HANDBOOK OF PHOTOCHEMISTRY

THIRD EDITION

Marco Montalti
Alberto Credi
Luca Prodi
M. Teresa Gandolfi

with introductory sections by Josef Michl and Vincenzo Balzani

CRC Press
Taylor & Francis Group
Boca Raton London New York

CRC Press is an imprint of the
Taylor & Francis Group, an **informa** business
A TAYLOR & FRANCIS BOOK

Published in 2006 by
CRC Press
Taylor & Francis Group
6000 Broken Sound Parkway NW, Suite 300
Boca Raton, FL 33487-2742

First issued in paperback 2020

ISBN 13: 978-0-367-57790-2 (pbk)
ISBN 13: 978-0-8247-2377-4 (hbk)

Visit the Taylor & Francis Web site at
http://www.taylorandfrancis.com

and the CRC Press Web site at
http://www.crcpress.com

Library of Congress Cataloging-in-Publication Data

Handbook of photochemistry.-- 3rd ed., rev. and expanded / Marco Montalti ...[et al.] ; with two
   introductory sections by Josef Michl, Vincenzo Balzani.
       p. cm.
       Rev. ed. of: Handbook of photochemistry. 2nd ed., rev. and expanded / Steven L. Murov,
       Ian Carmichael, Gordon L. Hug. c1993.
   Includes bibliographical references and index.
   ISBN 0-8247-2377-5 (alk. paper)
   1. Photochemistry--Tables. I. Montalti, Marco, 1971- II. Murov, Steven L. Handbook of
photochemistry.

QD719.M87 2005
541'.35'0212--dc22   2005051416

Foreword

Photochemists are always looking for information of all sorts. There is often an urgent demand for certain information that seemingly should be, but for some reason is not, conveniently available. All of us have a special reference book or two that we treasure that serve as portals to information that we need right away. Maybe it's a book with some good one liners for a speech we have to deliver or a thesaurus with the *mot juste* that we need to convince our colleagues of the beauty of our science in a lecture. Or maybe it's a reference book of prayers that we go to in desperation when experiments are not working. For a photochemist an ideal reference book provides rapid access to that rate constant, that quantum yield, that triplet energy, that fluorescence spectrum, that actinometer, etc. that suddenly is important in making an argument or checking a hypothesis. Where to find such data conveniently and authoritatively and quickly?

The Handbook of Photochemistry has been a very special reference book for photochemists since the first edition appeared in 1973. The collection of data complied in this edition was judiciously selected and presented in a user friendly manner such that it was easy to find the pertinent information. At that time important photochemical parameters such as excited state energies, rate constants, spectral information were not conveniently gathered. "The Handbook" was a break through in that regard and presented data with a high level of scholarship. The second edition, which appeared in 1993 built on the excellent scholarship of the first edition.

The third edition (2005) follows in the fine tradition of excellent, user-friendly scholarship, with some additional features. The new edition starts Chapter 1, an overview of the photophysics of organic molecules in solution by Josef Michl, followed by Chapter 2, an overview of the photophysics of transition metal complexes by Vincenzo Balzani. These are welcome essays, well worth reading by students and experts alike. Chapter 3 and the following chapter launch into the "meat" of the Handbook, large tables of data with references on the photophysical properties of organic and inorganic molecules.

The photochemical community will certainly be pleased to see the new updated arrival of The Handbook. Faculty should obtain a copy and put within easy reach of their students. When students ask where can I find....just point them to The Handbook. The sooner they become accustomed to using it as an information resource, the better off their research and understanding of photochemistry will be.

Enjoy.

Nicholas J. Turro
Columbia University (NY)

Preface to the third edition

In our everyday research and teaching activity as photochemists, the Handbook of Photochemistry has always been an invaluable reference. Therefore, needless to say, we embarked in the preparation of this revised and expanded edition of the Handbook with the greatest enthusiasm.

In the last thirteen years, since the release of the second edition of this Handbook, photochemical sciences have continued their vigorous expansion, as witnessed by the ever-increasing number of scientific papers dealing with photochemistry. Many spectroscopic and photochemical techniques that used to be prerogative of specialist research groups are now available in a wide range of laboratories. It has also become evident that chemistry as a whole has evolved rapidly. In many instances the traditional fields of chemistry – organic, inorganic, physical, analytical, biological – tend to overlap and merge together, giving rise to disciplines, such as those related to nanosciences. Light-induced processes play, indeed, an important role in these emerging fields of science. In such a context, we felt that times were ripe for a third edition of the Handbook of Photochemisty.

As for the previous editions, the goal of the third one is to provide a quick and simple access to the majority of the chemical and physical data that are crucial to photochemical investigations – from the planning and set up of experiments to the interpretation of the results. For the preparation of this Handbook, we could profit from an excellent starting material, that is, the second edition by S. L. Murov, I. Carmichael and G. L. Hug, published in 1993. We decided to maintain the format of most of the existing tables of data, not only because they were quite well organized, but also because many scientists got used to them during the years. By taking advantage of modern literature databases and related powerful search engines, we updated and expanded such tables with data on hundreds of new compounds. In particular, the section dealing with reduction potential values (Chapter 7), in the light of the importance of electron-transfer processes in photochemistry, was considerably enriched. In preparing the tabular and graphical material, we devoted a great effort to improve the readability and reach a style uniformity throughout the book.

The most relevant new entries of this third edition are indeed the tables gathering together the photophysical (Chapter 5), quenching (Section 6d) and reduction potential (Section 7b) data on metal complexes and organometallic compounds. Moreover, we found appropriate to expand and update, on the basis of our experience, some of the "technical" sections, and specifically light sources and filters (Chapter 11) and chemical actinometry (Chapter 12). We also introduced a section (Chapter 10) which describes the problems that are most frequently encountered in photoluminescence measurements, and illustrates a simple correction method to take into account the related effects. These sections will hopefully provide preliminary information at a glance, but are not intended to replace the many excellent books on photochemical methods and techniques that are available on the market and cited in the references.

Last but not least, the Handbook now features two introductory chapters written by two world leaders in photochemical sciences. Chapter 1, by Josef Michl, deals with the photophysics of organic molecules in solution. Chapter 2, by Vincenzo Balzani, describes the photophysical properties of transition metal complexes. Again, these sections cannot certainly be adopted in substitution of photochemistry textbooks; however, we felt that a concise overview of the most important light-induced processes that take place in organic and inorganic molecules would have been of help for students and for researchers that wish to get closer to the wonderful world of photochemistry.

In the preparation of this edition we benefited from the contribution of many people. First, we are grateful to all the members of our research group in Bologna, not only for fruitful discussions, but also for their friendship and support. We are particularly indebted to Vincenzo Balzani, Luca Moggi, Alberto Juris, Margherita Venturi of the Department of Chemistry of the University of Bologna, and to Roberto Ballardini of the ISOF Institute of CNR in Bologna. Mara Monari, Serena Silvi, Paolo Passaniti and Alberto Di Fabio gave us invaluable help for typing and checking several parts of the manuscript. Special thanks to Josef Michl for his contribution and Nick Turro for his fine foreword, and all the colleagues around the world that kindly sent us reprints and preprints of their papers or just simple, but precious information.

No book, of course, is free of errors and imperfections. We attempted to keep mistakes at a minimum by careful and repeated checks, and we apologize in advance for those that we could not avoid. We sincerely hope that readers will enjoy this new edition and that it will continue to be "The Handbook" of the photochemical community.

Photochemistry has come a long way since the pioneering work of Giacomo Ciamician, to whom our Department is named after. However, the ideas expressed by Ciamician in his visionary speech of 1912, entitled "The photochemistry of the future" are still the most appropriate to conclude this preface. The following sentence, quoted in the first two editions of the Handbook of Photochemistry, is worth repeating:

"On the arid lands there will spring up industrial colonies without smoke and without smokestacks; forests of glass tubes will extend over the plains and glass buildings will rise everywhere; inside of these will take place the photochemical processes that hitherto have been the guarded secret of the plants, but that will have been mastered by human industry which will know how to make them bear even more abundant fruit than nature, for nature is not in a hurry and mankind is."

G. Ciamician, *Science*, **1912**, *36*, 385-394.

Marco Montalti
Alberto Credi
Luca Prodi
Maria Teresa Gandolfi

Bologna, May 2005

Contents

1

Photophysics of Organic Molecules in Solution

By Josef Michl, Department of Chemistry and Biochemistry, University of Colorado, Boulder, Colorado 80309-0215, U.S.A.

1a INTRODUCTION

The following is a short overview of the principles of photophysics. We start by providing a brief survey of electronic excited states in Section 1b. This material can be found in textbooks of quantum chemistry but we have directed it to the specific needs of those wishing to learn the fundamentals of photophysics. We then proceed to the description of radiative (Section 1c) and non-radiative (Section 1d) transitions between electronic states. Strictly speaking, the material of Section 1c belongs to the discipline of electronic spectroscopy at least as much as it belongs to photophysics, but it was felt that it would be useful to outline the basics here instead of referring the reader elsewhere. Section 1e deals with the procedures that are in common use for the analysis of photophysical and photochemical kinetic data.

1b ELECTRONIC STATES

1b-1 Electronic Wave Functions

Because of their substantially smaller mass, electrons have much less inertia than nuclei and under most circumstances are able to adjust their positions and motion nearly instantaneously to any change in nuclear positions. It is therefore almost always acceptable to separate the problem of molecular structure into two parts, and to write the molecular wave function as a product of an electronic part, parametrically dependent on the nuclear geometry, and a nuclear part, different in each electronic state. The electronic wave function carries information about the motion of electrons within the molecule. Because of their light mass, electrons must be treated by quantum mechanics. The nuclear wave function contains information about molecular vibrational motion. Although strictly speaking nuclear motion must also be treated quantum mechanically, at times it is useful to approximate it by classical mechanics. The separation of electronic and nuclear motion is known as the *Born-Oppenheimer approximation*. The translation of a molecule as a whole is treated separately, almost always by classical mechanics, and need not concern us. Free rotation of an isolated molecule needs to be treated by quantum mechanics, but since we deal only with solutions, where it is severely hindered, we will be able to treat it classically if we need to consider it at all.

Electronic wave functions and their energies are found by solving the electronic Schrödinger equation, assuming stationary nuclear positions. In principle, an infinite number of solutions exists for any chosen geometry. Those at lower energies are quantized and their energy differences are on the order of tens of thousands of cm^{-1}. At higher energies, the energy differences decrease to thousands of cm^{-1} and less, and above the ionization potential, a solution exists at any energy. In the continuum regime, one or more electrons are unbound and the molecule is ionized (oxidized). For molecules with positive electron affinity, it is also possible to add an electron (reduction).

When only non-relativistic electrostatic energy terms are included in the potential energy part of the Hamiltonian operator contained in the Schrödinger equation, the resulting state wave functions are eigenfunctions of the total spin angular momentum operator and can be classified as singlets, triplets, etc., if the molecule contains an even number of electrons, or doublets, quartets, etc., if the number of electrons is odd. In organic molecules, only singlets and triplets are ordinarily of interest. Among the additional small terms normally neglected in the absence of atoms of high atomic number, spin-orbit coupling and electron spin-spin dipolar coupling are the most important, in that they cause the pure spin multiplet states to mix to a small degree. Also the hyperfine interaction term, which describes the coupling of electron and nuclear spin, can play this role. We shall return to these terms in Sections 1b-7, 1c-2, and 1d-3.

Often, we are only interested in the wave function of the lowest energy, which describes the ground electronic state. In ordinary organic molecules, this is the lowest singlet state (S_0). This wave function is the easiest one to solve for, but even it can only be found very approximately for molecules of any complexity. Most simply, the approximate solutions are expressed in the form of a spin symmetry adapted antisymmetrized product of one-electron wave functions, called an electron configuration state function (Fig. 1b-1). The antisymmetrization is needed to satisfy the Pauli principle, and is achieved by arranging the product of one-electron wave functions into a determinant (the Slater determinant). The one-electron wave functions used are referred to as occupied molecular spinorbitals, and those that could have been used, but were not, are known as virtual or unoccupied molecular spinorbitals, whose number is infinite. The best possible choice of molecular spinorbitals, defined as the one that gives the lowest ground-state energy, carries the name *Hartree-Fock* or *self-consistent spinorbitals*. Physically, a wave function approximated by a single configuration describes the motion of electrons in the field of stationary nuclei and the time-averaged field of the electrons.

Molecular spinorbitals are normally written as a product of an electron spin function (α, spin up, or β, spin down) and a space function, referred to as a molecular orbital. In a closed-shell configuration, each occupied molecular orbital is used twice, once with each choice of spin (Fig. 1b-1). Molecular orbital energies are related to reduction-oxidation properties of ground states of molecules. In an approximation developed by Koopmans, the energy of an occupied Hartree-Fock orbital is equal to minus the energy needed to remove an electron from that orbital to infinity (the ionization potential), and the energy of an unoccupied Hartree-Fock orbital is equal to minus the energy gained when an electron is brought from infinity and added to that orbital (electron affinity).

Molecular orbitals are ordinarily approximated as a linear combination of atomic orbitals centered at the atomic nuclei. These atomic orbitals are known as the basis set. Years of experience have revealed the basis set size and type that are needed to achieve a desired level of accuracy. Orbitals composed primarily from inner shell atomic orbitals are occupied in all low-energy states. Roughly half of the orbitals derived from atomic orbitals of the atomic valence shell are occupied and the other half are empty. Together, these orbitals span the so-called valence space. Rydberg orbitals are high-energy diffuse orbitals best expressed as combinations of atomic orbitals of higher principal quantum numbers.

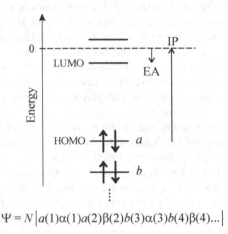

$$\Psi = N \left| a(1)\alpha(1)a(2)\beta(2)b(3)\alpha(3)b(4)\beta(4)... \right|$$

Fig. 1b-1. A symbolic representation of a closed-shell electronic wave function.

It is always better, but especially important at biradicaloid geometries (those with only two electrons in two approximately non-bonding orbitals in low-energy states), not to ignore the instantaneous as opposed to the time-averaged field of the other electrons. The energy lowering associated with this improvement is called electron correlation energy. The most common way to write a correlated wave function is to use a linear combination of many configurations instead of a single one. Depending on the details, this computationally much more demanding procedure is then called configuration interaction, coupled clusters, etc.

It is also possible to avoid molecular orbitals altogether and to construct the molecular electronic wave function directly from hybridized atomic orbitals (linear combinations of atomic orbitals located at the same nucleus). This so-called valence-bond method introduces correlation energy from the outset, but suffers from other difficulties. It has the intuitively appealing feature that the various contributions to the electronic wave function map readily onto the familiar Lewis structures of molecules. Carried to completion within a given starting atomic basis set, the molecular orbital and the valence-bond methods converge to the same result, known as the full configuration interaction wave function. Within the limits dictated by the use of a finite basis set, this is the exact solution of the Schrödinger equation, but present-day computer technology only permits its computation for very small molecules and limited basis sets, and this is not likely to change in the foreseeable future.

An altogether different approach is to give up the search for the molecular electronic wave function, which is a function of the space and spin coordinates of all electrons present in the molecule and contains far more information than is actually needed for any practical purpose, and to search for the total electron density function instead. Electron density within a molecule only depends on the three spatial variables, and is in principle sufficient for the evaluation of observable quantities. It is evaluated from the so-called Kohn-Sham determinant, built from Kohn-Sham orbitals. Although this determinant is analogous to the Slater determinant of wave function theories, it is not a wave function of the molecule under consideration, but only a construct used to make sure that the search for the optimal density is constrained to those densities for which an antisymmetric wave function in principle exists (it represents a wave function of a fictional molecule whose electrons do not mutually interact, and is chosen so as to produce the best total electron density in the variational sense). The various versions of this so called density functional method differ from each other in the functional used, i.e., in the assumptions they make in evaluating the total energy from electron density distribution in space. This relation is exact in principle and includes contributions from electron correlation energy, but the true form of the requisite functional is not known. The most popular functionals are semiempirical in that their general form agrees with first principles but the details have been adjusted empirically to yield optimal agreement with various experimental results.

Usually, the energy calculation is repeated for many stationary nuclear geometries and the one that yields the minimum total energy for the molecule is referred to as the optimized geometry of the ground state. The single-configuration approximation has the best chance of being adequate at geometries close to this optimized geometry, whereas at biradicaloid geometries the use of one of the methods that include electron correlation is mandatory. At these geometries, density functional methods usually have difficulties.

In photophysics and photochemistry, several of the lowest energy wave functions and their energies are normally needed. With the exception of the wave function of the lowest triplet state, and sometimes also one of the low-energy singlet states, it is only rarely possible to use the single-configuration approximation for excited states, and for accurate results, methods based on linear combinations of configuration state functions are always used. Optimization of geometries in excited states is possible, but again with the exception of the lowest triplet state, much more difficult than in the ground state. In density functional methods, electronic state energy differences are calculated directly from the ground state electron density. In the vicinity of ground state equilibrium geometries, these so called time-dependent density functional methods perform quite well.

The observable properties of a molecule in a particular electronic state, such as its permanent dipole moment, are obtained as the expectation value of the appropriate operator, such as the dipole moment operator M, over its electronic wave function, evaluated at the equilibrium geometry. They can change quite dramatically as a function of the electronic state. More properly, the calculation is repeated for many geometries in the vicinity of the equilibrium geometry and averaged over the molecular vibrational wave function, discussed below. Since the variation of molecular properties over a small range of geometries is usually small, this is rarely necessary unless the observable value vanishes by symmetry at equilibrium.

1b-2 Potential Energy Surfaces

The nuclear geometry of a molecule with N nuclei is specified by the values of $3N–6$ internal coordinates, since three of the total of $3N$ degrees of freedom are needed to describe the location of the center of mass and three to describe rotations relative to a laboratory frame. Only the geometry of a diatomic molecule, which is always linear and therefore has only two axes of rotation, is described by $3N–5$ internal coordinates, i.e., by the bond length alone.

A collection of $3N–6$ internal coordinates at a particular geometry represents a point in a $3N–6$ dimensional mathematical space. A surface produced in a $3N–5$ dimensional graph in which the total molecular electronic energy of the ground state is plotted against the geometry is known as the ground state potential energy surface. In spite of its name, the total electronic energy contains not only the kinetic and potential energy of the electrons, but also the potential energy of the nuclei. It is not easy to visualize multidimensional potential energy surfaces, and it is customary, albeit frequently misleading, to show limited portions of a surface in two-dimensional (Fig. 1b-2) or three-dimensional (Fig. 1b-3) cuts through the $3N–5$ dimensional plot.

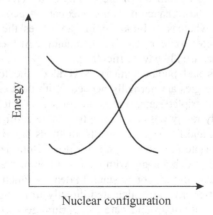

Fig. 1b-2. Two-dimensional cut through potential energy surfaces (schematic).

The ground state potential energy surface contains minima that correspond to the geometries and energies of more or less stable molecules and can be associated with their chemical (Lewis) structures. The minima are separated by barriers, and the lowest cols between minima correspond to transition states of chemical reactions. The height of a col above the minimum is the activation energy, and the width of the col is related to the activation entropy.

Connecting the set of points representing the next higher energy solutions yields the potential energy surface for the first excited state, and one can continue to define potential energy surfaces for as many states as needed (Fig. 1b-2). The excited surfaces also have minima and barriers separating them. Excited state minima located in the geometrical vicinity of a ground state minimum are called spectroscopic since their existence and shape can ordinarily be deduced from molecular electronic spectra. There usually are additional excited state minima and funnels at other geometries, particularly biradicaloid ones. Funnels are conical intersections, $(3N-8)$-dimensional subspaces in which one potential energy surface touches another, i.e., an electronic state has the same energy as the one just below or just above it. If the two touching states have equal multiplicity, their touching is avoided in the remaining two dimensions. Plotted in these two dimensions, the potential energy surfaces of the two states have the appearance of two conical funnels touching at a single point, the lower upside down and the upper right side up (Fig. 1b-3). Plotted along a single coordinate that passes through a touching point, they look like two crossing lines (Fig. 1b-2). If they have different multiplicity, potential energy surfaces cross freely. The minima and conical intersections in a potential energy surface are often separated by barriers with cols. Minima and funnels located at geometries far from any ground state minima can be referred to as reactive (see below).

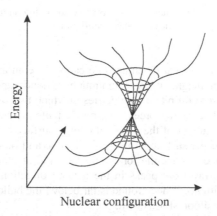

Fig. 1b-3. Three-dimensional cut through potential energy surfaces (perspective view, schematic).

1b-3 Vibrational Wave Functions

The total internal energy of a molecule is obtained by adding the internal nuclear kinetic energy to the electronic energy described by the potential energy surface. At any one time, the molecular geometry is represented by a point in the $3N–6$ dimensional nuclear configuration space, and thus by a point on the potential energy surface. The total energy is represented by a point located vertically above the latter, by an amount corresponding to the nuclear kinetic energy. In a molecule that is not exchanging energy with its environment, the total energy is constant, and the point representing the total energy moves in a horizontal plane, directly above the point that represents the electronic energy, which is located on the potential energy surface (Fig. 1b-4).

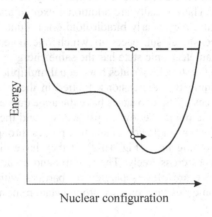

Fig. 1b-4. A schematic representation of the force vector acting on nuclei at a particular choice of nuclear configuration.

In contemplating nuclear dynamics one can invoke an analogy to a balloon soaring at a constant height above a mountainous landscape containing sharp peaks reaching almost to infinity at geometries in which two or more nuclei come very close together. The lateral force acting on the balloon above any one point is given by minus the gradient of the potential energy surface at that point. In completely flat regions of the surface, at its minima, cols, and other points where the gradient of the surface is zero, the force vanishes, and at other points, the force pushes the balloon away from peaks in the general direction of valleys leading toward a nearby minimum. When ground is far below, the balloon moves fast, and when it is close, the balloon slows down.

In the classical limit, and in an appropriately chosen coordinate system, the molecule behaves as if the shadow of the balloon, with the sun at zenith, were to follow the path of a ball that rolls on the surface without friction. In the quantum mechanical description, solutions of the Schrödinger equation for nuclear motion dictated by the potential energy surface need to be found. The resulting nuclear wave functions have a simple analytic form only in regions where the surface is harmonic, i.e. where a cut through the potential energy surface in any direction is a parabola. This generally occurs in the deep minima located in the vicinity of equilibrium geometries. In each such region, the molecule behaves as a $3N$–6 dimensional harmonic oscillator with frequencies v_1, v_2, ..., v_{3N-6}.

The nuclear eigenfunctions of a $3N$–6 dimensional harmonic oscillator are products of $3N$–6 wave functions, each describing motion along one of the normal modes q_i. Excitation of each of the normal modes is quantized. The stationary energy levels are equally spaced and are labeled by vibrational quantum numbers v, starting with v = 0 at the lowest energy. The energy separation is hv, with the v = 0 level located at hv/2 above the minimum of the parabola, at the so called zero point energy. The lowest energy wave function in each mode is Gaussian shaped and the higher energy ones have the form of a Gaussian multiplied by a polynomial. Each has v nodes. The total vibrational energy is the sum of the vibrational energies in each mode, and the energy in the lowest vibrational state is the sum of zero-point energies in all normal modes (Fig. 1b-5).

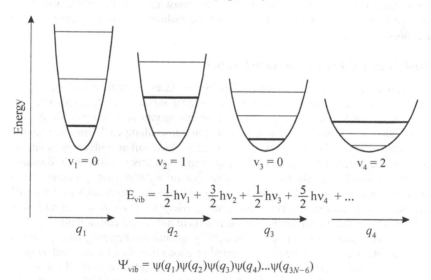

$$E_{vib} = \frac{1}{2}hv_1 + \frac{3}{2}hv_2 + \frac{1}{2}hv_3 + \frac{5}{2}hv_4 + ...$$

$$\Psi_{vib} = \psi(q_1)\psi(q_2)\psi(q_3)\psi(q_4)...\psi(q_{3N-6})$$

Fig. 1b-5. Contributions of the first few normal modes q_1–q_4 to vibrational energy (their total number is $3N$–6).

As one departs from the equilibrium geometry, deviations of the potential energy surface from a harmonic shape generally become more pronounced. At higher energies, vibrational wave functions and their spacing therefore deviate increasingly from the simple harmonic behavior. The effects of moderate anharmonicity can be treated as a perturbation of the harmonic case, in which the normal modes are only somewhat mixed and the spacing of levels within them is denser. At high nuclear kinetic energies, the normal mode picture breaks down altogether, and it is usually best to view the motion in terms of local modes, oscillators that correspond to individual bonds in the molecule. At the dissociation limit, the total energy is sufficient to break the weakest bond in the molecule. Above this point, the dissociation continuum is reached and vibrational energy is no longer quantized. Accurate quantum mechanical description of nuclear dynamics in highly vibrationally excited molecules is presently only possible in molecules containing very few atoms, and the standard procedure for larger molecules is to calculate a large number of classical trajectories and average the results, or to use statistical theories.

Vibrational eigenfunctions are calculated in the same manner for higher (electronically excited) potential energy surfaces. In regions of local minima, the harmonic approximation is again useful, but the location and shapes of minima are generally different than in the ground potential energy surface. Even in spectroscopic minima, there will be some displacement of the equilibrium geometry, some change in the vibrational frequencies, most often a decrease, and some change in the definition of the normal mode coordinates (which is known as the *Dushinsky effect*).

1b-4 Potential Energy Surface Shapes

An accurate determination of potential energy surfaces, even for small molecules, requires heavy computation, and very few global surfaces are known. After all, even in a tetraatomic molecule, the nuclear configuration space has six dimensions, and even if one wished to have only ten points along each direction, an energy calculation at 10^6 points would be needed. Most often, only very small regions of the space at low energies are explored. Even this is sufficiently demanding that it is often useful to derive qualitative information with minimal or no computations. This provides approximate answers to questions such as "*at which geometries are minima, barriers, and funnels likely to be located ?*", and it also helps with intuitive understanding of the results of numerical computations.

Some general answers are provided by qualitative bonding theory. In the ground state, minima are generally located at geometries that correspond to good Lewis structures, barriers are lower when weaker bonds are being broken and stronger ones made along a reaction path, and they are also lower for orbital symmetry allowed rather than forbidden reactions, etc. Much less chemical intuition is available for electronically excited potential energy surfaces.

Spectroscopic minima are to be expected when electronic excitation does not represent a major perturbation in chemical bonding, particularly in large conjugated systems. Often, they also occur at geometries at which two solute molecules stick to each other in the excited state, even if they do not in the ground state (*excimers* if the two molecules are alike and *exciplexes* if they are different). The increased intermolecular attraction in the excited state is due to a combination of exciton interactions and charge-transfer interactions in a proportion that depends on the nature of the partners. Since these minima have no ground-state counterpart, they are normally observed spectrally in emission rather than absorption.

Since by definition a spectroscopic minimum in an excited state is located at a geometry not too different from that of the ground state, radiative or radiationless vertical return to the ground state followed by vibrational relaxation typically results in no overall change in molecular structure, and the processes involved are considered a part of photophysics. Yet, spectroscopic minima in the lowest excited singlet and triplet states play an essential role in photochemistry, since they serve as holding reservoirs for excited molecules, with relatively long lifetimes, permitting thermally activated escape over small barriers to reactive minima and funnels located far from ground state geometries. Vertical return from the latter, followed by vibrational relaxation, has the potential to lead ultimately to different ground state species, hence to a photochemical event (Fig. 1b-6).

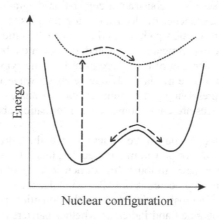

Fig. 1b-6. Motion on potential energy surfaces (schematic).

Reactive minima are to be expected at biradicaloid geometries, those in which the molecule, or a pair of molecules, have one fewer chemical bond than they could have according to the rules of valence. In the ground state biradicaloid geometries are energetically disadvantageous since two electrons in approximately non-bonding orbitals contribute little or nothing to bonding, while at other geometries they could be making a bond. In an excited state, one electron is typically excited from a bonding to an antibonding orbital anyway, more or less cancelling the bonding contribution of its erstwhile partner. There effectively is one bond fewer, and it is no great loss and often actually is an advantage to adopt a biradicaloid geometry and gain two non-bonding electrons. After all, in simple molecular orbital theory with overlap, an antibonding electron is more antibonding than its bonding partner is bonding, so it is better to have both of them non-bonding.

Exact degeneracy of the two non-bonding orbitals at a perfect biradical geometry leads to a zero energy difference between the lowest and next higher singlet state only in perfect biradicals of the axial type, which are relatively rare (methylnitrene, linear carbene, C_{5h} cyclopentadienyl cation). Ordinarily, and particularly in point biradicals (twisted ethylene, square cyclobutadiene, D_{8h} cyclooctatetraene), there is a considerable separation between S_0 and S_1. This can be reduced to zero by the introduction of a polarizing perturbation, which removes the exact degeneracy of the two most localized non-bonding orbitals of the biradical (e.g., by distorting square to a diamond cyclobutadiene). Similar recipes for approaching geometries of conical intersections can be formulated using valence-bond theory, and they depend on the similarity of atomic orbital interaction patterns to that observed in the simplest Jahn-Teller case, H_3.

Perhaps the simplest way to obtain rapid information about likely shapes of potential energy surfaces is to construct a correlation diagram. This tool depends on our ability to plot a sequence of electronic state energies at an initial and a final geometry of a chosen reaction path, or perhaps even at some points in between. These energies can be experimental, calculated, or estimated, but it is essential that an approximate understanding of the nature or at least the symmetry of the corresponding wave functions be available. If one relies on wave function symmetry, only reaction paths preserving that symmetry can be analyzed. However, since potential energy surfaces are continuous, some information about nearby paths is obtained as well.

Lines connecting similar wave functions are then drawn, and the non-crossing rule is invoked. For polyatomic molecules, this rule says that it is rare for two states of equal symmetry to touch (the dimensionality of the subspace of geometries at which touching occurs is only $3N–8$). After the crossings are avoided, a qualitative picture of a cross-section through the potential energy surfaces along the reaction path is obtained and indicates whether barriers or a minima are expected. For instance, if a locally excited and a charge-transfer excited state change their energy order along a reaction path, e.g., in the twisting of *p*-dimethylaminobenzonitrile around its C–N bond from planar to orthogonal geometry, one can expect a barrier in the lower excited surface, and the barrier will be smaller if the two state energies are closer to each other to start with. However, the non-crossing rule does not say how strongly the crossing is avoided, and this needs to be estimated from qualitative arguments or from a calculation. At times, a crossing is avoided so strongly that no trace of it seems to be left in the potential energy surface, and at other times, it is avoided only very weakly, causing the appearance of a barrier in the lower state and a minimum in the upper state. In the latter case, an appropriate reduction of symmetry will often lead from the point of nearest approach to a conical intersection, at which the crossing is not avoided at all, and as noted above, some guidelines are available for such searches. At times, state energies are not available, and then one can start with orbital correlation diagrams or valence bond structure correlation diagrams and proceed from these to state correlation diagrams and still obtain useful information.

1b-5 Singlet and Triplet States

A typical singlet wave function of a two-electron system can be written as a product of a function S that depends on the spatial coordinates of the two electrons, and a function that depends on their spin coordinates,

$$\Sigma(1,2) = N[\alpha(1)\beta(2) - \alpha(2)\beta(1)],$$

where $N = 2^{-1/2}$ is the normalization factor. Note that Σ is antisymmetric with respect to the exchange of electrons 1 and 2, so the function of space coordinates S must be symmetric in order for the overall wave function $S\Sigma$ to be antisymmetric and to satisfy the Pauli principle. Typical symmetric spatial functions S could be the closed-shell $a(1)a(2)$ or the open shell $N[a(1)b(2) + a(2)b(1)]$, where the two molecular orbitals a and b are orthogonal.

For a larger even number of electrons, the situation is more complicated in that a larger number of singlet spin functions is possible. For almost all organic molecules, the ground electronic state is a closed shell singlet with all bonding orbitals doubly occupied. Excited singlet states are generally of the open shell kind. The simplest among them can be approximated by a single open-shell configuration, with one unpaired electron in one of the antibonding orbitals and one in one of the bonding orbitals. This wave function can be written as an antisymmetrized product of the open shell wave function written above for two electrons and a closed shell wave function containing all other electrons paired in bonding orbitals. Often, the lowest excited singlet state is of this nature, with the two singly occupied orbitals a and b being the HOMO (highest occupied molecular orbital) and the LUMO (lowest unoccupied molecular orbital) of the molecule. In many aromatic molecules, such as benzene, naphthalene, and their derivatives, this is only the second excited singlet state and an excited singlet state with a more complicated wave function lies lower.

A triplet state consists of a collection of three states (sublevels) that have very similar energies. At room temperature, the populations of the three levels are in rapid equilibrium (equilibration occurs on the scale of ns), but at very low temperatures the equilibrium is established slowly (on the scale of seconds). For the non-relativistic electrostatic Hamiltonian that we have used so far, the three energies are identical. Typical triplet wave functions of a two-electron system can be written as a product of an open-shell function

$$T = N[a(1)b(2) - a(2)b(1)]$$

that depends on the spatial coordinates of the two unpaired electrons located in orbitals a and b and is common to all three sublevels, and one of three functions Θ discussed below that depend on the spin coordinates of the two electrons.

For a larger even number of electrons, the situation is more complicated in that a larger number of triplet spin functions is possible, and they need to be combined to the three functions appropriate for the three sublevels. The simplest among them can again be approximated by a single open-shell configuration, with one electron in one of the antibonding orbitals and the other in one of the bonding orbitals. As was the case for the singlet, this wave function can be written as an antisymmetrized product of the open shell wave function written above for two electrons and a closed shell wave function containing all other electrons paired in bonding orbitals. Almost invariably, the lowest excited triplet state is of this nature, with the two singly occupied orbitals a and b being the HOMO and the LUMO of the molecule.

Simple expressions are available for the energies of excited singlet (^1E) and triplet (^3E) states of a closed-shell ground state molecule that can be reasonably well described by a single configuration in which an electron has been promoted from an originally doubly occupied orbital a into an originally unoccupied orbital b. Their relation to the orbital energy difference is

$$\Delta E = E(b) - E(a)$$
$$^1E = \Delta E - J_{ab}$$
$$^3E = \Delta E - J_{ab} + 2K_{ab}$$

where J_{ab} is the Coulomb and K_{ab} the exchange integral between orbitals a and b. The integral J_{ab} can be thought of as the repulsion energy of the charge distribution produced by an electron occupying orbital a with the charge density produced by an electron in orbital b, and typical values are tens of thousands of cm^{-1}. The integral K_{ab} can be thought of as the self-repulsion energy of the overlap charge density (transition density) of orbitals a and b, and typical values are five thousand cm^{-1} or less. The energy of a triplet is therefore approximately $2K_{ab}$ below that of a singlet derived from the same configuration with singly occupied orbitals a and b. K_{ab} tends to be particularly small if the orbitals a and b avoid each other in space, as in charge-transfer or nπ* transitions.

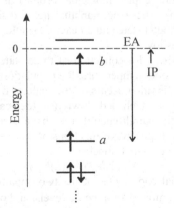

Fig. 1b-7. Electronically excited states have high electron affinities (EA) and low ionization potentials (IP).

Excited molecules are generally much easier to oxidize than ground state molecules, since the singlet or triplet excitation energy counts against the total energy needed to remove an electron to infinity. They are also much easier to reduce, since the excitation energy counts as a net energy gain when the ground state of the radical anion is produced by the addition of an electron. These relations are particularly easy to see for excited states that can be described by single configurations with singly occupied orbitals *a* and *b*. The easiest ionization of such an excited state corresponds to the removal of the electron that has been excited to the high-energy orbital *b*, and its easiest reduction corresponds to the addition of an electron to the low-energy orbital *a* (Fig. 1b-7).

For all three triplet spin functions the total length of the spin angular momentum vector is $2^{1/2}\hbar$, and the magnitudes of the projection of this vector into the quantization direction are $\hbar$, 0, and $-\hbar$, with spin quantum numbers 1, 0, and −1, respectively. Since the three sublevels are degenerate, an arbitrary linear combination can be used, but already a small perturbation that lifts the degeneracy will dictate the quantization direction and the proper choice of spin functions. In the presence of a strong outside magnetic field directed along the laboratory direction *Z*, as typically applied in EPR spectroscopy, the quantization direction is *Z*, and the three triplet spin functions are

$$\Theta[1] = \alpha(1)\alpha(2), \quad \Theta[0] = N[\alpha(1)\beta(2) + \alpha(2)\beta(1)], \quad \text{and} \quad \Theta[-1] = \beta(1)\beta(2).$$

The energies of the three sublevels then differ because of the Zeeman term in the molecular Hamiltonian (Fig. 1b-8).

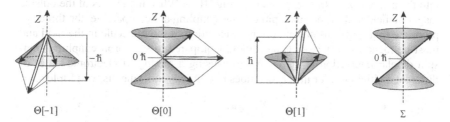

Fig. 1b-8. A symbolic representation of spin angular momenta associated with the triplet (Θ) and singlet (Σ) wave functions in the strong magnetic field limit. Contributions of each of two electrons are shown as simple arrows and their vector sums as double arrows. Projections into the magnetic field direction *Z* are shown.

If one includes small relativistic terms in the Hamiltonian, the weak interaction of the spin magnetic dipoles of the two unpaired electrons (*spin-spin dipolar coupling*) and the coupling interaction of the magnetic moment due to electron spin with the magnetic moment due to the orbital motion of this and other electrons in the attractive field of nuclei shielded by the other electrons (*spin-orbit coupling*), will cause the energies of the three sublevels to differ a little even in the absence of outside magnetic field. In organic molecules containing no heavy atoms, the resulting "*zero-field splitting*" generally is on the order of one cm^{-1} or less. In the presence of atoms of large atomic number, the effect of spin-orbit coupling can be in hundreds or even thousands of cm^{-1} (the "*heavy atom effect*"). In atoms and molecules of very high symmetry, such as octahedral or cubic, the effect of the spin dipole-dipole interaction vanishes by symmetry and spin-orbit coupling dominates. In low-symmetry organic molecules containing no atoms of high atomic number, the spin dipole-dipole interaction virtually always dominates the zero-field splitting and the spin-orbit coupling is a minor correction. The energy differences between the three triplet sublevels in the absence of outside magnetic field could be characterized by the energies of the three levels, X, Y, and Z, but it is customary to summarize them in the quantities D and E, defined by

$$D = (X + Y)/2 - Z$$
$$E = (Y - X)/2.$$

They are usually evaluated from EPR spectra.

When spin-orbit coupling is neglected relative to spin-spin dipolar coupling, and there is no outside magnetic field, the orientation of the magnetic axes is given by the principal axes x, y, and z of the spin dipolar tensor, which is fixed in the molecular frame, and the three spin functions are

$$\Theta[x] = N[\alpha(1)\alpha(2) - \beta(1)\beta(2)],$$
$$\Theta[y] = N[\alpha(1)\alpha(2) + \beta(1)\beta(2)],$$
$$\Theta[z] = \Theta[0] = N[\alpha(1)\beta(2) + \alpha(2)\beta(1)].$$

For the function $\Theta[u]$, the projection of the spin angular momentum vector into the axis u is zero ($u = x$, y, or z), cf. Fig. 1b-9. When the effects of the outside magnetic field and of the spin-spin dipolar coupling are comparable, the three appropriate spin functions depend on the orientation of the molecule in the field and must be found by matrix diagonalization. Compared to spin-orbit coupling, spin-spin dipolar interaction is ineffective in causing a mixing of electronic states of different multiplicity; for instance, it does not directly mix singlets with triplets.

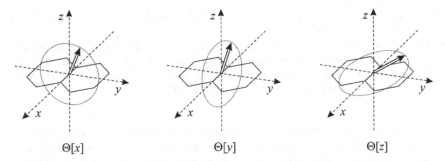

$\Theta[x]$ $\Theta[y]$ $\Theta[z]$

Fig. 1b-9. A symbolic representation of spin angular momenta associated with the triplet (Θ) wave functions in the absence of magnetic field (double arrows). The projection of the angular momentum into the molecular magnetic axis u is zero in the state $\Theta[u]$, $u=x$, y, or z.

   In organic molecules, the effect of spin-orbit coupling on the first-order description of the triplet sublevels provided above is often described by first order perturbation theory. The totally symmetric spin-orbit operator H^{SO} is approximated as a one-electron operator, and is expressed as a sum of three parts, each of which acts on one of the spin functions $\Theta[u]$ ($u = x, y, z$), which have the symmetry properties of rotation around the axis u. The operator mixes the three triplet sublevels with all singlet and other triplet (and quintet) states I to a degree that is dictated by its matrix elements $<I|H_u^{SO}|\Theta[u]>$ and by the energy separation between the states that are being mixed. Although this energy separation is virtually identical for all three sublevels of the triplet, their matrix elements are often vastly different (Fig. 1b-10). Because of the different symmetry properties of the spin functions of the three levels, in symmetric molecules one or more of the matrix elements vanish and different states I are admixed into the wave functions of each sublevel (*spin-orbit coupling selection rules*).

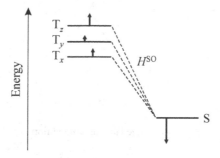

Fig. 1b-10. The arrows show the effect of the spin-orbit coupling operator H^{SO} on the energies of the sublevels of a triplet state and on the energy of a somewhat lower lying singlet state (schematic).

   As a result, it is no longer true that the three triplet sublevels share exactly the same spatial part of the wave function T. The introduction of spin-orbit coupling thus has two main effects: it modifies the energies of the three sublevels, changing the zero-field splitting, and it removes the strict separation into states of different spin multiplicities. Each singlet state will contain some triplet character, and each triplet will contain some singlet character. This weak admixture of singlet into triplet states and vice versa plays an essential role in organic photophysics and photochemistry.

1b-6 State Labels

It is frequently necessary to refer to singlet and triplet excited states by name, and several labeling schemes exist. The states are frequently referred to as S_n and T_n, respectively, and numbered in the order of increasing energy, starting with the

lowest singlet state S_0 and the lowest triplet state T_1. In ordinary organic molecules, S_0 lies below T_1 and represents the ground state, but in rare instances, T_1 lies
lower and is the ground state. The same labels can be used for potential energy
surfaces, and it is important to adhere to the convention that the states are labeled
by the order of their energies; i.e., S_1 cannot be above S_2 at any geometry, by definition. *Thus, states with different labels can touch, but they cannot cross, and at
points where they touch, they can change their slopes discontinuously* (Fig. 1b-
11).

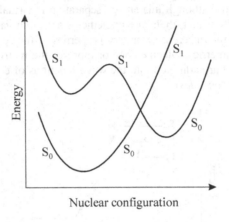

Fig. 1b-11. State labeling convention.

At nuclear geometries with elements of symmetry, such as reflection planes
and axes of rotation, it is common to label states by group theory symbols for irreducible representations to which their electronic wave functions belong. In deriving these labels, the symmetry of the triplet spin function is usually not included
and the symmetry of the space part of the triplet wave function is used to refer to
all three sublevels jointly. The total symmetry for each of the three separate sublevels $\Theta[x]$, $\Theta[y]$, and $\Theta[z]$ is then easily derived by multiplying with the irreducible representations appropriate for rotation about the axes x, y, and z, respectively.

Group theoretical labels are very useful for the derivation of selection rules
and in relating similar compounds of a larger family to each other. Although they
are strictly applicable only at special points on a potential energy surface, they are
often used in a loose sense even in areas near symmetrical geometries, where nuclear symmetry is only approximate.

There are, however, many circumstances in which molecular symmetry is
simply too low for any group theoretical classification to be useful. In such cases,
the best one can usually do is to provide another indication of the nature of the
electronic wave function. A common procedure is to approximate the excited state
wave function by a single configuration and to use the symbols of the orbital from

which an electron has been promoted, such as σ, π, or n (lone pair) for the initial orbital and π*, σ*, or Rydberg for the terminating orbital of the excitation. The symbols σ and π were originally introduced for linear molecules but it rapidly became common to use them for planar molecules and nowadays they are used in a more general sense to indicate the local symmetry of an orbital, σ for symmetric and π for antisymmetric relative to a local plane of symmetry. The resulting symbols such as ππ* are useful even for labeling states that cannot be represented by a single configuration, such as the states of benzene, as long as all of the important orbitals out of which promotion took place are of the same class (π) and all of the terminating orbitals are also of the same class (π*).

Another classification that is often useful is the distinction of locally excited and charge-transfer (CT) states. In the former, the starting and the terminating orbital of the transition are located in the same part of a molecule or pair of molecules, whereas in the latter, they are not. The part of the molecule that carries the starting orbital is referred to as the donor and the part that carries the terminating orbital is called the acceptor.

These types of nomenclature are useful in that they permit the formulation of sweeping statements about electronic transitions. For instance, nπ* transitions are never intense, whereas ππ* transitions can be. While nπ* transitions are shifted to higher energies (blue-shifted) in polar and particularly in hydrogen bonding solvents, ππ* transitions are generally shifted to lower energies (red-shifted). CT transitions are weak, have a small singlet-triplet splitting, and can produce a large change in the state dipole moment.

The labels discussed above are based on molecular orbital description of excited states. An alternative state characterization, based on the valence bond description of its wave function, identifies it as covalent or zwitterionic, depending on the nature of the dominant Lewis structures that symbolize the valence bond wave function. At biradicaloid gemetries of uncharged molecules, this designation coincides with the labels "dot-dot" and "hole-pair", respectively.

1b-7 Jablonski Diagram

A drawing of molecular electronic state energy levels, with singlet and triplet states in separate columns, is referred to as the Jablonski diagram (Fig. 1b-12). Often, vibrational sublevels are shown schematically as well. Radiative transitions from one level to another are indicated by straight arrows and non-radiative ones by wavy arrows. Most often, the levels correspond to vibrationally relaxed electronic states, i.e. to an equilibrium geometry of each individual state. Sometimes, an effort is made to show the state energies at two or more geometries, and as this representation becomes more elaborate, the diagram gradually turns into a drawing of a cut through potential energy surfaces (Fig. 1b-13).

The processes that change the nuclear geometry and/or modify the kinetic energy of the nuclei without changing the electronic potential energy surface that governs the nuclear motions are referred to as *adiabatic*, and are already familiar from ground state (thermal) chemistry. Those that change the potential energy surface are called *non-adiabatic* (or *diabatic*).

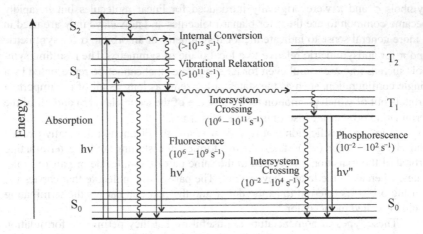

Fig. 1b-12. The Jablonski diagram.

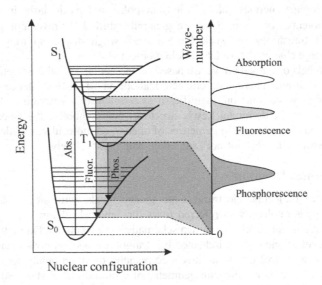

Fig. 1b-13. The relation of observed radiative transitions to potential energy curves (schematic).

1b-8 Adiabatic Processes

Under ordinary circumstances, molecules reside in potential energy minima. At the lowest temperatures, they are in a vibrational ground state. Ordinarily, they are in a state of dynamic equilibrium among all possible vibrational states, described by the Boltzmann distribution, i.e., they are in a thermal vibrational equilibrium. The shape of the potential energy surface in the vicinity of the minimum determines vibrational entropy and heat capacity, and ultimately, along with rotation and translation, the free energy.

In photochemistry and photophysics, we frequently deal with molecules that have vibrational energies much in excess of what would be expected from thermal equilibrium. This may be a result of electronic transition into a higher vibrational level of a state by photon absorption, emission, or energy transfer, or a result of radiationless transition from a higher electronic state. Such molecules remain vibrationally "hot" only for a very short time, since vibrational equilibration in condensed media is very rapid. Even in an isolated molecule, excitation in one normal mode is typically distributed statistically over all internal modes on a ps or sub-ps time scale, and in ordinary solvents, full thermal vibrational equilibrium with the environment is reached in a few ps or tens of ps. In solution, only very fast processes are capable of competing with vibrational equilibration.

In the following, we consider first monomolecular and subsequently bimolecular processes. Occasionally, the degree of vibrational excitation of a molecule exceeds the height of the lowest col in the barriers that surround the minimum (the activation energy). With a probability dictated by the frequency factor, the molecule will find itself going across the col (transition state), escaping into a neighboring minimum or funnel (Fig. 1b-14). The rate of the process is described by transition state theory, in which activation enthalpy and entropy correspond roughly to the energy of activation and the frequency factor, respectively. If the process does not involve the breaking of any chemical bonds, it corresponds to a change in conformation. If the process changes the chemical structure, it corresponds to a chemical reaction.

In the ground electronic state, the distinction between the two possibilities is quite unambiguous. In electronically excited states, it is much harder, and frequently is not made. The main difficulty is that electronic excitation in itself can be viewed as removing one bond, but its antibonding effect is frequently delocalized. To provide a very simple example, there is no doubt that in the ground state a cis and a trans alkene are two different compounds, whose interconversion involves the breaking of a π bond. In the triplet $\pi\pi^*$ state, there again are two minima, both at orthogonally twisted geometries. There is no π bond at any dihedral angle of rotation, and the minima clearly correspond to two conformers of the same molecule. But which bond is missing in the T_1 state of naphthalene? In most instances, the situation is not clear-cut, and we do not even have chemical names for many of the minima in excited state surfaces.

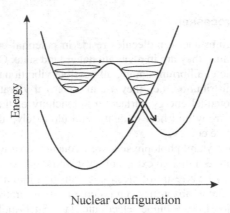

Fig. 1b-14. A symbolic representation of an excited state reaction followed by return to the ground state through a funnel (conical intersection).

Extensive adiabatic nuclear motion in an electronically excited state is quite common in T_1, and to a much lesser degree, in S_1. The reason for the difference is that the S_1 state is studded with funnels that can trap a passing molecule and return it to the S_0 state in an extremely rapid diabatic process. In this regard the situation is even less favorable for long-distance adiabatic travel in higher excited states of either multiplicity. Even in T_1, adiabatic motion typically represents only a fraction of the overall reaction path, since after eventual return to S_0, a chemical bond can usually be made before the final product geometry is reached.

Other than minor adjustments in internuclear distances and twisting motions, proton translocation is by far the most common type of adiabatic chemical reaction in the S_1 state (Fig. 1b-15), although a few examples of adiabatic C–C bond making and breaking processes have been reported. In organic photochemistry it is exceedingly rare to observe excited singlet product formation, and unusual to observe the formation of a product in its vertically excited triplet state.

Fig. 1b-15. An example of an adiabatic chemical reaction in the S_1 state.

The description of bimolecular adiabatic processes is similar to the one just given for monomolecular ones. One needs to treat both reaction partners as a "supermolecule" and to include all of their nuclei in constructing the potential energy surfaces. It is however necessary to recognize that motion along certain directions in the nuclear configuration space is relatively slow. One of the $3N–6$ dimensions corresponds to the intermolecular separation and motion in this direction is limited by the rate of translational diffusion. Five dimensions describe the relative orientation of the two molecules and motion in these directions is limited by the rate of rotational diffusion. The diffusion rates are generally slower than the intramolecular vibrational equilibration that we treated above, and depend strongly on solvent viscosity. In ordinary organic solvents, and at usual reactant concentrations, the typical scale for intermolecular encounters due to these motions is ns.

Under ordinary conditions, the two partners therefore are vibrationally equilibrated before they meet, except in neat or nearly neat liquids. In general, they are attracted to each other by dispersion (van der Waals) forces, and often also by dipole-induced dipole or dipole-dipole forces. At close separation, the dipole approximation to the description of electrostatic interactions is inadequate and the whole charge distribution needs to be considered. Hydrogen bonds are a common source of dimerization and aggregation.

In the gas phase, there is always at least a shallow minimum in the potential energy surface at the geometry of dimer, but in solution the intermolecular force may be purely repulsive if the interaction with the solvent is more favorable than interaction with another solute molecule. Often, however, dimerization or higher aggregation occurs even at relatively low concentrations, especially in poor solvents such as alkanes, and at low temperatures. If one of the partner molecules is an electron donor (reductant) and the other an acceptor (oxidant), intermolecular complex formation is often revealed by the presence of a new electronic transition in the absorption spectrum. This excitation is of charge-transfer character, in that an electron is transferred from the donor to the acceptor. Complexes that exhibit such a transition are called *charge-transfer complexes*, although in their ground state the amount of electron transfer from the donor to the acceptor is usually miniscule (see 1d.4).

As mentioned briefly above, electronically excited molecules are generally even more eager to form dimers and complexes with other molecules. These are referred to as *excimers* (*exci*ted di*mers*) and *mixed excimers*, respectively, if they do not exhibit a large degree of charge transfer from one component to the other, and *exciplexes* (*exci*ted com*plexes*), if they do. Their binding energies are frequently quite significant, comparable to those of hydrogen bonds. When the two partners form a solution charge-transfer complex already in the ground state, these names are not used, and instead, one uses the term *excited charge-transfer complex*.

The formation of excimers and exciplexes occurs on an excited potential energy surface, whether they are formed by an encounter of one excited and one ground state molecule, or, alternatively, by an encounter of a ground state radical anion of one partner with a ground state radical cation of the other (Fig. 1b-16). Although each of the ions is in its electronic ground state, the system of the two ions together is in an electronically highly excited state. Its ground state corresponds to a combination of two neutral ground-state molecules and its lowest two excited singlet states usually correspond to combinations of one ground-state and one excited-state neutral molecule, although in certain cases (polar solvents) the ion-pair state can lie below them. In these cases, an adiabatic process connects an exciplex with a pair of ions, whereas normally, an adiabatic process connects it with a pair of neutral molecules, one in its ground and the other in its excited state. Only in rare instances would a pair of dissolved organic molecules have an ionic ground state and be present as a salt.

The only other common case in which intermolecular reactions are adiabatic is proton transfer.

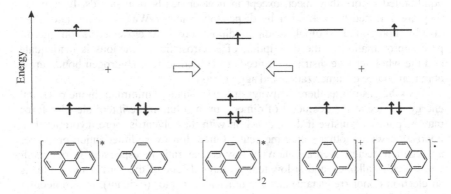

Fig. 1b-16. Generation of pyrene excimer by an encounter of an S_0 with an S_1 excited neutral pyrene molecule (left), and by an encounter of pyrene radical cation with pyrene anion (right).

1c RADIATIVE TRANSITIONS

1c-1 Electromagnetic Radiation

Electronic transitions are most commonly induced by electromagnetic radiation, characterized by its frequency ν and state of polarization, which provides information about the spatial direction of the electric field of the radiation.

The quantum of energy that the radiation field can exchange with matter is $E = h\nu$, which is usually expressed in units of kcal/mol, kJ/mol, or eV/molecule. It is also common to use a quantity proportional to E, such as the frequency ν of the radiation (in s^{-1}), or its wave number $\tilde{\nu} = \nu/c$ (in cm^{-1}). A less useful but also common practice is to use the wavelength of the light, $\lambda = c/\nu$.

Monochromatic light is an abstraction. In practice all light sources produce a distribution of frequencies and wavelengths, and in careful work it is necessary to recognize this and if necessary, integrate over the frequencies present. Lasers produce coherent light, in which the magnitude and spatial direction (polarization) of the electromagnetic wave are well defined at all times. There is a Fourier transform relation between the time dependence of the electric field strength of a pulse of coherent light and the frequency dependence of its spectral distribution, and the shorter is the pulse, the wider is its spectral distribution. Light intensity is proportional to the square of the electric field of the radiation. Other light sources usually produce incoherent light, whose electric field phase is random, but its polarization can still be well defined. For the most common light sources, see Chapter 12.

1c-2 Absorption and Emission

In absorption and emission spectra, light intensity is plotted as a function of one of the characteristics that identify photon energy (Fig. 1c-1). The relation $E = h\nu$ provides a bridge from the position of the observed spectral peaks to the energy difference between initial (G) and final (F) state involved in a spectroscopic transition. The vast majority of transitions are being studied under conditions in which a single photon is absorbed by a molecule at a time, and the *resonant condition* for absorption or emission then is that $h\nu$ be equal to the energy difference of F and G. Simultaneous absorption of two and at times an even larger number of photons is also possible with high laser light intensities. In two-photon absorption, if both photons are taken from the same beam, the resonance condition is that $2h\nu$ be equal to the energy difference of F and G, etc. Some of the advantages of two-photon absorption are a different set of selection rules, higher spatial resolution, and deeper penetration of the longer-wavelength exciting light into otherwise poorly transparent samples, such as biological tissue.

At room temperature equilibrium, organic molecules normally are in their electronic ground state S_0, and measurement of an absorption spectrum provides information about transitions from S_0 to electronically excited states. It is also possible to transfer a large fraction of molecules to an excited state, typically S_1 or T_1, usually with an intense laser pulse, and to measure the absorption spectrum of this excited state. Since the excited state population decays rapidly unless continually replenished, this is referred to as *transient absorption spectroscopy*.

Ordinary (one-photon) absorption spectroscopy relies on the *Lambert-Beer law*, which relates the intensity $I(\tilde{\nu})$ of monochromatic light of wave number $\tilde{\nu}$ transmitted through a sample to the intensity $I_0(\tilde{\nu})$ incident on the sample:

$$I = I_0 e^{-\sigma(\tilde{\nu})l} = I_0 10^{-\varepsilon(\tilde{\nu})cl}$$

where $\sigma(\tilde{\nu})$ is the absorption coefficient of the sample and l is its thickness. For solutions, it is common to use the molar decadic absorption coefficient (*molar absorptivity*) $\varepsilon(\tilde{\nu})$, obtained from the above expression by inserting the molar concentration c and the sample thickness in cm.

In order to observe emission spectra of organic molecules, their excited electronic states first need to be populated. When this is done by light absorption, the emission is referred to as photoluminescence. The measurement can be performed in a continuous mode of excitation (*steady state photoluminescence*) or with pulsed excitation (*pulsed photoluminescence*). Since both the wavelength of the exciting light and that of the detected emitted light can be varied, photoluminescence is intrinsically a two-dimensional spectroscopic technique, but the full two-dimensional spectra are rarely measured. A spectrum showing the dependence on the frequency (or wavelength) of the emitted light is referred to as an *emission spectrum*, and a spectrum showing the dependence on the frequency (or wavelength) of the exciting light of constant intensity is called the *excitation spectrum*. For a correct use of experimental emission and excitation spectra, see Chapter 10.

In the optical region, the wavelength of light is much larger than a molecule, and the electric vector of the radiation E is virtually the same at any point within the molecule. Moreover, the magnetic field carried by the radiation interacts with the molecule much more weakly than the electric field. As a result, the degree of interaction of electromagnetic radiation with the molecule is well described in the electric dipole approximation as $|E{\cdot}M|^2$, where M is the molecular transition dipole moment of the transition with which the radiation field is resonant, and all magnetic and higher order electric transition moments are neglected. In the absorption event, energy $h\nu$ is transferred from the field to the molecule, and in a stimulated emission event, it is transferred from the molecule to the field. Significant stimulated emission requires relatively intense electromagnetic fields and it is known from laser action. Under most circumstances, *spontaneous photon emission* from an electronically excited state, which takes place even in the absence of radiation imposed from the outside, is more probable. It is induced by the intrinsic zero-field fluctuation of the electromagnetic field that occurs even in vacuum. The rates of the three processes (absorption, stimulated and spontaneous emission) are related by the Einstein relations.

From these relations, Strickler and Berg derived an approximate expression relating the fluorescence rate constant k_F in units of s^{-1} to the integrated intensity of the first absorption band,

$$k_F = (\tilde{\nu}^2_{max} /3.47{\times}10^8) \int (\tilde{\nu}) d\tilde{\nu} \approx \tilde{\nu}^2_{max} f/1.5$$

where $\tilde{\nu}_{max}$ is the wave number of the absorption maximum and f is the oscillator strength, defined by $f = (4.319{\times}10^{-9}) \int \varepsilon(\tilde{\nu}) d\tilde{\nu}$.

In the absence of heavy atoms in the molecule, transitions interconnecting S and T states have small transition moments and are referred to as *spin-forbidden*. In absorption, singlet-triplet (and triplet-singlet) transitions have very small extinction coefficients, and are very difficult to observe. Such singlet-triplet absorption spectra consist of three spectrally nearly identical essentially unresolvable contributions, usually of widely different intensities, one into each of the triplet sublevels. In emission, such transitions have very small radiative constants and very long *natural radiative lifetimes*, and often have low quantum yields since competing radiationless processes can readily prevail (see Section 1e). The resulting emission is referred to as *phosphorescence*, whereas spin-allowed emission is referred to as *fluorescence* (Fig. 1c-1).

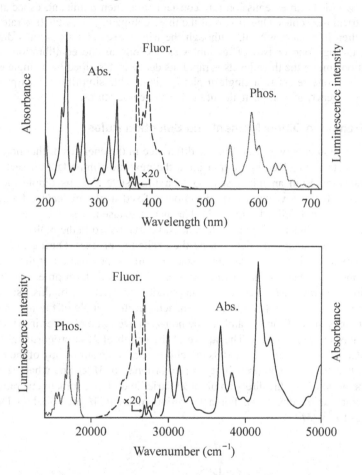

Fig. 1c-1. Absorption and emission spectra of pyrene. Absorption and fluorescence spectra recorded in acetonitrile solution at room temperature; phosphorescence spectrum recorded in acetonitrile rigid matrix at 77 K.

Since the sublevels of a triplet state are typically all populated to some degree, phosphorescence consists of three emissions, one from each sublevel. Their spectral positions ordinarily differ by less than a wave number, much less than the width of the individual peaks in the vibrational structure, and the overlapping emissions cannot be resolved. Commonly, one therefore talks about phosphorescence, its polarization degree, quantum yield, etc., as if it were a single emission. The three overlapping emissions can have widely different intensities, even though at equilibrium the three sublevels have nearly exactly the same populations, since one or two of the three emission rate constants are often dominant, especially in symmetrical molecules. One or two of the three competing non-radiative rate constants often dominate as well. Although the triplet excited state is thus drained mostly through one or two of its sublevels, as long as the equilibration of the population among the three levels is rapid, its decay is described by a single exponential, and one refers to a single triplet lifetime. This simplified description of phosphorescence is justified at all but the lowest temperatures.

1c-3 Transition Dipole Moment and Selection Rules

After transition energy, dictated by the difference in the energies of the molecular stationary states involved in the transition, the next most important characteristic of a transition from an initial state G to a final state F is its dipole moment $M(G \rightarrow F)$. This is a vector whose direction is fixed in the molecular frame but whose sense is not defined (Fig. 1c-2). This is so because it describes the transient electric dipole moment of the transition density, established in the molecule by the perturbing electromagnetic field during the excitation process. During the transition the molecule is not in a stationary state, but in a superposition of the states G and F, and its properties such as the dipole moment oscillate in time. The transition dipole moment M describes the amplitude and direction of this oscillating electric dipole. It is unrelated to the permanent electric dipole of the molecule in the stationary states G and F, and can be non-zero even in molecules in which the latter vanishes by symmetry. The square of the length of M is proportional to the integrated transition intensity, and its direction dictates the anisotropy of the transition, in that the transition probability is proportional to $|M|^2 \cos^2 \alpha$, where α is the angle between M and the direction of the electric field of electromagnetic radiation (light polarization direction). For a transition moment M measured in Debye, $f = 4.702 \times 10^{-7} \, \tilde{\nu} \, |M|^2$.

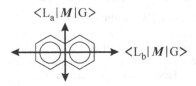

$$\langle L_a | M | G \rangle$$

$$\langle L_b | M | G \rangle$$

Fig. 1c-2. Transition dipole moments for the excitation of naphthalene from its ground state G to its excited states L_b (S$_1$) and L_a (S$_2$).

The quantum mechanical expression for the dipole of the transition electron density is $\langle F|M|G \rangle$, where $M = e\mathbf{R}$, e is the (negative) charge of the electron, and

R is the sum of the position operators of all the electrons in the molecule. None, one, two, or all three components of the transition dipole moment *M* may vanish by symmetry, depending on the point group that describes the symmetry of the molecule and on the irreducible representations to which the states G and F belong. Knowledge of state symmetries thus permits a prediction of transition moment directions, referred to as transition polarizations. Conversely, a measurement of transition polarizations is very helpful for the assignment of state symmetries. If all three components of *M* vanish, the transition is said to be forbidden by symmetry selection rules, or simply, *symmetry-forbidden*.

If the excitation can be described adequately as the promotion of an electron from orbital *a* to orbital *b*, *M* is equal to $<b|e\mathbf{R}|a>$, where *R* now is the position operator of a single electron. This can be thought of as the electric dipole moment of the overlap charge density of orbitals *a* and *b*, defined at any point in space by the product of the amplitudes of orbitals *a* and *b* at that point (Fig. 1c-3) times electron charge. Overlap charge density is positive in some parts of the molecule and negative in others, and its integral over all space vanishes since it is equal to the overlap of the orthogonal orbitals *a* and *b*. Transition moment directions are predictable from the knowledge of orbital symmetries. The general simple rule is that for a transition to be allowed, orbitals *a* and *b* should differ by the presence of a single nodal plane, and the polarization direction will be perpendicular to that plane.

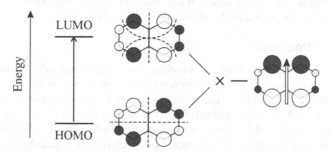

Fig. 1c-3. A schematic representation of the HOMO and LUMO orbitals of naphthalene, and of their product, the transition density for the HOMO → LUMO excitation (its dipole moment is indicated with a double arrow).

Even if a transition is allowed, the transition dipole is small if the orbitals *a* and *b* avoid each other in space, since then the overlap density is small everywhere and is unlikely to have a large dipole moment. This occurs for instance when *a* and *b* are an *s* and a *p* orbital on the same atom, as in certain nπ* transitions, or when one is mostly distributed over one set of *p* orbitals in a conjugated system and the other over a separate set, as can happen in non-alternant hydrocarbons (e.g., the first transition in azulene), and in molecules or supermolecules containing an electron donor with orbital a and an electron acceptor with orbital b separated in space (charge transfer transitions, as in absorption spectra of charge transfer complexes or emission spectra of exciplexes).

The transition moment between states G and F that differ in multiplicity, such as a singlet and a triplet, is generally extremely small in molecules composed of atoms with low atomic numbers. Due to orthogonality of the spin parts of the wave functions G and F, it would actually vanish if the states were not of slightly mixed multiplicity due to spin-orbit coupling. This statement is known as the *spin selection rule*. In molecules with some symmetry, the spin-orbit coupling selection rules mentioned above can be used to deduce the direction of the transition moment into or from each triplet sublevel separately, provided that the symmetries of the spatial parts of the wave functions S and T are known. These directions are normally different for each of the three sublevels.

When an atom of high atomic number Z is introduced into a molecule, spin-orbit coupling almost always increases, often considerably. The resulting increase in singlet-triplet mixing, which is usually quite different for the different sublevels of the triplet, is referred to as the *heavy-atom effect*. Its consequences are an increase in the absorption intensity and in the phosphorescence rate constant. The heavy atom can also be introduced into the solvent rather than the molecule (external as opposed to the internal heavy atom effect). Singlet-triplet mixing can also be induced by the presence of paramagnetic species in the solution, which has similar consequences. For instance, under high pressure of oxygen singlet-triplet absorption of solute molecules becomes quite easily observable.

1c-4 Linear Polarization

If the orientation of an ensemble of absorbing or emitting molecules is random, as is the case in isotropic liquid or solid solutions, light of any polarization is absorbed with equal probability, and emission occurs with equal probability in any direction and with any linear polarization. The former situation is common in absorption spectroscopy of isotropic samples and the latter situation is common in emission spectroscopy of isotropic samples in which excited molecules are equally likely to have any orientation because they were excited randomly, for instance by a chemical reaction (chemiluminescence) or, if they were excited by an anisotropic beam of light, they have had time to rotate into an arbitrary orientation either physically or because of excitation energy transfer (photoluminescence in fluid samples or in concentrated solid samples). The absorption or emission intensity observed on such samples is reduced by a factor of three relative to what would be observed if each molecule were lined up with its transition moment M along the direction of the electric vector of linearly polarized light that is being absorbed or observed in emission.

Fully isotropic measurements of this type provide information about the length of M but not about its orientation in the molecular frame. To obtain the latter, a fully or partially oriented molecular assembly is required. This is most easily understood by considering measurements on a single molecule and recalling that transition probability is proportional to $|M|^2\cos^2\alpha$, where α is the angle between M and the light polarization direction. Except in molecules of very high symmetry neither the initial state G nor the final state F will be degenerate, and at the wave length appropriate for the transition from G to F only a single electronic transition moment M will contribute to the absorption or emission process.

The absorption will be most likely if the light propagation direction is perpendicular to M and the light is linearly polarized parallel to M. Then, α is zero and $\cos\alpha$ has the value of one. A rotation of the light propagation direction by 90° to make it coincide with the direction of M, regardless of its state of polarization, or a rotation of its polarization direction by 90°, which also causes α to become 90° and hence $\cos\alpha$ to vanish, brings the absorption probability to zero. Thus, if the orientation of the molecule is known, a variation of either the light propagation direction or light polarization direction will provide information about the direction of M in the molecular frame. If the direction of M in the molecular frame is known, the measurements will provide information about the orientation of the molecule. The dependence of absorption properties on light direction or polarization is referred to as linear dichroism (LD).

Ordinarily, measurements are performed on molecular assemblies. Fully aligned molecular assemblies occur in crystals and partially aligned ones occur in numerous solutions, liquid (lyotropic or thermotropic liquid crystals, streaming samples, liquids in strong electric or magnetic fields, etc.) or solid (stretched polymers, flat surfaces, etc.). Integration over the orientation distribution function is then necessary to evaluate the results of measurements. The simplest situation prevails in uniaxial samples, which possess a single orientation axis Z, and all directions perpendicular to this axis are equivalent. The choice of the perpendicular axes X and Y is then arbitrary. In this case, only two linearly independent absorption spectra can be obtained and are usually chosen to be those with the light polarization direction either along Z or perpendicular to Z. As in single-molecule measurements, if information about the molecular orientation is available, conclusions about the direction of M in the molecular frame can be reached, and if the latter is known, information about molecular orientation is obtained.

In an entirely analogous fashion, the emission observed from a repeatedly excited oriented molecule through a linear polarizer will be most intense when observed in a direction perpendicular to M with the polarization direction parallel to M. A rotation of the light observation direction by 90°, which makes it coincide with the direction of M, regardless of the rotation of the polarizer, or a rotation of the polarizer by 90°, causes the emission detection probability to vanish. Once again, if the orientation of the molecule is known, a variation of either the light observation direction or polarizer orientation will provide information about the direction of M in the molecular frame. Conversely, if the direction of M in the molecular frame is known, the measurements will provide information about the orientation of the molecule.

Measurements of emission anisotropy or polarization can be performed on the same types of aligned molecular assemblies that were listed above for linear dichroism. Most often, however, it is simplest to produce partially aligned ensembles of excited molecules by photoselection on an isotropic sample in which the molecules are not free to rotate. The best procedure to use for this purpose is to excite the sample with a collimated beam of linearly polarized light. Then, the light electric vector selects molecules for excitation with probabilities given by $\cos^2\alpha$, as we have just seen in the discussion of absorption, and the direction of the light electric field serves as the Z axis of the resulting uniaxial partial oriented assembly. Since the orientation distribution function is then known, specific predictions can be made for the polarization directions to be expected for one or another value of the angle between the absorbing transition moment and the one responsible for the emission. Thus, polarized photoluminescence measurements on rigid solutions, such as organic glasses at low temperatures, produce information about relative polarization directions.

The simplest formulas result if the depopulation of the ground state is negligible and when the observed intensities are cast in the form of emission anisotropy r, defined by

$$r = (I_{\parallel} - I_{\perp})/(I_{\parallel} + 2I_{\perp})$$

where $I_{\parallel}$ is the emission intensity observed with the analyzing polarizer parallel to the polarization direction of the exciting light, and $I_{\perp}$ is the emission intensity observed with the analyzing polarizer perpendicular to it. Then, if differently polarized transitions do not overlap, $r = 0.4$ if the absorbing and emitting transition moments are parallel, and $r = -0.2$ if they are perpendicular to each other. These limiting values are not always observed in fluorescence measurements, for various reasons. They are less commonly reached in phosphorescence studies, because the differently polarized emissions from three sublevels overlap. Only if emission from one of the sublevels greatly dominates over emission from the other two will there be an opportunity to observe a simple result.

If polarized absorption or emission measurements are performed fast enough after the initial photoexcitation, using short laser pulses and observing the emission at times short relative to rotational diffusion times, similar information can be obtained even in fluid solutions. Moreover, a study of the decay of the polarization anisotropy in time provide information about the rate of molecular rotational diffusion.

1c-5 Circular Polarization

Unlike achiral molecules in an achiral environment, chiral ones, and more weakly, achiral ones in a chiral environment, distinguish left-handed from right-handed circularly polarized light, and absorb and emit the two with slightly different probabilities. This is one of the demonstrations of optical activity. One of two enantiomers has a somewhat larger extinction coefficient ε_L for left-handed light, and a somewhat smaller one, ε_R, for right-handed light. The mirror image molecule, the other enantiomer, has the same values of extinction coefficients, but L and R are interchanged. In absorption, the difference spectrum $\varepsilon_L - \varepsilon_R$ is known as natural circular dichroism (CD). It has positive and negative peaks, and the CD spectra of two enantiomers are mirror images of each other. Racemic samples, which contain equal amounts of the two enantiomers, show a zero CD spectrum, since their contributions cancel.

Each transition from an initial state G to a final state F can be characterized by a rotatory strength $R(G \to F)$, obtained by integration of the CD spectrum over the transition region. It is positive for one enantiomer and of equal size but negative for the other. Since the CD signal is sometimes easily observable for transitions that are too weak to be seen in absorption because they are covered up by other bands, it can be useful for spectral assignments, and R can be viewed as one of the useful characteristics of an electronic transition. However, the primary practical significance of CD measurements lies elsewhere, as it can be used for the assignment of absolute molecular chirality.

Circular dichroism occurs even in the absence of molecular alignment, and is more complicated and much harder to evaluate for aligned samples. In isotropic solutions, the rotatory strength is related to the scalar product of the electric (M) and magnetic (**M**) dipole transition moment vectors,

$$R(G \to F) = \text{Im}(<G|M|F> \cdot <F|\mathbf{M}|G>)$$

where Im stands for "imaginary part of" and assures a real value for R even though **M** is a pure imaginary operator. In molecules that are achiral by symmetry, the two vectors are mutually perpendicular, or one is zero, and the scalar product vanishes. For aligned samples, the expression for rotatory strength is more complicated, and contains a contribution from an electric quadrupole transition moment.

In the presence of magnetic field oriented parallel to the light propagation direction, all samples become optically active and exhibit circular dichroism to a degree proportional to the strength of the magnetic field. This effect is referred to as magnetic circular dichroism, MCD. It is related to the Zeeman effect and has nothing to do with chirality. Unlike natural CD, the MCD spectra of enantiomers are identical and equal to that of a racemic sample. Since MCD changes its sign when the direction of the magnetic field is reversed, the effect disappears when the measuring light is sent through the sample twice in opposite directions, and this permits an easy separation of CD and MCD of chiral samples.

In an MCD spectrum, each transition is characterized by up to three terms, called A, B, and C, normally evaluated by a particular type of integration of the MCD spectrum over the transition region. The first of these vanishes unless the initial or the final state of the transition is degenerate, and the third one vanishes unless the initial state is degenerate. The second one is always present unless the transition has zero intensity in absorption. The B and C terms contribute an identical spectral shape, similar to that of the absorption band, and are separable from each other because C is temperature dependent. The shape contributed by the A term is bisignate and looks like the letter S. The measurement of MCD terms is useful for spectral assignments, particularly for the recognition of degenerate transitions.

Similar differences between left-handed and right-handed circular polarized light also exist in emission probabilities, where the difference spectra are termed natural and magnetic circular polarized luminescence (CPL and MCPL, respectively). If a transition is somewhat more probable for left-handed circular polarized light in absorption, it is somewhat more probable for right-handed light in emission, and vice versa. CPL is observed for pure enantiomers, even when it is excited with natural or linearly polarized light. It is not observed for achiral molecules in an achiral environment, and only weakly in a chiral environment. It is not observed for racemic mixtures excited with natural or linearly polarized light, but it is observed weakly for racemic mixtures excited with circularly polarized light, since the ensemble of excited molecules is then photoselected to contain slightly more of one of the enantiomers and is no longer racemic. This can be used to distinguish achiral from racemic samples.

1c-6 Vibrational Fine Structure

Electronic transitions in molecules appear as groups of lines. The larger spacings between the individual transitions are due to vibrational structure and reveal the details of vibrational levels in the G and F states. In gas-phase spectra, finer rotational structure also appears. In solutions, only vibrational structure is seen, but its lines are considerably broadened. At low temperatures, and to a lesser degree, even at room temperature, only the lowest vibrational level of the initial state G is significantly populated (all vibrational quantum numbers equal to zero), and this simplifies the consideration of the vibrational fine structure (Fig. 1c-4).

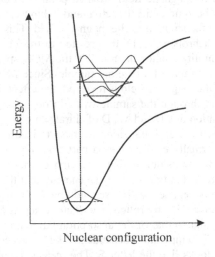

Fig. 1c-4. The origin of Franck-Condon factors (schematic).

Since even the lightest of nuclei are much heavier than electrons, electromagnetic radiation frequencies that are resonant with electronic excitation are much too high to drive an excitation of nuclear motion directly. During electronic excitation, even if it takes a relatively long time (e.g., the time between molecular collisions that destroy wave function coherence), the nuclear vibrational wave function therefore remains nearly intact (*Franck-Condon principle*). If the final electronic state F had a potential energy surface of the same shape as the initial state G (if they had identical vibrational frequencies), and if they were not displaced relative to each other (if they had identical equilibrium geometries), they would share the same set of vibrational wave functions, and only the transitions between levels characterized by identical quantum numbers would appear in the absorption or emission spectra (assuming that the electronic transition moment M is geometry-independent). This situation occurs rarely, since electronic excitation generally causes a modification of bonding conditions in the molecule and thus changes the equilibrium geometry and the force constants. An example is the excitation of inner-shell f electrons in complexes of rare earths.

More commonly, the vibrational wave function i of the initial state I differs from that of the final state F and has a significant non-vanishing projection into those (j) of several or many different vibrational levels of F. Then, transitions into all of the latter levels j will appear in the spectrum at appropriate energies, with probabilities proportional to the so-called Franck-Condon factors, $|\langle j|i\rangle|^2$ if the geometry dependence of M can be neglected. The resulting shape of intensity distribution in the absorption or emission band is often referred to as the Franck-Condon envelope (Fig. 1c-5). The vibrations that appear most prominently in the envelope are those whose normal modes need to be invoked to travel from the equilibrium geometry of the initial state to that of the final state, and those whose frequencies differ the most between the two states. The former are totally symmetric, and all components of the envelope have the same polarization as the zero-zero transition (the band origin). Asymmetric vibrations sometimes appear as double quanta and mark normal modes whose frequency changes strongly between the two states. An analysis of the vibrational fine structure permits a determination of the final state equilibrium geometry if that of the initial state is known. Often, the Franck-Condon envelopes of the absorption and the emission between two electronic states are approximate mirror images of each other in the spectra, since the vibrational frequencies in the two states differ little, and most of the envelope shape is dictated by the displacement of the equilibrium geometry (Fig. 1c-5).

If the change in equilibrium geometry upon excitation is small, the zero-zero transition between the two lowest vibrational levels of the two states is the most intense, and others appear weakly (Franck-Condon allowed transitions). If it is large, the zero-zero transition appears weakly, and sometimes is essentially inobservable (Franck-Condon forbidden transitions). In general, the peak of the absorption or emission band appears approximately at the energy of the so-called vertical transition, which corresponds to the difference between the initial and final state energy at the equilibrium geometry of the former (Fig. 1c-4). If transitions in both directions can be observed, one in absorption and the other in emission, it is possible to estimate the location of the zero-zero transition as the average of their peak energies, even when the transition itself is too weak to be observed (Fig. 1c-5).

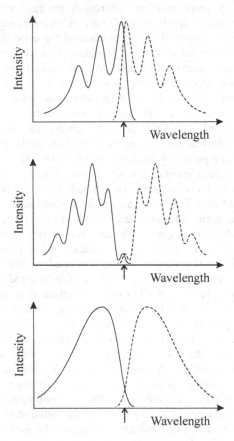

Fig. 1c-5. Examples of Franck-Condon allowed (top) and forbidden (center and bottom) absorption (left) and emission (right). The unresolved spectra shown at the bottom are typical of transitions in which many vibrational modes are active, especially when solvent-solute interactions are strong.

   At temperatures at which excited vibrational levels of the initial state are populated, transitions from these levels are observed as well. They are called hot bands and appear at the low-energy side of the transitions from the lowest vibrational level in absorption spectra and at the high-energy side in emission spectra.

In solution spectra, they are often not resolved and merely broaden the spectral bands observed. They are a common cause of thermochromism (dependence of the absorption spectrum on temperature).

1c-7 Vibronic Coupling

The description of the vibrational fine structure given in the preceding section applies to allowed transitions, i.e. those for which at least one of the three components of the transition moment M does not vanish. For transitions that are forbidden, or only very weakly allowed, it is not appropriate to neglect the geometry dependence of M, since in general there will be some symmetry-lowering normal mode that will cause M to differ significantly from zero. Although the origin of the forbidden transition remains forbidden, transitions into levels involving one or a higher odd number of quanta of such asymmetric mode or modes will then appear in the spectrum (*Herzberg-Teller* or *vibronic coupling*), each acting as a "false origin" on which a Franck-Condon envelope of symmetric modes of vibration will be built, using various totally symmetric vibrations. This transition intensity is said to be vibronically induced.

The distortion of the molecular geometry acts as a perturbation that permits mixing of allowed electronic states I that were of different symmetries at the more highly symmetrical equilibrium geometry. This effect introduces some of the nonvanishing transition moments $M(G,I)$ into the otherwise vanishing transition moment $M(G,F)$. Along with this "borrowed" or "stolen" intensity comes its polarization, which is that of the transition from G to I.

In solution, the molecular environment hardly ever has the full molecular symmetry, and it, too, can perturb the symmetry of the molecular wave function away from the one that would otherwise be present. Then, the origin of a forbidden transition will acquire weak but non-zero intensity. E.g., the origin of the lowest singlet-singlet transition in benzene, which is symmetry forbidden and not observed in the gas phase absorption spectrum, is clearly visible in the solution spectrum. The polarization of such solvent-induced transitions is again dictated by the symmetry of the states from which the intensity is borrowed. If the transition from G to F is not forbidden but merely very weak, the direction of its transition moment in solution will be intermediate between $M(G,F)$ and $M(G,I)$. The appearance of these solvent-induced or solvent-enhanced bands is referred to as the *Ham effect*. Their intensity is usually related to the polarizability of the solvent, and they can be used to probe molecular environment.

1d NON-RADIATIVE TRANSITIONS

Transitions between electronic states can occur in a radiationless manner as well as the radiative manner that we have discussed so far (Fig. 1b-12). In both cases, the jump is from a vibrational level of the electronic state associated with one potential energy surface to that of another. In a radiative process, conservation of the total energy was assured by making up for the acquired energy of electronic motion by absorbing a photon from the radiation field, or making up for the lost energy of electronic motion by donating a photon to the field. In a non-radiative process, conservation of the total energy is assured by compensating for an increase in the energy of electronic motion by taking up energy of nuclear motion, or compensating for the lost energy of electronic motion by releasing it to nuclear motion.

1d-1 Non-Born-Oppenheimer Terms

The coupling between electronic and vibrational motion is mediated by terms in the Hamiltonian that are neglected in the Born-Oppenheimer approximation, and is relatively inefficient. Its rate depends on several factors. The "Franck-Condon factor" drops in size as the number of quanta of vibrational excitation involved grows. This happens with increasing energy difference between the electronic states involved, and upon introduction of heavier isotopes, e.g. in compounds with C-D instead of C-H bonds. The "density of states factor" reflects the number of vibrational levels that are available for the conversion at the energy required. This increases as the energy difference between the electronic states grows, but in general not fast enough to compensate for the fall-off of the Franck-Condon factor. Radiationless transitions thus generally become exponentially slower as the energy difference increases (the *energy gap law*). This can be viewed as paradoxical since it means that a more exothermic process is slower. The "electronic matrix element factor" reflects the difference between the motion that the electrons execute as the molecular geometry changes during vibrational motion. The larger and more abrupt the changes, the harder it is for the electrons to follow the instantaneous changes in the positions of nuclei during vibrations, given that their moment of inertia may be small but is not zero (this is neglected in the Born-Oppenheimer approximation).

Under ordinary circumstances, it is rare for a molecule to have enough vibrational energy to be able to convert it into electronic excitation. It happens in energetic collisions at high temperatures, in intense IR fields, or upon bombardment with energetic particles. In organic molecules, it generally happens only when the energy required for the electronic excitation is fairly small. Thus, in molecules whose T_1 state is only a little below the S_1 state, thermal excitation at room temperature may be sufficient to populate the latter from the former. Because of the generally much longer lifetime of the T_1 state, fluorescence from S_1 can then be observed long after prompt fluorescence from the initially excited S_1 state has decayed (E-type *delayed fluorescence*). Another example is *chemiluminescence*, where thermal excitation from S_0 to S_1 or T_1 takes place in a potential energy surface region where the two are close in energy and is followed by geometry change to a region where S_0 has dropped far down below S_1 or T_1, such that the excited state produced has a substantial lifetime, and possible vertical emission occurs in the visible region.

Most often, non-radiative transitions convert electronic into vibrational energy, and the latter is lost to the environment within some tens of picoseconds as heat in the process of thermal equilibration. This is the major route of molecular return to the ground state equilibrium after electronic excitation (Fig. 1b-12).

The electronic matrix element that appears in the expression for the rate of a non-radiative process includes an integration over spin that yields unity if the two electronic states are of the same multiplicity. Such spin-allowed non-radiative transitions are known as *internal conversion*. Spin integration causes the matrix element to vanish if the two states involved are of pure spin and differ in multiplicity. The non-radiative transition would then be spin-forbidden. When spin-orbit coupling is included and the two states are of somewhat mixed multiplicity, a non-radiative process known as *intersystem crossing* takes place, but if other factors are the same, it is much slower than internal conversion.

1d-2 Internal Conversion

When the energy difference between the two states between which internal conversion takes place is large and corresponds to the visible of ultraviolet energy region, its rate is often competitive with that of fluorescence. This happens for the S_1 and S_0 states of many colored molecules. The quantum yields of fluorescence and internal conversion are then of comparable order of magnitude. Often, fluorescence dominates strongly when the energy gap is in the ultraviolet or blue whereas internal conversion dominates when the energy gap is in the red, and the lifetimes are on the order of nanoseconds.

When an electronically excited molecule is brought up to another solute, additional dimensions in the nuclear configuration space become accessible and nuclear motion in those directions may lead to internal conversion. Such a process is referred to as excited state quenching, and it can occur at a diffusion limited rate. A common example of a molecule that quenches many excited singlet states is molecular oxygen.

As the energy difference between the two states involved in internal conversion is reduced, its rate increases and the lifetime in the upper state shortens. Since the separations among the higher excited singlets generally are much smaller than the S_1-S_0 gap, their lifetimes are less than a picosecond with a few exceptions, and fluorescence from higher singlets is not competitive. *Kasha's rule states that only the lowest state of each multiplicity emits, and that all higher excited states convert to it faster than they can do anything else.*

The limit is reached when the potential surfaces of the two states actually or nearly touch (Fig. 1b-3). Then, the rate of internal conversion is normally limited only by the rate at which the molecule can reach this region of the potential energy surface, and this often takes as little as a few tens of femtoseconds. The region of surface touching is known as a conical intersection (see Section 1b-2).

1d-3 Intersystem Crossing

Under otherwise similar circumstances, intersystem crossing is many orders of magnitude slower than internal conversion. The efficiencies of S_1-S_0 fluorescence are often comparable to those of T_1-S_0 phosphorescence, reflecting a similar ratio of the rates of radiative and non-radiative processes, but whereas fluorescence lifetimes are typically measured in nanoseconds, those of phosphorescence are often measured in milliseconds or even seconds. This is only true in rigid solution, on surfaces, or in constrained media (cyclodextrins, carcerands, etc.), since in fluid solutions the long lifetimes make phosphorescence superbly sensitive to quenching.

In addition to quenching by impurities and oxygen, triplets also quench each other in a process known as *triplet-triplet annihilation*, in which two triplets encounter and combine their excitation energies to produce an excited singlet and excess vibrational energy. The excited singlet can then fluoresce. Because of the relatively long triplet lifetime, and the continued supply of excited singlets from the annihilation process, the emission continues long after the decay of prompt fluorescence and is known as P-type *delayed fluorescence*. A process inverse to triplet-triplet annihilation is also known and is referred to as *exciton splitting*. It occurs in crystals and polymers and leads to the formation of two triplet excited states from a high vibrational level of an excited singlet state. It is fast enough to compete with vibrational deactivation.

Another illustration of the spin forbiddenness of the intersystem crossing process is the fact that the crossing from S_1 to T_1 is often competitive with the internal conversion from S_1 to S_0 and occurs on a nanosecond time scale, although the energy gap involved is vastly smaller. In this process, the three sublevels of the triplet are generally populated at quite different rates, far from thermal equilibrium. This permits the use of time-resolved EPR spectroscopy to examine the transitions between the sublevels, and can also induce polarization of nuclear spin, permitting the use of CIDNP ("chemically induced dynamic nuclear polarization") for the study of photochemical reactions.

Intersystem crossing is subject to the same kind of internal and external heavy atom effects that have been discussed in connection with phosphorescence (Section 1c-3).

1d-4 Electron Transfer

We have already seen that one of the states participating in a radiative electronic transition can be locally excited while the other is of the charge-transfer type, particularly if the donor and the acceptor are located in different molecules located next to each other. These processes are referred to as charge-transfer absorption (Fig. 1d-1) and charge-transfer (exciplex) fluorescence or phosphorescence. The non-radiative analogs of these charge-transfer processes, internal conversion or intersystem crossing, are usually called electron transfer,

$$A^* + B \rightarrow A^+ + B^-$$

Their rates are subject to similar rules as other non-radiative processes, and have been studied in much detail. They are approximately described by the *Marcus theory*, which also leads to the conclusion that the rates of highly exothermic transfers are low (the "Marcus inverted region"). Because of the spatial separation of the donor and acceptor regions, the splitting between the singlet and triplet states of the dipolar species is particularly small.

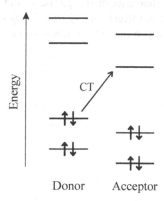

Fig. 1d-1. A schematic representation of a charge transfer transition in absorption.

Electron transfer processes play a crucial role in efforts to convert sunlight to electricity (photovoltaic cells) and to use it for the splitting of water. In both cases, the objective is to separate the generated charges spatially and to prevent their recombination by another non-radiative process to yield the ground state and heat. The charges are then led to electrodes or are used to reduce protons to molecular hydrogen and oxidize hydroxide anions to molecular oxygen.

1d-5 Energy Transfer

Another important class of internal conversion and intersystem crossing processes are of the energy transfer type. In these, both states involved are locally excited, but the excitation is localized in different parts of the molecule or on different molecules,

$$A* + B \rightarrow A + B^*$$

In addition to the trivial mechanism of energy transfer, in which one molecule emits a photon and another one absorbs it, there are two others (Fig. 1d-2). The *Förster mechanism* operates by an electrostatic coupling of the transition dipole moments of the two partners. It can be visualized as a simultaneous non-radiative deexcitation in the energy donor and excitation in the acceptor. Its rate falls off with the inverse sixth power of their distance, depends on their mutual orientation, and requires energy matching. It is resonant if the two chromophores involved have the same excitation energy, but normally proceeds downhill, from a chromophore with a higher excitation energy to one with a lower excitation energy. Then, some of the electronic energy is converted into vibrational excitation. Quantitatively, the match is expressed by the overlap of the emission spectrum of the donor and the absorption spectrum of the acceptor. Given the usual fluorescence lifetimes, the Förster mechanism is effective at distances of up to several nm. Since it depends on the transition moments of the partners, and singlet-triplet transition moments are very small, triplet-triplet energy transfer by the Förster mechanism is very slow. This may be compensated by the possibly very long lifetime of the donor triplet state.

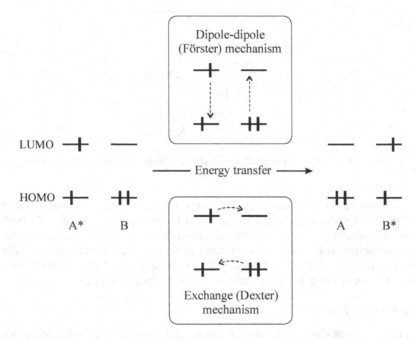

Fig. 1d-2. A schematic representation of the Förster and Dexter mechanisms for energy transfer from A* to B.

The *Dexter* or *exchange mechanism* is short-range and requires interpenetration of the wave functions of the two chromophores. It is best envisaged as a simultaneous transfer of an excited electron from one partner to the other and of its unexcited original counterpart in the opposite direction. As long as the partners effectively touch, it is competitive with the Förster mechanism for singlet excitation energy transfer and vastly faster than the Förster mechanism for triplet excitation energy transfer.

Singlet energy transfer has found much application in structural studies of biomolecules, where it is known under the acronym FRET (Förster resonant energy transfer). Triplet energy transfer has played an essential role in photochemistry, where it is used for excitation of substrates to their triplet state without going through their singlet states (sensitization), and for inducing the return of excited triplet states to the ground state (quenching).

1e EXCITED STATE KINETICS

A complete kinetic analysis of a photophysical or photochemical process requires the determination of the rate constants of all processes that occur, often as a function of their dependence on concentration, solvent, temperature, isotopic labels, and other variables. The basic time-independent observables from which this information is to be deduced are the fluorescence and phosphorescence emission and excitation spectra, quantum yields of fluorescence and phosphorescence of all emitting species, and quantum yields of starting material disappearance and product formation if photochemical transformations occur. The time-dependent observables are the decay curves of all observed emissions and transient absorptions. These basic data are often complemented by the investigation of the effect of quenchers and sensitizers, and sometimes by other measurements, such as photoacoustic calorimetry, CIDNP, etc. In the following, we only provide a brief survey of some of the basics.

The *quantum yield* of a species is measured by determining the number of moles of the species produced upon absorption of a mole of photons (an Einstein). In this sense, a photon of fluorescent or phosphorescent light is considered a species. The efficiency of an elementary step is defined as the fraction of molecules that undergo the step in competition with all other steps possible from the same starting species. The quantum yield of a species is the product of the efficiencies of all the steps that form the path by which it is produced. If a species is formed along more than one path, the contributions of all the participating paths need to be added.

The initial processes that one can anticipate after excitation of an organic molecule with light are formation of one of the excited singlet states and its rapid conversion to the thermally equilibrated lowest excited singlet state on the time scale of a few tens of ps, with a unit efficiency (*Kasha's rule*). Faster processes competitive with this scenario have been observed at times, but they are considered the exception rather than the rule (e.g., direct photochemical reactions that occur by return to the ground state through a conical intersection, or ultrafast intersystem crossing that takes place without prior vibrational equilibration in the S_1 state, especially in the presence of heavy atoms). In general, it is possible to use monochromatic radiation at a wavelength that assures that no state other than S_1 is excited.

The typical processes that compete for the thermally equilibrated S_1 state are fluorescence, internal conversion, intersystem crossing, bimolecular quenching with a quencher Q present at constant concentration [Q], a unimolecular photochemical transformation, and a bimolecular photochemical reaction with a partner B present at a constant concentration [B], with rate constants k_F, k_{IC}, k_{ISC}, $k_Q[Q]$, k_R, and $k_B[B]$, respectively (Figs. 1b-12 and 1d-3).

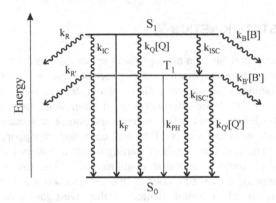

Fig. 1d-3. A list of processes that commonly contribute to a photochemical mechanism.

The *lifetime* of S_1, i.e. the time needed for its concentration to decay to $1/e$ of its initial value if it is not being replenished, is the inverse of the sum of the unimolecular rate constants of all the processes that depopulate it:

$$\tau(S_1) = 1/(k_F + k_{IC} + k_{ISC} + k_Q[Q] + k_R + k_B[B])$$

The lifetime can be measured by observing the decay of fluorescence intensity after a short pulse excitation, or by monitoring the decay of the transient absorption due to S_1 in time.

The *efficiencies* of the individual steps are

$$\eta_F = k_F\tau(S_1)$$
$$\eta_{IC} = k_{IC}\tau(S_1)$$
$$\eta_{ISC} = k_{ISC}\tau(S_1)$$
$$\eta_Q = k_Q[Q]\tau(S_1)$$
$$\eta_R = k_R\tau(S_1)$$
$$\eta_B = k_B[B]\tau(S_1)$$

Assuming that the efficiency with which S_1 is formed is unity, the quantum yield of fluorescence is $\phi_F = \eta_F$. Its measurement thus immediately yields the value of k_F. This can be checked against the value expected from the integrated intensity of the S_1 band in the absorption spectrum (Section 1c-2). The quantum yield of triplet formation is given by $\phi_{ISC} = \eta_{ISC}$ and is harder to determine, unless the transient T_1 absorption spectrum can be measured and the absorption coefficient is known or can be determined. Then, k_{ISC} can be found. If the product R is only formed from S_1, its quantum yield is $\phi_R = \eta_R$, and its measurement fixes the value of k_R.

Information on k_Q and k_B can be obtained from a so-called Stern-Volmer plot. For instance, when one measures the fluorescence quantum yield or lifetime for several choices of [Q] and labels them $\phi_F^{[Q]}$ and $\tau(S_1)^{[Q]}$

$$\phi_F^{[Q]} = k_F/(k_F + k_{IC} + k_{ISC} + k_Q[Q] + k_R + k_B[B])$$
$$\tau(S_1)^{[Q]} = 1/(k_F + k_{IC} + k_{ISC} + k_Q[Q] + k_R + k_B[B])$$

and uses the symbols ϕ_F^0 and $\tau(S_1)^0$ for the values measured in the absence of quencher, one obtains

$$\phi_F^0/\phi_F^{[Q]} = \tau(S_1)^0/\tau(S_1)^{[Q]} = 1 + \tau(S_1)^0 k_Q[Q]$$

Thus, when $\phi_F^0/\phi_F^{[Q]}$ or $\tau(S_1)^0/\tau(S_1)^{[Q]}$ is plotted against [Q], one obtains a straight line of slope $\tau(S_1)^0 k_Q$ and intercept 1. If $\tau(S_1)^0$ is known, k_Q can be determined, and vice versa. Similar results are obtained when ϕ_F or $\tau(S_1)$ is measured as a function of [B]. Often, the quenching can be assumed to be diffusion limited and its rate constant k_Q is of the order 10^{10} L mol^{-1} s^{-1}. When the Stern-Volmer plot is curved, the mechanism is more complicated. E.g., two excited states may be involved and only one of them is quenched, or static quenching in pre-formed complexes of the substrate with the quencher intervenes in addition to the dynamic quenching considered so far.

In general, with enough effort, all the rate constants of processes that start in S_1 can be obtained.

The molecules that reach T_1 can again branch. Some will proceed to S_0 by phosphorescent emission with a rate constant k_{PH}, others by intersystem crossing with a rate constant $k_{ISC'}$, some will be quenched with a rate constant $k_{Q'}[Q']$, some will react unimolecularly with a rate constant $k_{R'}$, and some bimolecularly with a rate constant $k_{B'}[B']$. The magnitudes of these rate constants will be obtained in similar ways. For the lifetime of phosphorescence we obtain

$$\tau(T_1) = 1/(k_{PH} + k_{ISC'} + k_{Q'}[Q'] + k_{R'} + k_{B'}[B'])$$

For the efficiency of phosphorescence, we obtain

$$\eta_{PH} = k_{PH}\tau(T_1)$$

and for its quantum yield,

$$\phi_{PH} = \eta_{ISC}\eta_{PH} = k_{ISC}\tau(S_1)k_{PH}\tau(T_1)$$

etc.

Acknowledgments

Support from the US National Science Foundation is gratefully acknowledged. The author thanks Prof. Alberto Credi for preparing the illustrations for this Chapter.

REFERENCES

Theoretical and computational aspects of photophysics

Michl, J. *Top. Curr. Chem.* **1974**, *46*, 1-59.
Bonacic-Koutecký, V.; Koutecký, J.; Michl, J. *Angew. Chem. Int. Ed. Engl.* **1987**, *26*, 170-190.
Michl, J.; Bonacic-Koutecký, V. *Electronic Aspects of Organic Photochemistry*; Wiley: New York (USA), 1990, 475p.
Domcke, W.; Yarkony, D. R.; Köppel, H. (eds.) *Conical Intersections: Electronic Structure, Dynamics, and Spectroscopy*; World Scientific: Singapore, 2004, 856p.
Kutateladze, A. G. (ed.) *Computational Methods in Photochemistry*; Marcel Dekker/CRC Press: Boca Raton (FL, USA), volume 13 in the series "Molecular and Supramolecular Photochemistry", 2005, 448p.
Olivucci, M. (ed.) *Computational Photochemistry*; Elsevier: Amsterdam (The Netherlands), part of the series "Theoretical and Computational Chemistry", 2005, 362p.

Experimental aspects of photophysics

Birks, J. B. *Photophysics of Aromatic Molecules*; Wiley-Interscience: New York (USA), 1970, 704p.
Murov, S. L.; Carmichael, I.; Gordon, L. H. *Handbook of Photochemistry*; Marcel Dekker: New York (USA), 1973, 420p.
Birks, J. B. *Organic Molecular Photophysics*; Wiley and Sons: New York (USA), volume 1, 1973, 600p.; volume 2, 1975, 653p.
Creation and Detection of the Excited State; Marcel Dekker: New York (USA), Lamola, A. A. (ed.), volume 1, 1971, parts A and B, 658p.; Ware, W. R. (ed.), volume 2, 1974, 230p., volume 3, 1974, 193p., volume 4, 1976, 320p.
Lim, E. C. (ed.) *Excited States*; Academic Press: New York (USA) volume 1, 1974, 347p.; volume 2, 1975, 403p.; volume 3, 1977, 351p.; volume 4, 1979, 400p.; volume 5, 1982, 204p.; volume 6, 1982, 224p.
Burgess, C.; Knowles, A. (eds.) *Standards in Absorption Spectrometry*; Chapman and Hall: London (UK), 1981, 141p.
Miller, J. N. (ed.) *Standards in Fluorescence Spectrometry*; Chapman and Hall: London (UK), 1981, 112p.
Rabek, J. F. *Experimental Methods in Photochemistry and Photophysics*; Wiley-Interscience: New York (USA), 1982, part 1, 592p., part 2, 506p.
Scaiano, J.-C. (ed.) *CRC Handbook of Organic Photochemistry*; CRC Press: Boca Raton (FL, USA), 1989, volume 1, 451p., volume 2, 481p.
Rabek, J. F. (ed.) *Photochemistry and Photophysics*; CRC Press: Boca Raton (FL, USA), 1991, volume 3, 202p., volume 4, 320p.
Turro, N. J. *Modern Molecular Photochemistry*; University Science Books: Mill Valley (CA, USA), 1991, 628p.
Gilbert, A.; Baggott, J. *Essentials of Molecular Photochemistry*; CRC Press: Boca Raton (FL, USA), 1991, 538p.
Perkampus, H.-H. *UV-VIS Spectroscopy and Its Applications*; Springer-Verlag: Berlin (Germany), 1992, 244p.

Klessinger, M.; Michl, J. *Excited States and Photochemistry of Organic Molecules*; VCH: New York (USA), 1995, 544p.

Michl, J.; Thulstrup, E. W. *Spectroscopy with Polarized Light. Solute Alignment by Photoselection*, in *Liquid Crystals, Polymers, and Membranes*; Revised soft-cover edition; VCH: Deerfield Beach (FL, USA), 1995, 573p.

Lakowitz, J. R. *Principles of Fluorescence Spectroscopy, 2nd Ed.*; Kluwer/Plenum: New York (USA), 1999, 698p.

Waluk, J. (ed.) *Conformational Analysis of Molecules in Excited States*; Wiley-VCH: New York (USA), 2000, 376p.

2

Photophysics of Transition Metal Complexes in Solution

By Vincenzo Balzani, Department of Chemistry "G. Ciamician", University of Bologna, Italy

2a Electronic Structure
    2a-1 Crystal Field Theory
    2a-2 Ligand Field Theory
    2a-3 Molecular Orbital Theory
2b Types of Excited States and Electronic Transitions
2c Absorption and Emission Bands
    2c-1 Ligand-Field Bands
    2c-2 Charge-Transfer Bands
    2c-3 Intra-Ligand Bands
    2c-4 A Remark
    2c-5 Band width
2d Jablonski Diagram
2e Photochemical Reactivity
    2e-1 Ligand-Field Excited States
    2e-2 Charge-Transfer Excited States
    2e-3 Intra-Ligand Excited States
    2e-4 Remarks
2f Electrochemical Behavior
    2f-1 Ligand Dissociation Processes
    2f-2 Structural Changes
    2f-3 Reversible Processes
2g Polynuclear Metal Complexes
    2g-1 Supramolecular Species or Large Molecules?
    2g-2 Electronic Interaction and Intervalence-Transfer Transitions

Transition metal complexes [9901, 0301], which are the most common coordination compounds, may be cationic, anionic or non-ionic species, depending on the charges carried by the central metal atom and the coordinated groups. These groups are usually called *ligands*. Ligands may be attached to the central atom through one or more atoms, i.e. they may be *mono-* or *poly-dentate*. The total number of attachments to the central atom is called *coordination number*. Coordination numbers of 2 to 10 are known, but most coordination compounds exhibit a coordination number of 6 or 4. A given metal ion does not necessarily have only one characteristic coordination number and, moreover, a given coordination number may give rise to several types of spatial arrangements. For example, Ni(II) forms both octahedral six-coordinated and tetrahedral or square-planar four-coordinated complexes. Coordination compounds that contain ligands of two or more types are called *mixed-ligand* complexes.

Many types of isomerism are encountered among transition metal complexes. The most common ones are *geometrical* and *optical isomerisms*. Another important type of isomerism is the *linkage isomerism* which occurs with ligands capable of coordinating in more than one way (ambidentate ligands). For example, the group NO_2^- may coordinate through the nitrogen or the oxygen atom giving rise to *nitro* isomers, such as $[(NH_3)_5Co-NO_2]^{2+}$ and to *nitrito* isomers, such as $[(NH_3)_5Co-ONO]^{2+}$.

Most of the transition metal complexes are electrically charged species (complex ions) and therefore, they are fairly soluble only in water or other polar solvents. Complex ions, like other charged particles, are particularly sensitive to their environment, i.e. to the nature of the solvent molecules and to the presence of other ionic species. For high ion concentrations, specific effects arise, such as the formation of "ion-pairs" (or "outer-sphere" complexes).

2a ELECTRONIC STRUCTURE

The most correct and complete approach to the electronic structure of metal complexes is that provided by the molecular orbital theory (MOT). Unfortunately, this theory requires a great deal of computation effort in order to provide quantitative results. Consequently, other less meaningful but more handy theories are often used in describing the bonding in complexes. These theories are the crystal field theory (CFT) and the ligand field theory (LFT). The valence bond theory, in its usual form, does not take into account the existence of electronic excited states and thus, it cannot offer any explanation for the photophysical and photochemical behavior.

2a-1 Crystal Field Theory

In this theory [6101, 9501] the ligands are treated as negative point charges which set up an electrostatic field. The effect of this field on the central metal ion is then investigated. Since in the transition-metal ions all but the outermost partially occupied d orbitals are either filled or completely empty, interesting effects, other than those dealt with in the classical ionic theory, arise only from the interaction of the electrostatic field of the ligands with the d electrons.

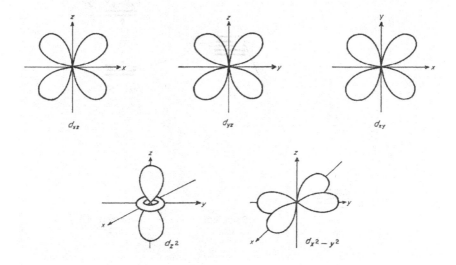

Fig. 2a-1. Pictures of the five d orbitals.

In a gaseous transition-metal ion, the five d orbitals (whose shapes are shown in Fig. 2a-1) are degenerate. In a complex where the ligand field is spherically symmetric, the five d orbitals would be higher in energy than in the free ion, because of the repulsion between the metal ion electron density and the spherical field of negative charge. In such a hypothetical environment, the d orbitals would be still five-fold degenerate. In an actual complex, however, a spherical field is never obtained. Therefore, the five d orbitals, because of their different orientation, are no longer equivalent and they are split according to the particular symmetry of the complex. For example, in a six-coordinated complex having the ligands located on the corners of a regular octahedron, simple pictorial arguments (or better, group-theoretical methods) show that the five d orbitals split into two sets: one set of three orbitals, d_{xy}, d_{xz} and d_{yz}, equivalent to one another and labeled t_{2g}, and another set of two orbitals, d_{z^2} and $d_{x^2-y^2}$, equivalent to each other but different from the previous set, labeled e_g. The e_g orbitals, which point directly toward the ligands, are higher in energy than the t_{2g} orbitals, which point between the ligands (Fig. 2a-2). The amount of splitting between e_g and t_{2g} orbitals is generally denoted by Δ or 10 Dq. The magnitude of Δ obviously depends on the central ion and ligands involved and, according to the CFT, it may be calculated using an electrostatic model. Using an analogous line of reasoning, the splitting patterns of the d orbitals can be obtained for the most important geometries of complexes (Fig. 2a-3).

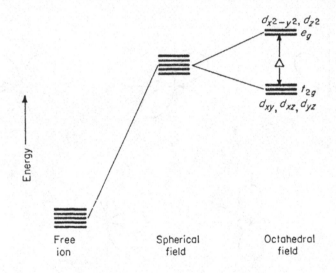

Fig. 2a-2. Splitting of the five d orbitals of a central metal ion in an octahedral field.

Now, let us consider a metal ion with n d-electrons. Owing to the interelectronic repulsions, the electrons are spread out in the five d orbitals so as to give the maximum number of unpaired spins (rule of maximum multiplicity, or Hund's rule). When the metal ion is introduced into an octahedral environment, the d orbitals split in the manner described above and there will now be a tendency for the electrons to fill the lowest orbitals, t_{2g}, before beginning to go into the upper orbitals, e_g. For a d^1 ion, the single electron will thus occupy one of the t_{2g} orbitals, and the complex can be said to have a t_{2g}^1 configuration. Similarly, for d^2 and d^3 ions, each electron will go into a different t_{2g} orbital and the electron spins will remain uncoupled, as required by Hund's rule. Therefore, the complexes of these ions have a t_{2g}^2 and t_{2g}^3 configuration, respectively. For a d^4 ion, a new and interesting situation arises, since the complex may choose between the following possibilities:

(i) *Low-spin configuration.* All the four electrons go into the three t_{2g} orbitals $(t_{2g}^4$ configuration); in such a case, two electrons must be coupled in the same orbital and this requires a certain amount of energy (P) in order to win the inte-relectronic repulsion.

(ii) *High-spin configuration.* All the four electrons remain unpaired, but one of them must then be promoted to the e_g orbitals $(t_{2g}^3 e_g^1$ configuration); this re-quires expenditure of the energy Δ.

Of course, the *ground state* of a complex will be represented by the low-spin configuration when $\Delta > P$ *(strong-field)* case), and by the high-spin configuration when $\Delta < P$ *(weak-field)* case). For complexes containing the same metal ion, P is nearly constant, while Δ strongly depends on the nature of the ligands. Therefore, a metal ion may give high-spin complexes with some ligands and low-spin com-plexes with other ligands. Octahedral complexes containing more than 4 d-electrons may be treated by the procedure outlined above, and complexes of symmetry other than the octahedral one may be treated by similar procedures.

Besides explaining the magnetic properties of complexes, the CFT enables us to predict, or at least to understand, a number of thermodynamic, kinetic, struc-tural, and spectroscopic features of complexes [6101, 9501].

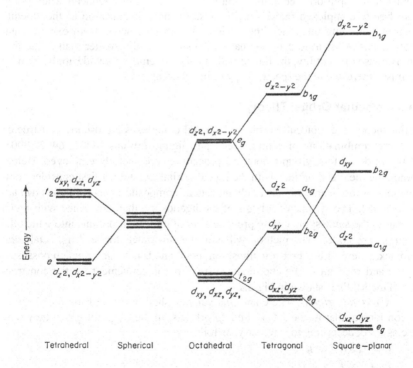

Fig. 2a-3. Splitting of the five d orbitals of a central metal ion in complexes of various sym-metry.

2a-2 Ligand Field Theory

A number of experimental data (electron spin resonance, nuclear magnetic resonance, and nuclear quadrupole resonance spectra, intensities of *d-d* transitions, nephelauxetic effect, antiferromagnetic coupling, etc.) prove that an appreciable degree of covalency does exist in metal complexes. Therefore, the pure electrostatic CFT model must be considered as a rough approximation of the electronic structure of complexes. A more real approach to this topic must take account of the existence of metal-ligand orbital overlap.

The theory which preserves all the conceptual and computational advantages of the simple CFT and which includes the possibility of accounting for the effects of covalent bonding is called ligand field theory (LFT) [6601, 9901, 0301]. According to this theory, the calculation of the energy level diagrams proceeds in the same manner as in CFT, except that one takes into account that the inter-electronic repulsion and spin-orbit coupling parameters are noticeably reduced when going from free to complexed metal ions, because of the delocalization of the "metal" electrons. In doing this, one either estimates such parameters from experimental observations or assumes that they have values reasonably smaller than in the free ion. Because of its utility, the ligand field theory has enjoyed a wide application by chemists interested in the spectroscopy of metal complexes.

2a-3 Molecular Orbital Theory

In this theory, the molecular orbitals for metal complexes are usually constructed as linear combinations of central metal and ligand orbitals [6401, 6402, 9901, 0301]. In doing this, group-theoretical procedures are usefully employed. Before combining the metal orbitals with the ligand orbitals, one must first consider combinations of the ligand orbitals to obtain sets of composite ligand orbitals (*symmetry orbitals*). The symmetry orbitals of the ligands are then combined with metal orbitals of the same symmetry to produce a set of molecular orbitals into which the electrons are added. The method will now be illustrated for an ML_6 octahedral complex, where M is a first-row transition metal and L is a ligand which possesses both σ and π orbitals. The coordinate system that is convenient for the construction of the MO's is shown in Fig. 2a-4.

(i) *Metal orbitals.* There are nine valence shell orbitals of the metal ion to be considered, that is, the 3*d,* 4*s,* and 4*p* orbitals. In the O_h point group they may be classified, according to symmetry, as follows:

$(3\,d_{z^2}, 3\,d_{x^2-y^2})$, e_g;

$(3\,d_{xy}, 3\,d_{xz}, 3\,d_{yz})$, t_{2g};

$(4s)$, a_{1g};

$(4\,p_x, 4\,p_y, 4\,p_z)$, t_{1u}.

The e_g and a_{1g} orbitals are suitable only for σ bonding, the t_{2g} orbitals are suitable only for π bonding, and the t_{1u} orbitals may give both σ and π bonding.

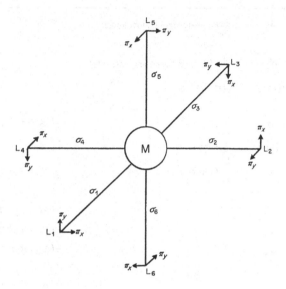

Fig. 2a-4. Coordinate system for σ and π bonding in an octahedral complex.

(ii) *Ligand orbitals.* The individual ligand σ orbitals are combined into six symmetry orbitals. Each one of these is constructed so as to overlap effectively with a particular one of the six metal orbitals which are suitable for σ bonding. This is done by choosing the linear combinations of ligand σ orbitals which have the same symmetry properties as the various metal σ orbitals. (For example, the linear combination of ligand σ orbitals which goes with the $4p_x$ metal orbital has a plus sign in the $+x$ direction and a minus sign in the $-x$ direction. This is the combination $\sigma_1 - \sigma_3$.) Similarly, the twelve individual ligand π orbitals (Fig. 2a-4) are combined into twelve symmetry orbitals; six of them will have appropriate symmetry to match the six metal π orbitals, whereas the other six have no metal orbital counterparts and therefore, they will be *non-bonding* with respect to the metal complex. All the σ and π metal and ligand orbitals are shown in Table 2a.

Table 2a Symmetry Classification of Orbitals for Octahedral Complexes

Symmetry represen-tations	Metal orbitals	Ligand orbital combinations	
		σ	π
a_{1g}	$4s$	$\frac{1}{\sqrt{6}}(\sigma_1+\sigma_2+\sigma_3+\sigma_4+\sigma_5+\sigma_6)$	
e_g	$3d_{z^2}$	$\frac{1}{2\sqrt{3}}(2\sigma_5+2\sigma_6-\sigma_1-\sigma_2-\sigma_3-\sigma_4)$	
	$3d_{x^2-y^2}$	$\frac{1}{2}(\sigma_1-\sigma_2+\sigma_3-\sigma_4)$	
t_{1u}	$4p_x$	$\frac{1}{\sqrt{2}}(\sigma_1-\sigma_3)$	$\frac{1}{2}(\pi_{2y}+\pi_{5x}-\pi_{4x}-\pi_{6y})$
	$4p_y$	$\frac{1}{\sqrt{2}}(\sigma_2-\sigma_4)$	$\frac{1}{2}(\pi_{1x}+\pi_{5y}-\pi_{3y}-\pi_{6x})$
	$4p_z$	$\frac{1}{\sqrt{2}}(\sigma_5-\sigma_6)$	$\frac{1}{2}(\pi_{1y}+\pi_{2x}-\pi_{3x}-\pi_{4y})$
t_{2g}	$3d_{xy}$		$\frac{1}{2}(\pi_{1x}+\pi_{2y}+\pi_{3y}+\pi_{4x})$
	$3d_{xz}$		$\frac{1}{2}(\pi_{1y}+\pi_{5x}+\pi_{3x}+\pi_{6y})$
	$3d_{yz}$		$\frac{1}{2}(\pi_{2x}+\pi_{5y}+\pi_{4y}+\pi_{6x})$
t_{2u}			$\frac{1}{2}(\pi_{2y}-\pi_{5x}-\pi_{4x}+\pi_{6y})$
			$\frac{1}{2}(\pi_{1x}-\pi_{5y}-\pi_{3y}+\pi_{6x})$
			$\frac{1}{2}(\pi_{1y}-\pi_{2x}-\pi_{3x}+\pi_{4y})$
t_{1g}			$\frac{1}{2}(\pi_{1y}-\pi_{5x}+\pi_{3x}-\pi_{6y})$
			$\frac{1}{2}(\pi_{2x}-\pi_{5y}+\pi_{4y}-\pi_{6x})$
			$\frac{1}{2}(\pi_{1x}-\pi_{2y}+\pi_{3y}-\pi_{4x})$

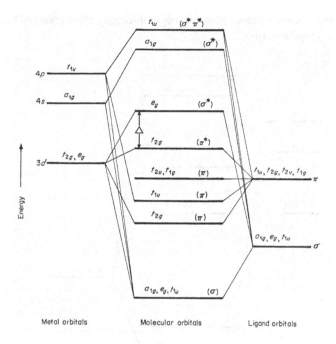

Fig. 2a-5. Molecular orbital energy level diagram for an octahedral complex containing ligands that possess σ and π orbitals.

(iii) *Molecular orbitals.* The bonding and antibonding molecular orbitals for the complex are then obtained by combining metal and ligand orbitals which have the same symmetry properties. The general formulation for MO's is

$$\psi = a_M \varphi_M + a_L \varphi_L \qquad (2\text{-}1)$$

where φ_M and φ_L are metal and ligand orbital combinations and a_M and a_L are co-efficients, whose values are restricted by conditions of normalization and orthogo-nality. Each MO so obtained is indicated with the appropriate symmetry label, using the asterisk to signify the antibonding character. The energy of each MO may be estimated using semiempirical methods and, in several cases, the relative molecular orbital energy order may be obtained. These results are conveniently expressed in schematic energy-level diagrams, such as that of Fig. 2a-5. A similar procedure may be applied to complexes containing ligands which do not possess π orbitals or which have two types of π orbitals (see later). Once the relative molecu-lar orbital energy order has been established, the valence electrons are assigned to the orbitals of lowest energy. It should be noted, however, that the energy differ-ence between the t_{2g} and e_g orbitals, Δ, may be lower than the inter-electronic repulsion energy; in such a case, electrons begin to occupy the e_g orbitals before the t_{2g} ones are completely filled.

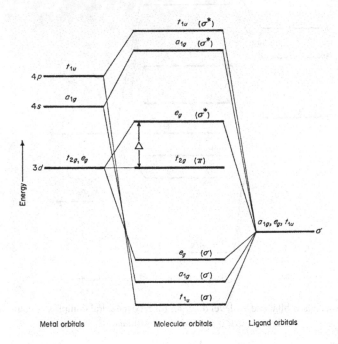

Fig. 2a-6. Molecular orbital energy diagram for an octahedral complex containing ligands which do not posses π-orbitals.

Some important features of energy-level diagrams are:

(i) The ligand σ orbitals are lower in energy than the metal σ orbitals. Therefore, in accordance with the theory which shows that a MO formed from two component orbitals includes a larger contribution from that component which is closer to it in energy, the six bonding σ MO's have more the character of ligand orbitals than they do of metal orbitals. Stated in another way, the electron density of bonding σ MO's is concentrated on the ligands rather than on the metal. Of course, the opposite is true for the antibonding σ* MO's.

(ii) The e_g and t_{2g} orbitals at the center of the diagrams are mainly localized on the metal and correspond, respectively, to the higher and lower energy sets of d orbitals described by the CFT.

(iii) When the ligands do not possess π orbitals (for example, NH_3), the t_{2g} MO's are purely metal d orbitals (non-bonding orbitals). The energy-level diagram corresponding to such a simple case is that of Fig. 2a-6.

(iv) If the ligands possess only π orbitals of lower energy than the metal t_{2g} orbitals (for example, ligands such as the fluoride and oxide ions), the π interaction destabilizes the metal t_{2g} orbitals and thus, the Δ value is lower than in the case of a σ interaction alone (Fig. 2a-7, *a* and *b*). The complete energy-level diagram for a complex of this type is that of Fig. 2a-5.

(v) If the ligands have only π orbitals of higher energy than the metal t_{2g} orbitals (for example, Ph_3P, Et_2S, etc.), the π interaction stabilizes the metal t_{2g} orbitals and thus, the Δ value is greater than in the case of a σ interaction alone (Fig. 2a-7, *a* and c).

(vi) If the ligands have both low energy and high energy π orbitals (examples are CN^-, CO, Cl^-, etc.), the net effect depends on the competition between the interaction of the two types of ligand π orbitals with the metal t_{2g} orbitals.

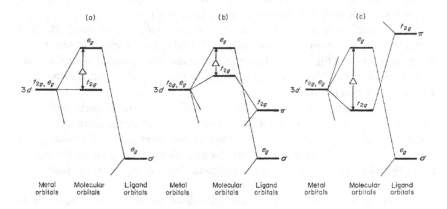

Fig. 2a-7. Effect of π bonding on the energy separation Δ between the e_g and t_{2g} "metal" orbitals (see text).

In principle, the molecular orbital treatment can be applied to complexes regardless of their symmetry. However, there is a rapid loss of simplicity going from octahedral complexes to less symmetric ones. In the T_d symmetry (tetrahedral complexes), the only pure metal σ orbital is $4s$ (a_1), the $3d_{z^2}$, $3d_{x^2-y^2}$ (e) are pure π orbitals, and the $3d_{xy}$, $3d_{xz}$, $3d_{yz}$ (t_2) and $4p_x$, $4p_y$, $4p_z$ (t_2) orbitals may give both σ and π bonding. For the symmetry D_{4h} (square-planar complexes), the $3d_{z^2}$ (a_{1g}), $4s$ (a_{1g}), $3d_{x^2-y^2}$ (b_{1g}) are pure σ orbitals, the $3d_{xz}$, $3d_{yz}$ (e_g) $3d_{xy}$ (b_{2g}) and $4p_z$ (a_{2u}) are pure π orbitals, and the $4p_x$, $4p_y$ (e_u) orbitals can be used in both σ and π bonding.

The MOT, using a much more realistic model, obtains the same *d*-orbital splitting as predicted by the CFT. Therefore, in principle, all the phenomena explained by the CFT (magnetism, visible spectra, etc.) may be accounted for equally well (or even better) by the MOT. In addition, a number of phenomena

(such as the charge-transfer spectra, see below), that are not taken into consideration by the CFT, can be interpreted by the MOT.

2b TYPES OF EXCITED STATES AND ELECTRONIC TRANSITIONS

The complexity of the electronic structure of coordination compounds is, of
course, reflected in their photophysical and photochemical properties [7001, 8401,
9301, 9401]. For example, the assignment of the various bands to transitions
which lead to specific excited states is a very difficult task. In order to make the
discussion easier, it is convenient to describe each state by using its preponderant
molecular orbital configuration and, then, to make a classification of the electronic
transitions according to the location of the MO's involved. Using this criterion,
three types of electronic transitions may be distinguished:

    1. *Transitions between MO's mainly localized on the central metal.* MO's of
such a type are those deriving from the metal d orbitals (for example, the $t_{2g}(\pi)$
and $e_g(\sigma^*)$ MO's in Fig. 2a-6). These transitions are usually called *d-d transitions*
or *metal centered* (MC) or *ligand-field transitions*, since they can also be dealt
with by the CF and LF theories.

    2. *Transitions between MO's mainly localized on the ligands and MO's
mainly localized on the central metal.* For obvious reasons, these transitions are
called *charge-transfer (CT)* or *electron-transfer transitions.* Depending on
whether the excited electron is originally located on the ligands or on the central
metal, ligand-to-metal (LM) or metal-to-ligand (ML) charge-transfer transitions
can occur.

    3. *Transitions between MO's mainly localized on the ligands.* These transi-
tions only involve ligand orbitals which are almost unaffected by coordination to
the metal. Transitions of such a type are called *ligand centered* (LC) or *intra-
ligand transitions.*

These kinds of electronic transitions for an octahedral complex are schematically
represented in the *orbital-level* diagram of Fig. 2b-1. More details about the three
types of bands [8401] will be reported below. Less frequently encountered types
of transitions (not shown in Fig. 2b-1) are those from a metal-centered orbital to a
solvent orbital (*charge-transfer to solvent*, CTTS), or between two orbitals predominantly localized on different ligands of the same metal center (*ligand-to-
ligand charge-transfer*, LLCT).

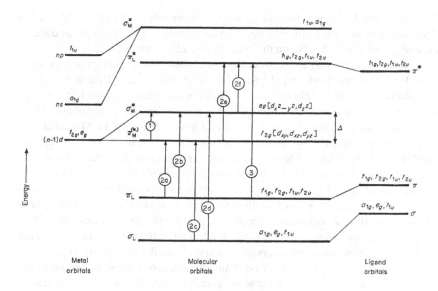

Fig. 2b-1. Schematic orbital energy diagram representing various types of electronic transitions in octahedral complexes. A line connects an atomic orbital to that MO in which it has the greatest participation. The asterisk of π_M is put into parentheses since this orbital may have bonding, non-bonding, or antibonding character, depending on the particular metal and ligands involved. 1: ligand field (or metal centered) transitions; 2a, 2b. 2c, 2d: ligand-to-metal charge-transfer transitions; 2e, 2f: metal-to-ligand charge-transfer transitions; 3: intra-ligand (or ligand centered) transitions.

2c ABSORPTION AND EMISSION BANDS

2c-1 Ligand-Field Bands

For complexes which originate from d^1 metal ions, the theoretical treatment of the ligand-field bands is very simple [6201, 7002, 8401]. Consider, for example, a d^1 octahedral complex, such as $Ti(H_2O)_6^{3+}$. In a MO energy-level diagram (Fig. 2a-6), all the orbitals lying below the $t_{2g}(\pi^*)$ orbitals are filled, the $t_{2g}(\pi^*)$ orbitals contain 1 electron, the $e_g(\sigma^*)$ orbitals are empty, and so are all the other upper orbitals. As previously noted, this picture coincides with that given by the ligand-field theory as far as the relevant orbitals, t_{2g} and e_g, are concerned. Since the t_{2g}^1 ground electronic configuration of a d^1 complex gives rise to only one state ($^2T_{2g}$), and the e_g excited electronic configuration also gives rise to only one state (^2E_g), only one d-d band may be present in the spectrum of $Ti(H_2O)_6^{3+}$. In an *orbital diagram* (Fig. 2b-1), this band corresponds to the metal centered transition, indicated by arrow 1, and in a *state diagram,* it corresponds to the $^2T_{2g} \rightarrow {}^2E_g$ transition. Note that in such a simple case, the wavenumber of the band gives directly the energy separation Δ between the "metal" t_{2g} and e_g orbitals.

For complexes containing more than one *d*-electron, the situation is complicated by the interelectronic repulsions. First of all, the promoted electron may either retain or change its spin orientation. As a consequence, each excited configuration gives rise to two distinct excited states, one having the same multiplicity as the ground state and the other of different multiplicity (Chapter 1). Moreover, additional complications arise because degenerate sets of orbitals (e_g and t_{2g}) are involved in the transition. The interelectronic repulsions, in fact, depend not only on the spin orientations, but also on the particular distribution of the electrons in the degenerate sets of orbitals. Thus, an electronic configuration can give rise to more than one state of a given multiplicity. For example, the ground electronic configuration of a d^2 octahedral complex, t_{2g}^2, gives rise to the states $^3T_{1g}$, $^1T_{2g}$, $^1A_{1g}$ and ^1E_g. According to Hund's rule, the lowest of these states, i.e. the ground state, is $^3T_{1g}$. Absorption bands could then arise from the transitions $^3T_{1g} \rightarrow {}^1T_{2g}$, $^3T_{1g} \rightarrow {}^1A_{1g}$, and $^3T_{1g} \rightarrow {}^1E_g$, i.e. from transitions between states belonging to the same configuration, t_{2g}^2. These intraconfigurational transitions, of course, cannot be shown in orbital-energy diagrams, such as that of Fig. 2b-1. Interestingly, they give rise to very narrow absorption and emission bands, as we will see later. The excitation of one electron from the ground electronic configuration t_{2g}^2 leads to the excited configuration $t_{2g}e_g$. This configuration gives rise to the following states: $^3T_{1g}$, $^3T_{2g}$, $^1T_{1g}$, $^1T_{2g}$. Therefore, four transitions arise when an electron undergoes the $t_{2g} \rightarrow e_g$ promotion. Then, there is the two-electron excited configuration e_g^2 which gives rise to the states $^3A_{2g}$, $^1A_{1g}$, and ^1E_g. Therefore, three transitions correspond to the $t_{2g}^2 \rightarrow e_g^2$ promotion. Transitions from the ground state $^3T_{1g}$ to each one of the excited states described above are symmetry forbidden since all the states have g character; moreover, transitions from the ground state (which is a triplet) to single excited states are spin forbidden (Section 1c-3), and the transitions from the ground state to the states arising from the e_g^2 excited configuration are also forbidden since they involve a two-electron promotion. Therefore, for a d^2 system, such as V(H₂O)₆³⁺, only two transitions, namely, $^3T_{1g}(t_{2g}^2) \rightarrow {}^3T_{2g}$ $(t_{2g}e_g)$ and $^3T_{1g}(t_{2g}^2) \rightarrow {}^3T_{1g}$ $(t_{2g}e_g)$ are expected (and found) to give bands of sufficient intensity to be observable.

In the above discussion, the assumption has been made that the t_{2g} and e_g levels are so far apart that the interactions between different configurations, such as t_{2g}^2 and $t_{2g}e_g$, can be ignored. This occurs in the limit of very strong perturbing fields [6601, 9501, 0301], i.e. when the ligand-field energy is much higher than the energy distance between terms of the same configuration.

From the actual spectra, however, it is found that, in most cases, such an approximation is not valid, and that a satisfactory interpretation of the band positions can be given only when configuration interactions are included. For example, in the two-electron case, $^3T_{1g}(t_{2g}^2)$ and $^3T_{1g}(t_{2g}\,e_g)$ levels "interact" giving rise to two levels, $^3T_{1g}(1)$ and $^3T_{1g}(2)$, which are, respectively, lower than $^3T_{1g}(t_{2g}^2)$ and higher than $^3T_{1g}(t_{2g}\,e_g)$. The $^3T_{1g}(1)$ wavefunction is a combination of the original $^3T_{1g}(t_{2g}^2)$ and $^3T_{1g}(t_{2g}\,e_g)$ wavefunctions, but it remains predominantly $^3T_{1g}(t_{2g}^2)$. Similarly, the $^3T_{1g}(2)$ wavefunction is a combination of the original ones but remains predominantly $^3T_{1g}(t_{2g}\,e_g)$. This means that in the $^3T_{1g}(1)$ state, which is the ground state, the electrons spend nearly- all their time in the configuration t_{2g}^2, but that a little is spent in the $t_{2g}\,e_g$, or that a certain fraction of the electrons is in the level t_{2g}, while the rest is in e_g. This may be thought of as occurring because the electrostatic repulsion between the two electrons is higher for t_{2g}^2 than for $t_{2g}\,e_g$, since they are moving in a larger space in the latter case and thus the energy of this configuration is lowered with respect to the simple one-electron t_{2g}^2 value.

In these cases (weak-field complexes) [6601, 9501, 0301], in order to calculate the energy of the various levels, it is convenient to start with the Russell-Saunders terms for the free ion, since the energies of these are known accurately from atomic spectra and to consider their splitting caused by the perturbing ligand field. For example, the configuration d^2 in the field free ion V^{3+} has these terms (in order of increasing energy): ^3F (ground state), ^1D, ^3P, ^1G, and ^1S. In the O_h symmetry, the free ion terms split as follows:

$$^3F: \ ^3T_{1g} \text{ (ground state)}, \ ^3T_{2g}, \ ^3A_{2g}$$

$$^1D: \ ^1T_{2g}, \ ^1E_g$$

$$^3P: \ ^3T_{1g}$$

$$^1G: \ ^1A_{1g}, \ ^1T_{2g}, \ ^1T_{1g}, \ ^1E_g$$

$$^1S: \ ^1A_{1g}$$

It can be easily verified that these states are exactly those obtained from the three possible configurations t_{2g}^2, $t_{2g}\,e_g$ and e_g^2 in an octahedral field (see above).

As previously mentioned, the ground state $^3T_{1g}$ (arising from the ^3F term and having t_{2g}^2 configuration in the limit of very strong perturbing fields) and the excited state having the same symmetry $^3T_{1g}$ (which arises from the ^3P term and has the $t_{2g}\,e_g$ strong field configuration) must interact. The states which arise from this interaction, which were previously called $^3T_{1g}(1)$ and $^3T_{1g}(2)$, are more commonly indicated with $^3T_{1g}(F)$ and $^3T_{1g}(P)$ to signify their origin.

The treatment of complexes which have more than two d-electrons or belong to other symmetries follows similar lines. From the above discussion, it appears that the number and position of the ligand field bands depend not only on the ligand field, but also on the interelectronic repulsion parameters.

Now, let us turn to the experimental features of the ligand-field bands. As mentioned above, these bands are weak since "pure" d-d transitions are symmetry forbidden and they may occur only by mechanisms (such as the coupling of vibrational and electronic motions) that permit the breakdown of the symmetry selection rule (see Chapter 1). For spin-allowed ligand-field bands, most of the experimentally observed ε_{max} values fall in the range 1-150 for octahedral complexes, but they may be much higher (up to 1000) for non-centrosymmetric complexes. Spin-forbidden ligand field bands are about 10-100 times weaker than the corresponding spin-allowed bands.

The large number of experimental studies carried out on ligand-field bands show that ligands may be arranged in a series according to their capacity to cause d-orbital splitting (i.e. according to their Δ values). This series, which is known as the *spectrochemical series* [6101, 6201, 7002, 9501], for the more important ligands is:

$$CN^- > NO_2^- > phen > bpy > SO_3^{2-} \sim en > NH_3 \sim py > NCS^- \gg H_2O \sim ox^{2-}$$
$$> ONO^- \sim OH^- > OC(NH_2)_2 > F^- \gg SCN^- > Cl^- > Br^- > I^-$$

It should be emphasized that this order is valid only for complexes having metal ions in their normal oxidation states.

According to the CFT, the energy separation Δ of the t_{2g} and e_g metal d orbitals in an octahedral complex depends on the strength of the electrostatic ligand field. But, in reality, it is impossible to understand the spectrochemical series in terms of any ionic model. Conversely, a semiquantitative explanation is possible on the basis of the MOT. In this theory, the splitting is attributed to the difference in the covalent interaction of the $d(\sigma)$ and $d(\pi)$ valence orbitals. As a consequence, Δ will depend on the σ- and π-bonding properties of the ligand. It seems reasonable to assume that *most of the ligands have similar σ-bonding abilities, while possessing widely ranging π-donor and π-acceptor properties*. The spectrochemical series should therefore reflect the π-bonding properties of the various ligands. In fact, strong π-acceptor ligands as CN^- and NO_2^-, which stabilize the "metal" $d(\pi^*)_M$ MO's, cause large Δ values, whereas strong π-donor ligands, such as I^- and Br^-, increase the energy of the "metal" $d(\pi^*)_M$ MO's and then, they cause small Δ values (Fig. 2a-7).

For the same ligand, the splitting between the most stable and the most unstable d orbitals generally increases with increasing the charge of the central metal and the principal quantum number of the metal d orbitals involved.

In six-coordinated mixed-ligand complexes, the actual symmetry is lower than the octahedral one, and thus, the t_{2g} and e_g sets of degenerate orbitals undergo further splitting (see, for example, Fig. 2a-3). While the pattern of the splitting depends on the symmetry of the mixed-ligand complex, the energy separation between orbitals which are degenerate in octahedral symmetry depends on the difference in the strength of the ligands involved, i.e. on their spectrochemical position. When the ligands are near to each other in the spectrochemical series, the energies of the levels which originate from the splitting of the octahedral levels are not different enough to cause splitting of the "octahedral" ligand-field bands.

When the ligands occupy sufficiently distant positions in the spectrochemical series, at least one of the "octahedral" bands splits. It should be pointed out that, for a number of reasons, electronic transitions can give rise to broad bands, so that, if the energy difference between two levels is less than about 1500 cm^{-1}, the corresponding absorption bands are not likely to be resolved.

As previously mentioned, in order to obtain a better agreement between the calculated energy levels and those experimentally obtained from the spectra, one must assume that the interelectronic repulsion parameters are smaller in the complexed than in the free ion. According to their tendency to reduce the interelectronic repulsion parameters, the ligands may be arranged in a series. Since the reduction of these parameters occurs because of the overlap of the metal d orbitals with ligand orbitals (i.e. because of the expansion of the d electronic cloud), this ligand arrangement has been named *nephelauxetic series* ("cloud-expanding" series) [6101, 6201, 7002, 9501]. A nephelauxetic series of the central ions can also be considered, but it is not so well defined as that of the ligands. The nephelauxetic effect may be taken as a measure of the tendency towards covalent bonding.

The position of a definite ligand in the spectrochemical and nephelauxetic series is important for determining whether its complexes are low spin or high spin. For example, CN^- gives rise to low-spin complexes since it occupies high positions in both the series, while H_2O, which is in a relatively low position in the spectrochemical series and in a low position in the nephelauxetic series, is likely to give rise to high-spin complexes.

2c-2 Charge-Transfer Bands

In order to discuss the charge-transfer (or, electron-transfer) bands, the simple CF and LF theories cannot be used, and a MO approach is thus required [6201, 7001, 7002, 8401].

The assignment of absorption bands to charge-transfer transitions is based on the generally valid assumption that the central atom and the ligands are separated systems which only interact weakly with each other. Under this condition, the ground state can be described by a preponderant electron configuration consisting of "localized" MO's and thus, a well defined oxidation state of the central metal can be determined. Charge-transfer transitions can then be defined as those adding (or removing) one electron from the partly filled shell of the metal and hence changing its oxidation state by -1 (or $+1$).

Ligand→ metal charge-transfer (LMCT) bands are exhibited in the u.v. and also in the visible spectral regions, particularly by complexes containing highly reducing ligands, such as I^-, Br^-, and ox^{2-}. It is a common observation that the wavenumber of these bands decreases as the central ion becomes more oxidizing and the ligands more reducing. When all of the ligands are equivalent, there is no reason for assuming that the electron transferred to the central metal comes from any particular one of them, so that the electron is described as coming from a de-localized ligand MO (Fig. 2b-1). For example, in the case of $[IrCl]_6^{2-}$, the preponderant configuration of the ground state corresponds to the oxidation state $+4$ for Ir and -1 for each ligand, while the excited states producing the LMCT bands have Ir(III) and a collectively oxidized set of ligands Cl_6^{5-}. When in a mixed-ligand complex one ligand is much more reducing than the others, it is possible to consider that the electron is transferred from *that* specific ligand to the central metal. Thus, the intense bands which appear in the near UV or visible spectrum of the monohalopentammine complexes of Co(III) are attributed to charge-transfer transitions from the halide ligand to the Co(III) ion [7001].

Using MO configurations as a classification, in an octahedral complex we may distinguish four types of LMCT bands. They are (Fig. 2b-1, arrows 2a, 2b, 2c, and 2d): $\pi_L \rightarrow \pi_M{}^{(*)}(t_{2g})$, $\pi_L \rightarrow \sigma_M{}^*(e_g)$, $\sigma_L \rightarrow \pi_M{}^{(*)}(t_{2g})$, and $\sigma_L \rightarrow \sigma_M{}^*(e_g)$, where π_L and σ_L are MO's mainly localized on the ligands, and $\pi_M{}^{(*)}(t_{2g})$ and $\sigma_M{}^*(e_g)$ are the MO's which receive the most important contribution from the two groups of metal d orbitals. The energy difference Δ between $\pi_M{}^{(*)}(t_{2g})$ and $\sigma_M{}^*(e_g)$ is known from the ligand-field spectra, and the energy difference between π_L and σ_L is a characteristic of the ligand involved. Note that in the low-spin d^6 octahedral complexes, such as those of Co(III), the "metal" t_{2g} orbitals are filled, so that the $\pi_L \rightarrow \pi_M{}^{(*)}(t_{2g})$ and $\sigma_L \rightarrow \pi_M{}^{(*)}(t_{2g})$ transitions cannot occur. As a consequence, these complexes cannot exhibit LMCT bands at low energy. On the contrary, in low-spin octahedral complexes containing fewer than six d-electrons (such as the complexes of (Ru(III), Ir(IV), etc.), the "metal" t_{2g} orbitals are available, and thus, LMCT bands can be present in the long wavelength region of the spectrum.

Charge transfers in the reverse direction, i.e. *metal→ligand charge-transfer* (MLCT) transitions (Fig. 2b-1, arrows 2e and 2f), are likely to happen in complexes with central atoms having small ionization potentials and ligands with easily available empty π^* orbitals. Ligands such as CN^-, CO, SCN^- and especially the conjugated carbon-containing molecules possess empty π^* orbitals suited for such transitions. Regarding the central atoms, it is obvious that MLCT transitions will be favored when the metal has a low oxidation state. For example, the bands in the visible of $[Ru(bpy)_3]^{2+}$ are due to MLCT transitions, while those of $[Ru(bpy)_3]^{3+}$ are due to LMCT transitions.

Another type of charge-transfer bands can occur in the ion pairs formed by a coordinatively saturated complex cation and a polarizable anion like iodide. These *ion-pair charge-transfer* (IPCT) bands are due to intermolecular charge-transfer transitions from the anion to the antibonding d orbitals of the central metal atom. The intensity of these bands depends on the formation constant of the ion-pair and on the concentration of the two ions. A typical example of an intermolecular charge-transfer band is given by the intense new band which appears in the spectrum of aqueous solutions of $[Co(NH_3)_6]^{3+}$ salts when the concentration of iodide ion is increased [8601, 9301].

Intermolecular charge-transfer transitions of the inverse type (i.e. from the complex to an outer species) are the so called *charge-transfer-to-solvent* (CTTS) transitions exhibited by some negative complex ions such as, for example, $[Fe(CN)_6]^{4-}$ [8601, 9301].

Finally, the charge-transfer-to-solvent transitions which may occur for hydrated cations in their low oxidation states (for example, Ce^{3+}) may be considered as intramolecular charge-transfer transitions.

2c-3 Intra-Ligand Bands

These bands (which may also be called "ligand-centered" (LC) bands) are due to transitions between two MO's both of which are principally localized on the ligand system [6201, 7001, 7002, 8401] (see, for example, arrow 3 in Fig. 2b-1). Such bands may be found at relatively low energy in complexes containing ligands which have π-systems of their own. Some bands of complexes containing aromatic ligands, such as pyridine, bipyridine and phenanthroline, surely belong to this class. Generally, it is possible to identify bands as intra-ligand or charge-transfer in nature by observing their energy shifts in a series of complexes of a given ligand with a variety of different metals. If a band has a rather fixed spectral position, independent of the oxidizing character of the central metal, it may be assigned to intra-ligand transitions.

2c-4 A Remark

It must be pointed out that the above classification of electronic transitions and electronically excited states is somewhat arbitrary and loses its meaning whenever the involved states cannot be described by localized MO configurations [6201, 7001, 7002, 8401]. It should also be noticed that the energy ordering of the various orbitals may be different from that shown in Fig. 2b-1. For example, in the case of $[Ru(bpy)_3]^{2+}$ (bpy=2,2'-bipyridine) the π^*_L orbital is thought to be lower in energy than the $\sigma^*_M(e_g)$ orbital. More generally, the excited state energy ordering is extremely sensitive to the type of the ligands and the nature and oxidation state of the metal. For example, the lowest excited state in $[Ir(phen)Cl_4]^-$, $[Ir(phen)_2Cl_2]^+$, and $[Ir(phen)_3]^{3+}$ (phen=1,10-phenanthroline) is MC, MLCT, and LC, respectively; the lowest excited state of $[Rh(bpy)_2Cl_2]^+$ is MC, whereas that of $[Ir(bpy)_2Cl_2]^+$ is MLCT; the lowest excited state of $[Ru(bpy)_3]^{2+}$ and $[Os(bpy)_3]^{2+}$ is MLCT, whereas that of $[Ru(bpy)_3]^{3+}$ and $[Os(bpy)_3]^{3+}$ is LMCT [8801]. Further complications arise from the fact that the energy splitting between the spin states (e.g., singlet and triplet) belonging to the same orbital configuration is very large for the MC and LC excited states, because in these cases the two interacting electrons occupy the same region of space, whereas it is smaller for the CT states. It follows that the excited state energy ordering may be different in the spin-allowed and spin-forbidden manifolds. Finally, it should be recalled that in transition-metal complexes there may be a considerable degree of spin-orbit coupling. This effect, being related to the central heavy metal ion, is different for different types of excited states. Almost pure LC states are scarcely affected, whereas for MC and CT states of metals belonging to the third transition row it may become almost meaningless to talk about discrete spin states. In the current literature, however, the spin label is generally used even when its meaning is not strict. Finally, it should be recalled that for compounds having open-shell ground-state configuration (e.g., Cr(III) and lanthanide complexes), intra-configuration MC excited states can be found at low energies.

2c-5 Band Width

The geometry of an electronically excited state is expected to be different from that of the ground state, whenever the bonding electron configuration is significantly different in the two states [5001, 6602, 8201]. In transition metal complexes the choice between different structures often depends on a delicate energy balance, so that the excitation of an electron from a bonding or non-bonding orbital to an antibonding orbital is likely to destabilize the ground-state geometry and cause the complex to undergo a more or less noticeable rearrangement. By contrast, intraconfiguration excited states (vide supra) are not distorted compared with the ground state geometry.

The width of the absorption bands can provide useful information regarding the difference in the equilibrium nuclear configuration between ground state and excited states. As a consequence of the Franck-Condon principle (see Chapter 1), the band width is proportional to the slope of the potential curve of the excited state as a function of the internuclear distances, taken at the equilibrium point of the ground state. Even though this slope will also depend on the force constant of the excited state (and its anharmonicity), it can certainly be expected that, for a given complex, the slope (and thus, the band width) will increase with increasing difference between the ground and excited state equilibrium nuclear configurations [7001].

A schematic representation of this relation is given in Fig. 2c-1. At the equilibrium nuclear configuration of the ground state, r_0, the slope of the curve of the intraconfiguration excited state E_1, whose equilibrium configuration distance is almost equal to r_0, is smaller than the slope of the curve corresponding to the state E_2 which, belonging to a different electronic configuration, has an equilibrium distance noticeably different from that of the ground state. Therefore, the $E_0 \rightarrow E_1$ transition will give a narrow band and the $E_0 \rightarrow E_2$ one a broad band.

In reality, the potential energy of an N-atomic molecule depends on $3N-6$ (or, for a linear molecule, $3N-5$) normal coordinates. Therefore, to represent the potential energy of a state, a $3N-6$ (or $3N-5$) dimensional hypersurface in a $3N-5$

(or $3N$–4) dimensional space should be used. Such hypersurfaces, unlike the simple potential curves of diatomic molecules, are difficult to visualize and still more difficult to represent graphically. However, keeping all but one coordinate fixed, one can draw curves like those of Fig. 2c-1. This figure can be considered as the section along the totally symmetrical coordinate of the hypersurfaces representing the potential energy functions of the ground and excited states.

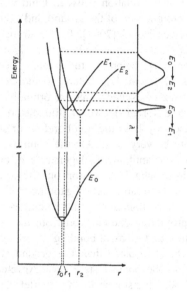

Fig. 2c-1. Schematic representation of the relation between band width and equilibrium nuclear configuration in the ground and excited state.

When the difference in the equilibrium geometry of the ground and excited state (i.e. the difference in the position of the minimum of the two corresponding hypersurfaces) only concerns the totally symmetrical coordinate, the photoexcitation will cause a change in the *size* but not in the shape of the complex. However, the displacement of the minimum of the excited state with respect to the minimum of the ground state can also (or only) concern vibrational coordinates other than the totally symmetrical one, and in this case, the photoexcitation will cause a change in the *shape* of the complex. For many excited states of coordination compounds, the occurrence of such distortions can be predicted on the basis of the Jahn-Teller theorem which states that, for an orbitally degenerate electronic state, the totally symmetric nuclear geometry is unstable except in the case of linear molecules. It can be shown, in fact, that for orbitally degenerate states the symmetrical conformation is *not* the position of minimum energy *if vibronic interaction is taken into consideration.* The Jahn-Teller effect is of basic importance in understanding the (distorted) structures of many ground state complexes [6201, 7002, 9501].

When the degeneracy is due to the not homogeneous occupation of the non-bonding or slightly π-antibonding t_{2g} orbitals, the distortion is expected to be smaller than in the case of the not homogeneous occupation of the σ-antibonding e_g orbitals.

As an example of the relation between band width and difference in the equilibrium nuclear configuration of the ground and excited states, let us consider the spectra of Cr(III) complexes [7001]. In the strong field approximation, the $^4A_{2g}$ ground state of Cr(III) corresponds to the t_{2g}^3 configuration. The excited doublet states ^2E_g and $^2T_{1g}$ belong to the same t_{2g}^3 configuration, whereas the excited quartet states $^4T_{2g}$ and $^4T_{1g}$ belong to the excited configuration $t_{2g}^2 e_g$ (Fig. 2c-2). All of the above excited states are orbitally degenerate and then, they should be Jahn-Teller distorted. However, the distortion of the doublet states, whose electronic anisotropy is in the non-bonding or slightly π-antibonding t_{2g} orbitals, is expected to be very small. On the contrary, the quartet states, which have one electron on the σ-antibonding degenerate e_g orbitals, should be noticeably distorted. Apart from Jahn-Teller distortion, the ground state and the doublet states should have almost the same metal-ligand bond distances since they all arise from the same electronic configuration; on the contrary, the excited quartet states, which are obtained promoting one electron from the non-bonding or slightly π-antibonding t_{2g} orbitals to the σ-antibonding e_g orbitals, are expected to have greater metal-ligand bond distance than the ground state. Therefore, the spin-forbidden transitions to the doublet states are expected to give narrow bands, whereas the spin-allowed transitions to the quartet states are expected to give broad bands, as schematically shown in Fig. 2c-1. Indeed, the doublet bands have half-widths of 100-200 cm^{-1}, and the quartet bands, 1500-2000 cm^{-1}.

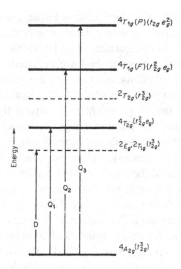

Fig. 2c-2. Schematic energy level diagram for octahedral Cr(III) complexes (d^3 electronic configuration).

The changes in size and shape of a complex in changing electronic state play an important role in determining not only the band widths of the radiative transitions, but also the rates of the radiationless conversions, which are favoured by crossing of the potential energy curves, and the occurrence of isomerization reactions.

2d JABLONSKI DIAGRAM

Before extending the rules and generalizations which have been formulated for radiative and radiationless transitions in organic molecules (see Chapter 1) to coordination compounds, it should be realized that these two classes of compounds differentiate for various reasons.

First of all, organic molecules usually have a closed-shell ground-electronic configuration, and therefore a singlet ground state, usually denoted by S_0 (see Chapter 1, Fig. 1b-12). When an electron is promoted from one of the low-energy occupied MO's to a high-energy unoccupied orbital, singlet and triplet excited states occur in pairs (S_1 and T_1, S_2 and T_2, etc). Each triplet state is invariably of lower energy than the corresponding singlet, since the interelectronic repulsions are less important when the orbitally-unpaired electrons have parallel spin. Thus, the lowest excited state of an organic molecule is usually a triplet state (T_1), i.e. a state having a *different* multiplicity from that of the ground state. For transition metal complexes, the situation is often quite different. Indeed, in most cases, the ground-electronic configuration contains degenerate orbitals which are not completely filled, and thus, as a consequence of the Hund rule, the ground state has a multiplicity higher than one. Moreover, the lowest excited state is not necessarily a state having a different multiplicity from that of the ground state. For example [7001], in the already discussed case of the Cr(III) complexes (Fig. 2c-2), the ground state is a quartet state, $^4A_{2g}$, which, in the strong field approximation, belongs to the t_{2g}^3 configuration. The low-lying excited states are either doublet states, which are obtained by spin-pairing within the t_{2g}^3 configuration, or quartet states, which are obtained by promotion of an electron from t_{2g} to e_g orbitals. The lowest doublet state is ^2E_g and the lowest excited quartet is $^4T_{2g}$. The energy of the ^2E_g state only depends on the interelectronic repulsion parameters and it is almost independent of the ligand field strength, while the energy of the $^4T_{2g}$ state depends on Δ (the ligand field parameter). Consequently, it results that the lowest excited state is the doublet state ^2E_g for strong ligand-field complexes, and the quartet state $^4T_{2g}$ for weak ligand-field ones (see also Fig. 2c-1). This difference is of the greatest importance for the photophysical behavior: when the lowest excited state is the spin forbidden and undistorted ^2E_g excited state, radiationless deactivation to the ground state is relatively slow and phosphorescence can be usually observed, whereas when the lowest excited state is the spin allowed and strongly distorted $^4T_{2g}$ one, radiationless decay is very fast and prevents fluorescence that, by the way, would be expected to exhibit a huge Stokes shift from the corresponding absorption band (Fig. 2c-1).

Another difference to be pointed out is that organic molecules, with a few exceptions, do not contain heavy atoms, and thus, they show a low degree of spin-orbit coupling. On the contrary, in transition-metal complexes, the presence of the metal atom brings about a considerable degree of spin-orbit coupling. This effect is particularly evident for metals belonging to the second or third transition series. For example, platinum complexes show formally spin forbidden bands which have similar intensities as the formally spin allowed ones. The lifetime of spin forbidden excited states is also strongly affected by spin-orbit coupling. As an example, a different amount of spin-orbit coupling is one of the reasons why in rigid matrix at 77 K (where bimolecular deactivation processes cannot occur) the lifetime of the lowest spin forbidden excited state of naphthalene is longer than a second, whereas that of the lowest spin forbidden excited state of $[Ru(bpy)_3]^{2+}$ is in the microsecond time scale.

2e PHOTOCHEMICAL REACTIVITY

The chemical reactivity of a molecule is determined principally by its electron distribution. In the case of transition-metal complexes, it is well known that the widely ranging rates of their thermal reactions may be qualitatively (or semi-quantitatively) explained on the basis of the *d*-electron configuration of the central atom. For example, the kinetic inertness of the t_{2g}^6 octahedral complexes can be accounted for on the basis of ligand field arguments, which show that the rearrangement of the ground-state octahedral structure to any transition state, whether of coordination number five or seven, would inevitably involve a large activation energy.

In the excited states, the electron distribution is more or less changed with respect to the ground state and, therefore, the reactivity is also changed. A full theory of excited state reactivity would begin with a calculation of the electron distribution in the excited state and would be followed by an account of the time-dependent nuclear configurations which follow as the molecule relaxes towards equilibrium with its environment. Such a possibility, however, seems to be a very remote one for polynuclear molecules in general, and even more so for coordination compounds. At the present time, we can only try to infer something on the reactivities of the excited states from an approximate and qualitative examination of their electron distribution [7001, 9301, 9401].

For this purpose, it is again convenient to classify the electronically excited states in terms of "localized" MO configurations (vide supra) as ligand-field, charge-transfer, and intra-ligand excited states.

2e-1 Ligand-Field Excited States

At a first approximation, the ligand-field (or metal-centered) excited states arise from electron transitions occurring between "metal" *d* orbitals (vide supra). With respect to the ground-state electron distribution, these transitions result in an *angular* electron rearrangement, which does *not* modify the charge distribution between ligands and metal. Therefore, since the ground state of complexes is usually stable towards intramolecular oxidation-reduction processes (i.e. towards the homolytic fission of metal-ligand bonds), it is to be expected that the ligand-field excited states will also be stable towards this type of reactivity.

The angular rearrangement of the metal electrons, however, can cause important consequences as far as the reactivity towards ligand substitution or isomerization reactions is concerned. Such consequences can be easily visualized, for example, in the case of strong field octahedral complexes, such as those of Cr(III). The ground state of these complexes, $^4A_{2g}$, belongs to the t_{2g}^3 configuration (Fig. 2c-2). As we have already discussed, the spin-allowed transitions, such as the $^4A_{2g}(t_{2g}^3) \rightarrow {}^4T_{2g}(t_{2g}^2 e_g)$ one, correspond to the promotion of an electron from the t_{2g} to the e_g orbitals. This promotion will strongly affect the metal-ligand bonds. In a crystal field picture, the promotion of an electron from a t_{2g} orbital, which is directed *away* from the ligands, to an e_g orbital, which is directed *towards* the ligands, is expected to cause: (i) an increase in the metal-ligand repulsion, and then, either some lengthening of metal-ligand distances (and, eventually the detachment of a ligand) or a rearrangement of the molecule towards a more stable structure (isomerization); (ii) a decrease of electron density in some directions between the ligands, which will facilitate a nucleophilic attack on the central metal by solvent molecules or other ligands present in the solution (substitution reaction). A molecular orbital approach leads to parallel results. The e_g orbitals are strongly σ-antibonding, so that their partial occupation is expected to weaken some metal-ligand bonds. On the other hand, the presence of an empty t_{2g} orbital, available for bond making, will facilitate the entering of a new ligand. Note that, in the case of intraconfiguration transitions, such as the $^4A_{2g}(t_{2g}^3) \rightarrow {}^2E_g(t_{2g}^3)$ one in Cr(III) complexes, the above considerations do not apply.

2e-2 Charge-Transfer Excited States

As we have seen above, these states arise from transitions between MO's principally localized on the metal and MO's principally localized on the ligands. Such transitions cause a *radial* redistribution of the electronic charge between the central metal and ligands and thus, a change in their oxidation state. It follows that the charge-transfer excited states will be particularly inclined to give *intramolecular* oxidation-reduction processes, resulting in the reduction of the metal and oxidation of a ligand, or viceversa. The tendency of the charge-transfer excited states to give such redox processes will be influenced by a number of factors, such as: (i) the stability of upper and lower oxidation states of the metal and ligands; (ii) the effective amount of charge transfer induced by irradiation; (iii) whether or not the charge-transfer is localized between *one* particular ligand and the metal (X → M charge-transfer states in $ML_{n-1}X$ complexes); (iv) environmental conditions (cage effect, solvent reactivity, etc.).

The above discussion shows that, going from the ground state to charge-transfer excited states, one must expect principally a strong increase in the reactivity towards oxidation-reduction reactions. However, other features of the charge-transfer excited states should be pointed out. First of all, one can expect that there will be an increase in the reactivity towards substitution reactions. In fact, because of the charge-transfer process, the oxidation state of the metal is changed and, as a consequence, an inert complex may be changed into a labile one. For example, LMCT transitions in Co(III) complexes lead to Co(II) species which are kinetically labile. More specifically, it should also be considered that: (i) the MLCT transitions increase the positive charge of the metal, so that a nucleophilic attack by outer ligands could be facilitated; (ii) in most LMCT transitions, the transferred electron goes into a metal orbital which points in the direction of the ligands (i.e. in a e_g σ-antibonding orbital): a metal-ligand repulsion could then result, analogously to what happens in the case of the ligand-field excited states. Moreover, charge-transfer transitions are expected to cause a strong change in the acid strength of complexes which contain "protogenic" ligands, such as H_2O, NH_3, en, etc.

As we have mentioned above, *intermolecular* charge-transfer transitions can also occur. For example, in ion-pair systems involving a complex cation, the excited states corresponding to such transitions are inclined to give redox processes which result in the formation of a reduced form of the complex that, if it is unstable, will undergo decomposition. Of course, the actual occurrence of the redox process will depend on intrinsic properties of the ion-pair and on environmental conditions. Similarly, charge-transfer-to-solvent transitions will lead to an oxidized form of the complex and a solvated electron.

Finally, it is worth-while emphasizing the relation between the position of the charge-transfer bands and the thermal stability towards redox reactions. If a system shows a charge-transfer band at sufficiently long wavelengths, it means that the energy which must be supplied to make an oxidation-reduction reaction possible is relatively small. Thus, a similar system, containing a more oxidizable (anionic) ligand, is expected to undergo a spontaneous (i.e. thermal) redox reaction. This is, in fact, true in a number of systems; for example, copper(II) bromide is dark red and copper(II) iodide, one can say, is so "black" that it does not exist at all [6201, 7002]. Of course, a similar behavior may be observed in a series of complexes which contain the same anionic ligand and different metal ions.

2e-3 Intra-Ligand Excited States

Intra-ligand (or ligand-centered) excited states arise from electron transitions occurring between two MO's which may be considered as mainly localized on the ligand system (Section 2c-3). It is difficult to make general predictions on the reactivity of intra-ligand excited states because these states may occur in a variety of different ligands. At a first approximation, the intra-ligand transitions neither change the charge distribution between the ligands and metal, nor affect the bonding structure of the complex in a direct way. However, the change in the electronic structure of the ligand can cause important changes in its donor or acceptor properties, equilibrium geometry, dipole moment, acid strength, etc., so that these excited states could be unstable towards metal-ligand dissociation. Moreover, the ligand itself may be photosensitive in the particular reaction medium, and this will certainly affect the stability of the complex.

2e-4 Remarks

On the basis of the above speculations on the inherent reactivities of the excited states of transition metal complexes, one could be tempted to formulate rules predicting which type of reaction should be obtained when the excitation is carried out in a given type of band. However, it must be pointed out again that to classify the electronically excited states in terms of the localized electronic configuration which is believed to make the major contribution to the state (using "metal" or "ligand" labeled orbitals) is only an approximation. More important, *no general relation must necessarily exist between type of excited band and type of photore-action observed* [7001]. It must be emphasized, in fact, that the excited state directly reached with the irradiation can undergo radiationless deactivation processes *before* having a chance to react. As a consequence, other (lower) excited states will be obtained which can be different in nature from the original one; for example, a ligand-field excited state can be obtained by deactivation of a charge-transfer state. Since the type of reaction is obviously dependent on the nature of the excited state from which the reaction originates, it follows that correlations such as "irradiation in charge-transfer bands → oxidation-reduction reactions" or "irradiation in ligand-field bands → substitution or related reactions" are devoid of any real significance. Similarly, the occurrence of an oxidation-reduction or ligand substitution photoreaction *cannot* be taken as a proof of the charge-transfer or ligand-field character of the excited band.

Indeed, the photochemical and photophysical behavior of a complex (as well as of any other molecule) is the result of a complicated balance between inherent chemical reactivities and radiative and nonradiative deactivation modes of the electronic states lying between the state directly reached with the irradiation and the ground state. Predictions concerning the photochemical behavior and formulation of rules will require detailed information on all the factors which govern this balance.

2f ELECTROCHEMICAL BEHAVIOR

In the localized MO approximation, oxidation and reduction processes are viewed as metal or ligand centered [7001]. The highest energy occupied molecular orbital (HOMO) is most usually metal centered, whereas the lowest empty molecular orbital (LUMO) is either metal or ligand centered, depending on the relative energy ordering. When the ligand field is sufficiently strong and/or the ligands can be easily reduced, reduction takes place on the ligand. When the ligand field is weak and/or the ligands cannot be easily reduced, the lowest empty orbital can be metal centered. Since the HOMO and LUMO orbitals are involved in oxidation and reduction processes, the electrochemical properties are related to the orbital nature of the lowest excited states and the electrochemical potentials are related to the spectroscopic energies [8101, 8801, 0302].

2f-1 Ligand Dissociation Processes

In several cases, oxidation and/or reduction processes lead to a decomposition of the complex. Typical examples are Co(III) complexes of amine-type ligands, e.g., $[Co(NH_3)_6]^{3+}$ and $[Co(en)_3]^{3+}$ (en = ethylenediamine) [9902]; decomposition occurs because the corresponding Co(II) complexes are labile:

$$[Co(NH_3)_6]^{3+} \xrightarrow{+e^-} [Co(NH_3)_6]^{2+} \xrightarrow{+6H_3O^+} Co^{2+}(aq) + 6NH_4^+ + 6H_2O$$

The electrochemical behavior changes substantially when the metal ion is encapsulated in a cage type ligand. This is the case of $[Co(sep)]^{3+}$ (sep = sepulchrate, 1,3,6,8,10,13,16,19-octaazabicyclo[6.6.6]eicosane) [9902], a cage version of Co(III) amine complexes, for which the Co(III)/Co(II) reduction process is reversible since the ligands units are kept around the reduced metal by the cage-like structure:

$$[Co(sep)]^{3+} + e^- \rightleftharpoons [Co(sep)]^{2+}$$

2f-2 Structural Changes

Other complexes undergo structural changes upon electrochemical reduction or oxidation. This is the case of Cu(I) four coordinated tetrahedral complexes like $Cu(phen)_2^+$ which exhibits a poorly reversible metal centered oxidation process. This behavior is attributed to the strong tendency of Cu(II) to form octahedral species by coordination of solvent molecules. When phen is replaced by its 2,9-dimethyl derivative, the metal centered oxidation process is reversible and strongly displaced towards more positive potential values. This change is due to the fact that the α-substituents to the coordinating nitrogen atoms prevent a change in geometry and protect the metal ion from solvent coordination.

2f-3 Reversible Processes

Typical examples of electrochemically well behaved compounds are the Ru(II)-polypyridine $[Ru(L)_3]^{n+}$ complexes [8801, 9402, 9502, 0101, 0302].

Oxidation: Their oxidation is metal centered and leads to Ru(III) compounds (low spin $\pi_M(t_{2g}^5)$ configuration) which are inert to ligand substitution:

$$[Ru^{II}(L)_3]^{2+} \rightleftharpoons [Ru^{III}(L)_3]^{3+} + e^-$$

Only this first metal centered oxidation process can be observed in the potential window available in the usual solvents (e.g., acetonitrile), but in SO_2 solution at -70 °C other processes involving ligand oxidation can be observed. [8802]. The Ru(III)/Ru(II) reduction potential in most complexes which contain only polypyridine-type ligands falls in a rather narrow range around +1.25 V (*vs* SCE, acetonitrile solution) [8801]. Substitution of one bpy ligand by two Cl⁻ ions to give $[Ru(bpy)_2Cl_2]$ lowers the potential to +0.32 V, whereas the strong π-acceptor CO ligand causes an increase of the reduction potential above +1.9 V [8801].

Reduction: Reduction of the Ru(II)-polypyridine complexes takes place on a π^* orbital of the polypyridine ligands. Therefore the reduced form, keeping the low-spin $\pi_M(t_{2g}^6)$ configuration, is usually inert and the reduction process is reversible:

$$[Ru^{II}(L)_3]^{2+} + e^- \rightleftarrows [Ru^{II}(L)_2(L^-)]^+$$

The added electron is substantially localized on a single ligand [9402, 9502, 9601, 0101]. Several reduction steps can often be observed in the accessible potential range. In DMF at −54 °C, up to six cyclic voltammetric waves can be observed for $[Ru(bpy)_3]^{2+}$ in the potential range between −1.33 and −2.85 V (*vs* SCE), which are assigned to successive first and second reduction of the three bpy ligands, yielding a complex that can be formulated as $[Ru^{2+}(bpy^{2-})_3]^{4-}$ [0102]. The localization of the acceptor orbitals in the reduction process is often particularly clear in mixed-ligand complexes involving polypyridine-type ligands with different energies of their π^* orbitals [8801].

As mentioned above, when the ligand field is weak and/or the ligands cannot be easily reduced, the lowest empty orbital in the reduction process could be metal centered ("parent" $\sigma^*_M(e_g)$ in octahedral symmetry). In such a case, reduction would lead to an unstable d^7 system which should give rise to a fast ligand dissociation, making the process electrochemically irreversible. Such a behavior has never been clearly observed for ruthenium complexes.

In the assumption that the molecular energy ordering does not change upon reduction or oxidation, the σ^*_M and π^*_L molecular orbitals involved in the reduction processes (redox orbitals) are the same orbitals which are involved in the MC and MLCT transitions, respectively (spectroscopic orbitals). Thus, reversibility of the first reduction step, indicating a ligand centered LUMO, also implies (to a first approximation) that the lowest excited state is MLCT. More generally, there are extended correlations between the electrochemical and spectroscopic data [8801, 9402, 9502, 0101, 0302].

2g POLYNUCLEAR METAL COMPLEXES

2g-1 Supramolecular Species or Large Molecules?

Polynuclear metal complexes, as well as any other multicomponent species, can fall into two quite different, limiting categories: (i) supramolecular species, and (ii) large molecules [9601].

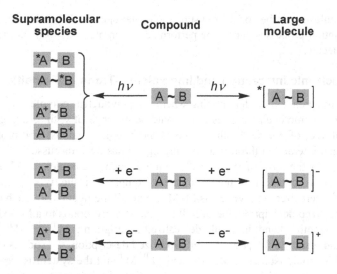

Fig. 2g-1. Schematic representation of the difference between a supramolecular system and a large molecule based on the effect caused by a photon or electron input.

From a photochemical and electrochemical viewpoint [9101, 9601, 0201, 0303], a supramolecular species may be defined as a complex system made of molecular components with definite individual properties. This happens when the interaction energy between components is small compared with other relevant energy parameters. As shown in Fig. 2g-1, light excitation of a supramolecular species A ~ B (~ indicates any type of bond or connection that keeps together the A and B components) leads to excited states that are substantially localized either on the A or on the B component (or causes directly an electron transfer from A to B or *vice versa*, see later). When the excited states are substantially delocalized on both A and B, the species is better considered as a large molecule. Similarly, oxidation and reduction of a supramolecular species can substantially be described as oxidation and reduction of specific components, whereas oxidation and reduction of a large molecule leads to species where the hole or the electron are delocalized on the entire species.

The above concepts can easily be applied to polynuclear metal complexes [9601]. Consider, for example, a dinuclear compound $[(L)_nM_aL–S–LM_b(L)_n]^{(x+y)+}$, where L-S-L is a bridging ligand in which the two coordination sites L are connected by a spacer S. Although the presence of the bridging ligand can create some ambiguities, in most cases $[(L)_nM_aL–S–L]^{x+}$ and $[L–S–LM_b(L)_n]^{y+}$ may be considered molecular components with well defined individual properties: for example, integral oxidation states can be assigned to M_a and M_b, and independent excitation of the $[(L)_nM_aL–S–L]^{x+}$ and $[L–S–LM_b(L)_n]^{y+}$ components is feasible.

Such complexes can be considered supramolecular species. When localized redox and excitation processes cannot be performed, a complex has to be considered a large molecule.

2g-2 Electronic Interaction and Intervalence-Transfer Transitions

A very important role in determining whether a polynuclear complex can be considered a supramolecular species or it should be better dealt with as a large molecule is played, of course, by the nature of the bridging ligands. The problem of metal-metal interaction through a bridging ligand has been discussed in considerable detail in the case of the so called mixed-valence complexes [6701, 6702, 8301]. Let us consider a homo-dinuclear complex of the type $[(L)_nML-S-LM(L)_n]^{5+}$, hereafter abbreviated as M–M, where all the ligands are uncharged. In a fully localized description, the overall 5+ charge corresponds to a $M^{II}-M^{III}$ complex. On the other hand, in a fully delocalized description a $M^{(II+1/2)}-M^{(II+1/2)}$ complex would result. In the case of a fully localized description, the species can exist as two "electronic isomers", $M^{II}-M^{III}$ and $M^{III}-M^{II}$ and the system is expected to exhibit properties which are an exact superposition of the properties of the isolated $[(L)_nML-S-L]^{2+}$ and $[L-S-LM(L)_n]^{3+}$ components. In the case of a fully delocalized description, the $M^{(II+1/2)}-M^{(II+1/2)}$ species will exhibit completely new properties compared with the properties of the separated components. In most cases, the electronic interaction between the two components is weak enough to perturb only slightly the properties of the two component units, but strong enough to allow the occurrence of the so called optical *intervalence-transfer* (IT) transition interconverting the two electronic isomers:

$$[(L)_nM^{II}L-S-LM^{III}(L)_n]^{5+} \xrightarrow{\ h\nu\ } [(L)_nM^{III}L-S-LM^{II}(L)_n]^{5+}$$

The arguments concerning the degree of electron delocalization are general and can be easily extended to systems which exhibit redox asymmetry because of the presence of different metals (e.g., $[(L)_nM_a^{II}L-S-LM_b^{III}(L)_n]^{5+}$) or different ligands (e.g., $[(L_a)_nM^{II}L-S-LM^{III}(L_b)_n]^{5+}$). The intervalence-transfer transitions are usually rather weak and appear in the visible or near infrared region of the spectrum; for a complete treatment, see refs. [6701, 6702, 8301, 0103].

REFERENCES

[5001] Herzberg, G. *Molecular Spectra and Molecular Structure. I. Spectra of Diatomic Molecules, 2nd ed.*; Van Nostrand: Princeton (NJ), 1950, 658 p.

[6101] Griffith, J. S. *The Theory of Transition-Metal Ions*; Cambridge University Press: Cambridge (UK), 1961, 455 p.

[6201] Jørgensen, C. K. *Absorption Spectra and Chemical Bonding in Complexes*; Pergamon Press: Oxford (UK), 1962, 359 p.

[6401] Ballhausen, C. J.; Gray, H. B. *Molecular Orbital Theory*; Benjamin: New York, 1964, 273 p.

[6402] Gray, H. B. *J. Chem. Educ.* **1964**, *41*, 2-12.

[6601] Orgel, L. E. *An Introduction to Transition-Metal Chemistry: Ligand-Field Theory, 2nd Ed.*; Methuen: London, 1966, 186 p.

[6602] Herzberg, G. *Molecular Spectra and Molecular Structure. III. Electronic Spectra and Electronic Structure of Polyatomic Molecules*; Van Nostrand: Princeton (NJ), 1966, 875 p.

[6701] Robin, M. B.; Day, P. *Adv. Inorg. Chem. Radiochem.* **1967**, *10*, 247-422.

[6702] Hush, N. S. *Prog. Inorg. Chem.*, **1967**, *8*, 391-444.

[7001] Balzani, V.; Carassiti, V. *Photochemistry of Coordination Compounds*; Accademic Press: London (UK), 1970, 432 p.
Wiley Interscience: USA, 1995, 856 p.

[7002] Jørgensen, C. K. *Prog. Inorg. Chem.* **1970**, *12*, 101-158.

[8101] De Armond M. K.; Carlin, C. M. *Coord. Chem. Rev.*, **1981**, *36*, 325-355.

[8201] Hollas, J., M. *High Resolution Spectroscopy*; Butterworths: London (UK), 1982, 638 p.

[8301] Creutz, C. *Prog. Inorg. Chem.*, **1983**, *30*, 1-73.

[8401] Lever A. B. P. *Inorganic Electronic Spectroscopy, 2nd Ed.*; Elsevier: Amsterdam (NL), 1984, 863 p.

[8601] Balzani, V.; Sabbatini, N.; Scandola, F. *Chem. Rev.* **1986**, *86*, 319-337.

[8801] Juris, A.; Balzani, V.; Barigelletti, F.; Campagna, S.; Belser, P.; von Zelewsky, A. *Coord. Chem. Rev.* **1988**, *84*, 85-277.

[8802] Garcia, E.; Kwak, J.; Bard, A. J. *Inorg. Chem.* **1988**, *27*, 4377-4382.

[9101] Balzani, V.; Scandola, F. *Supramolecular Photochemistry*; Horwood: Chichester (UK), 1991, 427 p.

[9301] Horwat, O.; Stevenson, K. L. *Charge-Transfer Photochemistry of Coordination Compounds*; VCH: New York (NY), 1993, 380 p.

[9401] Roundhill, D. M. *Photochemistry and Photophysics of Metal Complexes*; Plenum Press: New York (NY), 1994, 356 p.

[9402] Dodsworth, E. S.; Vlcek, A. A.; Lever, A. B. P. *Inorg. Chem.* **1994**, *33*, 1045-1049.

[9501] Cotton, F. A.; Wilkinson, G.; Gauss, P. L. *Basic Inorganic Chemistry*; Wiley Interscience: USA, 1995, 856 p.

[9502] Vlcek, A. A.; Dodsworth, E. S.; Pietro, W. J.; Lever, A. B. P.; *Inorg. Chem.* **1995**, *34*, 1906-1913.

[9601] Balzani, V.; Juris, A.; Venturi, M.; Campagna, S.; Serroni, S. *Chem. Rev.* **1996**, *96*, 759-833.

[9901] Cotton, F. A.; Wilkinson, G.; Murillo, C. A.; Bochmann, M. *Advanced Inorganic Chemistry, 6th Ed.*; Wiley-Interscience: New York (NY) 1999, 1376 p.

[9902] Venturi, M.; Credi, A.; Balzani, V. *Coord. Chem. Rev.* **1999**, *185-186*, 233-256.

[0101] Vlcek, A. Jr.; Heyrovsky, J. in *Electron Transfer in Chemistry*; Balzani, V. (Ed.); Wiley: Weinheim (D), 2001, vol. 2, 804-877.

[0102] Carano, M.; Ceroni, P.; Maggini, M.; Marcaccio, M.; Menna, E.; Paolucci, F.; Soffia, S.; Scorrano, G. *Collect. Czech. Chem. Commun.* **2001**, *66*, 276-290.

[0103] Launay, J. –P.; Coudret, C. in *Electron Transfer in Chemistry*; Balzani, V. Ed.; Wiley: Weinheim (D), 2001, vol. 5, 3-47.

[0201] Balzani, V., Credi, A.; Venturi, M. *Chem. Eur. J.* **2002**, *8*, 5525-5532.

[0301] McCleverty, J.; Meyer, T. J. (eds.) *Comprehensive Coordination Chemistry II*; Elsevier Pergamon (APS): USA, 2003, 9500 p.

[0302] Zanello, P. *Inorganic Electrochemistry*; RSC: Cambridge (UK), 2003, 615 p.

[0303] Balzani, V.; Venturi, M.; Credi, A. *Molecular Devices and Machines*; Wiley: Weinheim (D), 2003, 494 p.

3

Photophysical Properties of Organic Compounds

3a PHOTOPHYSICAL PARAMETERS IN SOLUTION

Three states play a dominant role in the photophysics of organic molecules in solution. They are the ground state, the lowest triplet state, and the lowest excited singlet state. In condensed media, vibrational relaxation is usually so fast that excited molecules quickly relax to one of the two excited states. For such systems, a three-state model, with the transitions between the states governed by first-order competing kinetics, forms an adequate framework to understand much of the data on unimolecular decays of the excited states.

In particular, radiative and radiationless transitions between these two excited states and the ground state delimit the extent to which photochemical processes can occur. For singlet photochemistry, the quantum yield can be written in terms of the fundamental kinetic parameters as

$$\Phi_{pc}^{S} = \frac{k_{pc}^{S}}{k_{S}} \tag{3-1}$$

where k_{pc}^{S} is the first-order rate constant for singlet photochemistry and k_{S} is the overall first-order rate constant for the singlet decay. Thus, in order to see singlet photochemistry, k_{pc}^{S} must be competitive with other decay channels of the lowest excited singlet state. Nonsensitized triplet photochemistry, on the other hand, is limited both by analogous processes to those in Eq. 3-1 and also by the triplet quantum yield.

In Table 3a, the energies of the two excited states are given. Both singlet and triplet energies are estimates of the lowest excited level of each multiplicity, namely the 0-0 vibronic band. For the singlet level, the fluorescence 0-0 band was chosen when it could be distinguished; otherwise, the 0-0 band in absorption was chosen. If neither was distinguishable, the midpoint in energy between the fluorescence and absorption maxima was taken as an estimate of the lowest excited singlet vibronic level. Triplet energy measurements are discussed in the introduction to Table 3b.

Table 3a displays two important quantum yields: the yield of fluorescence per photon absorbed (the fluorescence quantum yield, Φ_{fl}) and the triplet quantum yield, Φ_T (also called the intersystem crossing yield). They are related by

$$\Phi_{fl} + \Phi_T + \Phi_{ic}^S + \Phi_{pc}^S = 1 \qquad (3\text{-}2)$$

where the last two quantum yields in Eq. 3-2 are the internal conversion and the singlet photochemistry quantum yields, respectively. Since Φ_{ic}^S is often small due to the large singlet energy gaps, the sum of Φ_{fl} and Φ_T is often close to 1 when there is no singlet photochemistry. This rule-of-thumb breaks down often, even for relatively simple sets of compounds such as polyenes.

Lifetimes of the two excited states are also given in Table 3a. The overall fluorescence lifetime is the reciprocal of the decay rate constant of the excited singlet state which in the three-state model is given by

$$k_S = \frac{1}{\tau_S} = k_{isc}^S + k_{ic}^S + k_r^S + k_{pc}^S \qquad (3\text{-}3)$$

where k_{isc}^S, k_{ic}^S and k_r^S are the intersystem crossing, the internal conversion, and the natural radiative rate constants, respectively, from the excited singlet. The natural radiative rate constant can be calculated from

$$k_r^S = k_S \Phi_{fl} \qquad (3\text{-}4)$$

The other lifetime given in Table 3a is the triplet lifetime, τ_T, which is the reciprocal of the total decay rate constant for all processes out of the triplet state. The triplet lifetimes are usually several orders of magnitude longer than those of the singlet which makes triplet states susceptible to bimolecular quenching with impurities, cell walls, self-quenching with ground state species, or even weak reactions with solvents. For example, triplet lifetimes of aromatic ketones in hydroxylic solvents can reflect the progress of hydrogen-abstraction reactions between the triplet state and solvent. The triplet lifetimes in Table 3a should be taken as lower limits of a unimolecular lifetime. At room temperature, the unimolecular process contributing to the finite lifetime of triplet states is usually just the intersystem crossing to the ground state. However, in thioketones, the decay has a radiative component.

In Table 3a the molecules are listed primarily by inverted name. Many synonyms are listed in the Compound Name Index at the end of this Handbook. The photophysical measurements were initially classified by solvent media: nonpolar, polar, aromatic, gas, polar crystalline, nonpolar crystalline, etc. In Table 3a, these categories have been condensed into a simple bimodal (nonpolar and polar) classification, with only one value chosen for a particular excited state/solvent type pair. Some remnants of the more detailed media classification are retained as footnotes attached to the values themselves. The nonpolar/polar classification allows for a limited range of measurements to be displayed, yet it keeps the format succinct. Because of the widespread use of cyclohexane and ethanol, they were chosen whenever possible as the representative nonpolar and polar solvents, respectively.

The bimodal classification of solvents is not meant to imply that the unimolecular decays of molecules are solely, or even strongly, dependent on the solvent polarity. Polar solvents can blue-shift $n\pi^*$ transitions leading to energy shifts in the spectra. If these blue-shifts of the $n\pi^*$ transitions are large enough, they can lead to inversion of $n\pi^*$ and $\pi\pi^*$ transitions which can dramatically affect the kinetic parameters characterizing radiative and radiationless transitions. However, barring inversion of excited states, solvent couplings are generally weak enough that they are routinely ignored as a first approximation in theories of radia-tionless transitions. These considerations are the rationalization for choosing to display uni-molecular photophysical properties within a nonpolar/polar scheme. Caution should be exercised when using the scheme in systems where different ionization states are possible, when heterocyclic compounds (which have low-lying $n\pi^*$ transitions) are considered, or when there are strong specific solvent interactions (such as H-abstraction) with the excited molecule.

Table 3a Photophysical Parameters of Organic Species

No.	Solv.	E_S/kJ mol^{-1}	Ref.	τ_S/ns	Ref.	Φ_fl	Ref.	Φ_T	Ref.	E_T/kJ mol^{-1}	Ref.	τ_T/μs	Ref.
1	**Acenaphthene**												
	n	374	[7101]	46	[7101]	0.5	[7001]	0.46	[6901]	250	[6401]		
	p	376	[6801]			0.39	[6601]	0.58	[6802]	248	[6701]	3300	[6801]
2	**Acetone**												
	n			1.7	[7102]	0.0009	[7102]	0.9	[7002]			6.3	[8401]
	p			2.1	[7003]	0.01	[6602]	1	[6603]	332	[6602]	47	[7103]
3	**Acetophenone**												
	n	330	[6402]			$<1\times10^{-6}$	[7301]	1[b]	[8501]	310	[8101]	0.23	[7104]
	p	338	[8101]					1	[8501]	311	[8101]	0.14	[7104]
4	**Acetophenone, 2'-amino-**												
	n	312	[0101]	0.0094	[0101]	2.3×10^{-4}	[0101]	<0.02	[0101]				
	p	304	[0101]	0.386	[0101]	0.017	[0101]	0.07	[0101]	236	[0101]		
5	**Acetophenone, 4'-methoxy-**												
	n	340[b]	[6604]					1[b]	[8501]	300	[6702]		
	p							1	[8501]	299	[6703]		
6	**Acetylene, diphenyl-**												
	n					0.0028	[6704]	0.033	[6704]	262	[5701]		
	p	396	[5601]			0.0016	[6704]	0.056	[6704]	262	[5601]		

Table 3a Photophysical Parameters of Organic Species

No.	Solv.	E_S/kJ mol^{-1}	Ref.	τ_S/ns	Ref.	Φ_{fl}	Ref.	Φ_T	Ref.	E_T/kJ mol^{-1}	Ref.	τ_T/μs	Ref.
7	**Acridan**												
	p	356	[8801]	10	[7302]	0.32	[7302]			291	[7302]		
8	**Acridine**												
	n			0.045	[8402]	0.0001	[8402]	0.5	[7801]	190	[8102]	10000[b]	[6705]
	p	315	[7701]	0.35	[8402]	0.0079	[8402]	0.82[s]	[8502]	188	[8403]	14	[7802]
9	**Acridine-d_9**												
	n							0.75[b]	[7702]				
	p			11.5	[8103]	0.31	[8103]			190	[8103]		
10	**Acridine, 9-amino-**												
	p	280	[7105]	15.2	[6403]	0.99	[5702]			193	[8103]		
11	**1,8-Acridinedione, 3,4,6,7,9,10-hexahydro-3,3,6,6-tetramethyl-**												
	p			5.2	[9801]	0.91	[9801]			224	[9801]	1.5	[9801]
12	**1,8-Acridinedione, 3,4,6,7,9,10-hexahydro-3,3,6,6,10-pentamethyl-**												
	p			5.7	[9801]	0.83	[9801]			227	[9801]	16.2	[9801]
13	**Acridine Orange**												
	p	234	[6902]	4.4	[7101]	0.4	[8503]	<0.02	[8201]	206	[6902]	285	[6902]
14	**Acridine Orange, conjugate monoacid**												
	p	228	[7105]			0.46	[6801]	0.1	[7201]	191	[6801]	105	[7901]

Table 3a Photophysical Parameters of Organic Species

No.	Solv.	E_S/kJ mol⁻¹	Ref.	τ_S/ns	Ref.	Φ_{fl}	Ref.	Φ_T	Ref.	E_T/kJ mol⁻¹	Ref.	τ_T/μs	Ref.
15	**Acridine Yellow**												
	p	251	[7101]	5.1	[7101]	0.47	[7902]			220	[8301]		
16	**Acridinium**												
	p	270	[7105]	33.9	[8103]	0.66	[8103]					670	[8001]
17	**Acridinium, 3,6-diamino-**												
	p	250	[7105]			0.4	[6801]	0.22	[7201]	205	[6801]	20	[8002]
18	**Acridinium, 3,6-diamino-l0-methyl-**												
	p	247	[6902]	4.3	[8503]	0.16	[8503]			214	[6902]		
19	**9-Acridinone**												
	n	304	[7803]	0.78^b	[7601]	0.015	[7803]	0.99^b	[7601]	244	[7803]	20^b	[7601]
	p	290	[7803]	10.8	[8003]	0.97	[8003]			252	[7804]	9.2	[8901]
20	**9-Acridinone, 10-methyl-**												
	n	302	[7803]	9.9	[8003]	0.017	[8003]	0.96^b	[7903]	247	[7803]		
	p	283	[7803]	11	[8003]	0.98	[7803]	0.008	[7903]	252	[7803]	36	[8901]
21	**9-Acridinone, l0-phenyl-**												
	n	305	[7803]			0.018	[7803]			244	[7803]		
	p	289	[7803]	7.7	[8003]	0.99	[7803]			254	[7803]		
22	**Adenosine 5'-monophosphate**												
	p	421	[6706]			0.00005	[7401]			319	[6706]		

Table 3a Photophysical Parameters of Organic Species

No.	Solv.	E_S/kJ mol^{-1}	Ref.	τ_S/ns	Ref.	Φ_{fl}	Ref.	Φ_T	Ref.	E_T/kJ mol^{-1}	Ref.	τ_T/µs	Ref.
23	**Angelicin**												
	n							0.009[b]	[7805]				
	p	330	[7303]					0.031	[8404]	264	[7303]	2.6	[8404]
24	**Angelicin, 3,4'-dimethyl-**												
	n			<0.003	[9601]	$<1\times10^{-4}$	[9601]	0.005	[9601]				
	p			0.370	[9601]	0.018	[9601]	0.32	[9601]			8.8	[9601]
25	**Aniline**												
	n	398	[8202]	6.9	[8701]	0.17	[8701]	0.75	[8701]	297	[6707]	0.72	[8701]
	p	384	[8004]	16.6	[8701]	0.1	[8701]	0.9	[8004]	321	[4401]		
26	**Aniline, N,N-dimethyl-**												
	n	383	[7101]	2.4	[7101]	0.11	[7101]			317	[6102]		
	p	375	[7101]	2.8	[7101]								
27	**Aniline, N,N-diphenyl-**												
	n	362	[7101]			0.045[b]	[7004]	1.11	[7004]			0.06[b]	[7004]
	p									291	[6903]		
28	**Aniline, N-phenyl-**												
	n					0.05[b]	[7004]	0.32[b]	[7004]			0.3[b]	[7004]
	p	372	[7302]	2.4	[7302]	0.11	[7302]	0.47	[7004]	301	[7302]	0.5	[7004]

Table 3a Photophysical Parameters of Organic Species

No.	Solv.	E_S/kJ mol⁻¹	Ref.	τ_S/ns	Ref.	Φ_{fl}	Ref.	Φ_T	Ref.	E_T/kJ mol⁻¹	Ref.	τ_T/µs	Ref.
29	**Anthracene**												
	n	318	[8504]	5.3	[8504]	0.3	[7806]	0.71	[7501]	178	[5701]	670	[8505]
	p	319	[7701]	5.8	[6201]	0.27	[6803]	0.66	[8502]	178	[7701]	3300	[7402]
30	**Anthracene-d_{10}**												
	n	320	[7101]	4.9	[7101]	0.32	[7101]						
	p					0.24	[6803]						
31	**Anthracene, 1-amino-**												
	n	280	[7101]	22.8	[7101]	0.61	[7101]						
	p	259	[6904]	30.8	[6904]					184	[6904]		
32	**Anthracene, 2-amino-**												
	n			25.8^b	[7101]								
	p	264	[6904]	30.8	[7101]	0.57	[8701]			184	[6904]		
33	**Anthracene, 9-amino-**												
	p	257	[6904]	10	[5602]	0.29	[5602]			184	[6904]		
34	**Anthracene, 9-bromo-**												
	n	306	[7904]	1.1	[7904]	0.011	[7904]					43	[6202]
	p			1.1	[5602]	0.02	[6101]	0.98	[8902]	173	[8405]	19.5	[8405]
35	**Anthracene, 9-chloro-**												
	n	307	[7904]	2	[7806]	0.11	[7806]						

Table 3a Photophysical Parameters of Organic Species

No.	Solv.	E_S/kJ mol^{-1}	Ref.	τ_S/ns	Ref.	Φ_{fl}	Ref.	Φ_T	Ref.	E_T/kJ mol^{-1}	Ref.	τ_T/µs	Ref.
	p			2.8	[5602]	0.11	[5602]						
36	Anthracene, 9-cyano-												
	n	297	[8104]	15.6	[8504]	0.93	[8504]	0.04	[7304]			600	[7304]
	p	287	[7304]	11.9	[8802]			0.021	[7304]			1800	[7304]
37	Anthracene, 9,10-dibromo-												
	n	295[b]	[8406]	1.3	[8203]	0.094	[7806]	0.7[b]	[8406]	168	[5603]	36	[6202]
	p	295	[7502]	1.8	[8407]	0.15	[8407]	0.82	[8407]			11	[7802]
38	Anthracene, 9,10-dichloro-												
	n	298	[7101]	8.5	[7101]	0.46	[7001]	0.29[b]	[8406]	169	[5603]	100	[6202]
	p	297	[6605]	8.7	[6501]	0.56	[7005]	0.45	[7005]	169	[6605]		
39	Anthracene, 9,10-dicyano-												
	n	284	[8506]	11.7	[8506]	0.9	[8506]	0.23[b]	[8406]	175	[7807]	100	[7807]
	p	280	[8506]	15.1	[8506]	0.87	[8506]						
40	Anthracene, 9,10-dimethoxy-												
	n	296	[7904]	14.7	[7806]	0.87	[7806]						
	p			9.2	[5602]	0.41	[5602]						
41	Anthracene, 9,10-dimethyl-												
	n	300	[7105]	14	[7806]	0.93	[7806]	0.02[b]	[6804]				
	p	297	[6801]	11	[5602]	0.89	[5602]	0.032	[6801]			8000	[6708]

Table 3a Photophysical Parameters of Organic Species

No.	Solv.	E_S/kJ mol^{-1}	Ref.	τ_S/ns	Ref.	Φ_{fl}	Ref.	Φ_T	Ref.	E_T/kJ mol^{-1}	Ref.	τ_T/µs	Ref.
42	**Anthracene, 9,10-diphenyl-**												
	n	304	[8504]	7.7	[8504]	0.91	[8302]	0.02	[8303]	171	[6805]	2500	[7403]
	p	305	[7105]	8.2	[7603]	0.95	[7603]	0.02	[8303]			3000	[7403]
43	**Anthracene, 9-methoxy-**												
	n			4.9	[7806]	0.34	[7806]	0.26	[8903]				
	p			3.9	[5602]	0.17	[5602]						
44	**Anthracene, 9-methyl-**												
	n	310	[8504]	4.6	[7101]	0.29	[7001]	0.48	[7503]	173	[5701]		
	p	306	[6801]	5.8	[6605]	0.33	[6708]	0.67	[6708]	170	[6605]	10000	[6708]
45	**Anthracene, 9-(methylsulfinyl)-**												
	p	305	[0102]	1	[0102]	0.09	[0102]						
46	**Anthracene, 1,2,3,4,5,6,7,8-octahydro-**												
	n	414	[8204]	17.1	[8204]	0.27	[8204]			322	[8204]		
47	**Anthracene, 9-phenyl-**												
	n	309	[7101]	6.5	[7101]	0.41	[7101]	0.37	[6502]				
	p	304	[6801]	5.1	[6801]	0.49	[5602]	0.51	[6709]			15000	[6708]
48	**Anthracene, 9-[(1E)-2-phenylethenyl]-**												
	n	280	[0201]	3.6	[0201]	0.46	[0201]	0.4	[0201]			40	[0201]
	p	278	[0201]	4.2	[0201]	0.45	[0201]	0.27	[0201]			45	[0201]

Table 3a Photophysical Parameters of Organic Species

No.	Solv.	E_S/kJ mol⁻¹	Ref.	τ_S/ns	Ref.	Φ_{fl}	Ref.	Φ_T	Ref.	E_T/kJ mol⁻¹	Ref.	τ_T/μs	Ref.
49	**Anthracene, 9-[(1E)-2-(1-naphthalenyl)ethenyl]-**												
	n	269	[0201]	5.0	[0201]	0.43	[0201]	0.19	[0201]			37	[0201]
	p	269	[0201]	4.5	[0201]	0.35	[0201]	0.14	[0201]			36	[0201]
50	**Anthracene, 9-[2-(2-naphthalenyl)ethenyl]-**												
	n	274	[0201]	3.3	[0201]	0.72	[0201]	0.24	[0201]			48	[0201]
	p	273	[0201]	3.6	[0201]	0.68	[0201]	0.2	[0201]			49	[0201]
51	**9-Anthracenecarboxylic acid**												
	p			4.1	[5602]	0.04	[5602]			177	[8205]		
52	**1,8-Anthracenedisulfonate ion**												
	p			1.5	[8005]			0.87	[8507]			71	[8507]
53	**1-Anthracenesulfonate ion**												
	p			7.3	[8005]			0.98	[8507]			103	[8507]
54	**9,10-Anthraquinone**												
	n	284^g	[8508]					0.9^b	[6503]	261	[6404]	0.11^b	[8105]
	p									263	[6710]		
55	**9,10-Anthraquinone, 1-amino-**												
	n			1.8^b	[8206]	0.058^b	[8206]	0.02^b	[8601]			5^b	[7202]
	p	220	[7105]	0.46	[8206]	0.0082	[8206]						

93

Table 3a Photophysical Parameters of Organic Species

No.	Solv.	E_S/kJ mol⁻¹	Ref.	τ_S/ns	Ref.	Φ_{fl}	Ref.	Φ_T	Ref.	E_T/kJ mol⁻¹	Ref.	τ_T/μs	Ref.
56	**9,10-Anthraquinone, 2-amino-**												
	n			6.5[b]	[8206]	0.21[b]	[8206]	0.4[b]	[8601]			5[b]	[7202]
	p			0.054	[8206]	0.0006	[8206]						
57	**9,10-Anthraquinone, 2-(dimethylamino)-**												
	n	233[b]	[9401]	7.4	[9401]	0.21	[9401]	0.08	[9401]				
	p	218	[9401]	1.1	[9401]	0.0062	[9401]						
58	**9,10-Anthraquinone, 2,6-bis(dimethylamino)-**												
	n	242[b]	[9401]	2.8	[9401]	0.048	[9401]						
	p	232	[9401]	3.3	[9401]	0.049	[9401]	0.089	[9401]				
59	**3-Azafluorenone**												
	n	279	[0103]	<0.6	[0103]	4×10⁻⁴	[0103]	0.97	[0103]				
	p	264	[0103]	6.7	[0103]	2.4×10⁻²	[0103]	0.53	[0103]				
60	**Azulene**												
	n	170	[7105]	1.4 S₂	[7101]	0.02 S₂	[7203]			163[b]	[7504]	11[b]	[8408]
	p											3	[8106]
61	**Benz[a]acridine**												
	n			0.7	[8509]	0.05	[8509]	0.82	[8509]				
	p	295	[7808]	7.2	[8509]	0.32	[8509]	0.42	[8509]	209	[7808]		

Table 3a Photophysical Parameters of Organic Species

No.	Solv.	E_S/kJ mol^{-1}	Ref.	τ_S/ns	Ref.	Φ_{fl}	Ref.	Φ_T	Ref.	E_T/kJ mol^{-1}	Ref.	τ_T/µs	Ref.
62	**Benz[c]acridine**												
	n			3.6	[8509]	0.09	[8509]	0.83	[8509]				
	p	312	[7701]	5.6	[8509]	0.18	[8509]	0.65	[8509]	213	[7701]		
63	**Benz[a]acridinium**												
	p	276	[7105]	14.2	[8509]	0.5	[8509]	0.27	[8509]				
64	**Benz[b]acridin-12-one**												
	n									191	[8510]		
	p	245	[7804]	19.1	[7804]	0.14	[7804]			193	[7804]		
65	**Benzaldehyde**												
	n	323	[7204]			<1×10^{-6}	[7301]			301	[7204]		
	p									298	[7006]		
66	**Benz[a]antracene**												
	n	311	[7105]	45	[6405]	0.19	[6405]	0.79	[6905]	197	[5401]	100	[6406]
	p	307	[7808]	40	[7007]	0.22	[6802]	0.79	[7205]	198	[7808]	9400	[6407]
67	**Benzene**												
	n	459	[7105]	34	[7206]	0.06	[7206]	0.25	[6901]	353	[6702]		
	p	459	[5001]	28	[7206]	0.04	[7206]	0.15	[7305]	353	[6702]		
68	**Benzene-d_6**												
	n	449	[7101]	30	[7306]	0.042	[7306]	0.25	[7306]				

Table 3a Photophysical Parameters of Organic Species

No.	Solv.	E_S/kJ mol⁻¹	Ref.	τ_S/ns	Ref.	Φ_{fl}	Ref.	Φ_T	Ref.	E_T/kJ mol⁻¹	Ref.	τ_T/μs	Ref.
69	**Benzene, chloro-**												
	n	440	[7105]	0.74	[8409]	0.007	[8409]	0.6	[8304]	342[g]	[5901]	1.6	[8409]
	p			0.79	[8305]			0.7	[8511]		[8511]	0.715	[8511]
70	**Benzene, 1,4-dichloro-**												
	n	424	[8202]					0.8	[8304]	335	[6701]	430	[8512]
	p	427	[7905]					0.95	[8512]			330	[8512]
71	**Benzene, 1,4-dicyano-**												
	n	415	[8602]							294	[6203]		
	p	412	[8602]	9.7	[7604]					295	[7008]		
72	**Benzene, 1,4-dimethoxy-**												
	n	397	[7101]	2.9	[7101]	0.21	[7101]			314[b]	[7207]		
	p	396	[7101]	2.7	[7101]		[7101]						
73	**Benzene, fluoro-**												
	n	449	[7105]	7.6	[7101]	0.11	[7101]	0.8	[7001]	353[b]	[5701]	0.67	[7009]
74	**Benzene, hexachloro-**												
	n	397	[8202]					0.5		307	[6701]		
75	**Benzene, methoxy-**												
	n	437	[4901]	8.3	[7101]	0.24	[7001]	0.64	[8207]	338	[6702]	3.3	[7505]
	p	431	[8202]	7.5	[8207]	0.24	[8207]			338	[6702]		

Table 3a Photophysical Parameters of Organic Species

No.	Solv.	E_S/kJ mol⁻¹	Ref.	τ_S/ns	Ref.	Φ_{fl}	Ref.	Φ_T	Ref.	E_T/kJ mol⁻¹	Ref.	τ_T/μs	Ref.
76	**Benzene, 1-methoxy-4-methyl-**												
	n	416	[7101]	8.7	[7101]	0.26	[7101]			326[b]	[7809]		
77	**Benzene, nitro-**												
	n	372	[7105]					0.67[b]	[6806]	243	[8410]	0.0008	[8410]
	p									252	[4401]		
78	**Benzene, 1,2,3,4-tetramethyl-**												
	n	426	[8204]	36.2	[8204]	0.14	[8204]			331	[8204]		
79	**Benzene, 1,3,5-trichloro-**												
	n							0.8	[8304]	331	[6701]		
	p	427	[7905]										
80	**Benzene, 1,3,5-trimethyl-**												
	n	434	[7101]	36.5	[7101]	0.088	[7208]			335	[6103]		
	p	439	[7105]							336	[6408]		
81	**Benzene, 1,3,5-triphenyl-**												
	n	375	[7101]	42.6	[7101]	0.27	[7101]						
	p									269	[6701]		
82	**Benzene, 1,4-bis(phenylethynyl)-**												
	n	348	[0104]	0.64	[0104]	0.9	[0104]						

Table 3a Photophysical Parameters of Organic Species

No.	Solv.	E_S/kJ mol^{-1}	Ref.	τ_S/ns	Ref.	Φ_{fl}	Ref.	Φ_T	Ref.	E_T/kJ mol^{-1}	Ref.	τ_T/µs	Ref.
83	**Benzidine, N,N,N',N'-tetramethyl-**												
	n			10.1	[8411]	0.38	[8411]	0.52	[8411]	261	[7605]		
	p			9.4	[8411]	0.36	[8411]	0.41	[8411]			5	[8412]
84	**Benzil**												
	n	247^x	[6906]			0.0013	[6711]	0.92^b	[6503]	223	[6712]	150	[7106]
	p			2	[8513]					227	[6712]	1500	[6807]
85	**Benzimidazole**												
	n									321^x	[7506]		
	p	423	[6301]			0.67	[6606]			318	[8208]		
86	**Benzo[a]carbazole**												
	p	341	[7701]							256	[7701]		
87	**Benzo[b]carbazole**												
	p	296	[7808]							218	[7808]		
88	**Benzo[b]carbazole, N-methyl-**												
	n	302	[7101]	26.2	[7101]	0.6	[7101]						
	p	300	[7101]	23.5	[7101]								
89	**Benzo[def]carbazole**												
	p	317	[7808]							230	[7808]		

Table 3a Photophysical Parameters of Organic Species

No.	Solv.	E_S/kJ mol⁻¹	Ref.	τ_S/ns	Ref.	Φ_{fl}	Ref.	Φ_T	Ref.	E_T/kJ mol⁻¹	Ref.	τ_T/µs	Ref.
90	**Benzo[a]coronane**												
	n	277[b]	[5902]							215	[7810]		
	p					0.27	[6808]	0.55	[6808]				
91	**Benzo[3,4]cyclobuta[1,2-b]biphenylene**												
	p	203	[0202]	0.03	[0202]	<1×10⁻⁶	[0202]	<0.01	[0202]				
92	**Benzo[3,4]cyclobuta[1,2-a]biphenylene**												
	p	275	[0202]	20	[0202]	0.07	[0202]	0.3	[0202]				
93	**Benzo[b]fluoranthene**												
	n			44.3	[8209]	0.53	[8209]			228	[8209]		
	p	328	[7701]										
94	**Benzo[ghi]fluoranthene**												
	n	285	[7101]	45	[7101]	0.3	[7101]						
	p	282	[7701]							226	[7701]		
95	**Benzo[k]fluoranthene**												
	n			11.3	[8209]	1	[8209]						
	p	299	[7701]							211	[7701]		
96	**Benzo[a]fluorene**												
	n	346	[7101]	63	[7101]	0.51	[7101]						
	p	326	[7701]							241	[7701]		

Table 3a Photophysical Parameters of Organic Species

No.	Solv.	E_S/kJ mol^{-1}	Ref.	τ_S/ns	Ref.	Φ_{fl}	Ref.	Φ_T	Ref.	E_T/kJ mol^{-1}	Ref.	τ_T/μs	Ref.
97	**Benzo[b]fluorene**												
	n	352	[8202]										
	p	326	[7701]	32.3	[7307]					240	[7701]		
98	**Benzo[c]fluorene**												
	p	332	[7701]							231	[7701]		
99	**Benzofurazan, 5-(dimethylamino)-**												
	n	290	[0203]	6.8	[0203]	0.21	[0203]						
	p	261	[0203]	4.9	[0203]	0.04	[0203]						
100	**Benz[f]indole**												
	p	315	[0001]	19.1	[0001]	0.79	[0001]						
101	**Benzoic acid**												
	n					0.0068	[8603]			324	[6302]		
	p	428	[7105]							326	[7606]		
102	**Benzoic acid, 4-(dimethylamino)-, ethyl ester**												
	n			1	[8514]	0.29	[8514]	0.14	[8514]	284	[8306]		
103	**Benzonitrile, 4-amino-**												
	n	396	[8306]	3.3	[8306]	0.14	[8306]			298	[8306]		
	p									293	[7507]		

Table 3a Photophysical Parameters of Organic Species

No.	Solv.	E_S/kJ mol^{-1}	Ref.	τ_S/ns	Ref.	Φ_{fl}	Ref.	Φ_T	Ref.	E_T/kJ mol^{-1}	Ref.	τ_T/μs	Ref.
104	**Benzonitrile, 4-(diethylamino)-**												
	n			3.8	[8306]	0.12	[8306]			288	[8306]		
	p	365	[7906]							288	[7906]		
105	**Benzonitrile, 4-(dimethylamino)-**												
	n			2.9	[8307]	0.2	[8702]	0.18	[8006]	290	[8306]		
	p	368	[7906]							288	[7906]		
106	**Benzonitrile, 4-methoxy-**												
	n	422	[8306]	3.4	[8306]	0.1	[8306]			315	[8306]		
	p	421	[7508]							315	[7508]		
107	**Benzo[2'',1'':3,4;3'',4'':3',4']dicyclobuta[1,2-a:1',2'-a']bisbiphenylene**												
	p	248	[0202]	62	[0202]	0.21	[0202]	0.45	[0202]				
108	**Benzo[ghi]perylene**												
	n	293^b	[6907]	203^b	[6907]	0.38^b	[6907]	0.53	[6907]				
	p	294	[7701]	188	[6907]	0.25	[6803]			194	[5604]	150^b	[7107]
109	**Benzo[c]phenanthrene**												
	n	323	[7105]	76	[7101]	0.12	[7101]			237	[5401]		
	p	321	[7701]							239	[7701]		
110	**Benzo[f][4,7]phenanthroline**												
	n	352	[8107]							284	[8107]		

Table 3a Photophysical Parameters of Organic Species

No.	Solv.	E_S/kJ mol^{-1}	Ref.	τ_S/ns	Ref.	Φ_{fl}	Ref.	Φ_T	Ref.	E_T/kJ mol^{-1}	Ref.	τ_T/μs	Ref.
	p	350	[7509]			0.07	[7509]			285	[7509]		
111	**Benzo[a]phenazine**												
	n	295	[8108]			0.0008	[8509]	1.1	[8509]	209	[8515]		
	p	289	[8108]			0.02	[8509]	0.96	[8509]	202	[8108]		
112	**Benzo[a]phenazinium**												
	p	248	[8108]	2.7	[8509]	0.02	[8509]	0.81	[8509]	187	[8108]		
113	**Benzophenone**												
	n	316	[6402]	0.03[b]	[7404]	4×10^{-6}	[7301]	1	[6901]	287	[6702]	6.9[b]	[7703]
	p	311	[6303]	0.016	[7404]			1	[8501]	289	[8007]	50	[7607]
114	**Benzophenone, 4-amino-**												
	n	293[b]	[0002]	0.30	[0002]	1.2×10^{-3}	[0002]	0.6	[0002]				
	p	263	[0002]	0.15	[0002]	5×10^{-4}	[0002]	0.35	[0002]				
115	**Benzophenone, 4,4′-bis(dimethylamino)-**												
	n							0.91	[7704]	275	[6809]	25	[7705]
	p	295	[7907]					0.47	[8501]	255	[7907]	20	[7705]
116	**Benzophenone, 4,4′-dibromo-**												
	n	316[b]	[6604]	4×10^{-6}	[7301]								
	p											288	[6604]

Table 3a Photophysical Parameters of Organic Species

No.	Solv.	E_S/kJ mol⁻¹	Ref.	τ_S/ns	Ref.	Φ_{fl}	Ref.	Φ_T	Ref.	E_T/kJ mol⁻¹	Ref.	τ_T/μs	Ref.
117	**Benzophenone, 4,4'-dimethoxy-**												
	n	328[b]	[6604]			0.0065[b]	[7811]	1[b]	[8501]	293[b]	[8109]	3.6	[7703]
	p							1	[8501]	292	[8007]		
118	**Benzophenone, 4,4'-dimethyl-**												
	n	324	[6908]							288	[6702]	11	[7703]
	p	328	[6908]							290	[8007]		
119	**Benzophenone, 4-methoxy-**												
	n							1[b]	[8501]	287	[6810]		
	p							1	[8501]	290	[8007]	7.2	[8210]
120	**Benzophenone, 4-methyl-**												
	n	323	[6908]					1[b]	[8501]	287	[6702]		
	p	327	[6908]					1	[8501]	290	[8007]		
121	**Benzophenone, 4-phenyl-**												
	n												
	p	321	[6001]					1	[6901]	254	[6001]		
122	**Benzo[g]pteridine-2,4-dione**												
	p			1.0	[7706]	0.033	[7706]	0.31	[7706]			13	[7308]
123	**Benzo[g]pteridine-2,4-dione,7,8-dimethyl-**												
	p	282	[6902]	0.85	[7706]	0.036	[7706]	0.61	[7706]	232	[6902]	12	[7706]

Table 3a Photophysical Parameters of Organic Species

No.	Solv.	E_S/kJ mol^{-1}	Ref.	τ_S/ns	Ref.	Φ_{fl}	Ref.	Φ_T	Ref.	E_T/kJ mol^{-1}	Ref.	τ_T/μs	Ref.
124	**Benzo[g]pteridine-2,4-dione, 7,8,10-trimethyl-**												
	n	239	[6902]	5.5	[7706]	0.4	[7706]	0.3	[7706]	209	[6902]	30	[6811]
	p											320	[7010]
125	**Benzo[g]pteridine-2,4-dione, 7,8,10-trimethyl-, conjugate monoacid**												
	p			2.5	[7706]	0.16	[7706]	0.42	[7706]			29	[7510]
126	**Benzopyran-4-thione**												
	n	200	[9602]	0.024 S_2	[9602]	2.3×10^{-3} S_2	[9602]			190	[9602]	7.3	[9602]
	p	201	[9602]	0.210 S_2	[9602]	2.3×10^{-2} S_2	[9602]			191	[9602]	16	[9602]
127	**Benzo[a]pyrene**												
	n	295	[7105]	49	[7001]					175	[5701]	8700	[7812]
	p	297	[7701]	27.3	[8803]	0.42	[6801]			177	[7701]	8800	[6407]
128	**Benzo[e]pyrene**												
		325	[7105]									120^{b}	[7107]
		327	[7701]							221	[7701]		
129	**Benzo[f]quinoline**												
	n									262	[5901]		
	p	348	[4701]							262	[5903]		
130	**Benzo[g]quinoline**												
	n	307	[8110]	2.1	[8110]	0.1	[8110]	0.67	[8110]	180	[5901]		

Table 3a Photophysical Parameters of Organic Species

No.	Solv.	E_S/kJ mol^{-1}	Ref.	τ_S/ns	Ref.	Φ_{fl}	Ref.	Φ_T	Ref.	E_T/kJ mol^{-1}	Ref.	τ_T/µs	Ref.
	p	293	[8110]	11.6	[8110]	0.62	[8110]	0.32	[8110]	184	[8110]		
131	**Benzo[h]quinoline**												
	n	346	[7105]	4	[8604]	0.11	[8604]	0.88	[8604]	260	[5901]		
	p									261	[5903]		
132	**1,4-Benzoquinone**												
	n	262	[6909]							224	[6909]		
	p											0.53	[8008]
133	**1,4-Benzoquinone,tetrachloro-**												
	n	266	[6909]							206	[6909]	2	[6910]
	p						0.98	[7908]					1.2	[6910]
134	**1,4-Benzoquinone, tetramethyl-**												
	n							1	[7608]			21	[7608]
	p							1	[7608]			15	[7608]
135	**Benzo[f]quinoxaline**												
	n	329	[8308]							254	[8308]		
	p	323	[6304]							251	[6304]		
136	**Benzo[b]thiophene**												
	n									287	[5901]		
	p	394	[8516]							288	[8516]		

Table 3a Photophysical Parameters of Organic Species

No.	Solv.	E_S/kJ mol⁻¹	Ref.	τ_S/ns	Ref.	Φ_{fl}	Ref.	Φ_T	Ref.	E_T/kJ mol⁻¹	Ref.	τ_T/μs	Ref.
137	**Benzotriazole**												
	p	401	[6301]			0.02	[6606]			295	[6301]		
138	**Benzo[b]triphenylene**												
	n	319	[7105]	53.5	[7011]					213	[5701]	90[b]	[7107]
	p	320	[7701]	43	[7307]					213	[7701]		
139	**Benzoxazole, 2-phenyl-**												
	n					0.66	[8804]			260	[7309]		
	p	371	[8804]			0.71	[8804]			262	[8804]		
140	**Benzyl alcohol**												
	n	448	[7101]	29	[7101]	0.07	[8703]	0.51	[8703]				
	p	427	[8202]										
141	**Benzylamine**												
	n			27.8	[8703]	0.072	[8703]	0.33	[8703]	345	[6903]		
	p												
142	**Benzyl cyanide**												
	n			26.9	[8703]	0.046	[8703]	0.41	[8703]				
143	**Biacetyl**												
	n	267[x]	[5501]	11.5	[7209]	0.0027	[7209]	1[b]	[6002]			638	[7209]
	p			7.7	[6911]					236[d]	[5501]	145[e]	[8009]

Table 3a Photophysical Parameters of Organic Species

No.	Solv.	E_S/kJ mol⁻¹	Ref.	τ_S/ns	Ref.	Φ_{fl}	Ref.	Φ_T	Ref.	E_T/kJ mol⁻¹	Ref.	τ_T/μs	Ref.
144	**9,9'-Bianthryl**												
	n	304[b]	[7101]	10.61	[9802]	0.95	[9802]	0.06	[9802]				
	p			35.26	[9802]	0.33	[9802]	0.54	[9802]				
145	**Bibenzyl**												
	n	448	[7101]	35	[7101]	0.13	[7101]						
146	**2,2'-Bifuran**												
	n	401	[0003]	0.90	[0003]	0.63	[0003]	0.24	[0003]			102	[0003]
	p	413	[0003]	0.91	[0003]	0.51	[0003]					39	[0003]
147	**1,1'-Binaphthyl**												
	n	368	[7101]	3	[7101]	0.77	[7101]					14[b]	[7707]
	p	365	[7708]										
148	**2,2'-Binaphthyl**												
	n	359	[7101]	35.2	[7101]	0.41	[7101]						
	p									234	[5604]		
149	**Bisbenzo[3,4]cyclobuta[1,2-a:1',2'-c]biphenylene**												
	p	272	[0202]	81	[0202]	0.15	[0202]	0.03	[0202]				
150	**9,9'-Biphenanthryl**												
	p	340	[7708]							257	[7708]		

Table 3a Photophysical Parameters of Organic Species

No.	Solv.	E_S/kJ mol⁻¹	Ref.	τ_S/ns	Ref.	Φ_{fl}	Ref.	Φ_T	Ref.	E_T/kJ mol⁻¹	Ref.	τ_T/µs	Ref.
151	**Biphenyl**												
	n	418	[7101]	16	[7101]	0.15	[7001]	0.84	[7501]	274	[6401]	130	[6905]
	p	391	[7708]							4	[7310]		
152	**1,1'-Biphenyl, 4-(methylthio)-**												
	p	377	[0102]	1.9	[0102]	0.18	[0102]			255	[0102]		
153	**1,1'-Biphenyl, 4-(methylsulfinyl)-**												
	p	410	[0102]			0.005		0.12	[0102]	289	[0102]		
154	**1,1'-Biphenyl, 4-(methylsulfonyl)-**												
	p	402	[0102]	2.7	[0102]	0.26	[0102]			268	[0102]		
155	**Biphenyl, 4,4'-diamino-**												
	p	346	[7101]	13.5	[8411]	0.25	[8411]						
156	**Biphenyl, 4,4'-dibromo-**												
	n			0.03	[8203]							4[b]	[8010]
157	**Biphenyl, 4,4'-dimethoxy-**												
	n	386	[7101]	11	[7101]	0.21	[7101]						
	p	383	[7101]	11.7	[7101]					264	[7310]		
158	**Biphenyl, 3,3'-dimethyl-**												
	n	410	[7101]	13.2	[7101]	0.18	[7101]	0.32	[8517]	271	[8904]		
	p												

Table 3a Photophysical Parameters of Organic Species

No.	Solv.	E_S/kJ mol⁻¹	Ref.	τ_S/ns	Ref.	Φ_{fl}	Ref.	Φ_T	Ref.	E_T/kJ mol⁻¹	Ref.	τ_T/μs	Ref.
159	**Biphenyl, 4,4′-dimethyl-**												
	n	419	[7105]	16.5	[8517]	0.22	[8517]	0.47	[8517]				
	p									269	[7310]		
160	**Biphenyl, 4-methoxy-**												
	n	401	[7101]	9.4	[7101]	0.26	[7101]						
	p	399	[7101]	9.6	[7101]								
161	**Biphenylene**												
	p	287	[0202]	0.23	[0202]	2×10⁻⁴	[0202]	<0.01	[0202]				
162	**2,2′-Bipyridine, 5-(phenylethynyl)-**												
	n	351	[0104]	0.3	[0104]	0.23	[0104]		[0104]				
163	**4,4′-Bipyridinium, 1,1′-dimethyl-**												
	p	396	[0110]	1.0	[0110]	0.03	[0110]						
164	**α-Bithiophene**												
	n			0.046^b	[9603]	0.026	[9603]	0.99	[9603]			104	[9603]
	p	388	[9603]	0.04	[9603]	0.013	[9603]	0.93	[9603]			124	[9603]
165	**Buckminsterfullerene, 1,9-(4-hydroxycyclohexano)-**												
	n	168^b	[9402]			6×10⁻⁴	[9402]			140	[9402]	23	[9402]
166	**1,3-Butadiene, 1,4-diphenyl-**												
	n			0.6	[8211]	0.42	[8211]	0.02	[8211]	177	[7012]	1.6	[8211]

Table 3a Photophysical Parameters of Organic Species

No.	Solv.	E_S/kJ mol^{-1}	Ref.	τ_S/ns	Ref.	Φ_{fl}	Ref.	Φ_T	Ref.	E_T/kJ mol^{-1}	Ref.	τ_T/µs	Ref.
	p	334	[8011]	0.06	[8212]	0.042	[8212]	≤0.002	[8211]			5	[8413]
167	**1,3-Butadiene, 1,1,4,4-tetraphenyl-**												
	n	308	[7101]	1.8	[7101]	0.6	[7101]					0.665^b	[8414]
168	**1,3-Butadiene, di(2-thienyl)-**												
	n	308	[9901]	3.3	[9901]	0.29	[9901]	0.69	[9901]			7.5	[9901]
	p	297	[9901]			0.049	[9901]	0.07	[9901]			8.6	[9901]
169	**C$_{60}$**												
	n	193	[9101]	1.2^b	[9102]	1.5×10^{-4}	[9502]	1	[9201]	151^b	[9103]	250^b	[9202]
170	**C$_{70}$**												
	n	185	[9104]	0.67^b	[9105]			0.97^b	[9203]	148^b	[9106]	250^b	[9202]
171	**C$_{60}$H$_{18}$**												
	n			2.6^b	[9803]	6×10^{-2}	[9803]	0.15	[9803]			42	[9803]
172	**C$_{60}$H$_{36}$**												
	n			27^b	[9803]	0.37	[9803]	0.1	[9803]			25	[9803]
173	**C$_{120}$O**												
	n	174	[0105]	1.7	[0105]	8.7×10^{-4}	[0105]			121	[0105]		
174	**C$_{120}$**												
	n	164^b	[0106]	1.6	[0106]	7.9×10^{-4}	[0106]	0.7	[0106]			23	[0106]

Table 3a Photophysical Parameters of Organic Species

No.	Solv.	E_S/kJ mol^{-1}	Ref.	τ_S/ns	Ref.	Φ_{fl}	Ref.	Φ_T	Ref.	E_T/kJ mol^{-1}	Ref.	τ_T/µs	Ref.
175	**C$_{180}$**												
	n	164^b	[0106]	0.9	[0106]	5.5×10^{-4}	[0106]	0.7	[0106]			24	[0106]
176	**Carbazole**												
	n	361	[7101]	16.1	[7101]	0.31	[7001]	0.36^b	[6503]			170	[7709]
	p	347	[7808]	15.2	[7101]	0.42	[7302]			294	[5801]		
177	**Carbazole, N-methyl-**												
	n	349	[7101]	18.3	[7101]	0.51	[7101]			292	[8111]		
	p	347	[7101]	16	[7101]	0.42	[7609]			292	[6812]		
178	**Carbazole, N-phenyl-**												
	n	352	[7101]	10.3	[7101]	0.37	[7101]						
	p									294	[6812]		
179	**Carbazole, 3-chloro-**												
	p	335	[0204]	1.04	[0204]	0.033	[0204]	0.97	[0204]	284	[0204]		
180	**Carbazol-2-ol**												
	p	344	[0204]	12.3	[0204]	0.36	[0204]	0.64	[0204]	284	[0204]		
181	**Carbazole, 9-acetyl-**												
	p	366	[0204]	0.38	[0204]	0.003	[0204]	0.99	[0204]	288	[0204]		
182	**4,4'-Carbocyanine, 1,1'-diethyl-**												
	p					0.007	[7311]	<6×10^{-4}	[7311]			1100	[7311]

Table 3a Photophysical Parameters of Organic Species

No.	Solv.	E_S/kJ mol^{-1}	Ref.	τ_S/ns	Ref.	Φ_{fl}	Ref.	Φ_T	Ref.	E_T/kJ mol^{-1}	Ref.	τ_T/µs	Ref.
183	**β-apo-14'-Carotenal**												
	n	262	[7813]	0.8	[7813]	0.0098	[7813]	0.54	[7909]			8	[8309]
	p	257	[7813]			0.0015	[7813]	∼0.033	[7909]			10.3	[7909]
184	**β-Carotene**												
	n	228	[7105]	0.0084^b	[8605]			<0.001	[7710]	88^b	[7504]	70	[6607]
	p	170	[9204]							85	[9204]	9	[8112]
185	**Chlorophyll *a***												
	n	177^b	[8113]	7.8^b	[5703]	0.32^b	[5702]					1500^b	[5802]
	p	178	[7910]	5.5	[8606]	0.33	[8606]	0.53	[8606]	125		800	[6801]
186	**Chlorophyll *b***												
	n	181	[7910]	6.3^b	[5703]	0.11^b	[5702]			130	[7910]	2500^b	[5904]
	p	179	[6801]	3.5	[8606]	0.12	[8606]	0.81	[8606]	136	[6801]	1500	[6801]
187	**Chrysene**												
	n	331	[7105]	44.7	[7101]	0.12	[7001]	0.85	[7501]			710	[6905]
	p	332	[7701]	42.6	[7101]	0.17	[6601]	0.85	[6802]	239	[7701]		
188	**Cinnoline**												
	n	270	[7312]	0.24	[8607]	0.0018	[7312]						
189	**Coronene**												
	n	279^b	[6409]	307	[7011]								

Table 3a Photophysical Parameters of Organic Species

No.	Solv.	E_S/kJ mol^{-1}	Ref.	τ_S/ns	Ref.	Φ_{fl}	Ref.	Φ_T	Ref.	E_T/kJ mol^{-1}	Ref.	τ_T/μs	Ref.
190 Coumarin													
	p	279	[7701]	320	[6912]	0.23	[6802]	0.56	[6802]	228	[5604]		
	n					<0.0001	[7013]	0.054	[7911]	258[b]	[7711]	3.8[b]	[7911]
	p	350	[7303]			<0.0001	[7013]			261	[7303]	1.3	[7911]
191 Coumarin, 7-(diethylamino)-4-methyl-													
	n			2.8	[8012]	0.49	[8518]	0.3	[8518]				
	p			3.1	[8012]	0.73	[8518]	0.006	[8415]			3300	[7402]
192 Coumarin, 7-(diethylamino)-4-(trifluoromethyl)-													
	n			4.1	[8012]	1	[8518]	0.043	[8415]				
	p			0.85	[8518]	0.09	[8518]						
193 Coumarin, 5,7-dimethoxy-													
	n					0.003	[8704]	0.072	[8704]			10[b]	[7911]
	p	339	[7303]			0.65	[8704]			253	[7303]		
194 Curcumin													
	n	271	[0004]	0.239	[0004]	0.052	[0004]	0.012	[0004]			1.83	[0004]
	p	292	[0004]	0.130	[0004]	0.017	[0004]	0.03	[0004]			10.87	[0004]
195 1,3-Cyclohexadiene													
	n	410	[3901]							219	[6504]	30[b]	[8013]

Table 3a Photophysical Parameters of Organic Species

No.	Solv.	E_S/kJ mol^{-1}	Ref.	τ_S/ns	Ref.	Φ_{fl}	Ref.	Φ_T	Ref.	E_T/kJ mol^{-1}	Ref.	τ_T/μs	Ref.
196	**Cyclopentadiene**												
	n									243	[6504]	1.7[b]	[8114]
197	**4h-Cyclopenta[2,1-b:3,4-b']dithiophene**												
	p	384	[9804]	0.03	[9804]	0.0037	[9804]			215	[9804]	27	[9804]
198	**Cyclobuta[1,2-a:3,4-a']bisbiphenylene**												
	p	257	[0202]	49	[0202]	0.12	[0202]	0.25	[0202]				
199	**Dansylamide**												
	n	264	[7529]	17.8	[7529]	0.60	[7529]						
	p	252	[7529]	3.07	[7529]	0.028	[0011]						
200	**Cytidine 5'-monophosphate**												
	p	403	[6706]			0.0001	[7401]			334	[6706]		
201	**Deoxythymidine 5'-monophosphate**												
	p	408	[6706]			0.0001	[7401]	0.055	[7912]	315	[6706]	25	[7912]
202	**1,3-Diazaazulene**												
	n	265[b]	[7313]	0.8	[8519]	0.0024	[8519]	0.63[b]	[8519]	228[b]	[7313]		
	p			0.87	[8519]	0.0025	[8519]						
203	**2,3-Diazabicyclo[2.2.1]hept-2-ene**												
	n	357[b]	[9403]	0.15	[9403]	<0.001	[9403]			259	[9403]		

114

Table 3a Photophysical Parameters of Organic Species

No.	Solv.	E_S/kJ mol^{-1}	Ref.	τ_S/ns	Ref.	Φ_{fl}	Ref.	Φ_T	Ref.	E_T/kJ mol^{-1}	Ref.	τ_T/μs	Ref.
204	**2,3-Diazabicyclo[2.2.2]oct-2-ene**												
	n	318[b]	[9403]	682	[9403]	0.5	[9403]			222	[9403]		
205	**1,7-Diazaperylene**												
	n	270	[9501]	6.2	[9501]	0.97	[9501]	0.015	[9501]	145	[9501]	570	[9501]
206	**2,7-Diazapyrene, N,N'-dimethyl-, bis(tetrafluoroborate)**												
	p	280	[9107]	9.0	[9107]	0.63	[9107]	0.17	[9107]			0.8	[9107]
207	**Dibenz[a,h]acridine**												
	n	304	[7014]			0.24	[7014]			225	[7014]		
	p	306	[7701]			0.25	[7014]			229	[7701]		
208	**Dibenz[a,j]acridine**												
	n	305	[7014]	8.8	[7014]	0.42	[7014]						
	p	303	[7701]			0.51	[7014]			223	[7701]		
209	**Dibenz[a,h]anthracene**												
	n	303	[7105]	37[b]	[7101]			0.9[b]	[7205]				
	p	303	[7701]	27	[7701]	0.12	[6803]	0.9	[8310]	218	[7701]		
210	**Dibenz[a,j]antracene**												
	n	303[b]	[6410]										
	p	303	[7701]	80	[7307]					221	[7701]		

Table 3a Photophysical Parameters of Organic Species

No.	Solv.	E_S/kJ mol^{-1}	Ref.	τ_S/ns	Ref.	Φ_{fl}	Ref.	Φ_T	Ref.	E_T/kJ mol^{-1}	Ref.	τ_T/μs	Ref.
211	**Dibenzo[*def,mno*]chrysene**												
	n	279[b]	[7101]	5[b]	[7101]	0.62[b]	[6803]	0.21[b]	[6913]	141	[7210]		
	p	276	[7701]							143	[7701]		
212	**Dibenzofuran**												
	n	398	[8111]	7.3	[8111]	0.53	[7101]	0.39	[7015]	287	[8111]		
	p									293	[5801]		
213	**Dibenzothiophene**												
	n	367	[8111]	0.9	[8111]	0.09	[7101]	0.97	[7015]	285	[8111]		
	p									288	[7610]		
214	**2,2′-Dicarbocyanine, 1,1′-diethyl-**												
	p					0.0028	[7311]	<0.0003	[7311]			480	[7311]
215	**1,4-Dioxin, 2,3,5,6-tetraphenyl-**												
	n											0.63[b]	[7913]
	p									232		0.535	[7913]
216	**Dithieno[3,2-b:2′,3′-d]thiophene**												
	p	356	[9804]			0.001	[9804]			245	[9804]	45	[9804]
217	**4H-Dithieno[3,2-b:2′,3′-d]pyrrole**												
	p	358	[9804]			7.7×10^{-5}	[9804]			249	[9804]		

Table 3a Photophysical Parameters of Organic Species

No.	Solv.	E_S/kJ mol^{-1}	Ref.	τ_S/ns	Ref.	Φ_{fl}	Ref.	Φ_T	Ref.	E_T/kJ mol^{-1}	Ref.	τ_T/μs	Ref.
218	**Ethene, tetraphenyl-**												
	n	333	[7105]							209[b]	[8213]	0.18[b]	[8213]
219	**Ethene, 1,2-di(2-thienyl)-**												
	n	327	[9901]	0.16	[9901]	0.05	[9901]	0.27	[9901]			0.36	[9901]
	p	323	[9901]			0.009	[9901]	0.02	[9901]			0.25	[9901]
220	**Flavanone**												
	n					<0.001	[8608]			305	[8608]		
	p												
221	**Flavone, 4-thio-**												
	n	199	[9602]	0.009 S_2	[9602]	6×10^{-4} S_2	[9602]			183	[9602]	2.1	[9602]
	p			0.0412 S_2	[9602]	2.8×10^{-3} S_2	[9602]			184	[9602]	3.85	[9602]
222	**Flavone**												
	n							0.9[b]	[8609]	259	[8609]	4.5[b]	[8609]
223	**Fluoranthene**												
	n	295	[7611]	53	[7101]	0.35	[8209]			221	[5701]		
	p	295	[7701]			0.21	[6801]			221	[7701]	8500	[7701]
224	**4-Flavanthione**												
	n	268[b]	[0005]	0.008	[0005]	3.8×10^{-4}	[0005]	1	[0005]	190	[0005]	0.93	[0005]
	p	274	[0005]	0.0073	[0005]	1.8×10^{-4}	[0005]	0.9	[0005]	194	[0005]	0.04	[0005]

Table 3a Photophysical Parameters of Organic Species

No.	Solv.	E_S/kJ mol⁻¹	Ref.	τ_S/ns	Ref.	Φ_{fl}	Ref.	Φ_T	Ref.	E_T/kJ mol⁻¹	Ref.	τ_T/μs	Ref.
225	**4-Flavanthione, 6-hydroxy-**												
	n	267[b]	[0005]	<0.003	[0005]	3.7×10⁻⁴	[0005]	0.96	[0005]	191	[0005]	1.6	[0005]
	p	272	[0005]	0.0039	[0005]	8×10⁻⁵	[0005]			196	[0005]	0.1	[0005]
226	**Fluorene**												
	n	397	[7105]	10	[7101]	0.68	[8517]	0.22	[7015]	282	[6401]	150	[6905]
	p	397	[7701]			0.68	[6803]	0.32	[6802]	284	[7701]		
227	**Fluorene, 9,9-dioctyl-2,7-diphenyl-**												
	n	347	[0205]	0.88	[0205]	0.71	[0205]						
228	**9-Fluorenone**												
	n			2.8[b]	[7814]	0.0005	[6914]	0.94[b]	[7501]			500	[7814]
	p	266	[7006]	21.5	[7814]	0.0027	[6914]	0.48	[7814]	211	[7815]	100	[7814]
229	**9-Fluorenone, 1-amino-**												
	n	264	[0006]	1.3	[0006]	0.055	[0006]	0.86	[0006]	202	[0006]		
	p	258	[0006]	2.2	[0006]	0.057	[0006]	0.89	[0006]				
230	**9-Fluorenone, 2-amino-**												
	n	223	[0006]	0.8	[0006]	0.001	[0006]	0.02	[0006]				
231	**9-Fluorenone, 3-amino-**												
	n	262	[0006]	13.9	[0006]	0.16	[0006]	0.26	[0006]	193	[0006]		
	p	249	[0006]	4.2	[0006]	0.064	[0006]	0.08	[0006]				

Table 3a Photophysical Parameters of Organic Species

No.	Solv.	E_S/kJ mol⁻¹	Ref.	τ_S/ns	Ref.	Φ_{fl}	Ref.	Φ_T	Ref.	E_T/kJ mol⁻¹	Ref.	τ_T/µs	Ref.
232	**9-Fluorenone, 4-amino-**												
	n	243	[0006]	10.1	[0006]	0.08	[0006]	0.48	[0006]				
	p	227	[0006]	3.3	[0006]	0.012	[0006]	0.07	[0006]				
233	**9-Fluorenone, 2-methoxy-**												
	n			9.0	[9205]	0.024	[9205]	0.22	[9205]				
	p	237	[9205]	1.4	[9205]	0.0024	[9205]	0.05	[9205]				
234	**Fluorescein dianion**												
	p	230	[6902]	3.6	[8201]	0.97	[8201]	0.02	[8201]	197	[6902]	20000	[6003]
235	**Fluorescein dianion, 4',5'-dibromo-2',7'-dinitro-**												
	p	218	[6902]	4.5	[7101]					190	[6902]		
236	**Fluorescein dianion, 2',4',5',7'-tetrabromo-**												
	p	209	[6813]	3.6	[7712]	0.69	[7712]	0.33	[7712]	177	[8201]	1700^p	[6104]
237	**Fluorescein dianion, 2',4',5',7'-tetrabromo-3,6-dichloro-**												
	p			3.0	[8201]			0.19	[8201]				
238	**Fluorescein dianion, 2',4',5',7'-tetrabromo-3,4,5,6-tetrachloro-**												
	p	210	[6813]	3.3	[8201]			0.22	[8201]	167	[6813]		
239	**Fluorescein dianion, 3,4,5,6-tetrachloro-2',4',5',7'-tetraiodo-**												
	p	213	[7101]	0.82	[7712]	0.11	[7712]	0.61	[8201]	164	[6813]	30	[8416]

Table 3a Photophysical Parameters of Organic Species

No.	Solv.	E_S/kJ mol⁻¹	Ref.	τ_S/ns	Ref.	Φ_{fl}	Ref.	Φ_T	Ref.	E_T/kJ mol⁻¹	Ref.	τ_T/μs	Ref.
240		**Fluorescein dianion, 2',4',5',7'-tetraiodo-**											
	p	212	[6813]	0.56	[7712]	0.08	[7712]	0.83	[8201]	184	[6915]	630	[6411]
241		**Fluorescein monoanion**											
	p					0.45	[8520]	0.1	[8520]				
242		**Formaldehyde**											
	n	337^g	[8014]							303^g	[8014]		
243		**Fullero-1,2,5-triphenylpyrrolidine**											
	n	170	[9805]	1.2	[9805]	6×10⁻⁴	[9805]	0.95	[9805]			17	[9805]
244		**Furan, 2,5-diphenyl-**											
	n	349	[7101]	1.2	[7101]	1	[7101]						
	p	350	[7101]	1.6	[7101]								
245		**2,2'-Furil**											
	n	281	[0206]	3.7	[0206]			0.71	[0206]			2.8	[0206]
	p	292	[0206]	1.4	[0206]			0.24	[0206]			1	[0206]
246		**Guanosine 5'-monophosphate**											
	p	407	[6706]			8×10⁻⁵	[7401]			325	[6706]		
247		**(E,E,E)-2,4,6-Heptatrienal, 5-methyl-7-(2,6,6-trimethyl-1-cyclohexen-1-yl)-**											
	n	299	[7816]					0.66	[7909]	150	[8417]	6.2	[7909]
	p							0.41	[7909]			10.9	[7909]

Table 3a Photophysical Parameters of Organic Species

No.	Solv.	E_S/kJ mol^{-1}	Ref.	τ_S/ns	Ref.	Φ_{fl}	Ref.	Φ_T	Ref.	E_T/kJ mol^{-1}	Ref.	τ_T/μs	Ref.
248	**Hexahelicene**												
	n	291	[6204]	14.5	[6814]	0.041	[6814]			228	[6204]		
	p	290	[7701]							228	[7701]		
249	**1,3,5-Hexatriene, 1,6-diphenyl-**												
	n			12.9	[8211]	0.65	[8211]	0.029	[8211]	149	[7210]	20	[8211]
	p	300	[7817]	5.2	[8211]	0.27	[8211]	0.02	[7612]			30	[8211]
250	**1,3,5-Hexatriene, 1,6-di(2-thienyl)-**												
	n	281	[9901]	7.6	[9901]	0.077	[9901]	0.11	[9901]			9.1	[9901]
	p	280	[9901]			0.089	[9901]	0.14	[9901]			12.2	[9901]
251	**1,3,5-Hexatriene, 1,6-di(3-thienyl)-**												
	n	329	[9901]					0.27	[9901]			18	[9901]
	p	317	[9901]					0.24	[9901]			33	[9901]
252	**1-Indanone**												
	n	331	[8610]			<1×10^{-6}	[7301]			314	[8610]		
	p									317	[6608]		
253	**2-Indanone**												
	p	367^x	[8521]							345^x	[8521]		
254	**Indazole**												
	n									284^b	[7506]		

Table 3a Photophysical Parameters of Organic Species

No.	Solv.	E_S/kJ mol^{-1}	Ref.	τ_S/ns	Ref.	Φ_{fl}	Ref.	Φ_T	Ref.	E_T/kJ mol^{-1}	Ref.	τ_T/μs	Ref.
	p	396	[6301]			0.46	[6606]			284	[6301]		
255	**Indeno[2,1-*a*]indene**												
	n			2.1	[7314]	0.92	[7314]			199	[8015]		
	p									218	[8905]		
256	**Indole**												
	n	415	[5101]	7.9	[8016]	0.33	[8418]	0.43	[8115]	301	[8418]	16	[7713]
	p	401	[6301]	4.6	[7101]	0.42	[7101]	0.23	[8115]	296	[8418]	11.6	[7511]
257	**Indole, 1,2-diphenyl-**												
	n	364	[7101]	2	[7101]	0.9	[7101]						
	p	363	[7101]	2.4	[7101]								
258	**Indole, 1-methyl-**												
	n									289	[7714]	1.8^b	[7714]
	p			8.5	[6916]	0.38	[6917]			292	[7714]	29	[8311]
259	**Indole, 3-methyl-**												
	n			3.7	[8016]	0.31	[8016]						
	p	413	[7105]	8.5	[8419]	0.45	[8418]			285	[8418]		
260	**Indole, 2-phenyl-**												
	n	361	[7101]	2	[7101]	0.86	[7101]						
	p	354	[7101]	2.6	[7101]								

Table 3a Photophysical Parameters of Organic Species

No.	Solv.	E_S/kJ mol⁻¹	Ref.	τ_S/ns	Ref.	Φ_{fl}	Ref.	Φ_T	Ref.	E_T/kJ mol⁻¹	Ref.	τ_T/μs	Ref.
261	**Indolo[2,3-*b*]quinoxaline**												
	n			0.8	[8611]	0.02	[8611]	0.88	[8611]				
	p			1.4	[8611]	0.01	[8611]	0.92	[8611]				
262	**β-Ionone**												
	n	295[b]	[8522]					0.49[b]	[8522]	230[b]	[8522]	0.16	[7818]
263	**Isoquinoline**							0.21[b]	[8705]	254	[5901]		
	p	374	[5905]	0.25	[5905]	0.012	[7613]			254	[5905]		
264	**Lumichrome**												
	n	292	[0401]	0.64	[0401]	0.028	[0401]						
265	**Lumichrome, 1-methyl-**												
	n	292	[0402]	0.61	[0402]	0.021	[0402]					2.4	[0402]
	p	292	[0402]	0.63	[0402]	0.027	[0402]					6.9	[0402]
266	**Lumiflavin**												
	p	245	[0401]	7.7	[0401]	0.16	[0401]						
267	**Methylene Blue cation**												
	p	180	[6305]			0.04	[7902]	0.52	[6915]	138	[6713]	450	[7512]
268	**Naphthalene**												
	n	385	[7105]	96	[7101]	0.19	[7001]	0.75	[7501]	253	[5701]	175	[6202]

Table 3a Photophysical Parameters of Organic Species

No.	Solv.	E_S/kJ mol⁻¹	Ref.	τ_S/ns	Ref.	Φ_{fl}	Ref.	Φ_T	Ref.	E_T/kJ mol⁻¹	Ref.	τ_T/μs	Ref.
	p	384	[5001]	105	[6815]	0.21	[6601]	0.8	[6802]	255	[4401]	1800	[7402]
269	**Naphthalene-d_8**												
	n	382	[7101]	96	[7101]	0.27	[7101]	>0.38[b]	[6503]				
	p					0.22	[6803]						
270	**Naphthalene, 2-acetyl-**												
	n							0.84[b]	[6503]	249	[6702]	300[b]	[6609]
	p	325	[6001]							249	[6702]		
271	**Naphthalene, 1-amino-**												
	n	348	[7101]	6	[7101]	0.47	[8312]	>0.15[b]	[6503]				
	p	324	[7101]	19.6	[7101]	0.57	[7016]			229	[7614]		
272	**Naphthalene, 2-amino-**												
	n			6.9	[7016]	0.33	[7016]	0.58	[7205]				
	p	306	[7614]	16.6	[7016]	0.46	[7016]	0.32	[7205]	239	[7614]		
273	**Naphthalene, 1-bromo-**												
	n	373	[7105]	0.075	[8203]	0.003[b]	[8017]			247	[5701]	270	[6202]
	p									247	[6303]	830	[6105]
274	**Naphthalene, 2-bromo-**												
	n												
	p	372	[7105]	0.15	[8203]	0.004[b]	[8017]			252	[4401]	150	[6202]

Table 3a Photophysical Parameters of Organic Species

No.	Solv.	E_S/kJ mol^{-1}	Ref.	τ_S/ns	Ref.	Φ_{fl}	Ref.	Φ_T	Ref.	E_T/kJ mol^{-1}	Ref.	τ_T/µs	Ref.
275	**Naphthalene, 1-chloro-**												
	n			2.4	[8203]	0.014^b	[8017]	0.79	[7501]	245^b	[7819]	280	[6202]
	p	375	[7905]							248	[4401]		
276	**Naphthalene, 2-chloro-**												
	n	372	[7101]	4.2	[7101]					251		180	[6202]
	p	373	[7905]								[4401]		
277	**Naphthalene, 1-cyano-**												
	n	398^b	[8420]					>0.17^b	[6503]				
	p	373	[7105]	8.9	[7604]					240	[7604]		
278	**Naphthalene, 2-cyano-**												
	p	363	[7905]							248	[4401]		
279	**Naphthalene, 1,4-dicyano-**												
	n	356	[8420]	3.4	[8421]								
	p			10.1	[7604]			0.19	[8422]	232	[7604]	40	[8422]
280	**Naphthalene, 1,5-dimethoxy-**												
	p	385	[9808]	8.5	[9808]	0.26	[9808]						
281	**Naphthalene, 1,6-dimethyl-**												
	n	373	[8202]	50	[6610]	0.25	[6505]						
	p			55	[6611]	0.2	[6611]						

Table 3a Photophysical Parameters of Organic Species

No.	Solv.	E_S/kJ mol^{-1}	Ref.	τ_S/ns	Ref.	Φ_{fl}	Ref.	Φ_T	Ref.	E_T/kJ mol^{-1}	Ref.	τ_T/µs	Ref.
282	**Naphthalene, 2,6-dimethyl-**												
	n	370	[7101]	38	[7101]	0.37	[7001]						
	p			44	[6611]	0.3	[6611]						
283	**Naphthalene, 1-(dimethylamino)-**												
	n			0.13	[8312]	0.011	[8312]						
	p			2.1	[8312]	0.13	[8312]			243	[4401]		
284	**Naphthalene, 1,4-diphenyl-**												
	n	357	[7101]	1.3	[7101]	0.4	[7101]						
285	**Naphthalene, 1-hydroxy-**												
	n	372	[7101]	10.6	[7101]	0.17	[7001]	>0.27^b	[6503]	245	[4401]		
	p	371	[7105]	7.5	[7101]								
286	**Naphthalene, 2-hydroxy-**												
	n	362	[7101]	13.3	[7101]	0.27	[7001]			252	[4401]	67	[7315]
	p	362	[7105]	8.9	[7101]								
287	**Naphthalene, 1-methoxy-**												
	n					0.36^b	[6816]	0.45	[7501]				
	p	374	[7905]			0.53	[6601]	0.5	[6802]	250	[4401]	5500	[6801]
288	**Naphthalene, 1-methyl-**												
	n	377	[7101]	67	[7101]	0.21	[7206]	0.58	[7501]			25^b	[7615]

Table 3a Photophysical Parameters of Organic Species

No.	Solv.	E_S/kJ mol^{-1}	Ref.	τ_S/ns	Ref.	Φ_{fl}	Ref.	Φ_T	Ref.	E_T/kJ mol^{-1}	Ref.	τ_T/μs	Ref.
	p	377	[7105]	97	[7206]	0.19	[7206]			254	[4401]		
289	**Naphthalene, 2-methyl-**												
	n	376	[7101]	59	[7101]	0.27	[7001]	0.56	[7501]				
	p	374	[7105]	47	[6611]	0.16	[6611]			254	[6701]		
290	**Naphthalene, 1-nitro-**												
	n	313	[7105]	0.012[b]	[7405]			0.63[b]	[6806]	231	[7108]	0.93	[8116]
	p			0.008	[7405]							4.9	[8116]
291	**Naphthalene, 2-nitro-**												
	n	315	[7105]	0.01[b]	[7405]			0.83[b]	[7108]	238	[7108]	0.53	[7616]
	p			0.022	[7405]							1.7	[7616]
292	**Naphthalene, 1-phenyl-**												
	n	379	[7101]	13	[7101]	0.37	[7101]	0.52	[7501]	246	[6701]		
	p												
293	**Naphthalene, 2-phenyl-**												
	n	368	[7101]	114	[7101]	0.26	[7101]	0.43	[7501]	245	[6701]		
	p												
294	**Naphthalene, 1-styryl-, (E)-**												
	n			1.9	[7715]	0.64	[7715]	0.04[b]	[8423]			0.39[b]	[8423]
	p					0.25	[7715]						

127

Table 3a Photophysical Parameters of Organic Species

No.	Solv.	E_S/kJ mol⁻¹	Ref.	τ_S/ns	Ref.	Φ_{fl}	Ref.	Φ_T	Ref.	E_T/kJ mol⁻¹	Ref.	τ_T/µs	Ref.
295	**Naphthalene, 2-styryl-, (E)-**												
	n	372	[0102]	15	[7715]	0.71	[7715]					0.14[b]	[8423]
	p					0.51	[7715]						
296	**Naphthalene, 1-(methylsulfinyl)-**												
	p			13	[0102]	0.011	[0102]	0.42	[0102]	251	[0102]		
297	**Naphthalene, 2-(methylthio)-**												
	p	356	[0102]	13	[0102]	0.12	[0102]			243	[0102]		
298	**Naphthalene, 2-(methylsulfinyl)-**												
	p	381	[0102]			0.015	[0102]	0.18	[0102]	255	[0102]		
299	**Naphthalene, 2-(methylsulfonyl)-**												
	p	368	[0102]	11	[0102]	0.36	[0102]			247	[0102]		
300	**Naphthalene, 1,2,3,4-tetrahydro-**												
	n	429	[8204]	29.2	[8204]	0.2	[8204]			339	[8204]		
301	**1,2-Naphthalenedicarboximide, N-methyl-**												
	n			35.5	[9806]	0.26	[9806]	0.56	[9806]				
	p	305	[9404]	67	[9404]	0.75	[9404]	0.2	[9404]	218	[9404]		
302	**1,2-Naphthalenedicarboximide, N-phenyl-**												
	n	307	[9405]	0.45	[9405]	3×10^{-4}	[9405]	2.1×10^{-2}	[9405]				

Table 3a Photophysical Parameters of Organic Species

No.	Solv.	E_S/kJ mol^{-1}	Ref.	τ_S/ns	Ref.	Φ_{fl}	Ref.	Φ_T	Ref.	E_T/kJ mol^{-1}	Ref.	τ_T/μs	Ref.
303	**1,8-Naphthalenedicarboximide, N-methyl-**												
	p	335	[9404]	0.145	[9404]	0.027	[9404]	0.94	[9404]	222	[9404]		
304	**2,3-Naphthalenedicarboximide**												
	p	326	[9604]	7.8	[9604]	0.23	[9604]						
305	**2,3-Naphthalenedicarboximide, N-methyl-**												
	p	326	[9404]	8	[9404]	0.24	[9404]	0.71	[9404]	243	[9404]		
306	**2,3-Naphthalenedicarboximide, N-phenyl-**												
	p	322	[9604]	<0.05	[9604]	0.24	[9604]						
307	**(1-Naphthalenylcarbonyl)diphenyl-phosphine oxide**												
	n	269^b	[9902]			9.4×10^{-3}	[9902]	0.97	[9902]	229	[9902]	22	[9902]
	p	269	[9902]			8.1×10^{-4}	[9902]					18	[9902]
308	**1,5-Naphthyridine**												
	n	324^x	[8117]	0.27	[8607]			0.55	[8214]	278^x	[7316]		
	p			0.25	[8607]								
309	**1,8-Naphthyridine**												
	n	370	[7105]	0.21	[8607]			0.04	[8214]				
	p	388	[7105]	0.24	[8607]								
310	**1,3,5,7-Octatetraene, 1,8-diphenyl-**												
	n	270	[7211]	6.7	[7211]	0.085	[8211]	0.005	[8211]	132	[7210]	40	[8211]

Table 3a Photophysical Parameters of Organic Species

No.	Solv.	E_S/kJ mol^{-1}	Ref.	τ_S/ns	Ref.	Φ_{fl}	Ref.	Φ_T	Ref.	E_T/kJ mol^{-1}	Ref.	τ_T/µs	Ref.
	p			6.8	[8211]	0.091	[8211]	0.006	[7612]			34	[8211]
311	1,3,5,7-Octatetraene, 1,8-di(2-thienyl)-												
	n	260	[9901]	1.0	[9901]	0.0031	[9901]	0.01	[9901]			13.8	[9901]
	p	259	[9901]			0.0041	[9901]	<0.005	[9901]			4.4	[9901]
312	1,3,5,7-Octatetraene, 1,8-di(3-thienyl)-												
	n	289	[9901]					0.14	[9901]			4.8	[9901]
	p	288	[9901]					0.12	[9901]			42	[9901]
313	1,3,4-Oxadiazole, 2-(4-biphenylyl)-5-phenyl-												
	n	362	[7101]	1	[7101]	0.83	[7101]					0.46^b	[7716]
	p	360	[7101]	1.1	[7101]								
314	1,3,4-Oxadiazole, 2-(4-biphenylyl)-5-phenyl-												
	p	386	[7101]	1.5	[7101]								
315	2,2'-Oxadicarbocyanine, 3,3'-diethyl-												
	p		[8215]	1.1	[8215]	0.49	[8215]	<0.005	[7212]			5000	[7212]
316	Oxazole, 2,5-bis(4-biphenylyl)-												
	n	319^b	[7101]	1.2^b	[7101]							0.285^b	[7716]
	p			1.6	[8805]	0.84	[8805]						
317	Oxazole, 2,5-diphenyl-												
	n	357	[7101]	1.4	[7101]	0.85	[8018]	0.12	[8018]			1700	[8018]

Table 3a Photophysical Parameters of Organic Species

No.	Solv.	E_S/kJ mol^{-1}	Ref.	τ_S/ns	Ref.	Φ_{fl}	Ref.	Φ_T	Ref.	E_T/kJ mol^{-1}	Ref.	τ_T/μs	Ref.
	p	356	[7101]	1.6	[7101]	0.7	[8805]					2500	[7402]
318	**Oxazole, 2-(1-naphthyl)-5-phenyl-**												
	n	329	[7101]	2.1	[7101]	0.94	[7101]					0.215[b]	[7716]
	p	326	[7101]	2.3	[7101]	0.78	[8805]						
319	**Oxazole, 2,2′-(1,4-phenylene)bis (5-phenyl)-**												
	n	315	[7101]	1[b]	[8612]	0.98	[7717]	0.054[b]	[8612]			1750[b]	[8612]
	p	310	[7101]	1.3	[8806]	0.91	[7717]			232	[8612]		
320	**Pentacene**												
	n	205[b]	[7105]			0.08[b]	[7201]	0.16[b]	[7201]	75[b]	[7201]	110	[6105]
321	**(E,E)-2,4-Pentadienal, 3-methyl-5-(2,6,6-trimethyl-1-cyclohexen-1-yl)-**												
	n	334	[7816]					0.2	[8424]	188	[8424]	0.1	[7818]
	p							0.45	[7909]			0.19	[7909]
322	**Pentahelicene**												
	n			25	[7914]	0.07	[7914]					0.31	[7914]
	p	303	[6410]							237	[5604]		
323	**2,3-Pentanedione**												
	n	264[g]	[7213]							226[g]	[7213]		
	p									233	[6714]		

Table 3a Photophysical Parameters of Organic Species

No.	Solv.	E_S/kJ mol^{-1}	Ref.	τ_S/ns	Ref.	Φ_{fl}	Ref.	Φ_T	Ref.	E_T/kJ mol^{-1}	Ref.	τ_T/μs	Ref.
324	**2-Pentanone**												
	n	219^b	[7017]	1.8				0.59	[8425]			0.21	[8425]
	p		[8426]	3.3									
325	**Periflanthene**												
	n		[9206]	0.87	[9206]	0.015	[9206]	<0.01	[9206]	123	[9206]	0.6	[9206]
326	**Periflanthene, 1,16-benzo-**												
	n	239^b	[9206]	10.7	[9206]	0.52	[9206]	0.06	[9206]	170	[9206]	1.5	[9206]
327	**Perylene**												
	n	275	[7105]	6.4	[7101]	0.75	[7902]	0.014	[6612]	148^x	[6918]		
	p	273	[6801]	6	[6201]	0.87	[6106]	0.0088	[6612]	151	[6612]	5000	[6801]
328	**3,4:9,10-perylenebis(dicarboximide), N,N'-bis(1-hexylheptyl)-**												
	p	229	[9903]	4.9	[9903]	1.0	[9903]	0.005	[9903]	116	[9903]	200	[9903]
329	**3,4:9,10-perylenebis(dicarboximide), N,N'-ditridecyl-**												
	n	230	[0007]			0.94	[0007]						
	p	228	[0007]			0.96	[0007]						
330	**Phenanthrene**												
	n	346	[7105]	57.5	[7101]	0.14	[6715]	0.73	[7501]	260	[6306]	145	[6202]
	p	345	[7915]	60.7	[6817]	0.13	[6601]	0.85	[6802]	257	[8905]	910	[6105]

Table 3a Photophysical Parameters of Organic Species

No.	Solv.	E_S/kJ mol^{-1}	Ref.	τ_S/ns	Ref.	Φ_{fl}	Ref.	Φ_T	Ref.	E_T/kJ mol^{-1}	Ref.	τ_T/μs	Ref.
331	**Phenanthrene, 9,10-dihydro-**												
	n	395	[7101]	6.6	[7101]	0.55	[7101]	0.13	[6905]			120	[6905]
332	**Phenanthrene, 9-(methylsulfinyl)-**												
	p	356	[0102]	2	[0102]	0.013	[0102]	0.24	[0102]	289	[0102]		
333	**Phenanthridine**												
	p	342	[7808]							264	[7808]		
334	**6-Phenanthridone**												
	p	352	[8019]			0.25	[8019]			286	[8019]		
335	**6-Phenanthridone, *N*-methyl-**												
	n	351	[0107]	0.46	[0107]	0.056	[0107]	0.96	[0107]				
	p	347	[0107]	0.435	[0107]	0.059	[0107]	0.94	[0107]				
336	**1,10-Phenanthroline**												
	n	350	[8313]			0.001	[7718]					26	[7718]
	p	353	[6304]	2.1	[7718]	0.004	[7718]			264	[6304]	35	[7718]
337	**1,10-Phenanthroline, 3-(phenylethynyl)-**												
	n	345	[0104]	2	[0104]	0.025	[0104]						
338	**Phenazine**												
	n	273^x	[7916]	0.014	[7617]			0.21	[7110]	186	[6919]	42	[7110]
	p	299	[7701]	0.02	[7617]	0.0015	[8523]	0.45	[8523]	187	[8523]	770	[8523]

133

Table 3a Photophysical Parameters of Organic Species

No.	Solv.	E_S/kJ mol⁻¹	Ref.	τ_S/ns	Ref.	Φ_{fl}	Ref.	Φ_T	Ref.	E_T/kJ mol⁻¹	Ref.	τ_T/μs	Ref.
339	**Phenazinium, 3,7-diamino-5-phenyl-**												
	p			4.1		0.32	[8906]	0.1	[8906]			25	[8906]
340	**Phenol**												
	n	431	[7105]	2.1	[7101]	0.066	[7001]	0.32	[8207]				
	p	423	[8427]	7	[8207]	0.19	[6205]			342	[8427]	3.3	[7505]
341	**Phenol, 4-methyl-**												
	n	417	[7101]	2.3	[7101]	0.09	[7101]						
	p	412	[7101]	4.8	[7101]							3.4	[7505]
342	**Phenothiazine**												
	n									253	[7513]		
	p	323	[7105]	1.5	[8314]	0.0034	[8314]	0.54	[8314]				
343	**Phenoxazine**												
	n	330	[7214]	1.5	[7214]	0.027	[7214]			262	[7214]	32	[7214]
	p	327	[7214]	1.1	[7214]	0.023	[7214]			260	[7214]	44	[7018]
344	**Phenoxazine, 10-phenyl-**												
	n	328	[7214]	3.2	[7214]	0.047	[7214]			265	[7214]		
	p	327	[7214]	2.9	[7214]	0.04	[7214]			264	[7214]	49	[7214]
345	**Phenoxazinium, 3,7-diamino-, conjugate monoacid**												
	p							≤0.003	[7618]			55	[7619]

Table 3a Photophysical Parameters of Organic Species

No.	Solv.	E_S/kJ mol^{-1}	Ref.	τ_S/ns	Ref.	Φ_{fl}	Ref.	Φ_T	Ref.	E_T/kJ mol^{-1}	Ref.	τ_T/μs	Ref.
346	**p-Phenylenediamine, N,N,N′,N′-tetramethyl-**												
	n	334	[7101]	4.3	[7101]	0.18	[7101]					1.4	[8216]
	p	329	[7101]	7.1	[7101]							0.5	[8428]
347	**o-Phenylenepyrene**												
	n	260	[7101]	9.1	[7101]	0.17	[7101]						
348	**Phenyl ether**												
	n	426	[7101]	2	[7101]	0.03	[7101]						
	p	426								339	[8427]		
349	**Pheophytin a**												
	n	179	[7910]										
	p	179	[7019]					0.95	[7019]	130	[7910]	750	[7019]
350	**Pheophytin b**												
	n	182	[7910]										
	p	183	[7019]					0.75	[7019]	134	[7910]	1050	[7019]
351	**Phthalazine**												
	n			0.19	[8118]			0.29	[8214]	264^b	[8217]	2.7^b	[8705]
	p	309	[5905]					0.44	[8705]	275	[5905]	21.3	[7514]
352	**Phthalimide, N-methyl-**												
	p	368	[9404]	0.185	[9404]	0.0008	[9404]	0.7	[9404]	297	[9404]		

Table 3a Photophysical Parameters of Organic Species

No.	Solv.	E_S/kJ mol^{-1}	Ref.	τ_S/ns	Ref.	Φ_{fl}	Ref.	Φ_T	Ref.	E_T/kJ mol^{-1}	Ref.	τ_T/µs	Ref.
353	**Phthalocyanine**												
	n	170[b]	[6920]			0.67[b]	[7020]	0.14[b]	[7820]	120[b]	[7820]	130[b]	[7820]
354	**Phthalocyanine, magnesium(II)**												
	n	174[b]	[6920]	6.5	[5704]	0.48[b]	[6920]					100	[6506]
	p			7.6	[5704]	0.76	[7020]	0.23	[7020]			430	[8613]
355	**Phthalocyanine, zinc(II)**												
	n	175[b]	[7111]			0.3[b]	[6920]	0.65[b]	[8119]	109[b]	[7111]		
	p					0.45	[7020]	0.04	[8613]			270	[8613]
356	**Picene**												
	n	318[b]	[6410]									160[b]	[7107]
	p	318	[7701]							240	[7701]		
357	**Porphyrin**												
	n	195[b]	[7515]	12.5[b]	[7020]	0.055[b]	[7020]	0.9	[8218]	151[b]	[7515]		
	p	196	[7406]							152	[7406]		
358	**Porphyrin, magnesium(II)**												
	n	208	[7515]							164	[7515]		
	p	207	[7515]			0.066	[7114]			163	[7515]		
359	**Porphyrin, octaethyl-**												
	n	193	[7515]							155	[7515]		

Table 3a Photophysical Parameters of Organic Species

No.	Solv.	E_S/kJ mol^{-1}	Ref.	τ_S/ns	Ref.	Φ_{fl}	Ref.	Φ_T	Ref.	E_T/kJ mol^{-1}	Ref.	τ_T/μs	Ref.
	p	194	[7406]							156	[7406]		
360	Porphyrin, octaethyl-, zinc(II)												
	n			1.9[b]	[8614]	0.05[b]	[8614]					5000[b]	[8614]
	p	208	[7114]	2.1	[8614]	0.045	[8614]			170	[7114]	2100	[8614]
361	Porphyrin, 2,7,12,17-tetraethyl-3,8,13,18-tetramethyl-												
	n	192[b]	[7515]			0.09[b]	[6920]			155[b]	[7515]		
	p	194	[7515]							156	[7515]		
362	Porphyrin, 2,7,12,17-tetraethyl-3,8,13,18-tetramethyl-, magnesium(II)												
	n	207	[7515]			0.25[b]	[7515]			162	[7515]		
	p	203	[7515]										
363	Porphyrin, 2,7,12,17-tetraethyl-3,8,13,18-tetramethyl-, zinc(II)												
	n	209[b]	[6920]			0.04[b]	[6920]						
	p	208	[7114]			0.04	[7114]			170	[7114]		
364	Porphyrin, tetrakis(1-methylpyridinium-4-yl)-												
	p	177	[8429]	6.0	[8429]	0.047	[8429]	0.92	[8429]			170	[8429]
365	Porphyrin, tetrakis(1-methylpyridinium-4-yl)-, zinc(II)												
	p	191	[8429]	1.3	[8219]	0.025	[8429]	0.9	[8429]			2000	[8429]
366	Porphyrin, tetrakis(4-sulfonatophenyl)-												
	p			10.4	[8219]	0.16	[8430]	0.78	[8219]			420	[8219]

137

Table 3a Photophysical Parameters of Organic Species

No.	Solv.	E_S/kJ mol^{-1}	Ref.	τ_S/ns	Ref.	Φ_{fl}	Ref.	Φ_T	Ref.	E_T/kJ mol^{-1}	Ref.	τ_T/μs	Ref.
367	**Porphyrin, tetrakis(4-sulfonatophenyl)-, zinc(II)**												
	p			1.4	[8430]	0.041	[8430]	0.84	[8219]			80	[8220]
368	**Porphyrin, tetrakis(4-trimethylammoniophenyl)-**												
	p			9.3	[8315]	0.07	[8315]	0.8	[8315]			540	[8315]
369	**Porphyrin, tetrakis(4-trimethylammoniophenyl)-, zinc(II)**												
	p			1.8	[8315]			0.82	[8315]			1200	[8315]
370	**Porphyrin, tetraphenyl-**												
	n	179	[8221]	13.6	[8020]	0.11[b]	[8430]	0.82	[8218]	138	[8221]	1500[b]	[8430]
	p	185	[7406]	10.1	[7020]	0.15	[8120]	0.88	[8316]	140	[7406]		
371	**Porphyrin, tetraphenyl-, magnesium(II)**												
	n	196	[8221]	9.2	[8121]	0.15	[8121]			143	[8221]	1400	[8121]
	p	193	[7114]					0.85	[8316]	143	[7114]		
372	**Porphyrin, tetraphenyl-, zinc(II)**												
	n	198	[8221]	2.7	[8121]	0.04	[8121]	0.88	[8218]	153	[7516]	1200	[8121]
	p	199	[7114]					0.9	[8316]	153	[7114]		
373	**Porphyrin, zinc(II)**												
	n					0.022[b]	[7821]			166	[7114]		
	p	210	[7114]			0.023	[7114]						

Table 3a Photophysical Parameters of Organic Species

No.	Solv.	E_S/kJ mol⁻¹	Ref.	τ_S/ns	Ref.	Φ_{fl}	Ref.	Φ_T	Ref.	E_T/kJ mol⁻¹	Ref.	τ_T/μs	Ref.
374	**Porphyrin-2,12-dipropanoic acid, 7,17-diethyl-3,8,13,18-tetramethyl-, dimethyl ester**												
	n	191[b]	[6920]			0.08[b]	[6920]	0.81[b]	[8021]			220[b]	[8021]
375	**Porphyrin-2,18-dipropanoic acid, 7,12-diethenyl-3,8,13,17-tetramethyl-, dimethyl ester**												
	n			23[b]	[7719]	0.06	[7406]	0.8	[8218]			550[b]	[8022]
	p	191	[7406]							150	[7406]		
376	**Porphyrin-2,18-dipropanoic acid, 2,3-dihydro-3,3,7,12,17-pentamethyl-**												
	p	187	[8023]	6.3	[8023]	0.07	[8023]	0.85	[8023]	180	[8023]	430	[8023]
377	**Porphycene**												
	n			10.2[b]	[8615]	0.36[b]	[8615]	0.42[b]	[8615]			200[b]	[8615]
378	**Propiophenone**												
	n	336	[8610]							312	[8610]		
	p									313	[7006]		
379	**Psoralen**												
	n							0.034[b]	[7805]				
	p	327	[7303]	0.92[m]	[8431]	0.01	[8431]	0.06	[7917]	262	[7917]	5	[7303]
380	**Psoralen, 3-carbethoxy-**												
	n							0.3[b]	[8222]			20[b]	[8222]
	p			0.93	[8431]	0.025	[8431]	0.44	[8222]	254	[8317]	5.5	[8222]

Table 3a Photophysical Parameters of Organic Species

No.	Solv.	E_S/kJ mol⁻¹	Ref.	τ_S/ns	Ref.	Φ_{fl}	Ref.	Φ_T	Ref.	E_T/kJ mol⁻¹	Ref.	τ_T/μs	Ref.
381	**Psoralen, 5-methoxy-**												
	n			0.15	[8807]	0.0012	[8807]	0.067[b]	[8807]				
	p	319	[7303]	0.93	[8807]	0.01	[8807]	0.1	[8807]	254	[8317]	4.2	[8317]
382	**Psoralen, 8-methoxy-**												
	n							0.011[b]	[7805]			1.1[b]	[7805]
	p	309	[7303]			0.002	[8704]	0.04	[8317]	262	[7303]	10	[7917]
383	**Psoralen, 4,5',8-trimethyl-**												
	p	321	[7303]					0.093	[7918]	268	[7303]	7.1	[7918]
384	**Purine, 2-amino-**												
	n	362	[7407]	1.6	[7517]	0.11	[7517]						
	p	355	[7407]	6	[7517]	0.64	[7517]					83	[7517]
385	**Purine, 2-(dimethylamino)-**												
	n	345	[7407]	7.1	[7517]	0.13	[7407]						
	p	333	[7407]	9.4	[7517]	0.75	[7407]					59	[7517]
386	**Pyrazine**												
	n	365	[8318]			0.0004	[6716]	0.33	[6716]	315	[7822]		
	p							0.87	[7514]	311	[5803]	4.5	[7514]
387	**Pyrene**												
	n	322	[7105]			0.65	[7919]	0.37	[8223]	203	[5701]	180	[7010]

Table 3a Photophysical Parameters of Organic Species

No.	Solv.	E_S/kJ mol^{-1}	Ref.	τ_S/ns	Ref.	Φ_{fl}	Ref.	Φ_T	Ref.	E_T/kJ mol^{-1}	Ref.	τ_T/μs	Ref.
	p	321	[7808]	190	[8803]	0.72	[6601]	0.38	[6802]	202	[7808]	11000	[6801]
388	**Pyrene, 1-(methylsulfinyl)-**												
	p	326	[0102]	1.3	[0102]	0.008	[0102]	0.35	[0102]				
389	**1-Pyrenecarboxaldehyde**												
	n					0.0002	[8319]	0.78	[8319]	180[b]	[8319]	50	[8319]
	p			1.7	[8319]	0.084	[8319]	0.65	[8319]			38	[8319]
390	**Pyridazine**												
	n	318	[6716]	2.6	[6507]	0.0002	[6716]	0.07	[6716]	297	[9207]	0.065	[6716]
	p							0.08	[9207]	290[x]	[9207]	0.1	[6717]
391	**Pyridine, 2-amino-**												
	n			24.2	[8432]	0.04	[8224]						
	p			5.1	[8432]	0.19	[8224]						
392	**Pyrimidine**												
	n	360	[6716]	1.5	[6507]	0.0029	[6716]	0.12	[6716]	338	[7021]		
	p							1	[7514]			1.4	[7514]
393	**Pyronine cation**												
	p	215	[7720]	2.3	[8201]			<0.02	[8201]	178	[7720]		
394	**Pyrromethene 546**												
	p	240	[0301]	5.6	[0301]	0.99	[0301]						

Table 3a Photophysical Parameters of Organic Species

142

No.	Solv.	E_S/kJ mol^{-1}	Ref.	τ_S/ns	Ref.	Φ_{fl}	Ref.	Φ_T	Ref.	E_T/kJ mol^{-1}	Ref.	τ_T/μs	Ref.
395 Pyrromethene 560													
	p	227	[0301]	6.9	[0301]	0.7	[0301]						
396 Pyrromethene 567													
	p	227	[0301]	6.6	[0301]	0.83	[0301]						
397 Pyrromethene 597													
	p	221	[0301]	4.2	[0301]	0.77	[0301]						
398 Pyrromethene 650													
	p	200	[0301]	1.6	[0301]	0.54	[0301]						
399 Pyrromethene 556													
	p	230	[9904]	5.13	[0102]	0.84	[0102]						
400 Pyruvic acid													
	n							0.65^b	[8122]				
	p							0.88	[8122]			0.5	[8122]
401 2-Pyridyl 2-thienyl ketone													
	p	309	[9807]					1	[9807]	261	[9807]	30	[9807]
402 3-Pyridyl 2-thienyl ketone													
	p	324	[9807]					1	[9807]	262	[9807]	15	[9807]
403 4-Pyridyl 2-thienyl ketone													
	p	320	[9807]					1	[9807]	262	[9807]	24	[9807]

Table 3a Photophysical Parameters of Organic Species

No.	Solv.	E_S/kJ mol^{-1}	Ref.	τ_S/ns	Ref.	Φ_{fl}	Ref.	Φ_T	Ref.	E_T/kJ mol^{-1}	Ref.	τ_T/μs	Ref.
404	**2-Pyridyl 3-thienyl ketone**												
	p	314	[9807]					0.8	[9807]	291	[9807]	0.05	[9807]
405	**3-Pyridyl 3-thienyl ketone**												
	p	328	[9807]					0.9	[9807]	299	[9807]	10	[9807]
406	**4-Pyridyl 3-thienyl ketone**												
	p	327	[9807]					0.9	[9807]	299	[9807]	33	[9807]
407	**5h-Pyrido[3,2-b]indole**												
	n	353	[0207]	4	[0207]	0.145	[0207]			295	[0207]		
408	**Pyrylium, 2,6-dimethyl-4-phenyl-, perchlorate**												
	p	297	[9701]	12.1	[9701]	0.1	[9701]	0.5	[9701]	264	[9701]	0.3	[9701]
409	**Pyrylium, 2,6-dimethyl-4-(4-methylphenyl)-, perchlorate**												
	p	299	[9701]	8.7	[9701]	0.28	[9701]	0.2	[9701]	256	[9701]	2.4	[9701]
410	**Pyrylium, 4-[1,1'-biphenyl]-4-yl-2,6-dimethyl-**												
	p	266	[9905]	3.4	[9905]	0.78	[9905]	0.03	[9905]			2.2	[9905]
411	**Pyrylium, 2,6-dimethyl-4-(1-naphthalenyl)-, perchlorate**												
	p	246	[9905]	13.4	[9905]	0.22	[9905]	0.13	[9905]			6.1	[9905]
412	**Pyrylium, 2,6-dimethyl-4-(2-naphthalenyl)-, perchlorate**												
	p	249	[9905]	16.7	[9905]	0.64	[9905]	0.09	[9905]			3.5	[9905]

Table 3a Photophysical Parameters of Organic Species

No.	Solv.	E_S/kJ mol^{-1}	Ref.	τ_S/ns	Ref.	Φ_{fl}	Ref.	Φ_T	Ref.	E_T/kJ mol^{-1}	Ref.	τ_T/μs	Ref.
413	**Pyrylium, 4-(4-acetylphenyl)-2,6-dimethyl-, perchlorate**												
	p	326	[9905]	2	[9905]	0.03	[9905]	0.66	[9905]			8.3	[9905]
414	**2,2':5',2'':5'',2'''-Quaterfuran**												
	n	309[b]	[0003]	0.80	[0003]	0.8	[0003]	0.33	[0003]			67	[0003]
	p	318	[0003]	1.74	[0003]	0.79	[0003]	0.29	[0003]			14	[0003]
415	**p-Quaterphenyl**												
	n	362	[7101]	0.8	[7101]	0.92	[8320]						
416	**α-Quaterthiophene**												
	n			0.44[b]	[9603]	0.18	[9603]	0.73	[9603]			38	[9603]
	p			0.48	[9603]	0.16	[9603]	0.71	[9603]			40	[9603]
417	**Quinazoline**												
	n	326	[7823]	0.079	[8024]			0.7	[8214]	262	[7823]		
	p	327	[5905]							262	[5905]		
418	**Quinine bisulfate**												
	p	305	[7101]	19.2	[7101]	0.55	[6106]						
419	**Quinoline**												
	n	383	[8025]					0.31[b]	[6503]	258	[8025]		
	p	381	[8025]							261	[8321]		

Table 3a Photophysical Parameters of Organic Species

No.	Solv.	E_S/kJ mol^{-1}	Ref.	τ_S/ns	Ref.	Φ_{fl}	Ref.	Φ_T	Ref.	E_T/kJ mol^{-1}	Ref.	τ_T/µs	Ref.
420	**Quinoline, 4-[(1E)-2-(9-anthracenyl)ethenyl]-**												
	n	270	[0201]	2.6	[0201]	0.36	[0201]	0.14	[0201]			44	[0201]
	p	261	[0201]	0.088	[0201]	0.006	[0201]	0.009	[0201]				
421	**Quinoline, 2-[(1E)-2-(9-anthracenyl)ethenyl]-**												
	n	271	[0201]	3.4	[0201]	0.41	[0201]	0.13	[0201]			38	[0201]
	p	266	[0201]	0.33	[0201]	0.04	[0201]	0.014	[0201]			28	[0201]
422	**Quinoline, 3-[(1E)-2-(9-anthracenyl)ethenyl]-**												
	n	274	[0201]	3.2	[0201]	0.7	[0201]	0.24	[0201]			40	[0201]
	p	272	[0201]	3.0	[0201]	0.58	[0201]	0.16	[0201]			33	[0201]
423	**Quinoxaline**												
	n	314	[8225]	0.023	[8024]			0.99	[8214]	255	[8225]		
	p	319	[5905]					0.9	[8502]	254	[5905]	29.4	[7514]
424	**(all-E)-Retinal**												
	n	281	[7113]					0.43^w	[7518]	123	[8417]	9.3	[8226]
	p							0.12^w	[7824]			18	[6206]
425	**(all-E)-Retinol**												
	n	321	[6921]	5^b	[8524]	0.02	[6921]	0.017	[7710]	140	[8417]	25	[7114]
	p	327	[6921]	2.3	[8524]	0.006	[6921]	~0.003	[8524]				

Table 3a Photophysical Parameters of Organic Species

No.	Solv.	E_S/kJ mol⁻¹	Ref.	τ_S/ns	Ref.	Φ_{fl}	Ref.	Φ_T	Ref.	E_T/kJ mol⁻¹	Ref.	τ_T/μs	Ref.
426	**Rhodamine, inner salt**												
	p	230	[7720]	4.4	[6508]	0.88	[6508]	0.0024	[7721]	191	[7720]		
427	**Rhodamine, inner salt, *N,N'*-diethyl**												
	p	224	[7720]			0.94		0.005	[7722]	184	[7720]		
428	**Rhodamine B, inner salt**												
	p	213	[7720]	2.7	[8616]	0.65	[8616]	0.0024	[7721]	178	[7720]	1.6	[7722]
429	**Rhodamine, 4,5-dibromo, methyl ester**												
	p	233	[9605]	1.6	[9605]	0.34	[9605]	0.27	[9605]				
430	**Rhodamine 6G cation**												
	p	219	[7720]	3.8	[8201]	0.86[w]	[7902]	0.0021	[7408]	181	[7720]	3500	[7408]
431	**Riboflavin, conjugate monoacid**												
	p	263	[7105]	2.3	[7706]	0.12	[7706]	0.4	[7706]			19	[7706]
432	**Rubrene**												
	n	221[b]	[8123]	16.5[b]	[7101]	0.98[b]	[6818]	0.0092	[8617]	110[b]	[8124]	120	[8617]
	p			10.8	[8802]			0.023	[8902]			80	[6819]
433	**Spirilloxanthin**												
	n	219	[7105]					0.028	[7620]			6.2	[7620]
434	**Squaraine, bis[4-(dimethylamino)-phenyl]-**												
	p	187	[9208]	1.5	[9208]	0.45	[9208]					35.2	[9208]

Table 3a Photophysical Parameters of Organic Species

No.	Solv.	E_S/kJ mol^{-1}	Ref.	τ_S/ns	Ref.	Φ_{fl}	Ref.	Φ_T	Ref.	E_T/kJ mol^{-1}	Ref.	τ_T/µs	Ref.
435	**Squaraine, bis(3-ethylbenzothiazol-2-ylidene)-**												
	n	175[b]	[9209]	2.5	[9209]	0.54	[9208]						
	p	183	[9209]	1.35	[9209]	0.21	[9208]						
436	**(E)-Stilbene**												
	n	358[x]	[6207]	0.075	[7920]	0.036	[8026]			206	[8015]	14	[6820]
	p					0.016	[8227]			206	[8015]	62	[8125]
437	**(E)-Stilbene, 2-amino-**												
	n	321	[9906]	3.7	[9906]	0.88	[9906]						
	p	301	[9906]	5.4	[9906]	0.69	[9906]						
438	**(E)-Stilbene, 3-amino-**												
	n	336	[9906]	7.5	[9906]	0.78	[9906]	0.09	[0108]				
	p	312	[9906]	11.7	[9906]	0.40	[9906]	0.23	[0108]				
439	**(E)-Stilbene, 4-amino-**												
	n	338	[9906]	0.1	[9906]	0.05	[9906]						
	p	316	[9906]	0.1	[9906]	0.03	[9906]						
440	**(E)-Stilbene, 3,5-diamino-**												
	n	311	[0108]	23.4	[0108]	0.43	[0108]						
	p	286	[0108]	16.1	[0108]	0.1	[0108]						

Table 3a Photophysical Parameters of Organic Species

No.	Solv.	E_S/kJ mol⁻¹	Ref.	τ_S/ns	Ref.	Φ_{fl}	Ref.	Φ_T	Ref.	E_T/kJ mol⁻¹	Ref.	τ_T/µs	Ref.
441		**(E)-Stilbene, 4,4'-diamino-**											
	n	321	[0108]	1	[0108]	0.76	[0108]	0.1	[0108]				
	p	316	[0108]	0.57	[0108]	0.3	[0108]	0.34	[0108]				
442		**(E)-Stilbene, 4-bromo-**											
	n	347[x]	[6207]										
	p					0.038	[7921]			201	[8905]	0.03	[7922]
443		**(E)-Stilbene, 4,4'-dinitro-**											
	n					<0.0001[b]	[8433]	0.81[b]	[8433]			0.08	[7409]
	p					<0.0001	[8433]	0.69	[8433]	189	[8905]	0.13	[7923]
444		**(E)-Stilbene, 4,4'-diphenyl-**											
	n	322[b]	[7101]	1.1[b]	[7101]	0.81	[8808]						
	p					0.82	[8808]						
445		**(E)-Stilbene, 4-nitro-**											
	n					<0.0001[b]	[8433]	0.86[b]	[8433]	195	[8905]	0.063	[7409]
	p					<0.0001	[8433]	0.71	[8433]	208	[4401]	0.083	[7923]
446		**(Z)-Stilbene**											
	n	360[x]	[6207]							227[x]	[6922]	17	[6820]
447		**Styrene**											
	n	415	[7723]	13.9	[7924]	0.25	[7924]	0.4	[8228]	258	[8228]	0.025	[7723]

Table 3a Photophysical Parameters of Organic Species

No.	Solv.	E_S/kJ mol^{-1}	Ref.	τ_S/ns	Ref.	Φ_{fl}	Ref.	Φ_T	Ref.	E_T/kJ mol^{-1}	Ref.	τ_T/µs	Ref.
448	**Styrene, β-(2-pyridyl)-, (E)-**												
	n					0.0008	[8026]						
	p	349	[7105]			0.001	[8229]			205	[8905]		
449	**Styrene, β-(4-pyridyl)-, (E)-**												
	n					0.0016	[8026]			208	[8905]		
	p	355	[7105]			0.0015	[8229]			206	[8905]		
450	**2,2':5',2''-Terfuran**												
	n	342^b	[0003]	1.30	[0003]	0.77	[0003]	0.62	[0003]			34	[0003]
	p	352	[0003]	1.56	[0003]	0.74	[0003]	0.31	[0003]			13	[0003]
451	**m-Terphenyl**												
	n	393	[7101]	28.5	[7101]	0.29	[7101]	0.41	[8517]				
	p									269	[6701]		
452	**p-Terphenyl**												
	n	385	[7101]	0.95	[7101]	0.77	[7001]	0.11	[6905]			450	[6905]
	p									244	[6701]		
453	**α-Terthiophene**												
	n			0.16^b	[9603]	0.07	[9603]	0.95	[9603]			88	[9603]
	p			0.18	[9603]	0.056	[9603]	0.9	[9603]			62	[9603]

Table 3a Photophysical Parameters of Organic Species

No.	Solv.	E_S/kJ mol⁻¹	Ref.	τ_S/ns	Ref.	Φ_{fl}	Ref.	Φ_T	Ref.	E_T/kJ mol⁻¹	Ref.	τ_T/µs	Ref.
454	**Tetrabenzoporphyrin, zinc(II)**												
	n					0.23[b]	[7020]					525[b]	[7317]
	p	191	[7114]			0.11	[7114]			149	[7114]		
455	**Tetracene**												
	n	254	[7105]	6.4	[6509]	0.17	[7001]	0.62[b]	[7501]	123	[6412]	400	[8505]
	p	254	[7701]			0.16	[7115]	0.66	[7115]				
456	**1,4-Tetracenequinone**												
	n	243	[0109]	1.4	[0109]	0.03	[0109]	0.78	[0109]	179	[0109]	480	[0109]
457	**Thianthrene**												
	n	407	[7105]			0.036	[7724]	0.94	[7724]				
458	**Thieno[2,3-a]carbazole, 2,3,5-trimethyl-**												
	p	344	[0302]	1.08	[0302]	0.057	[0302]	0.94	[0302]			4	[0302]
459	**Thieno[2,3-b]carbazole, 2,3,4,10-tetramethyl-**												
	p	317	[0302]	1.52	[0302]	0.086	[0302]	0.34	[0302]			12	[0302]
460	**Thiobenzophenone**												
	n	191	[7519]					1[wb]	[8434]	165	[7519]	1.7[b]	[8434]
	p	191	[7215]							170	[7215]		
461	**Thiobenzophenone, 4,4'-bis(dimethylamino)-**												
	n	205	[7215]					1[b]	[8434]	176	[7215]	1.3[b]	[8434]

Table 3a Photophysical Parameters of Organic Species

No.	Solv.	E_S/kJ mol^{-1}	Ref.	τ_S/ns	Ref.	Φ_{fl}	Ref.	Φ_T	Ref.	E_T/kJ mol^{-1}	Ref.	τ_T/µs	Ref.
462	**Thiobenzophenone, 4,4'-dimethoxy-**												
	n							1^{wb}	[8434]	172	[7215]	1.4^b	[8434]
	p	197	[7215]							176			[7215]
463	**Thiocoumarin**												
	n							1^b	[8618]	212^b	[8618]	>2.5^b	[8618]
464	**Thiofluorenone**												
	n	173	[7519]							159	[7519]		
465	**Thiofosgene**												
	p	221	[9210]	0.04	[9210]	5.5×10^{-6}	[9210]	<0.04	[9210]	207	[9210]	0.09	[9210]
466	**Thioindigo**												
	n			11	[8435]	0.3	[8435]					0.158	[7825]
	p			7.5	[8435]	0.18	[8435]					0.2	[7925]
467	**Thionine cation**												
	p	196	[6923]					0.62	[6915]	163	[6923]	72	[6923]
468	**Thionine cation, conjugate monoacid**												
	p			0.37	[7725]	0.047	[7725]					16	[7726]
469	**Thiophene, 2,2'-(1,3-phenylene)bis-**												
	n	366	[8619]	6	[8619]								
	p									238		31	[8619]

Table 3a Photophysical Parameters of Organic Species

No.	Solv.	E_S/kJ mol^{-1}	Ref.	τ_S/ns	Ref.	Φ_{fl}	Ref.	Φ_T	Ref.	E_T/kJ mol^{-1}	Ref.	τ_T/µs	Ref.
470	Thiophene, 2,2'-(9,9-dioctyl-9H-fluorene-2,7-diyl)bis-												
	n	323	[0205]	0.96	[0205]	0.84	[0205]						
471	Thiopyrylium, 2,6-dimethyl-4-phenyl-, perchlorate												
	p	279	[9701]	12.3	[9701]	0.07	[9701]	0.4	[9701]	269	[9701]	0.3	[9701]
472	Thiopyrylium, 2,6-dimethyl-4-(4-methylphenyl)-, perchlorate												
	p	279	[9701]	5.1	[9701]	0.35	[9701]	0.3	[9701]	270	[9701]	3	[9701]
473	Thioxanthene												
	n					0.008	[7724]	0.78	[7724]				
474	Thioxanthen-9-one												
	n									265	[8620]	95^b	[8126]
	p			2	[8901]	0.12	[7410]					73	[7318]
475	Thioxanthione												
	n							1^{wb}	[8434]	166	[8230]	0.83^b	[8434]
476	Thymidine												
	n	421	[8027]										
	p	417	[8027]					0.069	[7912]			25	[7912]
477	Thymine												
	n	422	[8027]										
	p	423	[8027]					0.06	[7520]			10	[7520]

Table 3a Photophysical Parameters of Organic Species

No.	Solv.	E_S/kJ mol^{-1}	Ref.	τ_S/ns	Ref.	Φ_{fl}	Ref.	Φ_T	Ref.	E_T/kJ mol^{-1}	Ref.	τ_T/µs	Ref.
478	**α-Tocopherol**												
	n			0.8	[8907]	0.16	[8907]						
	p			1.8	[8907]	0.34	[8907]						
479	**Toluene**												
	n	445	[7105]	34	[7101]	0.14	[8204]	0.53	[6901]	346	[6103]		
	p			35	[7621]	0.13	[7411]			347	[6702]		
480	**Toluene-d_8**												
	n	447	[7101]	35	[7101]	0.21	[7101]						
	p			36	[7621]	0.14	[7621]						
481	**1,2,3-Triazolo[4,5-d]pyridazin-4-one, 2,5-dihydro-5-methyl-7-phenyl-**												
	n	354	[0208]	1.2	[0208]	0.031	[0208]	0.23	[0208]				
	p	353	[0208]	1.9	[0208]	0.033	[0208]	0.21	[0208]				
482	**Triphenylene**												
	n	349	[7101]	36.6	[7101]	0.066	[7001]	0.86	[6901]	280	[8436]	55	[6105]
	p	352	[7701]	37	[7607]	0.09	[6601]	0.89	[6601]			1000	[6105]
483	**Triphenylene-d_{12}**												
	n	351	[7101]	38	[7101]	0.11	[7101]						
	p					0.068	[6803]						

153

Table 3a Photophysical Parameters of Organic Species

No.	Solv.	E_S/kJ mol⁻¹	Ref.	τ_S/ns	Ref.	Φ_{fl}	Ref.	Φ_T	Ref.	E_T/kJ mol⁻¹	Ref.	τ_T/μs	Ref.
484	**Triphenylene,dodecahydro-**												
	n	415	[8204]	51.5	[8204]	0.21	[8204]			321	[8204]		
485	**Triphenylmethane**												
	n	444	[7105]			0.028	[6718]	0.16	[6718]				
486	**Tryptophan**												
	n					0.07	[8418]						
	p	399	[6208]	2.5	[7319]	0.13	[8418]	0.18	[7727]			14.3	[7511]
487	**Uracil**												
	n	430	[8027]										
	p	430	[8027]					0.1	[7912]			2	[7520]
488	**Uracil, 1,3-dimethyl-**												
	n	423	[8027]										
	p	423	[8027]					0.02	[8127]				
489	**Uridine**												
	p							0.078	[7912]			20	[7912]
490	**Uridine 5'-monophosphate**												
	p	417	[6706]			0.00003	[7401]	0.044	[7912]	328	[7320]	33	[7912]
491	**Xanthene**												
	p	408	[8801]							331	[8801]		

Table 3a Photophysical Parameters of Organic Species

No.	Solv.	E_S/kJ mol^{-1}	Ref.	τ_S/ns	Ref.	Φ_{fl}	Ref.	Φ_T	Ref.	E_T/kJ mol^{-1}	Ref.	$\tau_T/\mu s$	Ref.
492	**9-Xanthione**												
	n	193	[8128]					0.8^{wb}	[8434]	181	[8128]	1.8^b	[8434]
	p											0.7^p	[7412]
493	**Xanthone**												
	n	324	[7116]	0.013	[7926]					310	[7116]	0.02	[7622]
	p			0.008	[7811]					310	[7116]	17.9	[7622]
494	**m-Xylene**												
	n	439	[7105]	30.8	[7101]	0.14	[7001]			339	[6103]		
	p									336	[6004]		
495	**o-Xylene**												
	n	438	[8204]	32.2	[7101]	0.18	[8204]	0.28	[6924]	343	[6103]		
	p			38	[7206]	0.14	[7206]			344	[6004]		
496	**p-Xylene**												
	n	435	[7105]	30	[7101]	0.22	[7208]	0.63	[6901]	337	[6103]		
	p			33.7	[7621]	0.24	[7621]			337	[6004]		
497	**p-Xylene-d_{10}**												
	n	435	[7101]	30.3	[7101]	0.26	[7101]						
	p			32.9	[7621]	0.22	[7621]						

Table 3a Photophysical Parameters of Organic Species

No.	Solv.	E_S/kJ mol^{-1}	Ref.	τ_S/ns	Ref.	Φ_{fl}	Ref.	Φ_T	Ref.	E_T/kJ mol^{-1}	Ref.	τ_T/μs	Ref.
498	**Zinc(II) chlorophyll *a***												
	p	180	[7910]							131	[7910]		
499	**Zinc(II) chlorophyll *b***												
	p	186	[7910]							140	[7910]		

^{b} Aromatic solvent, benzene-like; ^{g} Gas-phase measurement; ^{m} Major component of multiexponential; ^{p} Phosphorescence decay; ^{s} Very solvent-dependent; ^{w} Wavelength dependent; ^{x} Crystalline medium.

3b TRIPLET-STATE ENERGIES: ORDERED

In Table 3b, the compounds are listed in order of increasing triplet energy. This table has been designed to facilitate selection of photosensitizers and quenchers. The energies are given in kJ mol^{-1}.

The triplet energies have been taken mainly from three types of experiments. First, triplet energies have been estimated from singlet-triplet absorption spectra. These radiative transitions are highly spin forbidden, but they can borrow intensity from singlet-singlet transitions through the use of oxygen as an external, paramagnetic perturber which enhances spin-orbit coupling in the molecules. A large fraction of the singlet-triplet absorption spectra show no clear 0-0 bands. However, empirical methods have been used to successfully estimate the location of the 0-0 bands. When triplet energies taken from singlet-triplet absorption have been chosen in Table 3b, either spectra showing distinct 0-0 bands or estimations of 0-0 bands have been quoted whenever possible.

Second, the short-wavelength band of the phosphorescence spectra usually represents the 0-0 band of a transition between the lowest triplet state and the ground state. Even though this is a spin-forbidden transition between the same vibronic levels as that seen in absorption, luminescence detection is much more sensitive than absorption methods. Phosphorescence is the most widely used method for determining triplet energies, and of the three common methods, it gives the most definitive value. However, phosphorescence spectra are almost always taken in rigid glasses or polymers, whereas one of the most common uses of the data is to design photosensitization experiments in fluid solution.

Third, photosensitized formation of triplet acceptors can be used to estimate the triplet energy of a molecule by measuring the energy-transfer rate constant of a series of triplet donors being quenched by (or acceptors quenching) the molecule in question.

$$^{3}D*+^{1}A\rightarrow^{1}D+^{3}A$$

When the triplet energy of the potential triplet donor is below the triplet energy of the potential triplet acceptor, the rate constant for energy transfer falls off very rapidly as the energy difference increases. Since triplet-triplet energy transfer is a spin-allowed process, as long as the triplet donor's energy is above (or within easy thermal assess to) that of the triplet acceptor's, the quenching normally proceeds at a rate constant approaching that of diffusion controlled. Of the three methods, this one best simulates applications to photosensitization, but it usually gives only a range of values.

When available, 0-0 bands of the phosphorescence are quoted in Table 3b in preference to the other two methods. Only as a last resort are the maxima of either singlet-triplet absorption or phosphorescence quoted. A spectral maximum from phosphorescence represents a lower limit to the lowest triplet energy, and a spectral maximum from singlet-triplet absorption represents an upper limit to the lowest triplet energy.

Table 3b Triplet State Energies of Organic Compounds

No.	Compound	$E_T(n)$			$E_T(p)$		
		kJ mol⁻¹	cm⁻¹	Ref.	kJ mol⁻¹	cm⁻¹	Ref.
1	Pentacene	75[b]	6300[b]	[7111]			
2	β-Carotene	88[b]	7300[b]	[7504]	85	7100	[9204]
3	3,4:9,10-Perylenebis(dicarboximide), N,N′-bis(1-hexylheptyl)-				116	9679	[9903]
4	Dibenzo[a,o]perylene, 7,16-diphenyl-	92	7700	[8505]			
5	Phthalocyanine, zinc(II)	109[b]	9150[b]	[7113]			
6	Rubrene	110[b]	9200[b]	[8124]			
7	Phthalocyanine	120[b]	10000[b]	[7820]			
8	(Z)-Azobenzene	121[b]	10100[b]	[8129]			
9	Tetracene, 5-methyl-	122	10170	[7210]			
10	Tetracene	123	10250	[6412]			
11	(all-E)-Retinal	123	10300	[8417]			
12	Periflanthen	123[b]	10300[b]	[9206]			
13	Chlorophyll a				125	10400	[7910]
14	Phthalocyanine, platinum(II)	127[b]	10590[b]	[7113]			
15	Pheophytin a	130	10800	[7910]			
16	Chlorophyll b	130	10900	[7910]	136	11400	[6801]

159

Table 3b Triplet State Energies of Organic Compounds

No.	Compound	$E_T(n)$			$E_T(p)$		
		kJ mol^{-1}	cm^{-1}	Ref.	kJ mol^{-1}	cm^{-1}	Ref.
17	Zinc(II) chlorophyll a				131	11000	[7910]
18	Benz[a]anthracene-3,4-dione	132[b]	11000[b]	[9301]			
19	Dibenz[a,j]anthracene-3,4-dione	132[b]	11000[b]	[9301]			
20	2,4,6,8,10-Dodecapentaenal	132	11050	[6005]			
21	1,3,5,7-Octatetraene, 1,8-diphenyl-	132	11050	[7210]			
22	Pheophytin b	134	11200	[7910]			
23	Methylene Blue cation				138	11500	[6713]
24	Porphyrin, tetraphenyl-	138	11500	[8221]	140	11700	[7406]
25	1,2-C$_{70}$H$_2$	139[b]	11610[b]	[9907]			
26	(all-E)-Retinol	140	11700	[8417]			
27	Zinc(II) chlorophyll b				140	11700	[7910]
28	1,9-(4-Hydroxycyclohexano)buckminsterfullerene	140[b]	11717[b]	[9402]			
29	Dibenzo[def,mno]chrysene	141	11800	[7210]	143	12000	[7701]
30	Isobenzofuran, 1,3-diphenyl-	142[b]	11900[b]	[8124]			
31	Porphyrin, tetraphenyl-, magnesium(II)	143	11900	[8221]	143	12000	[7114]
32	Benzo[c]cinnoline				144	12000	[6106]

Table 3b Triplet State Energies of Organic Compounds

No.	Compound	$E_T(n)$			$E_T(p)$		
		kJ mol^{-1}	cm^{-1}	Ref.	kJ mol^{-1}	cm^{-1}	Ref.
33	Dibenzo[b,def]chrysene	144[b]	12040[b]	[6209]			
34	1,7-Diazaperylene	145	12099	[9501]			
35	(E)-Azobenzene	146[b]	12200[b]	[8129]			
36	C$_{70}$	148[b]	12300[b]	[9106]			
37	Perylene	148[x]	12373[x]	[6918]	151	12600	[6801]
38	1,3,5-Hexatriene, 1,6-diphenyl-	149	12450	[7210]			
39	Tetrabenzoporphyrin, zinc(II)				149	12500	[7114]
40	(E,E,E)-2,4,6-Heptatrienal, 5-methyl-7-(2,6,6-trimethyl-1-cyclohexen-l-yl)-	150	12500	[8417]			
41	Porphyrin-2, 18-dipropanoic acid, 7,12-diethenyl-3,8,13,17-tetramethyl-, dimethyl ester				150	12600	[7406]
42	C$_{60}$	151[b]	12600[b]	[9103]			
43	Porphyrin	151[b]	12600[b]	[7515]	152	12700	[7406]
44	Porphyrin-2,18-dipropanoic acid, 7,12-diethenyl-3,8,13,17-tetramethyl-	151	12600	[7515]			
45	2,4,6,8-Decatetraenal	152	12700	[6005]			
46	Porphyrin, tetraphenyl-, zinc(II)	153	12800	[7516]	153	12800	[7114]
47	Porphyrin-2, 18-dipropanoic acid, 7,12-diethyl-3,8,13,17-tetramethyl-	154[b]	12900[b]	[7515]			

Table 3b Triplet State Energies of Organic Compounds

No.	Compound	$E_T(n)$			$E_T(p)$		
		kJ mol^{-1}	cm^{-1}	Ref.	kJ mol^{-1}	cm^{-1}	Ref.
48	Porphyrin-2,18-dipropanoic acid, 7,12-diethyl-3,8,13,17-tetramethyl-, dimethyl ester				155	12900	[7406]
49	Porphyrin, octaethyl-	155	13000	[7515]	156	13000	[7406]
50	Porphyrin, 2,7,12,17-tetraethyl-3,8,13,18-tetramethyl-	155^b	13000^b	[7515]	156	13000	[7515]
51	Porphyrin, tetrakis(4-chlorophenyl)-, zinc(II)	156	13000	[7516]			
52	Porphyrin, tetrakis(2-methylphenyl)-, zinc(II)	157	13100	[7516]			
53	Porphyrin, tetrakis(2-chlorophenyl)-, zinc(II)	158	13200	[7516]			
54	Cycloheptatriene	159^b	13300^b	[8525]			
55	Ferrocene	159^b	13300^b	[7504]			
56	Thiofluorenone	159	13300	[7519]			
57	5,6-Chrysenequinone	159^b	13300^b	[9301]			
58	Benzo[c]phenanthrene-5,6-dione	159^b	13300^b	[9301]			
59	Benzo[a]pyrene-7,8-dione	159^b	13300^b	[9301]			
60	Benzo[rst]pentaphene, 5-methyl-	161^b	13480^b	[6209]			
61	Porphyrin, 2,7,12,17-tetraethyl-3,8,13,18-tetramethyl-, magnesium(II)				162	13600	[7515]
62	Benzo[a]pyrene-4,5-dione	162^b	13500^b	[9301]			

Table 3b Triplet State Energies of Organic Compounds

No.	Compound	E_T(n)			E_T(p)		
		kJ mol^{-1}	cm^{-1}	Ref.	kJ mol^{-1}	cm^{-1}	Ref.
63	Azulene	163[b]	13600[b]	[7504]			
64	Thionine cation				163	13600	[6923]
65	Anthrathione, 9,9-dimethyl-	163	13640	[8322]		13700	[6813]
66	Fluorescein dianion, 3,4,5,6-tetrachloro-2',4',5',7'-tetraiodo-				164	13700	[6813]
67	Porphyrin, magnesium(II)	164	13700	[7515]	163	13700	[7515]
68	2,2'-Thiacarbocyanine, 3,3'-diethyl-				164	13750	[7826]
69	Thiobenzophenone	165	13760	[7519]	170	14200	[7215]
70	Anthracene, 1,5,10-trichloro-				165	13800	[5603]
71	Tetrabenzoporphyrin, magnesium(II)				165	13800	[7515]
72	Thioxanthione	166	13800	[8230]			
73	1-Aceanthrylenone, 2-methyl-				165	13830	[8205]
74	Porphyrin, zinc(II)				166	13900	[7114]
75	Fluorescein dianion, 2',4',5',7'-tetrabromo-3,4,5,6-tetrachloro-				167	13950	[6813]
76	Porphyrin, 2,8,12,18-tetraethyl-3,7,13,17-tetramethyl-, magnesium(II)	167	13990	[6307]			
77	Benzo[rst]pentaphene	168[b]	14050[b]	[6209]	169	14100	[7701]
78	Anthracene, 9,10-dibromo-	168	14060	[5603]			

Table 3b Triplet State Energies of Organic Compounds

No.	Compound	$E_T(n)$			$E_T(p)$		
		kJ mol⁻¹	cm⁻¹	Ref.	kJ mol⁻¹	cm⁻¹	Ref.
79	Dibenzo[a,d]cycloheptene-5-thione	169	14100	[8322]			
80	2,4,6-Octatriene, 2,6-dimethyl-	169[b]	14100[b]	[8323]			
81	Anthracene, 1,10-dichloro-				169	14128	[5603]
82	Anthracene, 9-benzoyl-10-bromo-				169	14140	[8205]
83	Anthracene, 9,10-dichloro-	169	14150	[5603]	169	14150	[6605]
84	Anthracene, 1,4,5,8-tetrachloro-				169	14155	[5603]
85	1,16-Benzoperiflanthene	170[b]	14200[b]	[9206]			
86	Anthracene, 9-benzoyl-10-chloro-				169	14160	[8205]
87	Porphyrin, octaethyl-, zinc(II)				170	14200	[7114]
88	Porphyrin, 2,7,12,17-tetraethyl-3,8,13,18-tetramethyl-, zinc(II)				170	14200	[7114]
89	Porphyrin-2, 18-dipropanoic acid, 7,12-diethyl-3,8,13,17-tetramethyl-, dimethyl ester, zinc(II)				171	14260	[6307]
90	Anthracene, 9,10-diphenyl-	171	14290	[6805]			
91	Quin[2,3-b]acridine-7,14-dione, 5,12-dihydro-2,9-dimethyl-				172	14388	[7804]
92	Thiobenzophenone, 4,4'-dimethoxy-	172	14400	[7215]	176	14700	[7215]
93	Anthracene, 9-bromo-				173	14400	[8405]

Table 3b Triplet State Energies of Organic Compounds

No.	Compound	$E_T(n)$				$E_T(p)$		
		kJ mol^{-1}	cm^{-1}	Ref.		kJ mol^{-1}	cm^{-1}	Ref.
94	Anthracene, 9-methyl-	173	14460	[5701]		170	14220	[6605]
95	Anthracene, 2-methyl-					173	14500	[6605]
96	Anthracene, 1,5-dichloro-					174	14568	[5603]
97	Anthracene, 9,10-dicyano-	175	14600	[7807]				
98	Pyronine B cation					175	14600	[7720]
99	Sulforhodamine B, inner salt					175	14600	[6902]
100	Anthracene, 9-nitro-	175	14630	[5701]				
101	Benzo[a]pyrene	175	14670	[5701]		177	14800	[7701]
102	Anthracene, 9-naphthoyl-					176	14690	[8205]
103	2,4-Hexadiene, 2,5-dimethyl-	176[b]	14700[b]	[8114]				
104	Thiobenzophenone, 4,4'-bis(dimethylamino)-	176	14700	[7215]				
105	Anthracene, 9-acetyl-					176	14720	[8205]
106	Anthracene, 9-benzoyl-					176	14730	[8205]
107	Anthracene, 9-propionyl-					176	14730	[8205]
108	Anthracene, 1-chloro-					176	14732	[5603]
109	Anthracene, 9-cinnamoyl-					176	14740	[8205]
110	9-Anthracenecarboxylic acid					177	14760	[8205]

165

Table 3b Triplet State Energies of Organic Compounds

No.	Compound	E_T(n)			E_T(p)		
		kJ mol⁻¹	cm⁻¹	Ref.	kJ mol⁻¹	cm⁻¹	Ref.
111	1,3-Butadiene, 1,4-diphenyl-	177	14800	[7012]			
112	Fluorescein dianion, 2',4',5',7'-tetrabromo-				177	14800	[6813]
113	Anthracene-9-carboxamide				177	14820	[8205]
114	Anthracene	178	14870	[5701]	178	14900	[7701]
115	Benz[g]isoquinoline	178	14870	[5901]			
116	Rhodamine 3G cation				178	14870	[7720]
117	Rhodamine B, inner salt				178	14890	[7720]
118	9-Acridinethione, 10-methyl-	178	14900	[8322]			
119	Pyronine cation				178	14900	[7720]
120	1,4-Tetracenequinone	179	14934	[0109]			
121	1-Pyrenecarboxaldehyde	180[b]	15000[b]	[8319]			
122	Thioketene, di-tert-butyl-	180[b]	15000[b]	[8621]			
123	Benz[g]quinoline	180	15070	[5901]	184	15400	[8110]
124	Rhodamine 6G cation				181	15100	[7720]
125	1-Benzothiopyran-4-thione, 2,6-dimethyl-	181	15110	[8322]			
126	9-Xanthione	181	15143	[8128]			

Table 3b Triplet State Energies of Organic Compounds

No.	Compound	$E_T(n)$			$E_T(p)$		
		kJ mol⁻¹	cm⁻¹	Ref.	kJ mol⁻¹	cm⁻¹	Ref.
127	Anthracene-9-carboxaldehyde	182[b]	15200[b]	[6822]	165	13790	[8205]
128	2,4,6-Octatrienal	182	15210	[6005]			
129	Rhodamine, inner salt, N,N'-diethyl				184	15350	[7720]
130	Flavone, 4-thio-	183	15319	[9602]			
131	Flavone, 4-thio-				184	15354	[9602]
132	Acridine, 9-chloro-				184	15400	[8622]
133	Anthracene, 1-amino-				184	15400	[6904]
134	Anthracene, 2-amino-				184	15400	[6904]
135	Anthracene, 9-amino-				184	15400	[6904]
136	Fluorescein dianion, 2',4',5',7'-tetraiodo-				184	15400	[6915]
137	Ovalene				184	15400	[7701]
138	Benz[a]anthracene, 7,12-dimethyl-	185	15500	[5401]	184	15400	[7701]
139	Dibenzo[b,g]phenanthrene	185	15500	[8130]			
140	4-Cholesten-3-thione	186	15500	[7215]	191	15900	[7215]
141	Phenazine	186	15561	[6919]	187	15600	[8523]
142	Indene-1-thione, 4-bromo-2,3-dihydro-2,2,3,3-tetramethyl-	187	15600	[8322]			

Table 3b Triplet State Energies of Organic Compounds

No.	Compound	$E_T(n)$			$E_T(p)$		
		kJ mol^{-1}	cm^{-1}	Ref.	kJ mol^{-1}	cm^{-1}	Ref.
143	2-Naphthalenethione, 1,1-dimethyl-	187[b]	15600[b]	[8618]			
144	Benzo[a]phenazinium				187	15650	[8108]
145	Naphtho[2,3-a]coronene	188	15676	[7810]			
146	Acridine, 9-methyl-				188	15690	[8103]
147	Acridine, 9-propyl-				188	15690	[8103]
148	Colchicine				188	15700	[7827]
149	(E,E)-2,4-Pentadienal, 3-methyl-5-(2,6,6-trimethyl-1-cyclohexen-1-yl)-	188	15700	[7818]			
150	(E)-Stilbene, 4,4'-dinitro-				189	15800	[8905]
151	Dinaphth[1,2-a;1',2'-h]anthracene	190	15852	[8706]			
152	Acridine	190	15870	[8102]	188	15740	[8403]
153	Acridine-d_9				190	15870	[8103]
154	Benzo[b]chrysene				190	15900	[7701]
155	Fluorescein dianion, 4',5'-dibromo-2',7'-dinitro-				190	15900	[6902]
156	3-Pentanethione, 2,2,4,4-tetramethyl-	190[b]	15900[b]	[8526]			
157	(E)-Stilbene, 4-cyano-4'-methoxy-				190	15900	[8905]
158	Benzopyran-4-thione	190	15914	[9602]	191	15949	[9602]

Table 3b Triplet State Energies of Organic Compounds

No.	Compound	$E_T(n)$			$E_T(p)$		
		kJ mol⁻¹	cm⁻¹	Ref.	kJ mol⁻¹	cm⁻¹	Ref.
159	4-Flavanthione	190	15900	[0005]	194	16200	[0005]
160	Indene-l-thione, 2,3-dihydro-2,2,3,3-tetramethyl-	190	15920	[8322]			
161	Rhodamine, inner salt				191	15930	[7720]
162	Benz[b]acridin-12-one	191	16000	[8510]	193	16130	[7804]
163	4-Flavanthione, 6-hydoxy-	191ᵇ	16000ᵇ	[0005]	196	16400	[0005]
164	3-Aminofluorenone	193ᵇ	16100ᵇ	[0006]			
165	Benzanthrone				192	16100	[7701]
166	Fluorescein dianion, 2',7'-dichloro-				192	16100	[6902]
167	(E)-Stilbene, 4,4'-dicyano-				192	16100	[8905]
168	Acridine, 9-amino-				193	16100	[8103]
169	Dibenzo[h,rst]pentaphene				193	16100	[7701]
170	Benzo[ghi]perylene				194	16180	[5604]
171	Hexabenzo[a,d,g,j,m,p]coronene	194	16200	[7623]			
172	Benz[a]anthracene, 12-methyl-	195	16260	[5401]	192	16100	[7701]
173	Benz[j]aceanthrylene, 1,2-dihydro-3-methyl-				195	16300	[7701]
174	2-Cyclohexenethione, 3,5,5-trimethyl-	195ᵇ	16300ᵇ	[8618]			
175	Porphyrin, platinum(II)	195	16300	[7022]			

Table 3b Triplet State Energies of Organic Compounds

No.	Compound	$E_T(n)$			$E_T(p)$		
		kJ mol^{-1}	cm^{-1}	Ref.	kJ mol^{-1}	cm^{-1}	Ref.
176	(E)-Stilbene, 4-nitro	195	16300	[8905]	208	17400	[4401]
177	Dibenz[a,e]aceanthrylene				196	16400	[7701]
178	(E)-1,3,5-Hexatriene	197	16450	[6005]			
179	Benz[a]anthracene	197	16500	[5401]	198	16520	[7808]
180	Fluorescein dianion				197	16500	[6902]
181	Naphtho[1,2,3,4-def]chrysene				197	16500	[7701]
182	(E)-Stilbene, 2-nitro-				197	16500	[8905]
183	(E)-Stilbene, 4-benzoyl-	198	16500	[8905]	198	16500	[8905]
184	(E)-Stilbene, 4-cyano-				198	16500	[8905]
185	Benz[a]anthracene, 9-methyl-	198	16530	[5401]	199	16600	[7701]
186	Benz[a]anthracene, 10-methyl-	198	16580	[5401]	199	16600	[7701]
187	Tetrabenzo[a,c,j,l]naphthacene				198	16600	[7623]
188	Indeno[2,1-a]indene	199	16600	[8015]	218	18300	[8905]
189	Benz[a]anthracene, 2-methyl-	199	16640	[5401]	200	16700	[7701]
190	Benz[a]anthracene, 8-methyl-	199	16650	[5401]	201	16800	[7701]
191	Benz[a]anthracene, 4-methyl-	199	16670	[5401]	199	16600	[7701]

Table 3b Triplet State Energies of Organic Compounds

No.	Compound	$E_T(n)$			$E_T(p)$		
		kJ mol[-1]	cm[-1]	Ref.	kJ mol[-1]	cm[-1]	Ref.
192	Dibenzo[a,j]coronene	199[b]	16700[b]	[7623]			
193	Tribenzo[a,d,g]coronene	199[b]	16700[b]	[7623]			
194	Benz[b]acridin-12-one, 5-methyl-	200	16700	[8510]			
195	(E)-Stilbene, 4-acetyl-	200	16700	[8905]	198	16500	[8905]
196	(E)-Stilbene, 4-iodo-	200	16700	[8905]	202	16900	[8015]
197	Benz[a]anthracene, 3-methyl-	200	16720	[5401]	201	16800	[7701]
198	Benz[a]anthracene, 11-methyl-	201	16780	[5401]	202	16900	[7701]
199	1,3-Cyclobutanedithione, 2,2,4,4-tetramethyl-				201[x]	16800[x]	[8324]
200	(E)-Stilbene-d_{12}	201	16800	[8015]	205	17100	[8905]
201	(E)-Stilbene, 4-bromo-				201	16800	[8905]
202	Benz[a]anthracene-7,12-dione	201[b]	16800[b]	[9301]			
203	Tropolone				201	16800	[7827]
204	Benz[a]anthracene, 5-methyl-	201	16810	[5401]	202	16900	[7701]
205	Thebenidine				202	16860	[7808]
206	1-Aminofluorenone	202[b]	16900[b]	[0006]			
207	(E)-Stilbene, 4-chloro-				203	16900	[8905]
208	Pentaphene				203	16930	[5604]

171

Table 3b Triplet State Energies of Organic Compounds

No.	Compound	$E_T(n)$			$E_T(p)$		
		kJ mol^{-1}	cm^{-1}	Ref.	kJ mol^{-1}	cm^{-1}	Ref.
209	Pyrene	203	16930	[5701]	202	16850	[7808]
210	α-Bithiophene				203	16938	[9804]
211	Benz[a]anthracene, 1-methyl-	203	16980	[5401]	203	17000	[7701]
212	Benz[a]anthracene, 6-methyl-	203	16980	[5401]	203	17000	[7701]
213	1-Naphthaleneacrylic acid, methyl ester	203^b	17000^b	[8622]			
214	(E)-Stilbene, 4,4'-dimethoxy-				203	17000	[8905]
215	(E)-Stilbene, 3-methoxy-				203	17000	[8905]
216	(E)-Stilbene, 3-nitro-				204	17000	[8905]
217	Styrene, β-(3-pyridyl)-, (E)-				204	17100	[8905]
218	2,4,6-Cycloheptatrien-l-one, 2-methoxy-				205	17100	[7827]
219	(E)-Stilbene, 3,3'-dibromo-				205	17100	[8015]
220	(E)-Stilbene, 4-fluoro-				205	17100	[8905]
221	(E)-Stilbene, 3-methyl-				205	17100	[8905]
222	Styrene, β-(2-pyridyl)-, (E)-				205	17100	[8905]
223	Octatetrayne, diphenyl-				205	17150	[5601]
224	Acridine Orange				206	17200	[6902]

Table 3b Triplet State Energies of Organic Compounds

No.	Compound	E_T(n) kJ mol⁻¹	cm⁻¹	Ref.	E_T(p) kJ mol⁻¹	cm⁻¹	Ref.
225	1,4-Benzoquinone, tetrachloro-	206	17200	[6909]			
226	Pyridine, 3,3'-(1,2-ethenediyl)bis-, (E)-				206	17200	[8905]
227	(E)-Stilbene	206	17200	[8015]	206	17200	[8015]
228	Benzo[g]pteridine-2,4-dione, 10-methyl-	206	17250	[7927]			
229	Dibenzo[g,p]chrysene				207	17300	[7701]
230	(E)-2-Indanthione, hexahydro-	207	17300	[7119]			
231	Pyran-4-thione	207	17300	[8325]			
232	Pyridine, 2,2'-(1,2-ethenediyl)bis-, (E)-				207	17300	[8905]
233	Thiofosgene				207	17300	[9210]
234	Thioacetone	207[g]	17300[g]	[8437]			
235	Benzo[g]pteridine-2,4-dione,3,10-dimethyl-	208	17350	[7927]			
236	Coumarin, 7-(diethylamino)-3,3'-carbonylbis-				208	17400	[8231]
237	Styrene, β-(4-pyridyl)-, (E)-	208	17400	[8905]	206	17200	[8905]
238	Benz[a]acridine				209	17480	[7808]
239	Benzo[g]pteridine-2,4-dione, 7,8,10-trimethyl-				209	17500	[6902]
240	Ethene, tetraphenyl-	209[b]	17500[b]	[8213]			
241	Riboflavin				209	17500	[6902]

Table 3b Triplet State Energies of Organic Compounds

No.	Compound	E_T(n)			E_T(p)		
		kJ mol^{-1}	cm^{-1}	Ref.	kJ mol^{-1}	cm^{-1}	Ref.
242	Triphenylethylene	209[b]	17500[b]	[8213]			
243	Benzo[a]phenazine	209	17512	[8515]	202	16900	[8108]
244	Diadamantylethanedione	210	17500	[8326]			
245	Dibenzo[hi,uv]hexacene	210	17538	[7928]			
246	Hexabenzo[bc,ef,hi,kl,no,qr]coronene	210[b]	17540[b]	[6812]			
247	(E)-Stilbene, 4-methoxy-	210	17550	[7929]	202	16900	[8905]
248	Benzo[k]fluoranthene				211	17600	[7701]
249	2,2'-Bibenzo[b]thiophene				211	17600	[8516]
250	Coumarin, 7-(diethylamino)-5',7'-dimethoxy-3,3'-carbonylbis-				211	17600	[8231]
251	Dibenzo[c,m]pentaphene				211	17600	[5604]
252	9-Fluorenone				211	17600	[7815]
253	Indene, 2-phenyl-				211	17600	[8905]
254	(E)-Stilbene, α-methyl-	211	17600	[8232]			
255	Benz[c]acridine, 9-methyl-				212	17700	[8403]
256	Pyridine, 4,4'-(1,2-ethenediyl)bis-, (E)-	212	17700	[8905]	211	17700	[8905]
257	Thiocoumarin	212[b]	17700[b]	[8618]			

Table 3b Triplet State Energies of Organic Compounds

No.	Compound	$E_T(n)$			$E_T(p)$		
		kJ mol^{-1}	cm^{-1}	Ref.	kJ mol^{-1}	cm^{-1}	Ref.
258	Benzo[*b*]triphenylene	213	17790	[5701]	213	17800	[7701]
259	Benz[*c*]acridine				213	17800	[7701]
260	Coumarin, 3-(2-benzofuroyl)-7-diethylamino-				213	17800	[8231]
261	Coumarin, 3-benzoyl-7-diethylamino-				213	17800	[8231]
262	Coumarin, 3,3'-carbonylbis(7-diethylamino)-				213	17800	[8231]
263	Pyran-4-thione, 2,6-dimethyl-	213	17800	[8325]	213	17800	[6719]
264	Pyrylium, 2,4,6-tris(4-methoxyphenyl)-				214	17900	[6902]
265	Acridine, 3,6-diamino-				214	17900	[6902]
266	Acridinium, 3,6-diamino-10-methyl-				215	18000	[6903]
267	Chrysene, 6-nitro-						
268	Benzo[*a*]coronene	215	18000	[7810]	216	18050	[6714]
269	Bicyclo[2.2.1]heptane-2,3-dione, 1,7,7-trimethyl-				216	18050	[6714]
270	3,4-Hexanedione, 2,2,5,5-tetramethyl-	216	18050	[8233]	215	17955	[8131]
271	Naphthalene, 1,4-dinitro-	216	18050	[8131]	215	17987	[9804]
272	4H-Cyclopenta[2,1-*b*:3,4-*b'*]dithiophene						
273	Ethanedione, dicyclohexyl-	216	18100	[8326]			
274	Tricycloquinazoline, 2-bromo-				217	18100	[8234]

Table 3b Triplet State Energies of Organic Compounds

No.	Compound	$E_T(n)$			$E_T(p)$		
		kJ mol^{-1}	cm^{-1}	Ref.	kJ mol^{-1}	cm^{-1}	Ref.
275	Naphthalene, 1,2-dinitro-				217	18181	[7728]
276	1,2-Naphthalenedicarboximide, N-methyl-				218	18187	[9404]
277	Benzo[b]carbazole				218	18200	[7808]
278	Coumarin, 7-(diethylamino)-3-thenoyl-				218	18200	[8231]
279	Dibenz[a,h]anthracene				218	18200	[7701]
280	Thiopyrylium, 2,4,6-triphenyl-				218	18200	[6719]
281	Coumarin, 3-phenyl-				219	18280	[7303]
282	1,3-Cyclohexadiene	219	18300	[6504]			
283	1,2-Propanedione, 1-phenyl-	219	18300	[8326]			
284	Acridine Yellow				220	18350	[8301]
285	Dinaphtho[2,1-b:1',2'-d]thiophene				220	18350	[7610]
286	Adamantanethione				220	18400	[7215]
287	Dibenzo[fg,ij]phenanthro[9,10,1,2,3-pqrst]pentaphene	220	18400	[7623]			
288	Fluoranthene	221	18450	[5701]	221	18500	[7701]
289	Dibenzo[a,g]coronene	221	18480	[7810]			
290	Benzo[e]pyrene				221	18500	[7701]

Table 3b Triplet State Energies of Organic Compounds

No.	Compound	$E_T(n)$ kJ mol^{-1}	cm^{-1}	Ref.	$E_T(p)$ kJ mol^{-1}	cm^{-1}	Ref.
291	Dibenz[a,j]anthracene				221	18500	[7701]
292	Tricycloquinazoline	221	18500	[8234]	224	18700	[8234]
293	Coumarin, 7-(diethylamino)-3-(4-dimethylaminobenzoyl)-				222	18500	[8231]
294	1,8(2H,5H)-Acridinedione, 3,4,6,7,9,10-hexahydro-3,3,6,6-tetramethyl-				224	18700	[9801]
295	2,3-Diazabicyclo[2.2.2]oct-2-ene	222[b]	18537[b]	[9403]			
296	Pyrylium, 2,4,6-triphenyl-				222	18500	[6719]
297	1,8-Naphthalenedicarboximide, N-methyl-				222	18537	[9404]
298	Dibenz[a,j]acridine				223	18600	[7701]
299	Tribenzo[b,n,pqr]perylene	223[b]	18620[b]	[6812]			
300	Dibenzo[a,c]phenazine	223	18630	[6925]	223	18600	[7701]
301	Benzo[b]naphtho[2,3-d]thiophene				224	18690	[6812]
302	Benzil	223	18700	[6712]	227	19000	[6712]
303	1,4-Benzoquinone	224	18740	[6909]			
304	1,3,5,7-Octatetraynediol				225	18790	[5601]
305	Bicyclo[2.2.1]heptane-2-thione, 1,3,3-trimethyl-				225	18800	[7215]
306	Dibenz[a,h]acridine	225	18800	[7014]	229	19100	[7701]

Table 3b Triplet State Energies of Organic Compounds

No.	Compound	$E_T(n)$			$E_T(p)$		
		kJ mol^{-1}	cm^{-1}	Ref.	kJ mol^{-1}	cm^{-1}	Ref.
307	Tricycloquinazoline, 2-methyl-				225	18800	[8234]
308	Dibenzo[b,h]fluoren-12-one				225	18817	[9702]
309	3,5,7,9-Dodecatetrayne				225	18820	[5601]
310	Naphthalene, 1-[(1-naphthyl)amino]-				225	18850	[7614]
311	2,3'-Bibenzo[b]thiophene				226	18870	[8516]
312	Benzo[g]chrysene				226	18900	[7701]
313	Benzo[ghi]fluoranthene				226	18900	[7701]
314	Chrysene, 6-benzoyl-				226	18900	[6926]
315	1,8(2H,5H)-Acridinedione, 3,4,6,7,9,10-hexahydro-3,3,6,6,10-pentamethyl-				227	18967	[9801]
316	2,3-Pentanedione	226[g]	18900[g]	[7213]	233	19400	[6714]
317	Benzo[c]phenanthrene, 1-methyl-	227	18940	[5401]			
318	Cinnoline, 4-methyl-				227	19000	[5905]
319	Dibenz[a,c]acridine				227	19000	[7701]
320	β-Naphthiazoline, 2-benzoyl-N-methyl-				227	19000	[7828]
321	(Z)-Stilbene	227[b]	19000[b]	[6922]			
322	Naphthalene, 1-anilino-				227	19010	[7614]

Table 3b Triplet State Energies of Organic Compounds

No.	Compound	$E_T(n)$			$E_T(p)$		
		kJ mol^{-1}	cm^{-1}	Ref.	kJ mol^{-1}	cm^{-1}	Ref.
323	Coronene				228	19040	[5604]
324	Benzo[b]fluoranthene				228	19050	[8209]
325	1,3-Diazaazulene	228[b]	19052[b]	[7313]			
326	Hexahelicene	228	19100	[6204]	228	19100	[7701]
327	1-Naphthalenecarbothioic acid, O-ethyl ester	229[b]	19100[b]	[7521]			
328	2,5-Thiophenedione, 3,4-dichloro-	229	19100	[7321]			
329	Naphthalene, 1-amino-				229	19150	[7614]
330	Tetrabenzo[g,lm,uv,a₁b₁]heptacene	229[b]	19160[b]	[6812]			
331	Phosphine oxide, (1-naphthalenylcarbonyl)diphenyl-	229[b]	19166[b]	[9902]			
332	Cinnamic acid, methyl ester	229[b]	19200[b]	[8622]			
333	Diazene, diethyl-, (E)-	230[g]	19200[g]	[7624]			
334	β-Ionone	230[b]	19200[b]	[8522]			
335	Naphthalene, 1,5-dinitro-				230	19210	[7110]
336	Aniline, N,N-dimethyl-4-nitro-				230	19230	[7906]
337	Benzo[def]carbazole				230	19230	[7808]
338	Benzo[b]naphtho[2,3-d]furan				230	19230	[6812]
339	Benzo[c]fluorene				231	19300	[7701]

Table 3b Triplet State Energies of Organic Compounds

No.	Compound	$E_T(n)$			$E_T(p)$		
		kJ mol⁻¹	cm⁻¹	Ref.	kJ mol⁻¹	cm⁻¹	Ref.
340	Cyclooctane-1,2-dione, 3,3,8,8,-tetramethyl-	231	19300	[8233]			
341	Naphthalene, 1,3-dinitro-				231	19300	[7110]
342	Naphthalene, 1-nitro-				231	19300	[7110]
343	Naphtho[1,2-b]triphenylene				232	19370	[5604]
344	Hexatriyne, diphenyl-				232	19380	[5601]
345	Aniline, 4-nitro-				232	19400	[5605]
346	Benzo[g]pteridine-2,4-dione, 7,8-dimethyl-				232	19400	[6902]
347	1,2-Cyclodecanedione				232	19400	[6714]
348	Diazene, dimethyl-, (E)-	232[g]	19400[g]	[7624]			
349	Diazene, dipropyl-, (E)-	232	19400	[7624]			
350	1,4-Dioxin, 2,3,5,6-tetraphenyl-				232	19400	[7913]
351	2,5-Furandione, 3,4-dichloro-	232	19400	[7321]			
352	Naphthalene, 1,4-dicyano-				232	19400	[7604]
353	Oxazole, 2,2'-(1,4-phenylene)bis(5-phenyl)-				232	19400	[8612]
354	Pyrrole-2,5-dione, 3,4-dichloro-				232	19400	[7321]
355	Pentacene-6,13-dione				232	19420	[6710]

Table 3b Triplet State Energies of Organic Compounds

No.	Compound	$E_T(n)$			$E_T(p)$		
		kJ mol^{-1}	cm^{-1}	Ref.	kJ mol^{-1}	cm^{-1}	Ref.
356	Anthrone, 1,8-dihydroxy-	233[b]	19500[b]	[8908]			
357	Benzo[c]chrysene				233	19500	[7701]
358	2,2'-Binaphthyl				234	19560	[5604]
359	Naphtho[1,8-de]-1,3,2-diazaborine, 2,3-dihydro-2-methyl-	234	19562	[8235]			
360	Tetrabenzo[a,c,hi,qr]pentacene	234[b]	19570[b]	[6812]			
361	Coumarin, 3-(4-cyanobenzoyl)-5,7-dimethoxy-				234	19600	[8231]
362	Coumarin, 5,7-dimethoxy-3,3'-carbonylbis-				234	19600	[8231]
363	Naphthalene, octafluoro-	234	19600	[6823]			
364	Pyrylium, 2-methyl-4,6-diphenyl-				234	19600	[6719]
365	1-Naphthaldehyde	235	19600	[6822]	236	19750	[6001]
366	Aniline, N-methyl-4-nitro-				235	19610	[7906]
367	Dibenzo[c,g]carbazole				235	19610	[7610]
368	5,12-Tetracenequinone				235	19610	[6710]
369	Benzo[c]phenanthrene, 6-methyl-	236	19690	[5401]			
370	Naphthalene, 2-[(2-naphthyl)amino]-				236	19690	[7614]
371	Coumarin, 3,3'-carbonylbis(5,7-dimethoxy)-				235	19700	[8231]

181

Table 3b Triplet State Energies of Organic Compounds

No.	Compound	$E_T(n)$			$E_T(p)$		
		kJ mol^{-1}	cm^{-1}	Ref.	kJ mol^{-1}	cm^{-1}	Ref.
372	Biacetyl				236[x]	19700[x]	[5501]
373	Coumarin, 5,7,7'-trimethoxy-3,3'-carbonylbis-				236	19700	[8231]
374	Naphthalene, l-acetyl-	236	19700	[6404]	236	19700	[6702]
375	Acetophenone, 2'-amino-				236	19700	[0104]
376	Benzo[c]phenanthrene, 4-methyl-	236	19720	[5401]			
377	1,1'-Dinaphthyl ketone				236	19761	[9702]
378	Benzo[c]phenanthrene, 5-methyl-	236	19720	[5401]			
379	Diindeno[1,2-a:2',1'-c]fluorene, 10,15-dihydro-				236	19720	[6006]
380	Benzo[c]phenanthrene, 2-methyl-	236	19760	[5401]			
381	Benzo[c]phenanthrene, 3-methyl-	236	19760	[5401]			
382	Diazene, diisopropyl-, (E)-	236	19800	[7624]			
383	Acenaphthene, 5-nitro-				237	19800	[6903]
384	1,3-Butadiene, 1-methoxy-	237	19800	[6504]			
385	Pentahelicene				237	19800	[5604]
386	Benzo[c]phenanthrene	237	19840	[5401]	239	20000	[7701]
387	1,3-Butadiene, 1-chloro-	238	19900	[6504]			

Table 3b Triplet State Energies of Organic Compounds

No.	Compound	$E_T(n)$			$E_T(p)$		
		kJ mol^{-1}	cm^{-1}	Ref.	kJ mol^{-1}	cm^{-1}	Ref.
388	Coumarin, 3-benzoyl-5,7-dimethoxy-				238	19900	[8231]
389	Naphthalene, 2-nitro-				238	19900	[7110]
390	Thiophene, 2,2'-(1,3-phenylene)bis-				238	19900	[8619]
391	Naphthalene, 1,5-dibenzoyl-				238	19925	[6001]
392	Naphthalene, 2-amino-				239	19960	[7614]
393	Chrysene				239	20000	[7701]
394	Coumarin, 3,3'-carbonylbis(7-methoxy)-				239	20000	[8231]
395	Coumarin, 3-(4-cyanobenzoyl)-7-methoxy-				239	20000	[8231]
396	(S)-Dinaphtho[2,1-d: 1',2'-f][1,3]dioxepin	239	20000	[8236]			
397	Naphthalene, 1,8-dinitro-				239	20000	[7110]
398	Coumarin, 5,7-dimethoxy-3-(4-methoxybenzoyl)-				240	20000	[8231]
399	Cycloheptane-1,2-dione, 3,3,7,7-tetramethyl-	240	20000	[8233]			
400	Dinaphtho[1,2-b: 1',2'-d]thiophene				240	20040	[7610]
401	Naphthalene, 2-anilino-				240	20040	[7614]
402	Naphtho[1,2-c][1,2,5]thiadiazole	240	20062	[7216]			
403	(Z)-Piperylene	240	20070	[8232]			
404	Benzo[b]fluorene				240	20100	[7701]

Table 3b Triplet State Energies of Organic Compounds

No.	Compound	$E_T(n)$ kJ mol⁻¹	$E_T(n)$ cm⁻¹	Ref.	$E_T(p)$ kJ mol⁻¹	$E_T(p)$ cm⁻¹	Ref.
405	Coumarin, 5,7-dimethoxy-3-thenoyl-				240	20100	[8231]
406	2-Furoic acid, 5-nitro-				240	20100	[8132]
407	Naphthalene, 1-cyano-				240	20100	[7604]
408	Picene				240	20100	[7701]
409	Benzo[a]fluorene				241	20100	[7701]
410	1,2-Cycloheptanedione				241	20100	[6714]
411	Naphthalene, 1-benzoyl-	241	20100	[6404]	240	20100	[6001]
412	1-Naphthalenecarboxylic acid				241	20100	[6701]
413	Coumarin, 3-acetyl-6-bromo-				241	20120	[7303]
414	1,4-Naphthoquinone				241	20160	[6710]
415	Naphthalene, 1,5-dihydroxy-				241	20200	[6701]
416	9-Acridinone, 2-bromo-				242	20200	[7804]
417	Biphenyl, 4,4'-dinitro-				242	20200	[4401]
418	1,2'-Dinaphthyl ketone				242	20216	[9702]
419	Biphenyl, 4,4'-dibenzoyl-	242ˣ	20222ˣ	[8028]			
420	2,6-Dithiocaffeine				242	20260	[7625]

Table 3b Triplet State Energies of Organic Compounds

No.	Compound	$E_T(n)$			$E_T(p)$		
		kJ mol⁻¹	cm⁻¹	Ref.	kJ mol⁻¹	cm⁻¹	Ref.
421	Dibenzothiophene, 5-oxide	243	20286	[9406]	255	21335	[9406]
422	Butadiyne, diphenyl-				242	20270	[5601]
423	Naphthalene, 2-(methylsulfinyl)-				255	21335	[0009]
424	1,1'-Biphenyl, 4-(methylthio)-				255	21335	[0009]
425	2,3-Naphthalenedicarboximide, N-methyl-				243	20286	[9404]
426	Naphthalene, 2-(methylthio)-				243	20286	[0009]
427	Coumarin, 3,3'-carbonylbis-				242	20300	[8231]
428	Benzene, nitro-	243	20300	[8410]	252	21100	[4401]
429	Carbazole, 9-(1-naphthoyl)-				243	20300	[6927]
430	Coumarin, 3-benzoyl-7-methoxy-				243	20300	[8231]
431	Cyclopentadiene	243	20300	[6504]			
432	Ethanedione, dicyclopropyl-	243	20300	[8326]			
433	Naphthalene, 1,4-dibromo-				243	20300	[6701]
434	Naphthalene, 1-(dimethylamino)-				243	20300	[4401]
435	Dibenzo[g,op]naphthacene				244	20360	[5604]
436	9-Acridinone	244	20370	[7803]	252	21050	[7804]
437	Coumarin, 3-(4-cyanobenzoyl)-				244	20400	[8231]

Table 3b Triplet State Energies of Organic Compounds

No.	Compound	E_T(n)			E_T(p)		
		kJ mol^{-1}	cm^{-1}	Ref.	kJ mol^{-1}	cm^{-1}	Ref.
438	Coumarin, 7-methoxy-3-(4-methoxybenzoyl)-				244	20400	[8231]
439	Coumarin, 3-thenoyl-7-methoxy-				244	20400	[8231]
440	2,5-Cyclohexadien-1-one,4,4-di(1-naphthyl)-	244	20400	[8527]			
441	Imidazole-1-ethanol, 2-methyl-5-nitro-				244	20400	[8707]
442	Phenanthrene, 9-acetyl-				244	20400	[6701]
443	Quinoline-4-carboxylic acid, ethyl ester	244	20400	[8708]	244	20400	[8708]
444	p-Terphenyl				244	20400	[6701]
445	9-Acridinone, 10-phenyl-	244	20430	[7803]	254	21220	[7803]
446	Naphthalene, 1-chloro-	245[b]	20490[b]	[7819]	248	20700	[4401]
447	Aziridine, 1-(2-naphthoyl)-	245	20500	[8327]			
448	Benzoxazole, 4',5-diamino-2-phenyl-				245	20500	[8804]
449	Biphenyl, 4-nitro-				245	20500	[4401]
450	1,3-Butadiene, 2-chloro-	245	20500	[6504]			
451	Carbazole, 9-(2-naphthoyl)-				245	20500	[6927]
452	2,4-Hexadiene	245	20500	[6504]			
453	Naphthalene, 1-hydroxy-				245	20500	[4401]

Table 3b Triplet State Energies of Organic Compounds

No.	Compound	$E_T(n)$			$E_T(p)$		
		kJ mol⁻¹	cm⁻¹	Ref.	kJ mol⁻¹	cm⁻¹	Ref.
454	Naphthalene, 1-iodo-				245	20500	[6303]
455	Naphthalene, 2-phenyl-				245	20500	[6701]
456	Thioxanthen-9-one-3-carboxylic acid, ethyl ester	245	20500	[8620]			
457	Phenanthrene, 9-chloro-				246	20530	[7915]
458	Tetrabenz[a,c,h,j]anthracene				246	20550	[5604]
459	Benzoxazole, 4',6-diamino-2-phenyl-				246	20600	[8804]
460	1,3-Butadiene, 2,3-dichloro-	246	20600	[6504]			
461	Coumarin, 3-benzoyl-				246	20600	[8231]
462	2,5-Cyclohexadien-1-one, 4,4-di(2-naphthyl)-	246	20600	[8527]			
463	Fluorene, 2-nitro-				246	20600	[4401]
464	Naphthalene, 1-[(methylsulfonyl)methyl]-				246	20600	[8528]
465	Naphthalene, 1-phenyl-				246	20600	[6701]
466	7,8-Benzoflavanone	247	20600	[8608]			
467	Flavanone, 6-methoxy-	247[b]	20600[b]	[8608]	259	21700	[8608]
468	Maleonitrile	247[b]	20600[b]	[8237]			
469	Phenanthrene, 3-acetyl-				247	20600	[6701]
470	Pyridinium, 2,4,6-tris(4-methoxyphenyl)-N-methyl-				247	20600	[6719]

Table 3b Triplet State Energies of Organic Compounds

No.	Compound	$E_T(n)$			$E_T(p)$		
		kJ mol⁻¹	cm⁻¹	Ref.	kJ mol⁻¹	cm⁻¹	Ref.
471	Styrene, α-phenyl-	247[b]	20600[b]	[8213]			
472	Thiophene, 2-nitro-				247	20600	[8238]
473	2,2'-Biquinoline	247	20614	[8239]			
474	9-Acridinone, 10-methyl-	247	20620	[7803]	252	21050	[7803]
475	Indolo[3,2-b]carbazole				247	20620	[6413]
476	Phenanthrene, 9-bromo-				247	20620	[7915]
477	Naphthalene, 1-bromo-	247	20650	[5701]	247	20600	[6303]
478	Naphthalene, 2-(methylsulfonyl)-				247	20635	[0009]
479	Dibenzo[a,g]carbazole				247	20660	[7610]
480	Dinaphtho[1,2-b:2',1'-d]thiophene				247	20660	[7610]
481	Benzo[c]carbazole				247	20700	[6720]
482	(E)-Piperylene	247	20700	[6504]			
483	Coumarin, 3-(4-methoxybenzoyl)-				248	20700	[8231]
484	1-Cyclopentene, 1-phenyl-				248	20700	[7729]
485	Cyclopentene, 1-phenyl-3-acetyl-				248	20700	[7729]
486	4-Cyclopentene- 1,3-dione, 4,5-dichloro-				248	20700	[7321]

Table 3b Triplet State Energies of Organic Compounds

No.	Compound	$E_T(n)$			$E_T(p)$		
		kJ mol^{-1}	cm^{-1}	Ref.	kJ mol^{-1}	cm^{-1}	Ref.
487	2-Naphthaldehyde				248	20700	[6702]
488	Naphthalene, 2-cyano-				248	20700	[4401]
489	Phenanthrene, 1-benzoyl-				248	20700	[6926]
490	2,2'-Dinaphthyl ketone				248	20705	[9702]
491	Dithiouracil				248	20715	[7829]
492	Dithieno[3,2-b:2',3'-d]thiophene				248	20729	[9804]
493	Benz[c]acridin-7-one	249	20800	[8510]			
494	Biphenyl, 2-nitro-				249	20800	[4401]
495	Coumarin, 3-thenoyl-				249	20800	[8231]
496	2,4-Hexadiene, 1-hydroxy-	249	20800	[6504]			
497	Naphthalene, 2-acetyl-	249	20800	[6702]	249	20800	[6702]
498	Naphthalene, 2-benzoyl-	249	20800	[6702]	249	20800	[6702]
499	2-Naphthalenecarboxylic acid				249	20800	[6701]
500	Styrene, β-methyl-, (E)-	249	20800	[7723]			
501	6-Thiocaffeine				249	20800	[7120]
502	4H-Dithieno[3,2-b:2',3'-d]pyrrole				249	20810	[9804]
503	Benzo[b]naphtho[2,1-d]thiophene				249	20830	[6812]

189

Table 3b Triplet State Energies of Organic Compounds

No.	Compound	$E_T(n)$ kJ mol^{-1}	cm^{-1}	Ref.	$E_T(p)$ kJ mol^{-1}	cm^{-1}	Ref.
504	4-Thiouracil, 1,3-dimethyl-				249	20830	[7625]
505	Quinoline, 8-chloro-	249[x]	20840[x]	[7819]	252	21100	[8321]
506	Acenaphthene	250	20872	[6401]	248	20700	[6701]
507	Psoralen, 4',5'-dihydro-				250	20876	[8704]
508	Benzoxazole, 4'-amino-2-phenyl-				250	20900	[8804]
509	Butadiene	250	20900	[6504]			
510	Naphthalene, 1-methoxy-				250	20900	[4401]
511	Benzothiazole, 2-phenyl-	251	20944	[7309]	250	20870	[7309]
512	Naphthalene, 1-fluoro-				251	20970	[4902]
513	Naphthalene, 1-(methylsulfinyl)-				251	20985	[0009]
514	Cambendazole				251	21000	[8208]
515	Isoprene	251	21000	[6504]			
516	Naphthalene, 2-chloro-				251	21000	[4401]
517	Quinoline, 5-chloro				251	21000	[8321]
518	Naphthalene, 2-iodo-				252	21040	[4401]
519	Tetrabenz[a,c,h,j]acridine				252	21097	[8403]

Table 3b Triplet State Energies of Organic Compounds

No.	Compound	$E_T(n)$			$E_T(p)$		
		kJ mol^{-1}	cm^{-1}	Ref.	kJ mol^{-1}	cm^{-1}	Ref.
520	Naphthalene, 2-bromo-				252	21100	[4401]
521	Naphthalene, 2-cyano-7-methoxy-				252	21100	[8909]
522	Naphthalene, 2-hydroxy-				252	21100	[4401]
523	2-Naphthalenecarboxylic acid, 7-methoxy-, methyl ester				252	21100	[8909]
524	9H-Carbazole, 3-nitro-				253	21132	[0204]
525	9H-Carbazole, 3-chloro-6-nitro-				253	21132	[0204]
526	1,3-Butadiene, 2,3-dimethyl-	253	21100	[6504]			
527	Phenothiazine	253	21100	[7513]			
528	Naphthalene	253	21180	[5701]	255	21300	[4401]
529	Coumarin, 5,7-dimethoxy-				253	21190	[7303]
530	Naphthalene, 2,7-dihydroxy-				254	21200	[6701]
531	Naphthalene, 1-methyl-				254	21200	[4401]
532	Psoralen, 3-carbethoxy-				254	21200	[8317]
533	Psoralen, 5-methoxy-				254	21200	[8317]
534	Isoquinoline	254	21210	[5901]	254	21200	[5905]
535	Benzophenone, 4-phenyl-				254	21225	[6001]
536	Benzo[f]quinoxaline	254	21262	[8308]	251	21000	[6304]

Table 3b Triplet State Energies of Organic Compounds

No.	Compound	$E_T(n)$			$E_T(p)$		
		kJ mol^{-1}	cm^{-1}	Ref.	kJ mol^{-1}	cm^{-1}	Ref.
537	Benzo[b]naphtho[2,1-d]furan				255	21280	[6812]
538	Coumarin-3-carboxylic acid				255	21280	[7303]
539	2-Cyclohexen-1-one,4,4-di(1-naphthyl)-	254	21300	[8529]			
540	Naphthalene, 2-methyl-				254	21300	[6701]
541	2-Cyclohexen-1-one	255^b	21300^b	[6721]			
542	Naphthalene, 2-cyano-6-methoxy-				255	21300	[8909]
543	2-Naphthalenecarboxylic acid, 7-methoxy-				255	21300	[8909]
544	Thioxanthen-9-one-1-carboxylic acid, 3-amino-, ethyl ester	255	21300	[8620]			
545	Coumarin, 7-hydroxy-				255	21320	[7303]
546	Coumarin, 3-methyl-				255	21320	[7303]
547	Quinoxaline	255	21325	[8225]	254	21250	[5905]
548	Quinoline, 4,6-dichloro-	255	21340	[8025]			
549	Quinoline, 4-chloro-2-methyl-				255	21350	[8321]
550	Acetophenone, 4'-phenyl-	255	21400	[6702]	254	21300	[6702]
551	Pyrylium, 2,6-dimethyl-4-(4-methylphenyl)-, perchlorate				256	21370	[9701]
552	Benzo[a]carbazole				256	21400	[7701]

Table 3b Triplet State Energies of Organic Compounds

No.	Compound	$E_T(n)$ kJ mol⁻¹	cm⁻¹	Ref.	$E_T(p)$ kJ mol⁻¹	cm⁻¹	Ref.
553	Indazole, 2-methyl-				256	21400	[7522]
554	2-Naphthalenecarboxylic acid, 6-methoxy-, methyl ester				256	21400	[8909]
555	1-Naphthalenemethanol acetate				256	21400	[8528]
556	Phosphoric acid, diethyl 1-naphthalenylmethyl ester				256	21400	[8528]
557	Quinoline, 3-chloro-				256	21400	[8321]
558	6-Phenanthridone, 8-nitro-				256	21415	[8709]
559	Quinoline, 4-chloro-	257	21450	[8025]	255	21300	[8321]
560	9,9'-Biphenanthryl				257	21460	[7708]
561	Dibenzo[a,j]carbazole				257	21460	[7610]
562	Psoralen, 5-hydroxy-				257	21460	[7303]
563	Naphtho[1,2-c][1,2,5]oxadiazole	257	21499	[7216]			[7303]
564	Coumarin, 6-methyl-				257	21500	[7303]
565	2-Naphthalenecarboxylic acid, 6-methoxy-				257	21500	[8909]
566	Quinoline, 6-methoxy-				257	21500	[6903]
567	2-Cyclohexen-1-one, 4,4-di(2-naphthyl)-	258	21500	[8529]			
568	o-Terphenyl				258	21500	[6701]
569	Quinoxaline, 2,3-dichloro-	257	21523	[8225]			

Table 3b Triplet State Energies of Organic Compounds

No.	Compound	$E_T(n)$			$E_T(p)$		
		kJ mol^{-1}	cm^{-1}	Ref.	kJ mol^{-1}	cm^{-1}	Ref.
570	Quinoline, 7-chloro-				258	21550	[8321]
571	Quinoline	258	21590	[8025]	261	21850	[8321]
572	Coumarin	258[b]	21600[b]	[7711]	261	21840	[7303]
573	Quinoline, 2-methyl-				258	21600	[6903]
574	Quinoline, 4-methyl-				258	21600	[6903]
575	Quinoline-3-carboxylic acid, ethyl ester	258	21600	[8708]	259	21700	[8708]
576	Styrene	258	21600	[7723]			
577	Naphthalene, 2-ethoxy-				259	21600	[6701]
578	Quinoline, 2,4-dichloro-	259	21640	[8025]			
579	Quinoline, 6-chloro-	259	21650	[8025]	255	21300	[8321]
580	Benzo[1,2-b:5,4-b]dipyran-2,8-dione				259	21670	[7303]
581	2,3-Diazabicyclo[2.2.1]hept-2-ene	259[b]	21685[b]	[9403]			
582	Flavone	259	21700	[8609]			
583	1-Naphthalenemethanminium chloride, N,N,N-trimethyl-				259	21700	[8528]
584	Styrene, α-methyl-	260	21700	[7723]			
585	Thioxanthen-9-one-4-carboxylic acid, ethyl ester	260	21700	[8620]			

Table 3b Triplet State Energies of Organic Compounds

No.	Compound	$E_T(n)$			$E_T(p)$		
		kJ mol^{-1}	cm^{-1}	Ref.	kJ mol^{-1}	cm^{-1}	Ref.
586	Benzo[*h*]quinoline	260	21740	[5901]	261	21790	[5903]
587	9-Bismafluorene, 9-phenyl-	260	21740	[8111]			
588	9-Stannafluorene, 9,9-diethyl-	260	21740	[8111]			
589	Benzoxazole,2-phenyl-	260	21746	[7309]	262	21900	[8804]
590	1,8-Phenanthroline	260	21755	[7830]			
591	Phenanthrene	260	21774	[6306]	257	21500	[8905]
592	9-Stibafluorene, 9-phenyl-	261	21790	[8111]			
593	9,10-Anthraquinone	261	21800	[6404]	263	21980	[6710]
594	Benzidine, *N,N,N',N'*-tetramethyl-	261	21800	[7605]			
595	Quinoline, 2,4-dimethyl-				261	21800	[6903]
596	2-Pyridyl 2-thienyl ketone				261	21818	[9807]
597	9-Germafluorene, 9,9-diphenyl-	261	21830	[8111]			
598	Acetylene, diphenyl-	262	21860	[5701]	262	21870	[5601]
599	Benzo[*f*]quinoline	262	21880	[5901]	262	21865	[5903]
600	Phenoxazine	262	21880	[7214]	260	21750	[7214]
601	Quinoxaline, 2-chloro-3-methyl-	262	21885	[8225]			
602	1,9-Phenanthroline	262	21886	[7830]			

Table 3b Triplet State Energies of Organic Compounds

No.	Compound	$E_T(n)$ kJ mol^{-1}	cm^{-1}	Ref.	$E_T(p)$ kJ mol^{-1}	cm^{-1}	Ref.
603	Psoralen, 8-methoxy-				262	21900	[7303]
604	Thiophene, 2-(4-cyanobenzoyl)-				262	21900	[7322]
605	3-Pyridyl 2-thienyl ketone				262	21901	[9807]
606	4-Pyridyl 2-thienyl ketone				262	21901	[9807]
607	Psoralen, 8-methyl-				262	21904	[7303]
608	Quinazoline	262	21925	[7823]	262	21900	[5905]
609	6-Phenanthridone, 8-amino-				262	21930	[8709]
610	Psoralen				262	21930	[7303]
611	6-Thiopurine				263	21980	[7625]
612	Biphenyl, 4,4'-dichloro-				263	22000	[7217]
613	Biphenyl, 4,4'-dihydroxy-				263	22000	[7310]
614	Biphenyl, 2-iodo-				263	22000	[6701]
615	Cambendazole, 1-amino-				263	22000	[8208]
616	Ethene, 1,2-dicyano-, (E)-	263	22000	[8232]			
617	Phenanthridine				264	22050	[7808]
618	Quinoline, 2-chloro-4-methyl-				264	22050	[8321]

Table 3b Triplet State Energies of Organic Compounds

No.	Compound	$E_T(n)$			$E_T(p)$		
		kJ mol^{-1}	cm^{-1}	Ref.	kJ mol^{-1}	cm^{-1}	Ref.
619	Pyrylium, 2,6-dimethyl-4-phenyl-, perchlorate				264	22069	[9701]
620	Indene	264[b]	22075[b]	[7506]			
621	Benz[b]arsindole, 5-phenyl-	264	22080	[8111]			
622	Coumarin, 7-hydroxy-4-methyl-				264	22080	[7303]
623	9-Phosphafluorene, 9-phenyl-	264	22090	[8111]			
624	Angelicin				264	22100	[7303]
625	Benzimidazole, 2-(4-thiazolyl)-				264	22100	[8208]
626	Biphenyl, 4,4'-dimethoxy-				264	22100	[7310]
627	1,10-Phenanthroline				264	22100	[6304]
628	Phthalazine	264[b]	22100[b]	[8217]	275	23000	[5905]
629	Thiophene, 2-benzoyl-				265	22100	[7322]
630	Thioxanthen-9-one	265	22100	[8620]			
631	Thioxanthen-9-one-l-carboxylic acid, ethyl ester	265	22100	[8620]			
632	Thioxanthen-9-one-2-carboxylic acid, ethyl ester	265	22100	[8620]			
633	Quinoline, 2-chloro-	265	22120	[8025]	262	21900	[8321]
634	9-Silafluorene, 9,9-diphenyl-	265	22120	[8111]			
635	Quinoxaline, 2,3-dimethyl-	265	22149	[8225]			

Table 3b Triplet State Energies of Organic Compounds

No.	Compound	$E_T(n)$			$E_T(p)$		
		kJ mol⁻¹	cm⁻¹	Ref.	kJ mol⁻¹	cm⁻¹	Ref.
636	4,7-Phenanthroline	265	22150	[7830]	266	22200	[6304]
637	Phenoxazine, 10-phenyl-	265	22150	[7214]	264	22070	[7214]
638	1,7-Phenanthroline	265	22154	[7830]	265	22150	[6304]
639	1,3,5-Hexatriynediol				265	22170	[5601]
640	2,4,6-Octatriyne	265	22170	[6005]	267	22320	[5601]
641	Biphenyl, 4-chloro-				266	22200	[7217]
642	Thiophene, 2-(4-methoxybenzoyl)-				266	22300	[7322]
643	Benzophenone, 4,4'-dithiomethoxy-				267	22300	[8133]
644	Biphenyl, 3-chloro-				267	22300	[7217]
645	Phenoxathiin	267	22300	[7724]			
646	Dibenzothiophene, 5,5-dioxide	268	22384	[9406]	268	22384	[9406]
647	Benzophenone, 4-thiomethoxy-				268	22400	[8133]
648	1,1'-Biphenyl, 4-(methylsulfonyl)-				268	22384	[0009]
649	Biphenyl, 4-hydroxy-				268	22400	[7310]
650	Biphenyl, 3,3',5,5'-tetrachloro-				268	22400	[7217]
651	Psoralen, 4,5',8-trimethyl-				268	22400	[7303]

Table 3b Triplet State Energies of Organic Compounds

No.	Compound	$E_T(n)$			$E_T(p)$		
		kJ mol^{-1}	cm^{-1}	Ref.	kJ mol^{-1}	cm^{-1}	Ref.
652	Benzo[2,1-b:3,4-b']bis[1]benzothiophene				268	22420	[8516]
653	9H-Carbazole, 1,3,6,8-tetrachloro-9-methyl-				268	22423	[0204]
654	Thiopyrylium, 2,6-dimethyl-4-phenyl-, perchlorate				269	22489	[9701]
655	Benzene, 1,3,5-triphenyl-				269	22500	[6701]
656	Biphenyl, 4,4'-bis(1,1-dimethylethyl)-				269	22500	[7310]
657	Biphenyl, 4,4'-bis(1-methylpropyl)-				269	22500	[7310]
658	Biphenyl, 3,3'-dichloro-				269	22500	[7217]
659	Biphenyl, 4,4'-diisopropyl-				269	22500	[7310]
660	Biphenyl, 4,4'-dimethyl-				269	22500	[7310]
661	m-Terphenyl				269	22500	[6701]
662	Thiopyrylium, 2,6-dimethyl-4-(4-methylphenyl)-, perchlorate				270	22559	[9701]
663	Benzo[b]tellurophene	270	22573	[8910]	272	22730	[8910]
664	Benzene, 1,2,4,5-tetracyano-	271	22650	[6722]			
665	Benzophenone, 4,4'-dicarbomethoxy-	271	22700	[8623]			
666	Biphenyl, decafluoro-	272	22700	[6823]			
667	Fluorene, 2-amino-				272	22700	[4401]

Table 3b Triplet State Energies of Organic Compounds

No.	Compound	$E_T(n)$			$E_T(p)$		
		kJ mol⁻¹	cm⁻¹	Ref.	kJ mol⁻¹	cm⁻¹	Ref.
668	Benzophenone, 4,4'-diiodo-	273[b]	22844[b]	[8109]			
669	Biphenyl	274	22871	[6401]	274	22900	[7310]
670	Pyrazole, 3-methyl-1,5-diphenyl-				274	22883	[8328]
671	Benzophenone, 3,4'-dicarbomethoxy-	274	22900	[8623]			
672	Dicumarol				274	22920	[7303]
673	Benzophenone, 4,4'-bis(dimethylamino)-	275	23000	[6809]	255	21300	[7907]
674	Biphenyl, 4,4'-difluoro-				275	23000	[7310]
675	Styrene, β-methyl-, (Z)-	275	23000	[7723]			
676	Acetophenone, 2,2,2-trifluoro-4'-methoxy-	276	23000	[8624]			
677	Benzophenone, 4,4'-dicyano-				276	23000	[8007]
678	Di-2-pyridyl ketone				275[x]	23005[x]	[8109]
679	Dibenzo[f,h]quinoxaline	276	23065	[6925]	275	23000	[6308]
680	9H-Carbazol-2-ol, 3-chloro-				276	23068	[0204]
681	Benzophenone, 2,2'-bis(trifluoromethyl)-	276	23100	[8438]			
682	Benzophenone, 4,4'-bis(trifluoromethyl)-	276	23100	[8438]			
683	Benzophenone, 4-cyano-2',4',6'-triisopropyl-				276	23100	[8329]

Table 3b Triplet State Energies of Organic Compounds

No.	Compound	$E_T(n)$			$E_T(p)$		
		kJ mol^{-1}	cm^{-1}	Ref.	kJ mol^{-1}	cm^{-1}	Ref.
684	Carbostyril	276	23100	[7121]			
685	Terephthaldicarboxaldehyde	277^b	23130^b	[7218]			
686	Quinoline-2-carboxylic acid, 4-hydroxy-				277	23188	[7930]
687	Acetophenone, 4'-amino-				278	23200	[6809]
688	Benzophenone, 4-cyano-	278	23200	[6404]	280	23400	[8007]
689	1,5-Naphthyridine	278^x	23215^x	[7316]			
690	6-Phenanthridone, 5-methyl-	278	23251	[8019]	282	23585	[8019]
691	6-Phenanthridone, 2-amino-				278	23255	[8709]
692	Benzo[1,2-b:3,4-b']bis[1]benzothiophene				278	23256	[8516]
693	Benzophenone, 4-carbomethoxy-	279	23300	[8134]			
694	Benzophenone, 4-cyano-4'-methoxy-				279	23300	[8007]
695	Indazole, 1-methyl-				279	23300	[7522]
696	Pyridine, 2-benzoyl-	279	23300	[6810]	273	22800	[7413]
697	9H-Carbazole, 9-acetyl-, 3,6-dichloro-				279	23310	[0204]
698	9H-Carbazole, 3,6-dichloro-2-methoxy-9-methyl-				280	23391	[0204]
699	Acetophenone, 4'-thiomethoxy-	279^x	23331^x	[7414]			
700	[1]Benzopyrano[5,4,3-cde][1]benzopyran-5,10-dione	279	23364	[8530]	280	23400	[8530]

Table 3b Triplet State Energies of Organic Compounds

No.	Compound	$E_T(n)$			$E_T(p)$		
		kJ mol^{-1}	cm^{-1}	Ref.	kJ mol^{-1}	cm^{-1}	Ref.
701	Pyrazole, 3,5-diphenyl-	280	23392	[8328]	285	23810	[8328]
702	Acetophenone, 2,2,2-trifluoro-3'-methyl-	279	23400	[8624]			
703	Benzophenone, 4-cyano-4'-methyl-				279	23400	[8007]
704	Benzophenone, 3,3'-bis(trifluoromethyl)-	280	23400	[8438]			
705	Benzophenone, 3,3'-dicarbomethoxy-	280	23400	[8623]			
706	Triphenylene				280	23400	[8436]
707	9H-Carbazole, 3-chloro-2-methoxy-9-methyl-				281	23471	[0204]
708	Pyrido[3,4-*b*]indole, 1-methyl-	281	23474	[8710]	285	23809	[8710]
709	Di-4-pyridyl ketone				281^x	23486^x	[8330]
710	Acetophenone, 4'-acetyl-	281	23500	[8624]			
711	Benzophenone,4-amino-	281	23500	[6809]	263	22000	[6809]
712	Phthalic anhydride, tetrachloro-	281	23500	[6722]			
713	Pyridine, 4-benzoyl-	281	23500	[6810]	278	23200	[7413]
714	Acetophenone, 2,2,2-trifluoro-4'-methyl-	282	23500	[8624]			
715	Fluorene, 9,9-dimethyl-	281	23520	[8111]			
716	Benzoic acid, 4-(methyl amino)-				281	23530	[7906]

Table 3b Triplet State Energies of Organic Compounds

No.	Compound	$E_T(n)$			$E_T(p)$		
		kJ mol[-1]	cm[-1]	Ref.	kJ mol[-1]	cm[-1]	Ref.
717	Benzo[b]selenophene	282	23585	[8910]	282	23585	[8910]
718	Dibenzotellurophene	282	23585	[8910]	281	23530	[8910]
719	Decafluorobenzophenone	282	23600	[8911]	292	24400	[8911]
720	Fumaric acid, dimethyl ester	282	23600	[8232]			
721	Acetophenone, 4'-tert-butyl-2,2,2-trifluoro-	283	23600	[8624]			
722	Acetophenone, 4'-chloro-2,2,2-trifluoro-	283	23600	[8624]			
723	Benzophenone, 4-(trifluoromethyl)-	283	23600	[6810]	285	23800	[6810]
724	Fluorene	282	23601	[6401]	284	23700	[7701]
725	Benzoic acid, 4-(diethylamino)-, ethyl ester	283	23640	[8306]			
726	Acetophenone, 3',4'-methylenedioxy-	283	23700	[6702]	275	23000	[6703]
727	Benzo[f][4,7]phenanthroline	284	23700	[8107]	285	23800	[7509]
728	Benzoic acid, 4-(dimethylamino)-, ethyl ester	284	23750	[8306]			
729	Indazole	284[b]	23753[b]	[7506]	284	23700	[6301]
730	Benzophenone, 4,4'-dichloro-	285	23800	[6404]	286	23900	[8711]
731	Benzophenone, 2,4,6-triethyl-				285	23800	[8329]
732	Flavanone, 6-methyl-	285	23800	[8608]	264	22000	[8608]
733	Indole, 3-methyl-				285	23800	[8418]

203

Table 3b Triplet State Energies of Organic Compounds

No.	Compound	$E_T(n)$			$E_T(p)$		
		kJ mol^{-1}	cm^{-1}	Ref.	kJ mol^{-1}	cm^{-1}	Ref.
734	Coumarin, 4-hydroxy-				285	23810	[7303]
735	9H-Carbazole, 3-chloro-9-phenyl-				286	23875	[0204]
736	9H-Carbazole, 1,6-dichloro-9-methyl-				286	23875	[0204]
737	Dibenzothiophene	285	23830	[8111]	288	24100	[7610]
738	Dibenzoselenophene	285	23866	[8910]	287	23980	[8910]
739	Benzophenone, 4'-(4-benzoylbenzyl)-2,4,6-triisopropyl-	286	23900	[8912]	287	24000	[8912]
740	Benzophenone, 3-chloro-	286	23900	[6810]	288	24100	[6810]
741	Benzophenone, 4-chloro-	286	23900	[6810]	288	24100	[6810]
742	Benzophenone, 4-hydroxy-				286	23900	[8711]
743	Benzophenone, 3-methoxy-	286	23900	[6810]			
744	Benzophenone, 3-(trifluoromethyl)-	286	23900	[8134]			
745	Benzophenone, 2,4,6-trimethyl-	286	23900	[8913]	284	23700	[8329]
746	Pyridine, 3-benzoyl-	286	23900	[6810]	282	23600	[7413]
747	9H-Carbazole, 1,3,6-trichloro-				284	23713	[0204]
748	9H-Carbazol-2-ol				284	23713	[0204]
749	9H-Carbazole, 9-acetyl-, 3-chloro-				284	23713	[0204]

Table 3b Triplet State Energies of Organic Compounds

No.	Compound	$E_T(n)$			$E_T(p)$		
		kJ mol⁻¹	cm⁻¹	Ref.	kJ mol⁻¹	cm⁻¹	Ref.
750	9H-Carbazole, 1,3,6,8-tetrachloro-				285	23794	[0204]
751	6-Phenanthridone				286	23920	[8019]
752	9H-Carbazole, 3-chloro-				287	23955	[0204]
753	Benzo[b]thiophene	287	23970	[5901]	288	24040	[8516]
754	Benzophenone, 3,3'-dibromo-	287ᵇ	23975ᵇ	[8109]			
755	Benzophenone	287	24000	[6702]	289	24200	[8007]
756	Benzophenone, 3-cyano-	287	24000	[8134]			
757	Benzophenone, 3,4-dimethyl-	287	24000	[6702]	289	24100	[6702]
758	Benzophenone, 4-methoxy-	287	24000	[6810]	290	24300	[8007]
759	Benzophenone, 4-methyl-	287	24000	[6702]	290	24300	[8007]
760	Benzophenone, 2,3,4,5,6-pentafluoro-	287	24000	[8911]			
761	Benzophenone, 2,4,6-triisopropyl-	287	24000	[8913]	288	24100	[8329]
762	Biphenyl, 4,4'-(trifluoromethyl)-				287	24000	[7310]
763	Dibenzofuran	287	24000	[8111]	293	24515	[5801]
764	Indole-3-carboxaldehyde	287	24000	[8418]	293	24500	[8418]
765	9H-Carbazole				293	24520	[0204]
766	9H-Carbazole, 9-phenyl-				293	24520	[0204]

205

Table 3b Triplet State Energies of Organic Compounds

No.	Compound	$E_T(n)$			$E_T(p)$		
		kJ mol^{-1}	cm^{-1}	Ref.	kJ mol^{-1}	cm^{-1}	Ref.
767	Aniline, N-ethyl-	288	24000	[6707]			
768	2,5-Cyclohexadien-l-one, 4,4-diphenyl-				288	24000	[6723]
769	Indole-5-carboxylic acid				288	24000	[8418]
770	Di-3-pyridyl ketone				287^x	24030^x	[8330]
771	9H-Carbazole, 2-methoxy-9-methyl-				288	24036	[0204]
772	9H-Carbazole, 3,6-dichloro-9-methyl-				288	24036	[0204]
773	Benzoic acid, 4-amino-				288	24040	[7906]
774	Benzonitrile, 4-(diethylamino)-	288	24060	[8306]	288	24100	[7906]
775	Pyrazole, 1,3,5-Biphenyl-				288	24096	[8328]
776	Acetophenone, 2',3',4',5',6'-pentafluoro-	288	24100	[8911]			
777	Aniline, N-methyl-	288	24100	[6707]			
778	Benzophenone, 3-carbomethoxy-	288	24100	[8134]			
779	Benzophenone, 4,4'-dibromo-				288	24100	[6604]
780	9H-Carbazole, 1,6-dichloro-				289	24117	[0204]
781	9H-Carbazole, 3,6-dichloro-				289	24117	[0204]
782	9H-Carbazole, 9-acetyl-				289	24117	[0204]

Table 3b Triplet State Energies of Organic Compounds

No.	Compound	$E_T(n)$			$E_T(p)$		
		kJ mol⁻¹	cm⁻¹	Ref.	kJ mol⁻¹	cm⁻¹	Ref.
783	1,1'-Biphenyl, 4-(methylsulfinyl)-				289	24133	[0009]
784	Phenanthrene, 9-(methylsulfinyl)-				289	24133	[0009]
785	9H-Carbazol-2-ol, 3,6-dichloro-, acetate				289	24197	[0204]
786	9H-Carbazole, 1,3,6-trichloro-2-methoxy-9-methyl-				289	24197	[0204]
787	9H-Carbazole, 3-bromo-9-methyl-				289	24197	[0204]
788	9H-Carbazol-2-ol, 3-bromo-6-chloro-				289	24197	[0204]
789	Benzophenone, 2,5-dimethyl-	288	24100	[6702]	290	24300	[6702]
790	Benzophenone, 4,4'-dimethyl-	288	24100	[6702]	290	24300	[8007]
791	Benzophenone, 4-fluoro-	288	24100	[6810]	292	24400	[8133]
792	Benzophenone, 2-(trifluoromethyl)-	288	24100	[6810]			
793	1,4-Naphthoquinone, 5-hydroxy-	288	24100	[8625]	297	24900	[8625]
794	Benzophenone, 2-carbomethoxy-	289	24100	[8134]			
795	Benzophenone, 2,4-dimethyl-	289	24100	[6702]	291	24300	[6702]
796	Phenoxathiin, 10-oxide	289	24133	[9406]	331	27630	[9406]
797	Acetophenone, 2-fluoro-	289	24200	[8626]			
798	Benzaldehyde, 2,3,5,6-tetramethyl-	289[b]	24200[b]	[7711]			
799	Benzophenone, 2-benzyl-	289	24200	[6702]	292	24400	[6702]

Table 3b Triplet State Energies of Organic Compounds

No.	Compound	$E_T(n)$			$E_T(p)$		
		kJ mol^{-1}	cm^{-1}	Ref.	kJ mol^{-1}	cm^{-1}	Ref.
800	Indole, 1-methyl-	289	24200	[7714]	292	24400	[7714]
801	Acetophenone, 4'-cyano-	290	24200	[8624]	291	24300	[6810]
802	Benzophenone, 4,4'-di-*tert*-butyl-	290	24200	[8624]			
803	Benzophenone, 2-methyl-	290	24200	[6810]			
804	Benzonitrile, 4-(dimethylamino)-	290	24210	[8306]	288	24100	[7906]
805	Pyridine, 4-acetyl-	290	24235	[8331]	295	24654	[8331]
806	Benzaldehyde, 2-chloro-	290	24240	[7204]			
807	Isoindole-1,3-dione, 2-[(dimethylamino)methyl]-	290	24300	[7831]	292	24400	[7831]
808	Aniline, *N,N'*-diphenyl-				291	24300	[6903]
809	2-Pyridyl 3-thienyl ketone				291	24326	[9807]
810	4-Pyridinecarboxaldehyde	291^x	24300^x	[7931]			
811	Acridan				291	24330	[7302]
812	9H-Carbazol-2-ol, 1,3-dichloro-				291	24358	[0204]
813	Benzoic acid, 4-amino-, methyl ester				292	24380	[7906]
814	9H-Carbazole, 9-methyl-				292	24439	[0204]
815	Acetophenone, 2,2,2-trifluoro-	292	24400	[8624]			

Table 3b Triplet State Energies of Organic Compounds

No.	Compound	$E_T(n)$			$E_T(p)$		
		kJ mol^{-1}	cm^{-1}	Ref.	kJ mol^{-1}	cm^{-1}	Ref.
816	Indole, 3-acetyl-	292	24400	[8418]	294	24600	[8418]
817	Indole, 5-fluoro-	292	24400	[8418]	291	24300	[8418]
818	Pyrido[3,4-b]indole, 7-methoxy-1-methyl-				292	24400	[8439]
819	1,3,5-Triazine, 2,4-diphenyl-	292	24400	[6613]			
820	Carbazole, N-methyl-	292	24450	[8111]	292	24450	[6812]
821	Benzophenone, 4,4'-dimethoxy-	293[b]	24470[b]	[8109]	292	24400	[8007]
822	Acetophenone, 2-chloro-	293	24500	[6810]			
823	Acetophenone, 2,2,2,4-tetrafluoro-	293	24500	[8624]			
824	Pyridine, 2-acetyl-	294	24538	[8331]	298	24899	[8331]
825	Carbazole				294	24540	[5801]
826	Benzonitrile, 2-amino-	294[b]	24552[b]	[7506]			
827	Benzene, 1,4-dicyano-	294	24560	[6203]	295	24700	[7008]
828	Benzophenone, 4,4'-difluoro-	294	24600	[8624]	294	24600	[8133]
829	Benzophenone, 4-methoxy-2',4',6'-triisopropyl-				294	24600	[8329]
830	Carbazole, N-phenyl-				294	24600	[6812]
831	1,3,5-Triazine, 2,4,6-triphenyl-	294	24600	[6613]			
832	Acetophenone, 3'-cyano-	295	24600	[6810]	307	25600	[6810]

Table 3b Triplet State Energies of Organic Compounds

No.	Compound	$E_T(n)$			$E_T(p)$		
		kJ mol^{-1}	cm^{-1}	Ref.	kJ mol^{-1}	cm^{-1}	Ref.
833	Benzophenone, 4-methyl-2',4',6'-triisopropyl-				295	24600	[8329]
834	Benzonitrile, 4-(methylamino)-	295	24660	[8306]			
835	Benzaldehyde, 2-fluoro-	295	24670	[7204]			
836	Benzoic acid, 4-amino-, ethyl ester	295	24690	[8306]			
837	Acetophenone, 4'-hydroxy-	295	24700	[6822]	303	25300	[6604]
838	5H-Pyrido[3,2-b]indole	295	24692	[0207]			
839	Acetophenone, 2,2,2-trifluoro-4'-(trifluoromethyl)-	295	24700	[8624]			
840	Benzotriazole				295	24700	[6301]
841	Ketone, phenyl 3-thienyl				296	24744	[0008]
842	Limonene	295	24700	[8627]			
843	Acetophenone, 2,2,2-triphenyl-	296	24700	[6702]	300	25000	[6702]
844	1,4-Naphthoquinone, 5,8-dihydroxy-	296	24700	[8625]	282	23600	[8625]
845	2-Pyridinecarboxaldehyde	296^x	24760^x	[7931]			
846	Dibenz[b,f]azepine, 10,11-dihydro-				296	24770	[7302]
847	Acetophenone, 2,2-difluoro-	297	24800	[8626]			
848	Aniline	297	24800	[6707]	321	26800	[4401]

Table 3b Triplet State Energies of Organic Compounds

No.	Compound	$E_T(n)$			$E_T(p)$		
		kJ mol⁻¹	cm⁻¹	Ref.	kJ mol⁻¹	cm⁻¹	Ref.
849	Dimethyl phthalate				297	24800	[7219]
850	Phthalimide, N-methyl-				297	24832	[9404]
851	3-Pyridinecarboxaldehyde	297[x]	24810[x]	[7931]			
852	Pyridazine	297	24850	[6716]	290[x]	24251[x]	[6717]
853	9H-Carbazol-2-ol, acetate				290	24278	[0204]
854	Pyridine, 3-acetyl-	298	24899	[8331]	307	25633	[8331]
855	Acetophenone, 4'-bromo-	297	24900	[6702]	299	25000	[6702]
856	2-Cyclohexen-1-one, 4,4-diphenyl-				298	24900	[6723]
857	2-Thiouracil				298	24932	[7829]
858	Benzaldehyde, 3-chloro-	298	24940	[7204]			
859	Benzonitrile, 4-amino-	298	24940	[8306]	293	24500	[7507]
860	Acetophenone, 3',5'-dimethyl-	299	25000	[6702]	298	24900	[6703]
861	Benzaldehyde, 3-fluoro-	299	25000	[7204]			
862	1,2,3-Triazole, 4-benzoyl-5-methyl-				299	25000	[7122]
863	Benzaldehyde, 4-chloro-	300	25090	[7204]			
864	Acetophenone, 4'-methoxy-	300	25100	[6702]	299	25000	[6703]
865	Acetophenone, 4'-(trifluoromethyl)-	300	25100	[6702]	301	25200	[6702]

Table 3b Triplet State Energies of Organic Compounds

No.	Compound	$E_T(n)$			$E_T(p)$		
		kJ mol^{-1}	cm^{-1}	Ref.	kJ mol^{-1}	cm^{-1}	Ref.
866	Benzaldehyde, 4-methoxy-	300	25100	[6702]	295	24700	[6925]
867	Anthrone, 10,10-dimethyl-	301	25100		301	25100	[8914]
868	Benzene, 1,1'-sulfonylbis(4-chloro)-	301	25100	[7730]			
869	Aniline, N-phenyl-				301	25140	[7302]
870	Anthrone				301	25150	[7006]
871	Benzofuran	301	25157	[8910]	301	25130	[8910]
872	Benzaldehyde, 2-methyl-	301	25160	[7204]			
873	Acetophenone, 3',4'-dimethyl-	301	25190	[7123]	300	25060	[7123]
874	Benzene, 1-methoxy-4-(methylthio)-	301	25182	[9406]	305	25532	[9406]
875	Acetylene, phenyl-	301	25190	[5901]			
876	Benzonitrile, 3-methoxy-	301	25190	[7508]			
877	Benzaldehyde	301	25200	[7204]	298	24950	[7006]
878	3-Pyridyl 3-thienyl ketone				299	24994	[9807]
879	4-Pyridyl 3-thienyl ketone				299	24994	[9807]
880	3,3'-Dithienyl ketone				299	24994	[0008]
881	Benzoic acid, 4-cyano-, methyl ester				301	25200	[7219]

Table 3b Triplet State Energies of Organic Compounds

No.	Compound	$E_T(n)$			$E_T(p)$		
		kJ mol^{-1}	cm^{-1}	Ref.	kJ mol^{-1}	cm^{-1}	Ref.
882	Indole	301	25200	[8418]	296	24800	[8418]
883	Methanone, (1,2-dimethyl-2-cyclopenten-l-yl)phenyl-				301	25200	[7626]
884	Acetophenone, 4′-chloro-	302	25200	[6810]	301	25100	[6810]
885	Acetophenone, 2′,4′,6′-trimethyl-	302	25200	[8913]	303	25300	[6702]
886	Benzaldehyde, 3-methyl-	302	25280	[7204]			
887	Acetophenone, 4′-acetyl-2,2,2-trifluoro-	303	25300	[8624]			
888	Acetophenone, 3′-methoxy-	303	25300	[6702]	303	25300	[6703]
889	Benzene, 1,1′-sulfonylbis(4-methoxy)-	303	25300	[7730]			
890	Caffeine				303	25300	[7120]
891	Formaldehyde	303^g	25316^g	[8014]			
892	Phenol, 3-cyano-				303	25320	[7906]
893	α-Tetralone	303	25320	[8610]	304	25400	[6702]
894	Benzene, 1,2-dicyano-	303	25340	[6203]	305	25500	[7008]
895	Acetophenone, 3′-bromo-	303	25400	[6702]	304	25400	[6702]
896	Acetophenone, 4′-tert-butyl-	304	25400	[8624]			
897	Acetophenone, 3′-chloro-	304	25400	[8624]			
898	Acetophenone, 3′-(trifluoromethyl)-	304	25400	[6702]	307	25600	[6702]

213

Table 3b Triplet State Energies of Organic Compounds

No.	Compound	$E_T(n)$			$E_T(p)$		
		kJ mol^{-1}	cm^{-1}	Ref.	kJ mol^{-1}	cm^{-1}	Ref.
899	Dibenzo[b,e][1,4]dioxin	304	25400	[7724]			
900	4-Chromanone	304	25445	[7123]	304	25445	[7123]
901	Benzaldehyde, 4-methyl-	305	25490	[7204]	298	24900	[6702]
902	Acetophenone, 4'-methyl-	305	25500	[7220]	305	25500	[6703]
903	Benzene, 1,1'-sulfonylbis(4-methyl)-	305	25500	[7730]			
904	Dimethyl terephthalate				305	25500	[7219]
905	Flavanone	305	25500	[8608]			
906	Benzaldehyde, 4-fluoro-	305	25520	[7204]			
907	Benzene, 1,1'-thiobis-	305	25532	[9406]	305	25532	[9406]
908	2-Thiouracil, 1-methyl-	306	25562	[8135]			
909	Acetophenone, 3'-methyl-	306	25600	[6702]	303	25400	[6703]
910	Benzene, (methylthio)-				305	25532	[9406]
911	2-Δ^2-Thiazoline, (4'-chlorobenzoyl)amino-	306	25600	[8332]			
912	Benzene, hexachloro-	307	25600	[6701]			
913	4-Cholesten-3-one	307	25600	[7124]			
914	Testosterone acetate	307	25600	[6824]			

Table 3b Triplet State Energies of Organic Compounds

No.	Compound	$E_T(n)$			$E_T(p)$		
		kJ mol⁻¹	cm⁻¹	Ref.	kJ mol⁻¹	cm⁻¹	Ref.
915	Benzoic acid, 2-methoxy-				307	25640	[7906]
916	Benzoic acid, 2-cyano-, methyl ester				307	25700	[7219]
917	Benzonitrile, 4-bromo-	308[b]	25777[b]	[7809]	308	25770	[7906]
918	Benzene, 1,1'-sulfonylbis-	309	25800	[7730]			
919	Phthalic anhydride	309	25800	[6722]			
920	2-Δ²-Thiazoline, (4'-methylbenzoyl)amino-	309	25800	[8332]			
921	Benzonitrile, 4-chloro-	309[x]	25840[x]	[7809]	313	26136	[8531]
922	Benzene, 1-methoxy-4-(methylsulfonyl)-	310	25882	[9406]	314	26231	[9406]
923	Phenoxathiin, 10,10-dioxide	310	25882	[9406]	331	27630	[9406]
924	1-Indanone, 2,2-dimethyl-	310	25900	[8610]			
925	Methanone, phenyl(1,2,3-trimethyl-2-cyclopenten-1-yl)-				310	25900	[7626]
926	Bicyclo[4.3.0]non-1(6)-en-2-one				314	26231	[9211]
927	Progesterone	310	25900	[7124]			
928	Xanthone	310	25906	[7118]	310	25905	[7118]
929	Acetophenone	310	25933	[8101]	311	26034	[8101]
930	Benzene, pentachloro-	310[b]	25944[b]	[8440]			
931	Benzonitrile, 2-bromo-				311	25970	[7906]

215

Table 3b Triplet State Energies of Organic Compounds

No.	Compound	E_T(n)			E_T(p)		
		kJ mol^{-1}	cm^{-1}	Ref.	kJ mol^{-1}	cm^{-1}	Ref.
932	Benzonitrile, 2-methoxy-	311	25970	[7508]			
933	Acetophenone, 2-allyl-	311	26000	[6702]	311	26000	[6702]
934	Acetophenone, 2-(2'-phenylethyl)-	311	26000	[6702]	311	26000	[6702]
935	Acetophenone, 2-(phenylmethyl)-				311	26000	[6702]
936	Acetophenone, 2-propyl-	311	26000	[6702]	312	26100	[6702]
937	Benzoic acid, 3,5,-dimethyl-				311	26000	[7606]
938	Propiophenone	312	26080	[8610]	313	26150	[7006]
939	Acetophenone, 2-ethyl-				313	26100	[6825]
940	Benzonitrile, 2-(trifluoromethyl)-				313	26100	[8134]
941	Pyrazine-d_4	313^b	26146^b	[8240]			
942	Aniline, 4-chloro-	313^x	26154^x	[7809]			
943	Borine, triphenyl-	313	26171	[8333]			
944	2-Δ^2-Thiazoline, (2'-chlorobenzoyl)amino-	313	26200	[8332]			
945	Benzene, 1-methoxy-4-(methylsulfinyl)-	314	26231	[9406]	335	27980	[9406]
946	Benzene, 1,4-dimethoxy-	314^b	26250^b	[7207]			
947	Benzoic acid, 4-methoxy-, ethyl ester	314	26250	[8306]			

Table 3b Triplet State Energies of Organic Compounds

No.	Compound	$E_T(n)$			$E_T(p)$		
		kJ mol^{-1}	cm^{-1}	Ref.	kJ mol^{-1}	cm^{-1}	Ref.
948	1-Indanone	314	26250	[8610]	317	26500	[6608]
949	Benzonitrile, 3-methyl-	314	26270	[6203]	315	26400	[7606]
950	Benzoyl chloride	314	26280	[5901]			
951	Benzonitrile, 2-chloro-	315	26291	[8531]	315	26317	[8531]
952	Benzonitrile, 3-chloro-	315	26291	[8531]	316	26395	[8531]
953	Benzoic acid, 3,4-dimethyl-				315	26300	[7606]
954	Deoxythymidine 5'-monophosphate				315	26300	[6706]
955	2-Δ^2-Thiazoline, benzoylamino-	315	26300	[8332]			
956	2-Δ^2-Thiazoline, (2'-methylbenzoyl)amino-	315	26300	[8332]			
957	Pyrazine, 2,5-dimethyl-				315	26318	[5803]
958	Benzoic acid, 3-methoxy-				315	26320	[7906]
959	Benzonitrile, 3-bromo-				315	26320	[7906]
960	Benzonitrile, 4-methoxy-	315	26320	[8306]	315	26300	[7508]
961	Phenol, 2-cyano-				315	26320	[7906]
962	Phenol, 4-cyano-				315	26320	[7906]
963	Borane, dichlorophenyl-	315	26347	[8532]			
964	Pyrazine	315	26361	[7822]	311	25991	[5803]

Table 3b Triplet State Energies of Organic Compounds

No.	Compound	$E_T(n)$			$E_T(p)$		
		kJ mol⁻¹	cm⁻¹	Ref.	kJ mol⁻¹	cm⁻¹	Ref.
965	Chromone	316	26385	[7123]	313	26180	[7123]
966	Benzoic acid, 4-methoxy-, methyl ester				315	26400	[7606]
967	Benzene, 1,3-dicyano-				316	26400	[7008]
968	1,3,5-Triazine				316	26400	[6107]
969	Benzonitrile, 4-methyl-	316	26410	[6203]	317	26500	[6203]
970	Purine	316[b]	26434[b]	[7506]			
971	Benzonitrile, 2-methyl-	316	26450	[6203]	318	26550	[6203]
972	Benzoic acid, 4-methoxy-	317	26460	[6005]	317	26460	[7906]
973	2,4-Hexadiyne, 1,6-dichloro-	317[b]	26460[b]	[7207]			
974	Pyridine, 3-cyano-				322	26946	[6928]
975	Aniline,N,N-dimethyl-				317	26500	[6102]
976	Benzoic acid, 3-methyl-, methyl ester				318	26500	[7606]
977	Pyridine, 2-cyano-	317[b]	26510[b]	[7207]			
978	Pyrazine, 2-methyl-	317	26511	[7822]			
979	Benzene, 1,2,3,4-tetrachloro-				317[x]	26519[x]	[8440]
980	Benzoic acid, 3-methyl-	318	26580	[7731]	316	26400	[7606]

Table 3b Triplet State Energies of Organic Compounds

No.	Compound	$E_T(n)$			$E_T(p)$		
		kJ mol^{-1}	cm^{-1}	Ref.	kJ mol^{-1}	cm^{-1}	Ref.
981	Aniline, 4-bromo-	318[x]	26600[x]	[7809]			
982	Benzonitrile, 3-fluoro-				318	26600	[7523]
983	Benzonitrile, 4-(trifluoromethyl)-				318	26600	[8134]
984	Borinane,1-phenyl-	319	26630	[8029]			
985	Pyridine, 4-cyano-				319	26650	[6928]
986	Adenosine 5'-monophosphate				319	26700	[6706]
987	Benzoic acid, 3-cyano-, methyl ester				319	26700	[7219]
988	Pyrazine, tetramethyl-				319[x]	26700[x]	[8441]
989	Benzene, 1,2,4,5-tetrabromo-	320[x]	26700[x]	[6701]			
990	Benzoic acid, 4-methyl-, methyl ester				320	26700	[7606]
991	Benzene, 1,2,3,5-tetrachloro-				320[x]	26710[x]	[8440]
992	Pyridine, 2-methoxy-				320	26738	[7732]
993	Pyrazine, 2,6-dimethyl-	320	26758	[7822]			
994	Benzonitrile	320	26780	[6203]			
995	Aniline, 4-methyl-	321[b]	26795[b]	[7809]	323	27000	[7606]
996	Benzene, 1,2,4,5-tetrachloro-	320[x]	26800[x]	[6701]			
997	Benzonitrile, 3-(trifluoromethyl)-				320	26800	[8134]

219

Table 3b Triplet State Energies of Organic Compounds

No.	Compound	$E_T(n)$			$E_T(p)$		
		kJ mol⁻¹	cm⁻¹	Ref.	kJ mol⁻¹	cm⁻¹	Ref.
998	Trimethyl 1,3,5-benzenetricarboxylate				320	26800	[7219]
999	Benzimidazole	321[x]	26810[x]	[7506]	318	26600	[8208]
1000	Aniline, 4-fluoro-	321[b]	26838[b]	[7809]			
1001	Benzonitrile, 2-fluoro-				321	26850	[7523]
1002	Triphenylene, dodecahydro-	321	26850	[8204]			
1003	Benzamide, 3-methyl-				321	26870	[7906]
1004	Benzoic acid, 4-methyl-	322	26880	[7731]	320	26800	[7606]
1005	Dimethyl isophthalate				322	26900	[7219]
1006	Anthracene, 1,2,3,4,5,6,7,8-octahydro-	322	26950	[8204]			
1007	Mesitylpentamethyldisilane				322	26950	[8442]
1008	Aniline,N,N-diethyl-				323	27000	[6102]
1009	Phenanthrene, 1,2,3,4,5,6,7,8-octahydro-	323	27000	[8204]			
1010	Silane, tris(trimethylsilyl)mesityl-				323	27000	[8442]
1011	Benzamide, 4-methyl-				323	27010	[7906]
1012	Benzoic acid, ethyl ester	324	27100	[8306]			
1013	Benzoic acid, 2-methyl-	324	27100	[7731]			

Table 3b Triplet State Energies of Organic Compounds

No.	Compound	E_T(n)			E_T(p)		
		kJ mol^{-1}	cm^{-1}	Ref.	kJ mol^{-1}	cm^{-1}	Ref.
1014	Disilane, (2,5-dimethylphenyl)pentamethyl-				324	27100	[8442]
1015	Benzoic acid	324	27110	[6302]	326	27200	[7606]
1016	1,3-Benzodioxole	325[b]	27189[b]	[7506]			
1017	Guanosine 5'-monophosphate				325	27200	[6706]
1018	Pyrazole, 3,5-dimethyl-l-phenyl-				325	27210	[8328]
1019	Benzonitrile, 4-fluoro-	326[b]	27229[b]	[7809]	325	27200	[7523]
1020	Benzene, 1,1'-sulfinylbis-				326	27281	[9406]
1021	Benzoic acid, methyl ester	326	27240	[6302]	326	27200	[7219]
1022	Benzene, 1-methoxy-4-methyl-	326[b]	27280[b]	[7809]			
1023	1,2,3-Triazole, 4-acetyl-5-methyl-				326	27300	[7122]
1024	Butadiyne	327	27300	[6005]			
1025	Benzene, 1-bromo-4-methoxy-	327[b]	27350[b]	[7809]			
1026	(Z)-2-Butene	327[g]	27360[g]	[7125]			
1027	Benzene, pentamethyldisilyl-				327	27370	[8442]
1028	Benzene, 1-chloro-4-methoxy-	327[b]	27377[b]	[7809]			
1029	2,4-Hexadiyne, 1,6-dihydroxy-	328	27380	[6005]			
1030	Uridine 5'-monophosphate				328	27400	[7320]

Table 3b Triplet State Energies of Organic Compounds

No.	Compound	$E_T(n)$			$E_T(p)$		
		kJ mol^{-1}	cm^{-1}	Ref.	kJ mol^{-1}	cm^{-1}	Ref.
1031	1,3-Octadiyne	328	27420	[6005]			
1032	Benzene, hexamethyl-	328^x	27423^x	[6408]			
1033	Ethylene	330	27550	[8232]			
1034	Benzene, 1-methoxy-4-fluoro-	330^b	27563^b	[7809]			
1035	Benzene, 1,3,5-tribromo-	330	27600	[6701]			
1036	Benzene, 1,3,5-trichloro-				331	27600	[6701]
1037	Benzamide, 2-methyl-				331	27690	[7906]
1038	Benzene, 1,2,3,4-tetramethyl-	331	27700	[8204]			
1039	Xanthene				331	27700	[8801]
1040	Pyridine	332^b	27770^b	[8443]			
1041	Benzamide	332^b	27785^b	[6302]	330	27550	[6302]
1042	Acetone				332	27800	[6602]
1043	Benzene, 1,4-dibromo-	332^x	27800^x	[6701]			
1044	Tetrazole, 5-phenyl-, ion(1-)				332	27800	[6929]
1045	Ethene, 1,2-dichloro-, (E)-	333	27800	[8232]			
1046	5,7-Dodecadiyne	333	27820	[6005]			

Table 3b Triplet State Energies of Organic Compounds

No.	Compound	$E_T(n)$			$E_T(p)$		
		kJ mol^{-1}	cm^{-1}	Ref.	kJ mol^{-1}	cm^{-1}	Ref.
1047	Benzene, 1-bromo-4-chloro-				333	27900	[6701]
1048	Pyridine, 2,4,6-trimethyl-				333	27900	[8136]
1049	Cytidine 5′-monophosphate				334	27900	[6706]
1050	2,4-Hexadiyne	334	27910	[6005]			
1051	Benzene, 1,2,4,5-tetramethyl-	334	27920	[8204]	335	28000	[6702]
1052	Benzene, 1,4-dichloro-	335	28000	[6701]			
1053	Benzene, 1,3,5-trimethyl-	335	28010	[6103]	336	28075	[6408]
1054	p-Xylene	337	28135	[6103]	337	28145	[6004]
1055	Pyridine, 2,6-dimethyl-				337	28160	[6928]
1056	Benzene, methoxy-	338	28200	[6702]	338	28200	[6702]
1057	Pyrimidine	338	28214	[7021]			
1058	1,3,2-Benzothiazaborolidine, 2-methyl-	338	28215	[8235]			
1059	Benzene, cyclopropyl-	339	28300	[5901]			
1060	Naphthalene, 1,2,3,4-tetrahydro-	339	28300	[8204]			
1061	Phenyl ether				339	28320	[8427]
1062	m-Xylene	339	28325	[6103]	336	28120	[6004]
1063	Benzene, (methylsulfinyl)-				339	28330	[9406]

Table 3b Triplet State Energies of Organic Compounds

No.	Compound	$E_T(n)$			$E_T(p)$		
		kJ mol^{-1}	cm^{-1}	Ref.	kJ mol^{-1}	cm^{-1}	Ref.
1064	Benzene, 1-fluoro-4-methyl-	341[g]	28500[g]	[7415]			
1065	Phenol				342	28563	[8427]
1066	Benzene, chloro-	342[g]	28570[g]	[5901]			
1067	Pyridinium, 2,6-dimethyl-				342	28600	[8136]
1068	Benzene, (methylsulfonyl)-				343	28680	[9406]
1069	1,3,2-Benzodiazaborolidine, 2-methyl-	343	28678	[8235]			
1070	o-Xylene	343	28705	[6103]	344	28760	[6004]
1071	Indan	344	28750	[7416]			
1072	Benzylamine				345	28800	[6903]
1073	Pyrimidine, 2-chloro-	345	28800	[7021]			
1074	2-Indanone				345[x]	28853[x]	[8521]
1075	Anthracene, 9,10-dihydro-				346	28900	[8801]
1076	Toluene	346	28920	[6103]	347	29000	[6702]
1077	Pyridine, 4-amino-				347	28985	[7323]
1078	Benzene, 1,4-difluoro-	347[g]	29000[g]	[7415]			
1079	Benzene, 1-fluoro-2-methyl-	347[g]	29000[g]	[7415]			

Table 3b Triplet State Energies of Organic Compounds

No.	Compound	$E_T(n)$			$E_T(p)$		
		kJ mol^{-1}	cm^{-1}	Ref.	kJ mol^{-1}	cm^{-1}	Ref.
1080	Benzene, 1,2,4,5-tetrafluoro-	347[g]	29000[g]	[7415]			
1081	Benzene, 1-fluoro-3-methyl-	348[g]	29100[g]	[7415]			
1082	Benzene, hexafluoro-	348[g]	29100[g]	[7415]			
1083	Benzene, (trifluoromethyl)-	349[g]	29150[g]	[5901]			
1084	Benzene, pentafluoro-	349[g]	29200[g]	[7415]			
1085	Benzene, 1,2,3,4-tetrafluoro-	349[g]	29200[g]	[7415]			
1086	Octanoic acid, 8-phenyl-				350	29200	[7023]
1087	Benzene, 1,2-difluoro-	351[g]	29300[g]	[7415]			
1088	Benzene, 1,2,4-trifluoro-	351[g]	29300[g]	[7415]	352	29450	[7125]
1089	Butanoic acid, 4-phenyl-				351	29300	[7023]
1090	Phenylacetic acid				351	29300	[7023]
1091	3-Phenylpropionic acid				351	29300	[7023]
1092	Benzene	353	29500	[6702]	353	29500	[6702]
1093	Benzene, 1,3-difluoro-	353[g]	29500[g]	[7415]			
1094	Benzene, fluoro-	353[b]	29500[b]	[5701]			
1095	Benzene, 1,2,3-trifluoro-	353	29500	[7125]	349	29150	[7125]
1096	Benzene, 1,2,3,5-tetrafluoro-	354[g]	29600[g]	[7415]			

Table 3b Triplet State Energies of Organic Compounds

No.	Compound	$E_T(n)$			$E_T(p)$		
		kJ mol^{-1}	cm^{-1}	Ref.	kJ mol^{-1}	cm^{-1}	Ref.
1097	Pyridinium				355	29700	[8136]
1098	Benzene, 1,3,5-trifluoro-	358[g]	29900[g]	[7415]	358	29890	[7125]
1099	Methyl methacrylate	358[g]	29900[g]	[8915]			
1100	Pyridine, 4-hydroxy-				358	29940	[7627]
1101	Methyl acrylate	372[g]	31100[g]	[8915]			
1102	Pyridinium, 2-amino-				392	32800	[8136]
1103	Acetylene	502[g]	41900[g]	[7832]			
1104	1-Butyne	502[g]	41900[g]	[7832]			

[b] Aromatic solvent, benzene-like; [g] Gas-phase measurement; [x] Crystalline medium

3c FLASH PHOTOLYSIS PARAMETERS

Table 3c is intended to bring together information pertinent to designing triplet-triplet energy-transfer experiments in flash photolysis. These experiments are central to triplet photosensitized photochemistry, measurements of triplet quenching rates, and estimations of many triplet-triplet extinction coefficients. Table 3 can also be useful for the design of steady-state photochemistry, but triplet extinction coefficients and corresponding wavelengths are included which are needed to quickly monitor time-resolved experiments.

The data for four of the columns is already presented in Table 3a in SI units, although the criteria used for the selection of compounds in Table 3a differ from that adopted here. For convenience, the triplet energy, E_T, in Table 3c is given in units of kcal mol^{-1}, and the singlet energy is given as an excitation wavelength, λ_S^{0-0}. The other two items from Table 3a, namely the triplet quantum yields, Φ_T, and triplet lifetimes, τ_T, are carried along unmodified.

The central new item of Table 3c is the information on triplet-triplet absorption. Whenever possible, this absorption is characterized by both a wavelength, λ_T, and a molar absorption coefficient, ε_T, at that wavelength. When an absorption maximum exists, the wavelength refers to that maximum. In cases where molar absorption coefficients are not available, but the triplets are important donors or acceptors, only a wavelength is given, and this isolated wavelength refers to an absorption maximum. Molar absorption coefficients are often called decadic extinction coefficients. A review of the primary literature was presented previously, led to the publication of a set of recommended values for, and procedures for the determination of the molar absorptivity of transients. [9108]

In Table 3c, the solvent type is a dual one. Benzene and some aromatic solvents have a large influence on triplet-triplet absorption. There is a strong tendency to broaden the peaks such that the extinction coefficient at the maximum is lowered, but the oscillator strength (related to the total area) in the transition remains constant. To distinguish aromatic solvents for the (λ_T, ε_T) pair of a given compound from non aromatic solvents, "b" for "benzene-like" is used for aromatic solvents, and "nb" is used for other solvents. One or the other of these two symbols follow the "/" in the solvent column and refer only to the (λ_T, ε_T) column. The symbol preceding the "/" in the solvent column is a nonpolar/polar classification (discussed in 3a) for the four properties, λ_{0-0}, E_T, Φ_T, and τ_T.

Table 3c Flash Photolysis Parameters

No.	Solv.	λ_S^{0-0}/nm	Ref.	E_T/kcal mol^{-1}	Ref.	Φ_T	Ref.	λ_T/nm, ε_T/M^{-1}cm^{-1}	Ref.	τ_T/μs	Ref.
1	**Acenaphthene**										
	n/b	319	[7101]	59.7	[6401]	0.46	[6901]	422	[8241]		
	p/nb	318	[6801]	59.2	[6701]	0.58	[6802]	430, 6000	[8712]	3300	[6801]
2	**Acetone**										
	n					0.90	[7002]			6.3	[8401]
	p/nb			79.4	[6602]	1	[6603]	300, 600	[8712]	47	[7103]
3	**Acetophenone**										
	n	363	[6402]	74.1	[8101]	1	[8501]			0.23	[7104]
	p/nb	354	[8101]	74.4	[8101]	1	[8501]	330, 7160	[8712]	0.14	[7104]
4	**Acetophenone, 4'-methoxy-**										
	n	352	[6604]	71.8	[6702]	1	[8501]				
	p/nb			71.5	[6703]	1	[8501]	360, 8840	[8712]		
5	**Acetophenone, 4'-methyl-**										
	n	352	[6826]	72.9	[7220]	1	[8501]				
	p/nb			72.8	[6703]	1	[8501]	331, 11400	[7324]		
6	**Acetophenone, 4'-phenyl**										
	n			61.1	[6702]						

Table 3c Flash Photolysis Parameters

No.	Solv	λ_S^{0-0}/nm	Ref.	E_T/kcal mol^{-1}	Ref.	Φ_T	Ref.	λ_T/nm, ε_T/M^{-1}cm^{-1}	Ref.	τ_T/µs	Ref.
	p/nb			60.8	[6702]			435, 130000	[8712]		
7	**Acetophenone, 4'-(trifluoromethyl)-**										
	n	360	[6826]	71.7	[6702]	1	[8501]				
	p/nb	360	[6908]	72.0	[6702]	1	[8501]	455, 2290	[8712]		
8	**Acridine**										
	n/b			45.4	[8102]	0.5	[7801]	440, 24300	[7126]	10000	[6705]
	p/nb	380	[7701]	45.0	[8403]	0.82^s	[8502]	432.5, 31500	[7126]	14	[7802]
9	**Acridine, 3,6-diamino-**										
	p/nb	485	[6902]	51.1	[6902]			550, 9510	[8712]		
10	**Acridine Orange, conjugate monoacid**										
	p/nb	524	[7105]	45.7	[6801]	0.10	[7201]	540, 9570	[8712]	105	[7901]
11	**9-Acridinethione, 10-methyl**										
	n/b	633	[8322]	42.6	[8322]	0.95	[8444]	520, 9300	[8444]	2.6	[8444]
	p/nb					0.90	[8444]	520, 8790	[8712]	2.3	[8444]
12	**Acridinium, 3,6-diamino-**										
	p/nb	478	[7105]	48.9	[6801]	0.22	[7201]	550, 8270	[8712]	20	[8002]
13	**Acridinium, 3,6-diamino-l0-methyl-**										
	p/nb	485	[6902]	51.1	[6902]			620, 8600	[8712]		

229

Table 3c Flash Photolysis Parameters

No.	Solv	λ_S^{0-0}/nm	Ref.	E_T/kcal mol^{-1}	Ref.	Φ_T	Ref.	λ_T/nm, ε_T/M^{-1}cm^{-1}	Ref.	τ_T/µs	Ref.
14	**9-Acridinone**										
	n/b	393	[7803]	58.2	[7803]	0.99	[7601]	620, 37800	[7601]	20	[7601]
	p/nb	413	[7803]	60.2	[7804]			620, 41400	[8712]	9.2	[8901]
15	**Angelicin**										
	n					0.009	[7805]				
	p/nb	362	[7303]	63.2	[7303]	0.031	[8404]	450, 4330	[8712]	2.6	[8404]
16	**Angelicin, 3,4'-dimethyl-**										
	n					0.005	[9601]	450	[9601]		
	p/nb					0.32	[9601]	450, 5900	[9601]	8.8	[9601]
17	**Aniline**										
	n	300	[8202]	71.0	[6707]	0.75	[8004]	320	[6930]	0.72	[8701]
	p/nb	312	[8004]	76.6	[4401]	0.90	[8004]				
18	**Aniline, N,N-dimethyl-**										
	n	313	[7101]								
	p/nb	319	[7101]	75.8	[6102]			460, 4000	[8533]		
19	**Aniline, N-phenyl-**										
	n					0.32	[7004]			0.3	[7004]

Table 3c Flash Photolysis Parameters

No.	Solv	λ_S^{0-0}/nm	Ref.	E_T/kcal mol^{-1}	Ref.	Φ_T	Ref.	λ_T/nm, ε_T/M^{-1}cm^{-1}	Ref.	τ_T/μs	Ref.
	p/nb	322	[7302]	71.9	[7302]	0.47	[7004]	530, 10400	[7126]	0.5	[7004]
20	**Anthracene**										
	n/b	376	[8504]	42.5	[7501]	0.71	[5701]	432.5, 45500	[7126]	670	[8505]
	p/nb	375	[7701]	42.5	[8502]	0.66	[7701]	425, 64700	[7126]	3300	[7402]
21	**Anthracene, 9-bromo-**										
	n/b	390	[7904]					430, 48000	[6931]	43	[6202]
	p/nb			41.3	[8902]	0.98	[8405]	430, 47500	[8712]	19.5	[8405]
22	**Anthracene, 9-cyano-**										
	n/b	403	[8104]			0.040	[7304]	435, 10300	[6931]	600	[7304]
	p/nb	417	[7304]			0.021	[7304]	435, 8490	[8712]	1800	[7304]
23	**Anthracene, 9,10-dibromo-**										
	n/b	406	[8406]	40.2	[8406]	0.70	[5603]	427.5, 48000	[8406]	36	[6202]
	p/nb	405	[7502]			0.82	[6605]	425, 46300	[8712]	11	[7802]
24	**Anthracene, 9,10-dichloro-**										
	n/b	401	[7101]	40.5	[8406]	0.29	[5603]	425, 46000	[8406]	100	[6202]
	p/nb	403	[6605]	40.5	[7005]	0.45	[6605]	425, 42500	[8712]		
25	**Anthracene, 9,10-dicyano-**										
	n/b	421	[8506]	41.7	[8407]	0.23	[7807]	440, 9000	[8406]	100	[7807]

Table 3c Flash Photolysis Parameters

No.	Solv	λ_S^{0-0}/nm	Ref.	E_T/kcal mol^{-1}	Ref.	Φ_T	Ref.	λ_T/nm, ε_T/M^{-1}cm^{-1}	Ref.	τ_T/μs	Ref.
	p/nb	428	[8506]					440, 9180	[8712]		
26	**Anthracene, 9,10-dimethyl-**										
	n/b	398	[7105]			0.02	[6804]	435, 35300	[8712]		
	p/nb	403	[6801]			0.032	[6801]	435, 35300	[8712]	8000	[6708]
27	**Anthracene, 9,10-diphenyl-**										
	n/b	393	[8504]	40.9	[6805]	0.02	[8303]	445, 14500	[8712]	2500	[7403]
	p/nb	392	[7105]			0.02	[8303]	445, 15600	[8712]	3000	[7403]
28	**Anthracene, 9-methyl-**										
	n/b	387	[8504]	41.3	[5701]	0.48	[7503]	430, 42000	[6931]		
	p/nb	391	[6801]	40.7	[6605]	0.67	[6708]	430, 45900	[8712]	10000	[6708]
29	**Anthracene, 9-[(1E)-2-(1-naphthalenyl)ethenyl]-**										
	n	444	[0201]			0.19	[0201]	507	[0201]	37	[0201]
	p/nb	445	[0201]			0.14	[0201]	504	[0201]	36	[0201]
30	**Anthracene, 9-[2-(2-naphthalenyl)ethenyl]-**										
	n	437	[0201]			0.24	[0201]	492	[0201]	48	[0201]
	p/nb	439	[0201]			0.2	[0201]	487	[0201]	49	[0201]

Table 3c Flash Photolysis Parameters

No.	Solv	λ_S^{0-0}/nm	Ref.	E_T/kcal mol^{-1}	Ref.	Φ_T	Ref.	λ_T/nm, ε_T/M^{-1}cm^{-1}	Ref.	τ_T/μs	Ref.
31	Anthracene, 9-phenyl-										
	n	387	[7101]			0.37	[6502]				
	p/nb	394	[6801]			0.51	[6709]	428, 14600	[8712]	15000	[6708]
32	Anthracene, 9-[(1E)-2-phenylethenyl]-										
	n	427	[0201]			0.4	[0201]	445	[0201]	40	[0201]
	p/nb	431	[0201]			0.27	[0201]	450	[0201]	45	[0201]
33	1-Anthracenesulfonate ion										
	p/nb					0.98	[8507]	440, 20000	[8507]	103	[8507]
34	2-Anthracenesulfonate ion										
	p/nb					0.65	[8507]	425, 30000	[8507]	83	[8507]
35	9,10-Anthraquinone										
	n/b			62.4	[6404]	0.90	[6503]	390, 10300	[7202]	0.11	[8105]
	p			62.8	[6710]						
36	Anthrone										
	n									0.170	[7628]
	p/nb			71.9	[7006]			341, 74000	[8712]		
37	Azulene										
	n/b	704	[7105]	39	[7504]			360, 4000	[8106]	11	[8408]

233

Table 3c Flash Photolysis Parameters

No.	Solv	λ_S^{0-0}/nm	Ref.	E_T/kcal mol^{-1}	Ref.	Φ_T	Ref.	λ_T/nm, ε_T/M^{-1}cm^{-1}	Ref.	τ_T/μs	Ref.
	p/nb							360, 4140	[8712]	3	[8106]
38 Benz[b]acridin-12-one											
	n			45.7	[8510]						
	p/nb	488	[7804]	46.1	[7804]			590, 53600	[8712]		
39 Benz[a]anthracene											
	n/b	385	[7105]	47.2	[5401]	0.79	[6905]	490, 20500	[7126]	100	[6406]
	p/nb	390	[7808]	47.2	[7808]	0.79	[7205]	480, 28800	[7126]	9400	[6407]
40 Benzene											
	n	260	[7105]	84.3	[6702]	0.25	[6901]				
	p/nb	261	[5001]	84.3	[6702]	0.15	[7305]	235, 11000	[8712]		
41 Benzene, chloro-											
	n	272	[7105]	81.7	[5901]	0.6	[8304]			1.6	[8409]
	p/nb					0.7	[8511]	300, 6150	[8712]	0.715	[8511]
42 Benzene, 1,4-dichloro-											
	n	282	[8202]	80.1	[6701]	0.8	[8304]			430	[8512]
	p/nb	280	[7905]			0.95	[8512]	310, 3800	[8512]	330	[8512]

Table 3c Flash Photolysis Parameters

No.	Solv	λ_S^{0-0}/nm	Ref.	E_T/kcal mol^{-1}	Ref.	Φ_T	Ref.	λ_T/nm, ε_T/M^{-1} cm^{-1}	Ref.	τ_T/µs	Ref.
43	Benzene, methoxy-										
	n	274	[4901]	80.7	[6702]	0.64	[8207]				
	p/nb	278	[8202]	80.7	[6702]			252	[7505]	3.3	[7505]
44	Benzidine, N,N,N',N'-tetramethyl-										
	n			62.3	[7605]	0.52	[8411]				
	p/nb					0.41	[8411]	475, 38700	[8712]	5	[8412]
45	Benzil										
	n	485	[6906]	53.4	[6712]	0.92	[6503]			150	[7106]
	p/nb	279	[7105]	54.3	[6712]			480	[7932]	1500	[6807]
46	Benzo[a]coronene										
	n	432	[5902]	51.5	[7810]						
	p/nb					0.55	[6808]	570, 22300	[8712]		
47	Benzoic acid										
	n			77.5	[6302]						
	p/nb	279		77.9	[7606]			320, 1000	[8712]		
48	Benzo[rst]pentaphene										
	n/b	433	[6409]	40.2	[6209]			490	[7108]	170	[7108]
	p/nb	433	[7701]	40.4	[7701]			490	[7108]		

Table 3c Flash Photolysis Parameters

No.	Solv	λ_S^{0-0}/nm	Ref.	E_T/kcal mol^{-1}	Ref.	Φ_T	Ref.	λ_T/nm, ε_T/M^{-1} cm^{-1}	Ref.	τ_T/μs	Ref.
49	**Benzo[*ghi*]perylene**										
	n/b	408	[6907]					470	[7108]	150	[7108]
	p/nb	407	[7701]	46.3	[5604]	0.53	[6907]	465, 39300	[8712]		
50	**Benzo[c]phenanthrene**										
	n	370	[7105]	56.7	[5401]						
	p/nb	373	[7701]	57.1	[7701]			517, 4800	[5804]		
51	**Benzo[3,4]cyclobuta[1,2-*b*]biphenylene**										
	p/nb	588	[0202]			<0.01	[0202]				
52	**Benzo[3,4]cyclobuta[1,2-*a*]biphenylene**										
	p/nb	438	[0202]			0.3	[0202]	430, 16000	[0202]		
53	**Benzo[2'',1'':3,4;3'',4'':3',4']dicyclobuta[1,2-*a*:1',2'-*a*']bisbiphenylene**										
	p/nb	483	[0202]			0.45	[0202]	685, 75000	[0202]		
54	**Benzophenone**										
	n/b	379	[6402]	68.6	[6702]	1.0	[6901]	530, 7220	[8334]	6.9	[7703]
	p/nb	384	[6303]	69.2	[8007]	1	[8501]	525, 6250	[8712]	50	[7607]
55	**Benzophenone, 4,4'-bis(dimethylamino)-**										
	n			65.8	[6809]	0.91	[7704]			25	[7705]

Table 3c Flash Photolysis Parameters

No.	Solv	λ_S^{0-0}/nm	Ref.	E_T/kcal mol^{-1}	Ref.	Φ_T	Ref.	λ_T/nm, ε_T/M^{-1}cm^{-1}	Ref.	τ_T/µs	Ref.
	p/nb	405	[7907]	60.8	[7907]	0.47	[8501]	500	[7705]	20	[7705]
56	**Benzophenone, 4-chloro-**										
	n			68.3	[6810]	1	[8501]				
	p/nb			68.8	[6810]	1	[8501]	320, 12800	[8445]		
57	**Benzophenone, 4,4'-dichloro-**										
	n			68.0	[6404]	1	[8501]			2.2	[7703]
	p/nb			68.4	[8711]	1	[8501]	320, 23300	[8712]		
58	**Benzophenone, 4,4'-dimethoxy-**										
	n	365	[6604]	70.0	[8109]	1	[8501]			3.6	[7703]
	p/nb			69.8	[8007]	1	[8501]	350, 14100	[8712]		
59	**Benzophenone, 4-fluoro-**										
	n			68.9	[6810]	1	[8501]				
	p/nb			69.8	[8133]	1	[8501]	315, 21900	[8712]		
60	**Benzophenone, 4-methoxy-**										
	n			68.6	[6810]	1	[8501]				
	p/nb			69.4	[8007]	1	[8501]	335, 10100	[8712]	7.2	[8210]
61	**Benzophenone, 4-methyl-**										
	n	370	[6908]	68.6	[6702]	1	[8501]				

Table 3c Flash Photolysis Parameters

No.	Solv	λ_S^{0-0}/nm	Ref.	E_T/kcal mol^{-1}	Ref.	Φ_T	Ref.	λ_T/nm, ε_T/M^{-1}cm^{-1}	Ref.	τ_T/µs	Ref.
	p/nb	366	[6908]	69.4	[8007]	1	[8501]	315, 21100	[8712]		
62	Benzophenone, 4-(trifluoromethyl)-										
	n	373	[6908]	67.6	[6810]	1	[8501]				
	p/nb	368	[6908]	68.1	[6810]	1	[8501]	320, 22400	[8712]		
63	Benzo[g]pteridine-2,4-dione, 7,8,10-trimethyl-										
	n									30	[6811]
	p/nb	500	[6902]	50.0	[6902]	0.30	[7706]	650, 6090	[8712]	320	[7010]
64	Benzo[a]pyrene										
	n/b	405	[7105]	41.9	[5701]			475	[7108]	8700	[7812]
	p/nb	403	[7701]	42.3	[7701]			465	[7812]	8800	[6407]
65	Benzo[e]pyrene										
	n/b	368	[7105]					560	[7108]	120	[7108]
	p/nb	366	[7701]	52.8	[7701]			555, 17000	[8712]		
66	1,4-Benzoquinone										
	n	457	[6909]	53.6	[6909]						
	p/nb							450	[7926]	0.53	[8008]

Table 3c Flash Photolysis Parameters

No.	Solv	λ_S^{0-0}/nm	Ref.	E_T/kcal mol^{-1}	Ref.	Φ_T	Ref.	λ_T/nm, ε_T/M^{-1}cm^{-1}	Ref.	τ_T/μs	Ref.
67	**1,4-Benzoquinone, tetrachloro-**										
	n/b	450	[6909]	49.2	[6909]			516	[7221]	2.0	[6910]
	p/nb					0.98	[7908]	510, 6990	[8712]	1.2	[6910]
68	**1,4-Benzoquinone, tetramethyl-**										
	n/b					1.0	[7608]	490, 6950	[7126]	21	[7608]
	p/nb					1.0	[7608]	490, 5330	[7126]	15	[7608]
69	**Benzo[b]triphenylene**										
	n/b	375	[7105]	50.9	[5701]			450	[7108]	90	[7108]
	p/nb	374	[7701]	50.9	[7701]			450, 29900	[8712]		
70	**Benzoxazole, 2,2'-(1,4-phenylene)bis-**										
	n										
	p/nb							480, 18600	[8712]	0.48	[8242]
71	**Biacetyl**										
	n/b					1.0	[6002]	315, 5160	[7126]	638	[7209]
	p/nb							315, 4580	[8712]		
72	**1,1'-Binaphthyl**										
	n	325	[7101]								
	p/nb	328	[7708]					615, 12400	[8712]	14	[7707]

Table 3c Flash Photolysis Parameters

No.	Solv	λ_S^{0-0}/nm	Ref.	E_T/kcal mol^{-1}	Ref.	Φ_T	Ref.	λ_T/nm, ε_T/M^{-1}cm^{-1}	Ref.	τ_T/μs	Ref.
73	**2,2'-Binaphthyl**										
	n	333	[7101]								
	p/nb			55.9	[5604]			450, 24800	[8712]		
74	**Biphenyl**										
	n/b	286	[7101]	65.4	[6401]	0.84	[7501]	367.3, 27100	[7126]	130	[6905]
	p/nb	306	[7708]	65.5	[7310]			361.3, 42800	[7126]		
75	**Biphenyl, 4,4'-diamino-**										
	p/nb	346	[7101]					460, 35500	[8712]		
76	**Biphenylene**										
	n	392	[6409]			<0.01	[0202]	350, 10500	[8712]	100	[7222]
	p/nb	417	[0202]								
77	**2,2'-Bifuran**										
	p/nb	290	[0003]					360	[0003]	39	[0003]
	n	298	[0003]			0.24	[0003]			102	[0003]
78	**Bisbenzo[3,4]cyclobuta[1,2-a:1',2'-c]biphenylene**										
	p/nb	441	[0202]			0.03	[0202]	715, 12000	[0202]		

Table 3c Flash Photolysis Parameters

No.	Solv	λ_S^{0-0}/nm	Ref.	E_T/kcal mol^{-1}	Ref.	Φ_T	Ref.	λ_T/nm, ε_T/M^{-1}cm^{-1}	Ref.	τ_T/μs	Ref.
79	**α-Bithiophene**										
	n/b					0.99	[9603]	385	[9603]	104	[9603]
	p/nb	358	[8011]			0.93	[9603]			124	[9603]
80	**1,3-Butadiene, 1,4-diphenyl-**										
	n/b		[7012]	42.3		0.020	[8211]	390, 45000	[8712]	1.6	[8211]
	p/nb					<0.002	[8211]	390, 54500	[8712]	5.0	[8413]
81	**1,3-Butadiene, di(2-thienyl)-**										
	n	389	[9901]			0.69	[9901]	425, 30000	[9901]	7.5	[9901]
	p/nb	403	[9901]			0.07	[9901]	425, 23000	[9901]	8.6	[9901]
82	**C$_{60}$**										
	n/b	620	[9101]	36	[9103]	1	[9201]	740, 12000	[9202]	250	[9202]
	p/nb							740, 14000	[9201]		
83	**C$_{70}$**										
	n/b	648	[9104]	35.3	[9106]	0.97	[9203]	490, 12000	[9202]	250	[9202]
	p/nb							470, 19000	[9201]		
84	**C$_{120}$**										
	n/b	730	[0106]			0.7	[0106]	700, 14000	[0106]	23	[0106]

Table 3c Flash Photolysis Parameters

No.	Solv	λ_S^{0-0}/nm	Ref.	E_T/kcal mol^{-1}	Ref.	Φ_T	Ref.	λ_T/nm, ε_T/M^{-1} cm^{-1}	Ref.	τ_T/µs	Ref.
85	**C$_{180}$**										
	n/b	730	[0106]			0.7	[0106]	700, 2700	[0106]	24	[0106]
86	**C$_{60}$H$_{18}$**										
	n/b	480	[9803]			0.15	[9803]	640, 26000	[9803]	42	[9803]
87	**C$_{60}$H$_{36}$**										
	n/b	480	[9803]			0.1	[9803]	600	[9803]	25	[9803]
88	**C$_{120}$O**										
	n/b	690	[0105]	28.8	[0105]			700	[0105]	0.16	[0105]
89	**Carbazole**										
	n/b	332	[7101]			0.36	[6503]	418	[8446]	170	[7709]
	p/nb	345	[7808]	70.2	[5801]			425, 14000	[8712]		
90	**2,2'-Carbocyanine, 1,1'-diethyl-**										
	p/nb					0.0029	[7933]	635, 58000	[8712]	190	[7933]
91	**4,4'-Carbocyanine, 1,1'-diethyl-**										
	p/nb					<6×10^{-4}	[7311]	778, 35600	[8712]	1100	[7311]
92	**β-*apo*-8'-Carotenal**										
	n	518	[7816]			0.003	[7818]			10	[7325]

Table 3c Flash Photolysis Parameters

No.	Solv	λ_S^{0-0}/nm	Ref.	E_T/kcal mol^{-1}	Ref.	Φ_T	Ref.	λ_T/nm, ε_T/M^{-1}cm^{-1}	Ref.	τ_T/µs	Ref.
93	**β-apo-14'-Carotenal**										
	p/nb							520, 223000	[8712]		[...]
	n/b	457	[7813]			0.54	[7909]	480, 114000	[8712]	8	[8309]
	p/nb	465	[7813]			−0.033	[7909]	480, 116000	[8712]	10.3	[7909]
94	**β-Carotene**										
	n/b	524	[7105]	21	[7504]	<0.001	[7710]	520	[7629]	70	[6607]
	p/nb	704	[9204]	20.3	[9204]			515, 187000	[8712]	9	[8112]
95	**Chlorophyll *a***										
	n/b	677	[8113]					460	[5805]	1500	[5802]
	p/nb	671	[7910]	29.8	[7910]	0.53	[8606]	460, 48000	[7934]	800	[6801]
96	**Chlorophyll *b***										
	n/b	660	[7910]	31.1	[7910]			316, 36500	[5904]	2500	[5904]
	p/nb	667	[6801]	32.6	[6801]	0.81	[8606]	450, 24300	[8712]	1500	[6801]
97	**Chrysene**										
	n/b	361	[7105]			0.85	[7501]	575	[7630]	710	[6905]
	p/nb	360	[7701]	57.1	[7701]	0.85	[6802]	580, 29800	[8712]		
98	**Coronene**										
	n/b	428	[6409]					480	[7630]		

Table 3c Flash Photolysis Parameters

No.	Solv	λ_S^{0-0}/nm	Ref.	E_T/kcal mol^{-1}	Ref.	Φ_T	Ref.	λ_T/nm, ε_T/M^{-1} cm^{-1}	Ref.	τ_T/μs	Ref.
	p/nb	429	[7701]	54.4	[5604]	0.56	[6802]	480, 15000	[8712]		
99	**Coumarin**										
	n/b			61.7	[7711]			400, 11000	[7911]	3.8	[7911]
	p/nb	341	[7303]	62.4	[7303]	0.054	[7911]	400, 10100	[8712]	1.3	[7911]
100	**Coumarin, 5,7-dimethoxy-**										
	p/nb	352	[7303]	60.6	[7303]	0.072	[7911]	450, 10500	[8712]	10	[7911]
101	**Cyclobuta[1,2-a:3,4-a']bisbiphenylene**										
	p/nb	465	[0202]			0.25	[0202]	585, 28000	[0202]		
102	**1,3-Cyclohexadiene**										
	n/b	292	[3901]	52.4	[6504]			310	[8447]	30	[8013]
	p/nb							303, 2300	[8712]		
103	**4H-Cyclopenta[2,1-b:3,4-b']dithiophene**										
	p/nb	312	[9804]	51.4	[9804]			375, 23000	[9804]	27	[9804]
104	**Deoxythymidine 5'-monophosphate**										
	p/nb	293	[6706]	75.2	[6706]	0.055	[7912]	370, 4000	[7912]	25	[7912]

Table 3c Flash Photolysis Parameters

No.	Solv	λ_S^{0-0}/nm	Ref.	E_T/kcal mol^{-1}	Ref.	Φ_T	Ref.	λ_T/nm, ε_T/M^{-1}cm^{-1}	Ref.	τ_T/µs	Ref.
105	**Dibenz[a,h]anthracene**										
	n/b	395	[7105]			0.90	[7205]	584, 13000	[8310]		
	p/nb	395	[7701]	52.1	[7701]	0.9	[8310]	580, 25100	[8712]		
106	**2,2'-Dicarbocyanine, 1,1'-diethyl-**										
	p/nb					$<3\times10^{-4}$	[7311]	780, 69300	[8712]	480	[7311]
107	**Dithieno[3,2-b:2',3'-d]thiophene**										
	p/nb	336	[9804]	59.3	[9804]			384, 22000	[9804]	45	[9804]
108	**4H-Dithieno[3,2-b:2',3'-d]pyrrole**										
	p/nb	344	[9804]	59.5	[9804]			400, 14000	[9804]		
109	**Ethene, 1,2-di(2-thienyl)-**										
	n	366	[9901]			0.27	[9901]	410, 16000	[9901]	0.36	[9901]
	p/nb	370	[9901]			0.02	[9901]	410, 17500	[9901]	0.25	[9901]
110	**4-Flavanthione**										
	n	446	[0005]	45.5	[0005]	1	[0005]	360, 7600	[0005]	0.93	[0005]
	p/nb	436	[0005]	46.3	[0005]	0.9	[0005]	355, 7450	[0005]	0.04	[0005]
111	**4-Flavanthione, 6-hydoxy-**										
	n/b	448	[0005]	45.7	[0005]	0.96	[0005]	370, 4400	[0005]	1.6	[0005]
	p/nb	440	[0005]	46.9	[0005]		[0005]	350, 4800	[0005]	0.1	[0005]

Table 3c Flash Photolysis Parameters

No.	Solv	λ_S^{0-0}/nm	Ref.	E_T/kcal mol^{-1}	Ref.	Φ_T	Ref.	λ_T/nm, ε_T/M^{-1}cm^{-1}	Ref.	τ_T/μs	Ref.
112	**Fluorene**										
	n	301	[7105]	67.5	[6401]	0.22	[7015]			150	[6905]
	p/nb	301	[7701]	67.9	[7701]	0.32	[6802]	380, 22700	[8712]		
113	**9-Fluorenone**										
	n/b					0.94	[7501]	430, 5900	[7501]	500	[7814]
	p/nb	450	[7006]	50.3	[7815]	0.48	[7814]	425, 6040	[8712]	100	[7814]
114	**Fluorescein dianion**										
	p/nb	520	[6902]	47.2	[6902]	0.02	[8201]	535, 8700	[6003]	20000	[6003]
115	**Fluorescein dianion, 2′,4′,5′,7′-tetrabromo-**										
	p/nb	571	[6813]	42.3	[6813]	0.33	[8201]	580, 10200	[8712]	30	[8416]
116	**Fluorescein dianion, 2′,4′,5′,7′-tetraiodo-**										
	p/nb	565	[6813]	44	[6915]	0.83	[8201]	526, 26000	[6715]	630	[6411]
117	**2,2′-Furil**										
	n	425	[0206]			0.71	[0206]	390	[0206]	2.8	[0206]
	p/nb	410	[0206]			0.24	[0206]	400	[0206]	1	[0206]
118	**(E,E,E)-2,4,6-Heptatrienal, 5-methyl-7-(2,6,6-trimethyl-1-cyclohexen-1-yl)-**										
	n/b	400	[7816]	35.9	[8417]	0.66	[7909]	430, 63000	[7909]	6.2	[7909]

Table 3c Flash Photolysis Parameters

No.	Solv	λ_S^{0-0}/nm	Ref.	E_T/kcal mol^{-1}	Ref.	Φ_T	Ref.	λ_T/nm, ε_T/M^{-1}cm^{-1}	Ref.	τ_T/µs	Ref.
	p/nb		[7909]			0.41	[7909]	440, 52000	[8712]	10.9	[7909]
119	1,3,5-Hexatriene, 1,6-diphenyl-										
	n/b		[7210]	35.6		0.029	[8211]	420, 105000	[8712]	20	[8211]
	p/nb	399	[7817]			0.020	[7612]	420, 114000	[8712]	30	[8211]
120	1,3,5-Hexatriene, 1,6-di(2-thienyl)-										
	n	426	[9901]			0.11	[9901]	445, 41000	[9901]	9.1	[9901]
	p/nb	428	[9901]			0.14	[9901]	445, 36000	[9901]	12.2	[9901]
121	1,3,5-Hexatriene, 1,6-di(3-thienyl)-										
	n	364	[9901]			0.27	[9901]	410, 10500	[9901]	18	[9901]
	p/nb	377	[9901]			0.24	[9901]	405, 8600	[9901]	33	[9901]
122	Indole										
	n/b	288	[5101]	72	[8418]	0.43	[8115]	430	[7714]	16	[7713]
	p/nb	299	[6301]	70.8	[8418]	0.23	[8115]	430, 4260	[8712]	11.6	[7511]
123	β-Ionone										
	n/b	405	[8522]	55	[8522]	0.49	[8522]	345	[8522]	0.16	[7818]
	p/nb							330, 85300	[8712]		
124	Isoquinoline										
	n		[5901]	60.6		0.21	[8705]				

Table 3c Flash Photolysis Parameters

No.	Solv	λ_S^{0-0}/nm	Ref.	E_T/kcal mol^{-1}	Ref.	Φ_T	Ref.	λ_T/nm, ε_T/M^{-1}cm^{-1}	Ref.	τ_T/μs	Ref.
125 Lumichrome	p/nb	320	[5905]	60.6	[5905]			418, 11900	[8712]		
	p/nb	409	[0401]					441, 21000	[0401]		
126 Methylene Blue cation	p/nb	664	[6305]	33	[6713]	0.52	[6915]	420, 14400	[8712]	450	[7512]
127 Naphthalene	n/b	311	[7105]	60.6	[5701]	0.75	[7501]	425, 13200	[7126]	175	[6202]
	p/nb	311	[5001]	60.9	[4401]	0.80	[6802]	415, 24500	[7126]	1800	[7402]
128 Naphthalene, 1-bromo-	n	321	[7105]	59	[5701]					270	[6202]
	p/nb			59.0	[6303]			425, 11500	[6105]	830	[6105]
129 Naphthalene, 1-chloro-	n			58.6	[7819]	0.79	[7501]				
	p/nb	319	[7905]	59.2	[4401]			420, 29500	[8712]	280	[6202]
130 Naphthalene, 1,4-dicyano-	n	336	[8420]								
	p/nb			55.5	[7604]	0.19	[8422]	455, 6730	[8712]	40	[8422]

Table 3c Flash Photolysis Parameters

No.	Solv	λ_S^{0-0}/nm	Ref.	E_T/kcal mol^{-1}	Ref.	Φ_T	Ref.	λ_T/nm, ε_T/M^{-1}cm^{-1}	Ref.	τ_T/μs	Ref.
131	**Naphthalene, 1,4-diphenyl-**										
	n	336	[7101]								
	p/nb							444, 32500	[8712]		
132	**Naphthalene, 1-hydroxy-**										
	n	321	[7101]			>0.27	[6503]				
	p/nb	323	[7105]	58.6	[4401]			430, 9000	[8534]		
133	**Naphthalene, 2-hydroxy-**										
	n	330	[7101]							67	[7315]
	p/nb	330	[7105]	60.3	[4401]			435, 6680	[8712]		
134	**Naphthalene, 1-methoxy-**										
	n					0.45	[7501]				
	p/nb	320	[7905]	59.8	[4401]	0.50	[6802]	440, 9980	[8712]	5500	[6801]
135	**Naphthalene, 2-methoxy-**										
	n					0.50	[7501]			50	[7733]
	p/nb							435, 21400	[8712]		
136	**Naphthalene, 1-methyl-**										
	n/b	317	[7101]			0.58	[7501]	425	[8448]	25	[7615]
	p/nb	317	[7105]	60.6	[4401]			420, 20200	[8712]		

Table 3c Flash Photolysis Parameters

No.	Solv	λ_S^{0-0}/nm	Ref.	E_T/kcal mol^{-1}	Ref.	Φ_T	Ref.	λ_T/nm, ε_T/M^{-1}cm^{-1}	Ref.	τ_T/μs	Ref.
137	**Naphthalene, 2-methyl-**										
	n	319	[7101]								
	p/nb	319	[7105]	60.8	[6701]	0.56	[7501]	420, 30600	[8712]		
138	**Naphthalene, 1-nitro-**										
	n	382	[7105]								
	p/nb			55.2	[7110]	0.63	[6806]	525	[8116]	0.93	[8116]
										4.9	[8116]
139	**Naphthalene, 2-nitro-**										
	n	380	[7105]								
	p/nb			56.9	[7110]	0.83	[7110]	360, 3600	[8712]	0.53	[7616]
										1.70	[7616]
140	**Naphthalene, 1-phenyl-**										
	n	315	[7101]								
	p/nb			58.8	[6701]	0.52	[7501]	490, 21700	[8712]		
141	**Naphthalene, 2-phenyl-**										
	n	325	[7101]								
	p/nb			58.7	[6701]	0.43	[7501]	430, 43000	[8712]		
142	**1,2-Naphthalenedicarboximide, *N*-methyl-**										
	p/nb	391	[9404]	52	[9404]	0.2	[9404]	510, 8400	[9404]		

Table 3c Flash Photolysis Parameters

No.	Solv	λ_S^{0-0}/nm	Ref.	E_T/kcal mol^{-1}	Ref.	Φ_T	Ref.	λ_T/nm, ε_T/M^{-1}cm^{-1}	Ref.	τ_T/μs	Ref.
143	**2,3-Naphthalenedicarboximide, *N*-methyl-**										
	p/nb	367	[9404]	58		0.71	[9404]	440, 10000	[9404]		
144	**1,8-Naphthalenedicarboximide, *N*-methy-**										
	p/nb	357	[9404]	53		0.94	[9404]	475, 10600	[9404]		
145	**1,3,5,7-Octatetraene, 1,8-diphenyl-**										
	n/b	443	[7211]	31.6	[7210]	0.005	[8211]	440, 178000	[8712]	40	[8211]
	p/nb					0.006	[7612]	440,198000	[8712]	34	[8211]
146	**1,3,5,7-Octatetraene, 1,8-di(2-thienyl)-**										
	n	461	[9901]			0.01	[9901]	465, 53000	[9901]	13.8	[9901]
	p/nb	462	[9901]			<0.005	[9901]	465, 43000	[9901]	4.4	[9901]
147	**1,3,5,7-Octatetraene, 1,8-di(3-thienyl)-**										
	n	414	[9901]			0.14	[9901]	420, 14000	[9901]	4.8	[9901]
	p/nb	415	[9901]			0.12	[9901]	420, 10000	[9901]	42	[9901]
148	**1,3,4-Oxadiazole, 2,5-diphenyl-**										
	n	311	[7101]								
	p/nb	310	[7101]					425, 980	[8712]	0.300	[7716]
149	**2,2′-Oxadicarbocyanine, 3,3′-diethyl-**										
	p/nb					<0.005	[7212]	650, 81400	[8712]	5000	[7212]

Table 3c Flash Photolysis Parameters

No.	Solv	λ_S^{0-0}/nm	Ref.	E_T/kcal mol^{-1}	Ref.	Φ_T	Ref.	λ_T/nm, ε_T/M^{-1} cm^{-1}	Ref.	τ_T/μs	Ref.
150	Oxazole, 2,5-bis(4-biphenylyl)-										
	n	375	[7101]								
	p/nb							560, 110000	[8712]	0.285	[7716]
151	Oxazole, 2,5-diphenyl-										
	n	335	[7101]			0.12	[8018]			1700	[8018]
	p/nb	336	[7101]					500, 14800	[8712]	2500	[7402]
152	Oxazole, 2,2'-(1,4-phenylene)bis(5-phenyl)-										
	n/b	379	[7101]			0.054	[8612]	550, 37600	[8612]	1750	[8612]
	p	385	[7101]	55.5	[8612]						
153	Pentacene										
	n/b	585	[7105]	18.0		0.16	[7201]	505, 120000	[7223]	110	[6105]
	p/nb							305, 595000	[8712]		
154	(E,E)-2,4-Pentadienal, 3-methyl-5-(2,6,6-trimethyl-1-cyclohexen-1-yl)-										
	n	358	[7816]	45.0	[7818]	0.20	[8424]			0.1	[7818]
	p/nb					0.45	[7909]	385, 32300	[8712]	0.19	[7909]
155	Pentaphene										
	p/nb	424	[6410]	48.4	[5604]			493, 45900	[8449]		

Table 3c Flash Photolysis Parameters

No.	Solv	λ_S^{0-0}/nm	Ref.	E_T/kcal mol^{-1}	Ref.	Φ_T	Ref.	λ_{TT}/nm, ε_T/M^{-1}cm^{-1}	Ref.	τ_T/μs	Ref.
156	**Perylene**										
	n/b	435	[7105]	35.4	[6918]	0.014	[6612]	490, 14300	[7126]		
	p/nb	439	[6801]	36.0	[6801]	0.0088	[6612]	485, 13400	[8712]	5000	[6801]
157	**3,4:9,10-perylenebis(dicarboximide), *N*,*N*'-bis(1-hexylheptyl)-**										
	p/nb	523	[9903]	27.7	[9903]	0.005	[9903]	494, 63000	[9903]	200	[9903]
158	**Phenanthrene**										
	n/b	346	[7105]	62.3	[6306]	0.73	[7501]	492.5, 15700	[7126]	145	[6202]
	p/nb	347	[7915]	61.4	[8905]	0.85	[6802]	482.5, 25200	[7126]	910	[6105]
159	**1,10-Phenanthroline**										
	n/b	342	[8313]					440	[8243]	26	[7718]
	p/nb	339	[6304]	63.2	[6304]			445	[7718]	35	[7718]
160	**Phenazine**										
	n/b			44.5	[6919]	0.21	[7112]	440	[7127]	42	[7112]
	p/nb	438	[7916]	44.6	[8523]	0.45	[8523]	355, 37700	[8712]	770	[8523]
161	**Phenazinium, 3,7-diamino-2,8-dimethyl-5-phenyl-**										
	p/nb					0.50	[8201]	420, 10000	[8712]	67	[6724]
162	**Phenazinium, 3,7-diamino-5-phenyl-**										
	p/nb					0.10	[8906]	440, 29000	[8906]	25	[8906]

Table 3c Flash Photolysis Parameters

No.	Solv	λ_S^{0-0}/nm	Ref.	E_T/kcal mol^{-1}	Ref.	Φ_T	Ref.	λ_T/nm, ε_T/M^{-1} cm^{-1}	Ref.	τ_T/μs	Ref.
163	**Phenol**										
	n	278	[7105]			0.32	[8207]				
	p/nb	283	[8427]	81.7	[8427]			250	[7505]	3.3	[7505]
164	**Phenothiazine**										
	n			60.4	[7513]						
	p/nb	370	[7105]			0.54	[8314]	460, 27000	[8712]		
165	**Phenoxazine**										
	n/b	362	[7214]	62.6	[7214]			465	[7018]	32	[7214]
	p/nb	366	[7214]	62.2	[7214]			465	[7018]	44	[7018]
166	**Phenoxazinium, 3,7-diamino-, conjugate monoacid**										
	p/nb					≤0.003	[7618]	650, 16000	[7619]	55	[7619]
167	***p*-Phenylenediamine, *N,N,N′,N′*-tetramethyl-**										
	n/b	358	[7101]					620, 12200	[8712]	1.4	[8216]
	p/nb	364	[7101]					605, 12200	[7126]	0.5	[8428]
168	**Pheophytin *a***										
	n	670	[7910]	31.0	[7910]						
	p/nb	667	[7019]			0.95	[7019]	407, 62800	[8712]	750	[7019]

Table 3c Flash Photolysis Parameters

No.	Solv	λ_S^{0-0}/nm	Ref.	E_T/kcal mol^{-1}	Ref.	Φ_T	Ref.	λ_T/nm, ε_T/M^{-1}cm^{-1}	Ref.	τ_T/μs	Ref.
169	**Pheophytin b**										
	n	658	[7910]	32.1	[7910]						
	p/nb	654	[7019]			0.75	[7019]	423, 71200	[8712]	1050	[7019]
170	**Phthalazine**										
	n/b			63.1	[8217]	0.29	[8214]	396	[7417]	2.7	[8705]
	p/nb	387	[5905]	65.8	[5905]	0.44	[8705]	421, 4450	[8712]	21.3	[7514]
171	**Phthalocyanine**										
	n	704	[6920]	28.7	[6920]	0.14	[7820]				
	p/nb							480, 29900	[8712]	130	[7820]
172	**Phthalocyanine, magnesium(II)**										
	n	687	[6920]							100	[6506]
	p/nb					0.23	[7020]	480, 32300	[8712]	430	[8613]
173	**Phthalocyanine, zinc(II)**										
	n/b	683	[7113]	26.2	[7113]	0.65	[8119]	480, 51000	[8712]		
	p/nb					0.04	[8613]	480, 28900	[8712]	270	[8613]
174	**Picene**										
	n/b	376	[6410]					560	[7108]	160	[7108]
	p/nb	376	[7701]	57.4	[7701]			630, 45500	[8712]		

Table 3c Flash Photolysis Parameters

No.	Solv	λ_S^{0-0}/nm	Ref.	E_T/kcal mol^{-1}	Ref.	Φ_T	Ref.	λ_T/nm, ε_T/M^{-1} cm^{-1}	Ref.	τ_T/μs	Ref.
175	**Porphyrin**										
	n	615	[7515]	36.0	[7515]	0.90	[8218]				
	p/nb	610	[7406]	36.4	[7406]			419, 98600	[8712]		
176	**Porphyrin, 2,7,12,17-tetraethyl-3,8,13,18-tetramethyl-, zinc(II)**										
	n	573	[6920]								
	p/nb	576	[7114]	40.6	[7114]			440, 99000	[8712]		
177	**Porphyrin, tetrakis(4-sulfonatophenyl)-**										
	p/nb					0.78	[8219]	790, 3400	[8430]	420	[8219]
178	**Porphyrin, tetrakis(4-sulfonatophenyl)-, zinc(II)**										
	p/nb					0.84	[8219]	840, 6000	[8430]	80	[8220]
179	**Porphyrin, tetrakis(4-trimethylammoniophenyl)-**										
	p/nb					0.80	[8315]	800, 3200	[8315]	540	[8315]
180	**Porphyrin, tetrakis(4-trimethylammoniophenyl)-, zinc(II)**										
	p/nb					0.82	[8315]	840, 5000	[8315]	1200	[8315]
181	**Porphyrin, tetraphenyl-**										
	n/b	667	[8221]	33	[8221]	0.82	[8218]	790, 6000	[8430]	1500	[8430]
	p/nb	645	[7406]	33.5	[7406]	0.88	[8316]	785, 6000	[8712]		

Table 3c Flash Photolysis Parameters

No.	Solv	λ_S^{0-0}/nm	Ref.	E_T/kcal mol^{-1}	Ref.	Φ_T	Ref.	λ_T/nm, ε_T/M^{-1}cm^{-1}	Ref.	τ_T/μs	Ref.
182	**Porphyrin, tetraphenyl-, magnesium(II)**										
	n	611	[8221]	34.1						1400	[8121]
	p/nb	620	[7114]	34.2		0.85	[7114]	485, 72000	[8712]		
183	**Porphyrin, tetraphenyl-, zinc(II)**										
	n/b	605	[8221]	36.6	[7516]	0.88	[8218]	845, 8200	[8430]	1200	[8121]
	p/nb	602	[7114]	36.6	[7114]	0.9	[8316]	470, 71000	[8121]		
184	**Porphyrin, zinc(II)**										
	p/nb	569	[7114]	39.6	[7114]			840, 7000	[8030]		
185	**Porphyrin-2,18-dipropanoic acid, 7,12-diethenyl-3,8,13,17-tetramethyl-, dimethyl ester**										
	n/b					0.8	[8218]	710, 9000	[7719]	550	[8022]
	p	627	[7406]	36.0	[7406]						
186	**Porphyrin-2,18-dipropanoic acid, 3,7,12,17-tetramethyl-, dimethyl ester**										
	n					0.63	[8021]			210	[8021]
	p/nb							440, 23900	[8712]		
187	**Porphycene**										
	n/b					0.42	[8615]	380, 66000	[8615]	200	[8615]
188	**Psoralen**										
	n/b					0.034	[7805]	450, 8100	[7805]		

Table 3c Flash Photolysis Parameters

No.	Solv	λ_S^{0-0}/nm	Ref.	E_T/kcal mol^{-1}	Ref.	Φ_T	Ref.	λ_T/nm, ε_T/M^{-1} cm^{-1}	Ref.	τ_T/μs	Ref.
	p/nb	365	[7303]	62.7	[7303]	0.06	[7917]	450, 11200	[8712]	5	[7917]
189	**Psoralen, 5-methoxy-**										
	n/b					0.067	[7805]	450, 10200	[7805]		
	p/nb	375	[7303]	60.7	[7303]	0.1	[8317]	450, 9450	[8712]	4.2	[8317]
190	**Psoralen, 8-methoxy-**										
	n/b					0.011	[7805]	480, 10000	[7805]	1.1	[7805]
	p/nb	387	[7303]	62.6	[7303]	0.04	[8317]	370, 17700	[8712]	10	[7917]
191	**Psoralen, 4,5',8-trimethyl-**										
	p/nb	372	[7303]	64.0	[7303]	0.093	[7918]	470, 25700	[8712]	7.1	[7918]
192	**Purine**										
	n			75.6	[7506]						
	p/nb	286	[7105]					390, 4100	[8712]		
193	**Pyranthrene**										
	n					0.55	[8310]				
	p/nb					0.52	[8310]	500, 20600	[8712]		
194	**Pyrazine**										
	n	327	[8318]	75.4	[7822]	0.33	[6716]				

Table 3c Flash Photolysis Parameters

No.	Solv	$\lambda_S^{0\text{-}0}$/nm	Ref.	E_T/kcal mol^{-1}	Ref.	Φ_T	Ref.	λ_T/nm, ε_T/M^{-1}cm^{-1}	Ref.	τ_T/µs	Ref.
	p/nb			74.3	[5803]	0.87	[7514]	260, 3600	[8712]	4.5	[7514]
195	**Pyrene**										
	n/b	372	[7105]	48.4	[5701]	0.37	[8223]	420, 20900	[7126]	180	[7010]
	p/nb	373	[7808]	48.2	[7808]	0.38	[6802]	412.5, 30400	[7126]	11000	[6801]
196	**1-Pyrenecarboxaldehyde**										
	n/b			43	[8319]	0.78	[8319]	440, 20100	[8319]	50	[8319]
	p/nb					0.65	[8319]	440, 18400	[8712]	38	[8319]
197	**Pyrylium, 2,6-dimethyl-4-phenyl-, perchlorate**										
	p/nb	402	[9701]	63.1		0.5	[9701]	440, 27000	[9701]	0.3	[9701]
198	**Pyrylium, 2,6-dimethyl-4-(4-methylphenyl)-, perchlorate**										
	p/nb	400	[9701]	61.1		0.2	[9701]	430, 52000	[9701]	2.4	[9701]
199	**Pyrylium, 4-[1,1'-biphenyl]-4-yl-2,6-dimethyl-, perchlorate**										
	p/nb	450	[9905]			0.03	[9905]	500, 40000	[9905]	2.2	[9905]
200	**Pyrylium, 2,6-dimethyl-4-(1-naphthalenyl)-, perchlorate**										
	p/nb	485	[9905]			0.13	[9905]	500, 55000	[9905]	6.1	[9905]
201	**Pyrylium, 2,6-dimethyl-4-(2-naphthalenyl)-, perchlorate**										
	p/nb	480	[9905]			0.09	[9905]	520, 15000	[9905]	3.5	[9905]

Table 3c Flash Photolysis Parameters

No.	Solv	λ_S^{0-0}/nm	Ref.	E_T/kcal mol^{-1}	Ref.	Φ_T	Ref.	λ_T/nm, ε_T/M^{-1}cm^{-1}	Ref.	τ_T/μs	Ref.
202	**Pyrylium, 4-(4-acetylphenyl)-2,6-dimethyl-, perchlorate**										
	p/nb	367	[9905]			0.66	[9905]	460, 27000	[9905]	8.3	[9905]
203	**2,2':5',2'':5'',2'''-Quaterfuran**										
	p/nb	376	[0003]			0.29	[0003]	530, 12000	[0003]	14	[0003]
	n/b	387	[0003]			0.33	[0003]			67	[0003]
204	**α-Quaterthiophene**										
	n/b					0.73	[9603]	560	[9603]	38	[9603]
	p/nb					0.71	[9603]			40	[9603]
205	**Quinoline**										
	n	313	[8025]	61.7	[8025]	0.31	[6503]		[8712]		
	p/nb	314	[8025]	62.5	[8025]			425, 6750			
206	**Quinoline, 2-[(1E)-2-(9-anthracenyl)ethenyl]-**										
	n	441	[0201]			0.13	[0201]	502	[0201]	38	[0201]
	p/nb	450	[0201]			0.014	[0201]	499	[0201]	28	[0201]
207	**Quinoline, 3-[(1E)-2-(9-anthracenyl)ethenyl]-**										
	n	437	[0201]			0.24	[0201]	495	[0201]	40	[0201]
	p/nb	440	[0201]			0.16	[0201]	484	[0201]	33	[0201]

Table 3c Flash Photolysis Parameters

No.	Solv	λ_S^{0-0}/nm	Ref.	E_T/kcal mol^{-1}	Ref.	Φ_T	Ref.	λ_T/nm, ε_T/M^{-1} cm^{-1}	Ref.	τ_T/µs	Ref.
208	**Quinoline, 4-[(1E)-2-(9-anthracenyl)ethenyl]-**										
	n	443	[0201]			0.14	[0201]	493	[0201]	44	[0201]
	p/nb	459	[0201]			0.009	[0201]				
208	**Quinoxaline**										
	n	381	[8225]	61.0	[8225]	0.99	[8214]				
	p/nb	375	[5905]	60.8	[5905]	0.90	[8502]	425, 8100	[7024]	29.4	[7514]
209	**(all-E)-Retinal**										
	n/b	426	[7115]	29.4	[7115]	0.43^w	[8417]	450, 58400	[8712]	9.3	[8226]
	p/nb					0.12^w	[7824]	450, 69300	[8712]	18	[6206]
210	**(all-E)-Retinol**										
	n	373	[6921]	33.5	[8417]	0.017	[7710]	405, 80000	[7710]	25	[7116]
	p/nb	366	[6921]			~0.003	[8524]				
211	**Rhodamine, inner salt, N,N'-diethyl**										
	p/nb	534	[7720]	43.9	[7720]	0.005	[7722]	615, 12000	[7833]		
212	**Rhodamine 6G cation**										
	p/nb	546	[7720]	43.2	[7720]	0.0021	[7408]	620, 16000	[8712]	3500	[7408]
213	**Riboflavin, conjugate monoacid**										
	p/nb	455	[7105]			0.40	[7706]	415, 7560	[8712]	19	[7706]

Table 3c Flash Photolysis Parameters

No.	Solv	λ_S^{0-0}/nm	Ref.	E_T/kcal mol^{-1}	Ref.	Φ_T	Ref.	λ_T/nm, ε_T/M^{-1} cm^{-1}	Ref.	τ_T/μs	Ref.
214	**Rubrene**										
	n/b	542	[8123]	26.3	[8124]	0.0092	[8617]	495, 26000	[8123]	120	[8617]
	p/nb					0.023	[8902]	495, 26500	[8712]	80	[6820]
215	**2H-seleno[3,2-g]-1-benzopyran-2-one**										
	n/b					0.42	[0010]	560, 16600	[0010]	2	[0010]
	p/nb					1	[0010]	470, 6000	[0010]	6	[0010]
216	**Squaraine, bis[4-(dimethylamino)-phenyl]-**										
	p/nb	641	[9208]					540, 30500	[9208]	35.2	[9208]
217	**(E)-Stilbene**										
	n/b	334	[6207]	49.3	[8015]			360	[7224]	14	[6821]
	p/nb			49.3	[8015]			378, 34000	[8712]	62	[8125]
218	**Styrene**										
	n	288	[7723]	61.8	[7723]	0.4	[8228]	325, 2210	[8712]	0.025	[8228]
	p/nb										
219	**2,2':5',2''-Terfuran**										
	p/nb	340	[0003]			0.31	[0003]	470, 9000	[0003]	13	[0003]
	n/b	350	[0003]			0.62	[0003]			34	[0003]

Table 3c Flash Photolysis Parameters

No.	Solv	λ_S^{0-0}/nm	Ref.	E_T/kcal mol^{-1}	Ref.	Φ_T	Ref.	λ_T/nm, ε_T/M^{-1}cm^{-1}	Ref.	τ_T/μs	Ref.
220	**p-Terphenyl**										
	n/b	310	[7101]			0.11	[6905]	460, 90000	[7126]	450	[6905]
	p/nb			58.3	[6701]			460, 72700	[8712]		
221	**α-Terthiophene**										
	n/b					0.95	[9603]	460	[9603]	88	[9603]
	p/nb					0.9	[9603]			62	[9603]
222	**Tetracene**										
	n/b	472	[7105]	29.3	[6412]	0.62	[7501]	465, 31200	[7126]	400	[8505]
	p/nb	471	[7701]			0.66	[7117]	465, 57900	[8712]		
223	**1,4-Tetracenequinone**										
	n	460	[0109]	42.7	[0109]	0.78	[0109]	440, 19500	[0109]	480	[0109]
224	**Thiobenzophenone**										
	n/b	627	[7519]	39.3	[7519]	1.0^w	[8434]	400, 4800	[8434]	1.7	[8434]
	p/nb	626	[7215]	40.6	[7215]			400, 4950	[8712]		
225	**Thiobenzophenone, 4,4'-bis(dimethylamino)-**										
	n/b	583	[7215]	42.2	[7215]	1.0	[8434]	335, 14400	[8434]	1.3	[8434]
	p/nb							335, 14900	[8712]		

Table 3c Flash Photolysis Parameters

No.	Solv	λ_S^{0-0}/nm	Ref.	E_T/kcal mol^{-1}	Ref.	Φ_T	Ref.	λ_T/nm, ε_T/M^{-1} cm^{-1}	Ref.	τ_T/μs	Ref.
226	Thiobenzophenone, 4,4'-dimethoxy-										
	n			41.2		1.0^w	[8434]			1.4	[8434]
	p/nb	606	[7215]	42.1	[7215]			295, 20500	[8712]		
227	Thionine cation										
	p/nb	612	[6923]	39	[6923]	0.62	[6915]	770, 10900	[8712]	72	[6923]
228	Thionine cation, conjugate monoacid										
	p/nb							650, 18000	[8712]	16	[7726]
229	7H-thiopyran[3,2-g]-1-benzofuran-7-one										
	n/b					0.37	[0010]	510, 13100	[0010]	2	[0010]
	p/nb					0.6	[0010]	560, 7500	[0010]	6	[0010]
230	Thiopyrylium, 2,6-dimethyl-4-phenyl-, perchlorate										
	p/nb	429	[9701]	64.3	[9701]	0.4	[9701]	460, 22000	[9701]	0.3	[9701]
231	Thiopyrylium, 2,6-dimethyl-4-(4-methylphenyl)-, perchlorate										
	p/nb	429	[9701]	64.5	[9701]	0.3	[9701]	470, 14000	[9701]	3	[9701]
232	Thioxanthen-9-one										
	n/b			63.3	[8620]			650, 30000	[7935]	95	[8126]
	p/nb							650, 26200	[8712]	73	[7318]

Table 3c Flash Photolysis Parameters

No.	Solv	λ_S^{0-0}/nm	Ref.	E_T/kcal mol^{-1}	Ref.	Φ_T	Ref.	λ_T/nm, ε_T/M^{-1}cm^{-1}	Ref.	τ_T/μs	Ref.
233	**Thioxanthione**										
	n/b			39.6	[8230]	1.0[w]	[8434]	505, 2500	[8434]	0.83	[8434]
	p/nb							505, 2580	[8712]		
234	**Thymidine**										
	n	284	[8027]								
	p/nb	287	[8027]			0.069	[7912]	370, 2320	[8712]	25	[7912]
235	**Thymine**										
	n	283	[8027]								
	p/nb	283	[8027]			0.06	[7520]	340, 4000	[7520]	10	[7520]
236	**1,3,5-Triazine**										
	n	327	[7105]								
	p/nb			75.5	[6107]			245, 6000	[8712]	0.91	[7524]
237	**Triphenylene**										
	n/b	343	[7101]			0.86	[6901]	430, 6190	[8712]	55	[6105]
	p/nb	340	[7701]	67.0	[8436]	0.89	[6601]	430, 13500	[8712]	1000	[6105]
238	**Tryptophan**										
	p/nb	300	[6208]			0.18	[7727]	460, 5000	[8712]	14.3	[7511]

Table 3c Flash Photolysis Parameters

No.	Solv	λ_S^{0-0}/nm	Ref.	E_T/kcal mol^{-1}	Ref.	Φ_T	Ref.	λ_T/nm, ε_T/M^{-1} cm^{-1}	Ref.	τ_T/μs	Ref.
239	**Uracil**										
	n	278	[8027]								
	p/nb	278	[8027]			0.1	[7912]	350, 1730	[8712]	2	[7520]
240	**Uracil, 1,3-dimethyl-**										
	n	283	[8027]								
	p/nb	283	[8027]			0.02	[8127]	380, 8000	[8712]		
241	**Uridine**										
	p/nb					0.078	[7912]	370, 4130	[8712]	20	[7912]
242	**Uridine 5′-monophosphate**										
	p/nb	287	[6706]	78.3	[7320]	0.044	[7912]	390, 5810	[8712]	33	[7912]
243	**9-Xanthione**										
	n/b	621	[8128]	43.3	[8128]	0.8^w	[8434]	345, 15400	[8434]	1.8	[8434]
	p/nb							345, 15900	[8712]		
244	**Xanthone**										
	n/b	370	[7118]	74.1	[7118]			610, 5300	[7622]	0.02	[7622]
	p/nb			74.1	[7118]			605, 6480	[8712]	17.9	[7622]

^{s} Very solvent-dependent; ^{w} Wavelength dependent.

3d LOW-TEMPERATURE PHOTOPHYSICAL PARAMETERS

The same excited states that dominate the photophysics of organic molecules at room temperature are also dominant at low temperature. However, by immobilizing a molecule in glasses, crystalline media, or polymer matrices, the radiative processes of the triplet state (phosphorescence) can not only compete with nonradiative decay channels but also with deactivation due to extrinsic agents such as impurities, residual oxygen, etc. Some information from low temperature that is relevant for room temperature photochemical experiments is listed below.

First, as discussed in the introduction to Table 3b, triplet energies are routinely estimated from phosphorescence spectra. Triplet energies have several uses. One of the main uses is for assessing potential photosensitizers and quenchers. Triplet energies, whether from low temperature or not, are tabulated in Tables 3a and 3b.

Second, when no triplet quantum yields, Φ_T, measurements have been done, the phosphorescence quantum yield, Φ_{ph}, gives a lower bound to the Φ_T because

$$\Phi_{ph} = \Phi_T k_r^T \tau_{ph} \qquad (3\text{-}5)$$

where k_r^T is the natural radiative rate constant of the triplet state and τ_{ph} is the measured phosphorescence lifetime given by

$$\frac{1}{\tau_{ph}} = k_r^T + k_{isc}^T + k_{pc}^T \qquad (3\text{-}6)$$

corresponding to all decay channels out of the lowest triplet state. See Chapter 1 for a definition of the rate constants in terms of the transitions underlying them.

Third, the nature of the triplet can often be guessed from the natural lifetime of the phosphorescence. For a rule-of-thumb $n\pi^*$ triplets have natural phosphorescence lifetimes $(1/k_r^T)$ that are 0.1 to 0.001 times less than those of $\pi\pi^*$ triplets. The natural phosphorescence lifetimes can be computed from Eq. 3-5 when the quantities, Φ_T, Φ_{ph}, and τ_{ph} are known. These latter three quantities are among the five parameters presented in Table 3d.

For completeness, low-temperature fluorescence data is also presented in Table 3d. Singlet states, which are not easy to study at room temperature because of temperature activated radiationless transitions, can often be investigated at low temperature. Even excited singlet states which fluoresce at room temperature can often be analyzed more conveniently at low temperature where the fluorescence lifetimes are usually longer and thus easier to measure.

Table 3d Low Temperature Data

No.	Solv.	Φ_T	Ref.	τ_{fl}/ns	Ref.	Φ_h	Ref.	τ_{ph}/s	Ref.	Φ_{ph}	Ref.	E_T/kcal mol⁻¹	Ref.
1 Acenaphthene													
	n												
	p	0.23	[7326]									20872	[6401]
2 Acetanilide													
	p					0.57	[7936]	2.6	[7418]	0.05	[7936]	20700	[6701]
3 Acetone													
	n							3.6	[6210]	0.05	[6210]		
	p							6×10^{-4}	[4902]	0.042	[7631]	27800	[6602]
4 Acetophenone													
	n	0.90	[8331]					2.1×10^{-3}	[7220]	0.65	[8331]	25933	[8101]
	p					0	[6211]	5.0×10^{-3}	[8101]	0.6	[8331]	26034	[8101]
5 Acetophenone, 4'-amino-													
	n												
	p					0.11	[7834]	0.72	[7834]	0.14	[7834]	23200	[6809]
6 Acetophenone, 4'-hydroxy-													
	n											24700	[6822]
	p							0.26	[7025]	0.55	[7025]	25300	[6604]
7 Acetophenone, 4'-methoxy-													
	n							0.35	[6703]			25100	[6702]
	p							0.26		0.72	[7025]	25000	[6703]

Table 3d Low Temperature Data

No.	Solv.	Φ_T	Ref.	τ_{fl}/ns	Ref.	Φ_{fl}	Ref.	τ_{ph}/s	Ref.	Φ_{ph}	Ref.	E_T/ kcal mol^{-1}	Ref.
8	**Acetophenone, 4'-methyl-**												
	n							0.027	[6702]			25500	[7220]
	p							0.084	[6703]	0.61	[6703]	25500	[6703]
9	**Acridan**												
	p			12	[7302]	0.38	[7302]	3.4	[7302]	0.51	[7302]	24330	[7302]
10	**Acridine**												
	n	0.15	[7525]									15870	[8102]
	p							0.155	[7701]			15740	[8403]
11	**9-Acridinone**												
	n											20370	[7803]
	p			12.9	[7804]	0.78	[7804]	2.50	[7804]	0.014	[7804]	21050	[7804]
12	**9-Acridinone, 10-phenyl-**												
	n											20430	[7803]
	p			7.3	[7804]	0.78	[7804]	2.58	[7804]	0.021	[7804]	21220	[7803]
13	**Adenine**												
	p	0.17	[8628]			0.008	[9001]	2.0	[9001]	0.011	[9001]		
14	**Adenine, conjugate acid**												
	p					0.010	[9001]	3.35	[6606]	0.038	[9001]		

269

Table 3d Low Temperature Data

No.	Solv.	Φ_T	Ref.	τ_{fl}/ns	Ref.	Φ_{fl}	Ref.	τ_{ph}/s	Ref.	Φ_{ph}	Ref.	E_T/ kcal mol^{-1}	Ref.
15	**Adenine ion(1−)**												
	p					0.003	[9001]	2.8	[9001]	0.004	[6606]		
16	**Adenosine**												
	p					0.005	[9001]	2.8	[9001]	0.035	[9001]		
17	**Adenosine, conjugate acid**												
	p					0.002	[9001]			0.008	[9001]		
18	**Adenosine ion(1−)**												
	p					0.004	[9001]			0.032	[9001]		
19	**Adenosine 5′-monophosphate**												
	p	0.02	[6706]	2.8	[7320]	0.006	[9001]	2.3	[9001]	0.025	[9001]		
20	**Adenosine 5′-monophosphate, conjugate acid**												
	p					<0.001	[6606]			0.005	[9001]		
21	**Adenosine 5′-monophosphate ion(1−)**												
	p					0.008	[9001]			0.028	[9001]		
22	**Aniline**												
	n					0.17	[8004]	4.2	[8004]	0.65	[8004]	24800	[6707]
	p					0.21	[8427]	4.0	[8427]	0.67	[8427]	26800	[4401]

Table 3d Low Temperature Data

No.	Solv.	Φ_T	Ref.	τ_{fl}/ns	Ref.	Φ_{fl}	Ref.	τ_{ph}/s	Ref.	Φ_{ph}	Ref.	E_T/ kcal mol^{-1}	Ref.
23	**Aniline, *N*-phenyl**												
	p	0.87	[7004]	2.6	[7327]	0.12	[7327]	1.9	[7327]	0.74	[7327]	25140	[7302]
24	**Anthracene**												
	n											14870	[5701]
	p			6.5	[6510]	0.34	[6510]	~0.04	[7936]	0.0003	[7936]	14900	[7701]
25	**Anthracene-*d*₁₀**												
	p	0.53	[6827]	5.7	[6510]	0.32	[6510]	0.12	[7936]	0.0007	[7936]		
26	**Antracene, 9,10-diphenyl-**												
	n			7.9	[7603]							14290	[6805]
	p			7.95	[7603]	1.0	[7403]						
27	**Anthracene, 9-(methylsulfinyl)-**												
	p					0.19	[0102]			0.0003	[0102]		
28	**9,10-Anthraquinone**												
	n							0.0033	[8137]			21800	[6404]
	p							0.0033	[8137]	0.29	[8713]	21980	[6710]
29	**Anthrone**												
	p							0.0015	[7006]	0.51	[7006]	25150	[7006]

Table 3d Low Temperature Data

No.	Solv.	Φ_T	Ref.	τ_{fl}/ns	Ref.	Φ_{fl}	Ref.	τ_{ph}/s	Ref.	Φ_{ph}	Ref.	E_T/ kcal mol⁻¹	Ref.
30	**Benzaldehyde**												
	n							0.0015				25200	[7204]
	p					0	[6211]	0.0023	[6702]	0.49	[7006]	24950	[7006]
31	**Benzaldehyde, 4-hydroxy-**												
	p							0.10	[7025]	0.32	[7025]		
32	**Benzaldehyde, 4-methoxy-**												
	n							0.094	[6702]			25100	[6702]
	p							0.10	[7225]	0.32	[7225]	24700	[6925]
33	**Benz[a]anthracene**												
	n											16500	[5401]
	p	0.67	[6808]	60	[6510]			0.50	[7808]			16520	[7808]
34	**Benzene**												
	n					0.17	[7419]	4.50	[7419]	0.15	[7419]	29500	[6702]
	p					0.19	[7225]	6.3	[7225]	0.18	[7225]	29500	[6702]
35	**Benzene, 1,4-dicyano-**												
	n											24560	[6203]
	p					0.42	[8602]	2.1	[7008]	0.25	[8602]	24700	[7008]

Table 3d Low Temperature Data

No.	Solv.	Φ_T	Ref.	τ_{fl}/ns	Ref.	Φ_{fl}	Ref.	τ_{ph}/s	Ref.	Φ_{ph}	Ref.	E_T/ kcal mol^{-1}	Ref.
36	**Benzil**												
	n							0.0051	[6712]			18700	[6712]
	p							0.0056	[6712]	0.67	[6210]	19000	[6712]
37	**Benzo[a]coronene**												
	n											18003	[7810]
	p	0.64	[6808]			0.39	[6912]	5.65^a	[6808]	0.07	[6912]		
38	**Benzofuran**												
	n											25157	[8910]
	p	0.37	[8910]			0.63	[8910]	2.35	[8910]	0.24	[8910]	25130	[8910]
39	**Benzoicacid**												
	n							3.06	[7526]			27110	[6302]
	p							2.1	[7023]	0.27	[6210]	27200	[7606]
40	**Benzoicacid, 4-(dimethylamino)-, ethylester**												
	n			2.3	[8306]	0.58	[8306]	2.3	[8306]	0.26	[8306]	23750	[8306]
41	**Benzoicacid, 4-methyl**												
	n							2.41	[7731]	0.90	[7731]	26880	[7731]
	p							1.7	[7906]			26800	[7606]

273

Table 3d Low Temperature Data

No.	Solv.	Φ_T	Ref.	τ_{fl}/ns	Ref.	Φ_{fl}	Ref.	τ_{ph}/s	Ref.	Φ_{ph}	Ref.	$E_T/\text{kcal mol}^{-1}$	Ref.
42	**Benzonitrile**												
	n											26780	[6203]
	p					0.30		3.10	[8602]	0.43	[8602]	27000	[7606]
43	**Benzonitrile, 4-amino-**												
	n							2.1		0.64	[8306]	24940	[8306]
	p			3.3	[7507]	0.15		2.45	[7507]	0.11	[7507]	24500	[7507]
44	**Benzonitrile, 4-chloro-**												
	n					0.004				0.10	[8531]	25840	[7809]
	p					0.003		0.15	[7906]	0.11	[8531]	26136	[8531]
45	**Benzonitrile, 4-methoxy-**												
	n					0.20		1.9	[8306]	0.73	[8306]	26320	[8306]
	p			7	[7508]	0.22		1.6	[7508]	0.24	[7508]	26300	[7508]
46	**Benzonitrile, 4-methyl-**												
	n							3.4	[8450]			26410	[6203]
	p					0.36		3.80	[8450]	0.33	[8450]	26500	[6203]
47	**Benzo[*ghi*]perylene**												
	p	0.59	[6827]	240	[6907]	0.35	[6907]	0.438	[7701]			16180	[5604]

Table 3d Low Temperature Data

No.	Solv.	Φ_T	Ref.	τ_{fl}/ns	Ref.	Φ_{fl}	Ref.	τ_{ph}/s	Ref.	Φ_{ph}	Ref.	E_T/kcal mol^{-1}	Ref.
48	**Benzo[c]phenanthrene**												
	n			96	[8706]	0.097	[8706]	3.1	[8706]	0.092	[8706]	19840	[5401]
	p							3.34	[7701]			20000	[7701]
49	**Benzo[f][4,7]phenanthroline**												
	n					0.11	[8107]	4.95	[8107]	0.43	[8107]	23700	[8107]
	p					0.15	[8107]	6.3	[8107]	0.50	[8107]	23800	[7509]
50	**Benzo[a]phenazine**												
	n											17512	[8515]
	p					0.063		0.110	[8108]	0.02	[8108]	16900	[8108]
51	**Benzophenone**												
	n							~0.001[ab]	[7835]			24000	[6702]
	p					0	[6211]	0.0060	[6702]	0.84	[6702]	24200	[8007]
52	**Benzophenone, 4,4'-dimethyl-**												
	n							0.0056	[6702]			24100	[6702]
	p							0.0064	[6702]	0.86	[6702]	24300	[8007]
53	**Benzo[g]pteridine-2,4-dione, 3,10-dimethyl-**												
	n			10.12	[7927]	0.62	[7927]	0.168	[7927]	0.0048	[7927]	17350	[7927]

Table 3d Low Temperature Data

No.	Solv.	Φ_T	Ref.	τ_{fl}/ns	Ref.	Φ_{fl}	Ref.	τ_{ph}/s	Ref.	Φ_{ph}	Ref.	E_T/ kcal mol^{-1}	Ref.
54	**Benzo[g]pteridine-2,4-dione, 10-methyl-**												
	n			10.4	[7927]	0.7	[7927]	0.134	[7927]	0.0025	[7927]	17250	[7927]
55	**Benzo[b]thiophene**												
	n											23970	[5901]
	p	0.98	[8910]			0.035	[8516]	0.35	[8516]	0.42	[8910]	24040	[8516]
56	**Biacetyl**												
	p							0.00225	[4902]	0.23	[6309]	19700	[5501]
57	**1,1'-Binaphthyl**												
	p					0.70	[7708]	2.6	[7708]	0.11	[7708]		
58	**Biphenyl**												
	n											22871	[6401]
	p					0.14	[7418]	4.6	[7225]	0.24	[7225]	22900	[7310]
59	**Biphenyl, 4,4'-dichloro-**												
	p							0.28	[7217]	0.63	[7217]	22000	[7217]
60	**1,1'-Biphenyl, 4-(methylsulfinyl)-**												
	p							0.79	[0102]	0.53	[0102]	21430	[0102]

Table 3d Low Temperature Data

No.	Solv.	Φ_T	Ref.	τ_{fl}/ns	Ref.	Φ_{fl}	Ref.	τ_{ph}/s	Ref.	Φ_{ph}	Ref.	E_T/ kcal mol^{-1}	Ref.
61	**1,1′-Biphenyl, 4-(methylsulfonyl)-**												
	p					0.04	[0102]	>1	[0102]	0.41	[0102]	22380	[0102]
62	**1,1′-Biphenyl, 4-(methylthio)-**												
	p					0.08	[0102]	>1	[0102]	0.59	[0102]	21335	[0102]
63	**C$_{70}$**												
	n							53000^b	[9106]	0.0013^b	[9106]	12300	[9106]
64	**Carbazole**												
	p	0.23	[7302]	15	[7326]	0.44	[7302]	6.8	[7302]	0.24	[7808]	24540	[5801]
65	**Carbazole, 3-chloro-**												
	p	0.94	[0204]			0.059	[0204]	1.85	[0204]	0.66	[0204]	23955	[0204]
66	**Carbazole, 1,6-dichloro-**												
	p	0.97	[0204]			0.026	[0204]	0.49	[0204]	0.7	[0204]	24115	[0204]
67	**Carbazole, 1,3,6-trichloro-**												
	p	0.99	[0204]			0.013	[0204]	0.25	[0204]	0.73	[0204]	23710	[0204]
68	**Carbazole, 1,3,6,8-tetrachloro-**												
	p	1	[0204]			>0.001	[0204]	0.03	[0204]	0.73	[0204]	23790	[0204]
69	**Carbazol, 2-hydroxy-**												
	p	0.54	[0204]			0.46	[0204]	4.19	[0204]	0.22	[0204]	23710	[0204]

Table 3d Low Temperature Data

No.	Solv.	Φ_T	Ref.	τ_{fl}/ns	Ref.	Φ_{fl}	Ref.	τ_{ph}/s	Ref.	Φ_{ph}	Ref.	E_T/ kcal mol^{-1}	Ref.
70	**Carbazole, 3-nitro-**												
	p	1	[0204]					0.16		0.35	[0204]	21130	[0204]
71	**Carbazole, N-methyl-**												
	n					0.5	[8111]	7	[8111]	0.1	[8111]	24450	[8111]
	p							8	[7609]			24450	[6812]
72	**Carbazole, N-phenyl-**												
	p	0.62	[0204]			0.34	[0204]	7.57	[0204]	0.35	[0204]	24520	[0204]
73	**Carbazole, N-acetoxy-**												
	p	0.99	[0204]			0.003	[0204]	6.93	[0204]	0.4	[0204]	24115	[0204]
74	**Chlorophyll *a***												
	n	0.65	[8031]			0.35	[8031]	0.002^a	[8244]				
	p					0.54	[7910]	0.0014	[7910]	1×10^{-5}	[7910]	10400	[7910]
75	**Chlorophyll *b***												
	n					0.14	[7910]	0.0028	[7910]	3×10^{-5}	[7910]	10900	[7910]
	p											11400	[6801]
76	**Chrysene**												
	n							2.2	[6310]				

Table 3d Low Temperature Data

No.	Solv.	Φ_T	Ref.	τ_{fl}/ns	Ref.	Φ_{fl}	Ref.	τ_{ph}/s	Ref.	Φ_{ph}	Ref.	E_T/kcal mol^{-1}	Ref.
	p	0.7	[7525]	55	[7418]	0.23	[7225]	2.54	[7701]	0.05	[7418]	20000	[7701]
77	**Chrysene-d_{12}**												
	p			56	[7418]	0.23	[7418]	13.4	[7418]	0.2	[7418]		
78	**Coronene**												
	n							8.45	[6310]				
	p	0.68	[6912]	320	[6912]	0.27	[6912]	9.5	[7701]	0.12	[6912]	19040	[5604]
79	**Coronene-d_{12}**												
	p	0.67	[6912]	355	[6912]	0.28	[6912]	35	[6912]	0.4	[6912]		
80	**Coumarin**												
	n											21600	[7711]
	p					0.009	[7303]	0.45	[7303]	0.055	[7303]	21840	[7303]
81	**Cytidine**												
	p					0.002	[9001]	0.66	[9001]	0.005	[9001]		
82	**Cytidine 5′-monophosphate**												
	p	0.03	[7320]			0.018	[9001]	0.4	[9001]	0.015	[9001]	27900	[6706]
83	**Cytidine 5′-monophosphate, conjugate acid**												
	p					0.04	[9001]			0.005	[9001]		

Table 3d Low Temperature Data

No.	Solv.	Φ_T	Ref.	τ_{fl}/ns	Ref.	Φ_{fl}	Ref.	τ_{ph}/s	Ref.	Φ_{ph}	Ref.	E_T/kcal mol^{-1}	Ref.
84	Cytidine 5'-monophosphate ion (1−)												
	p					0.005	[9001]			0.018	[9001]		
85	Cytosine												
	p					0.002	[9001]	0.8	[9001]	0.006	[9001]		
86	Cytosineion (1−)												
	p					0.1	[6606]			0.03	[6606]		
87	Deoxythymidine 5'-monophosphate												
	p	<0.003	[8245]	3.2	[7320]	0.16	[7320]	0.35	[6725]	~0	[7320]	26300	[6706]
88	2'-Deoxyuridine												
	p					4×10^{-4}	[9001]			0.002	[9001]		
89	Dibenz[*a,h*]acridine												
	n			7	[7014]			0.84	[7014]			18800	[7014]
	p							2.31	[7701]			19100	[7701]
90	Dibenz[*a,j*]acridine												
	n			15	[7014]			1.6	[7014]				
	p							1.04	[7701]			18600	[7701]

Table 3d Low Temperature Data

No.	Solv.	Φ_T	Ref.	τ_{fl}/ns	Ref.	Φ_{fl}	Ref.	τ_{ph}/s	Ref.	Φ_{ph}	Ref.	E_T/kcal mol^{-1}	Ref.
91	**Dibenz[a,h]anthracene**												
	p	0.98	[6827]	45.5	[7418]	0.19	[7701]	1.6	[7418]	0.021	[7701]	18200	[7701]
92	**Dibenzofuran**												
	n					0.31	[8111]	4.1	[8111]	0.31	[8111]	24000	[8111]
	p	0.6	[8910]			0.4	[8910]	5.6	[8910]			24515	[5801]
93	**Dibenzothiophene**												
	n					0.027	[8111]	1.3	[8111]	0.97	[8111]	23830	[8111]
	p	0.97	[8910]			0.025	[8910]	1.5	[8910]	0.47	[7610]	24100	[7610]
94	**Diphenylmethane**												
	n			38	[7128]	0.18	[7128]	7.8	[7128]	0.41	[7128]		
95	**Ethene, tetraphenyl-**												
	n					0.9	[8213]					17500	[8213]
	p	<0.01	[8213]			0.95	[8213]						
96	**Fluoranthene**												
	n							0.84	[6310]			18450	[5701]
	p			58	[7632]			0.99	[7701]			18500	[7701]
97	**Fluorene**												
	n					0.83	[7226]	5.1	[6310]			23601	[6401]

Table 3d Low Temperature Data

No.	Solv.	Φ_T	Ref.	τ_{fl}/ns	Ref.	Φ_{fl}	Ref.	τ_{ph}/s	Ref.	Φ_{ph}	Ref.	E_T/kcal mol^{-1}	Ref.
98 9-Fluorenone													
	p					0.8	[7226]	5.7	[6310]	0.07	[6210]	23700	[7701]
99 Fluoresceindianion													
	p	0.0078	[8031]	20	[7006]	0.09	[7006]	0.0019	[7815]	3×10^{-5}	[7815]	17600	[7815]
100 Fluoresceindianion, 2′,4′,5′,7′-tetrabromo-													
	p	0.022	[6104]			0.97	[8031]	0.3	[7720]	0.0003	[7720]	16500	[6902]
	p							0.0093		0.022	[6104]	14800	[6813]
101 Guanine													
	p					0.06		1.42	[6606]	0.07	[6606]		
102 Guanine ion (1–)													
	p					0.2		1.35	[6606]	0.06	[6606]		
103 Guanosine													
	p					0.006		1.2	[9001]	0.042	[9001]		
104 Guanosine, conjugate acid													
	p					0.5	[9001]			<0.03	[9001]		
105 Guanosine ion (1–)													
	p					0.04	[9001]	1.3	[9001]	0.025	[9001]		

Table 3d Low Temperature Data

No.	Solv.	Φ_T	Ref.	τ_{fl}/ns	Ref.	Φ_{fl}	Ref.	τ_{ph}/s	Ref.	Φ_{ph}	Ref.	E_T/kcal mol^{-1}	Ref.
106	**Guanosine 5′-monophosphate**												
	p	0.15	[6706]			0.028	[9001]	1.2	[9001]	0.095	[9001]	27200	[6706]
107	**Guanosine 5′-monophosphate, conjugate acid**												
	p					0.04	[9001]			<0.02	[9001]		
108	**Guanosine 5′-monophosphate ion (1–)**												
	p					0.04	[9001]	1.25	[6606]	0.06	[9001]		
109	**Indan**												
	n							7.6	[7416]			28750	[7416]
	p					0.43	[7411]	8	[7411]	0.16	[7411]		
110	**Indole**												
	n	0.34	[7525]									25200	[8418]
	p					0.6	[8418]	2.4	[6917]			24800	[8418]
111	**Isoorotic acid**												
	n			2.4	[8335]	0.44				<0.0005	[8335]		
	p					0.04	[9001]	~0.2	[9001]	0.04	[9001]		
112	**Isoquinoline**												
	n	0.19	[7129]					0.71	[7734]			21210	[5901]
	p	0.24	[7129]	9	[7613]	0.61	[7613]	1	[7613]	0.038	[7613]	21200	[5905]

Table 3d Low Temperature Data

No.	Solv.	Φ_T	Ref.	τ_{fl}/ns	Ref.	Φ_{fl}	Ref.	τ_{ph}/s	Ref.	Φ_{ph}	Ref.	E_T/kcal mol^{-1}	Ref.
113	**Naphthalene**												
	n			262	[8451]	0.45	[7418]	2.3	[7418]	0.033	[7418]	21180	[5701]
	p	0.29	[7026]	273	[6815]	0.45	[7633]	2.6	[7936]	0.039	[7633]	21300	[4401]
114	**Naphthalene-d_8**												
	n			304	[7027]	0.5	[7418]	18.5	[7418]	0.16	[7418]		
	p	0.25	[6932]	250	[7027]	0.41	[7633]	20.4	[7633]	0.34	[7633]		
115	**Naphthalene, 1-amino-**												
	n			10.9	[7016]								
	p			15.1	[7016]			1.5	[4902]			19150	[7614]
116	**Naphthalene, 2-amino-**												
	n			20.6	[7016]								
	p			20.5	[7016]	0.08		1.2	[8138]	0.0024	[8246]	19960	[7614]
117	**Naphthalene, 1-chloro-**												
	n											20490	[7819]
	p			12.9	[6815]	0.03	[6211]	0.305	[8321]	0.16	[6211]	20700	[4401]
118	**Naphthalene, 1,4-dichloro**												
	p					0.04	[7633]	0.11	[7633]	0.25	[7633]		

Table 3d Low Temperature Data

No.	Solv.	Φ_T	Ref.	τ_{fl}/ns	Ref.	Φ_{fl}	Ref.	τ_{ph}/s	Ref.	Φ_{ph}	Ref.	E_T/kcal mol^{-1}	Ref.
119	**Naphthalene, 1-(methylsulfinyl)-**												
	p					0.031	[0102]	>1	[0102]	0.21	[0102]	20985	[0102]
120	**Naphthalene, 2-(methylsulfinyl)-**												
	p					0.044	[0102]	>1	[0102]	0.22	[0102]	21330	[0102]
121	**Naphthalene, 2-(methylsulfonyl)-**												
	p							>1	[0102]			20635	[0102]
122	**Naphthalene, 2-(methylthio)-**												
	p					0.027	[0102]	>1	[0102]	0.6	[0102]	20285	[0102]
123	**Orotic acid**												
	n					0.002	[8335]	0.11	[8335]	0.017	[8335]		
	p					<0.008	[9001]	0.3	[9001]				
124	**Phenanthrene**												
	n											21774	[6306]
	p	0.46	[7326]	63	[7418]	0.2	[7936]	3.6	[7808]	0.16	[7936]	21500	[8905]
125	**Phenanthrene-d_{10}**												
	p	0.45	[6932]	65	[7418]	0.2	[7936]	15.4	[7936]	0.66	[7936]		
126	**Phenanthrene, 9-(methylsulfinyl)-**												
	p			37	[0102]	0.066	[0102]	>1	[0102]	0.41	[0102]	24130	[0102]

Table 3d Low Temperature Data

No.	Solv.	Φ_T	Ref.	τ_{fl}/ns	Ref.	Φ_{fl}	Ref.	τ_{ph}/s	Ref.	Φ_{ph}	Ref.	E_T/kcal mol^{-1}	Ref.
127	**6-Phenanthridone**												
	p					0.31	[8709]	4.9	[8709]	0.51	[8019]	23920	[8019]
128	**Phenol**												
	n					0.44	[8207]			0	[8207]		
	p					0.45	[8427]	2.5	[8427]	0.4	[8427]	28563	[8427]
129	**Phenothiazine**												
	n											21100	[7513]
	p							0.049		0.34	[8032]		
130	**Phenoxazine**												
	n					0.064	[7214]	2.6	[7214]	0.29	[7214]	21880	[7214]
	p					0.11	[7214]	2.76	[7214]	0.41	[7214]	21750	[7214]
131	**Phenoxazine, 10-phenyl-**												
	n					0.082	[7214]	2.31	[7214]	0.94	[7214]	22150	[7214]
	p					0.12	[7214]	2.65	[7214]	0.72	[7214]	22070	[7214]
132	**Phenylacetic acid**												
	p					0.17	[7411]	5	[7411]	0.35	[7411]	29300	[7023]

Table 3d Low Temperature Data

No.	Solv.	Φ_T	Ref.	τ_{fl}/ns	Ref.	Φ_{fl}	Ref.	τ_{ph}/s	Ref.	Φ_{ph}	Ref.	E_T/ kcal mol⁻¹	Ref.
133	**Phenyl ether**												
	p					0.14	[8427]	1	[8427]	0.67	[8427]	28320	[8427]
134	**Pheophytin *a***												
	n	0.79	[8031]			0.21	[8031]	0.001	[7910]	3×10⁻⁵	[7910]	10800	[7910]
135	**Pheophytin *b***												
	n					0.2	[7910]			5×10⁻⁵	[7910]	11200	[7910]
136	**Phthalazine**												
	n											22100	[8217]
	p	0.49	[7227]			0	[7228]	0.42	[7228]			23000	[5905]
137	**Phthalimide, *N*-methyl-**												
	p							0.75	[9908]	0.6	[9908]		
138	**Phthalocyanine, zinc(II)**												
	n					0.3ᵇ	[7113]	1.1×10⁻³ᵇ	[7113]	1×10⁻⁴	[7113]	9150	[7113]
139	**Picene**												
	p	0.36	[6827]					2.7	[7701]			20100	[7701]
140	**Porphyrin**												
	n					0.055ᵇ	[7515]	0.011ᵇ	[7515]	0.0001ᵇ	[7515]	12600	[7515]
	p							0.009	[7821]	5×10⁻⁵	[7821]	12700	[7406]

Table 3d Low Temperature Data

No.	Solv.	Φ_T	Ref.	τ_{fl}/ns	Ref.	Φ_{fl}	Ref.	τ_{ph}/s	Ref.	Φ_{ph}	Ref.	E_T/kcal mol^{-1}	Ref.
141	**Porphyrin, magnesium(II)**												
	n					0.07	[7515]	0.1	[7515]	0.0006	[7515]	13700	[7515]
	p					0.07	[7515]	0.1	[7515]	0.0006	[7515]	13700	[7515]
142	**Porphyrin, octaethyl-**												
	n					0.16	[7515]	0.016	[7515]	0.0006	[7515]	13000	[7515]
	p											13000	[7406]
143	**Porphyrin, octaethyl-, zinc(II)**												
	p							0.057	[7114]	0.065	[7114]	14200	[7114]
144	**Porphyrin, 2,7,12,17-tetraethyl-3,8,13,18-tetramethyl-**												
	n					0.17[b]	[7515]	0.015[b]	[7515]	0.0006[b]	[7515]	13000	[7515]
	p					0.17	[7515]	0.02	[7515]	0.0003	[7515]	13000	[7515]
145	**Porphyrin, 2,7,12,17-tetraethyl-3,8,13,18-tetramethyl-, magnesium(II)**												
	n					0.25	[7515]			0.006	[7515]		
	p					0.25	[7515]	0.085	[7515]	0.0051	[7515]	13600	[7515]
146	**Porphyrin, 2,7,12,17-tetraethyl-3,8,13,18-tetramethyl-, zinc(II)**												
	p							0.057	[7114]	0.07	[7114]	14200	[7114]

Table 3d Low Temperature Data

No.	Solv.	Φ_T	Ref.	τ_{fl}/ns	Ref.	Φ_{fl}	Ref.	τ_{ph}/s	Ref.	Φ_{ph}	Ref.	E_T/kcal mol^{-1}	Ref.
147	**Porphyrin, tetraphenyl**												
	n							0.006	[8020]	4×10^{-5}	[8020]	11500	[8221]
	p							0.0066	[7406]	2×10^{-5}	[7406]	11700	[7406]
148	**Porphyrin, tetraphenyl-, magnesium(II)**												
	n							0.045	[8121]	0.015	[8121]	11900	[8221]
	p							0.046	[7114]	0.015	[7114]	12000	[7114]
149	**Porphyrin, tetraphenyl-, zinc(II)**												
	n							0.026	[8121]	0.012	[8121]	12800	[7516]
	p							0.025	[7114]	0.014	[7114]	12800	[7114]
150	**Porphyrin, zinc(II)**												
	p							0.0425	[7821]	0.01	[7114]	13900	[7114]
151	**Porphyrin-2,18-dipropanoicacid, 7,12-diethenyl-3,8,13,17-tetramethyl-, dimethyl ester**												
	p							0.0046	[7406]	5×10^{-5}	[7406]	12600	[7406]
152	**Propiophenone**												
	n							0.0063	[7420]			26080	[8610]
	p							0.0038	[7006]	0.7	[7006]	26150	[7006]
153	**Psoralen**												
	p					0.019	[7303]	0.66	[7303]	0.13	[7303]	21930	[7303]

Table 3d Low Temperature Data

No.	Solv.	Φ_T	Ref.	τ_{fl}/ns	Ref.	Φ_{fl}	Ref.	τ_{ph}/s	Ref.	Φ_{ph}	Ref.	E_T/kcal mol^{-1}	Ref.
154	Psoralen, 5-methoxy-												
	p					0.019	[7303]	1.42	[8317]	0.22	[7303]	21200	[8317]
155	Psoralen, 8-methoxy-												
	p					0.013	[7303]	0.72	[7303]	0.17	[7303]	21900	[7303]
156	Purine												
	n	0.51	[7525]									26434	[7506]
	p	0.35	[8247]					3.6	[8033]				
157	Pyrazine												
	n					6×10^{-4}	[6716]	0.0185	[7822]	0.3	[6716]	26361	[7822]
	p							0.02	[4902]			25991	[5803]
158	Pyrene												
	n					0.88	[7418]	0.55	[7418]	0.0021	[7418]	16930	[5701]
	p	0.22	[6933]	515	[6933]	0.92	[7418]	0.58	[7418]	0.0022	[7418]	16850	[7808]
159	Pyrene d_{10}												
	n					0.8	[7418]	3.35	[7418]	0.009	[7418]		
	p	0.15	[6933]	460	[7418]	0.87	[7936]	3.95	[7936]	0.012	[7936]		

Table 3d Low Temperature Data

No.	Solv.	Φ_T	Ref.	τ_{fl}/ns	Ref.	Φ_{fl}	Ref.	τ_{ph}/s	Ref.	Φ_{ph}	Ref.	E_T/ kcal mol^{-1}	Ref.
160	**Pyrene, 1-(methylsulfinyl)-**												
	p			82	[0102]	0.11	[0102]			0.0022	[0102]		
161	**Pyridazine**												
	n	0.66	[8809]			0.01	[6716]			<1×10^{-5}	[6716]	24850	[6716]
	p	0.24	[7227]									24251	[6717]
162	**Pyridine**												
	n					0.005	[8443]	0.7	[8443]	0.004	[8443]	27770	[8443]
	p					0.025	[8443]	2.4	[8443]	0.015	[8443]		
163	**Pyridine, 2-amino-**												
	p							1.9	[7323]	0.068	[7323]		
164	**Pyridine, 4,4'-(1,2-ethenediyl)bis-,(*E*)-**												
	n											17700	[8905]
	p					0.03	[8905]	0.012	[8905]	<0.0001	[8905]	17700	[8905]
165	**Pyridine, 2-hydroxy-**												
	p							0.1	[7732]	0.009	[7732]		
166	**Pyridinium, 2-amino-**												
	p					0.07	[6606]			0.33	[6606]		
167	**Pyrimidine**												

Table 3d Low Temperature Data

No.	Solv.	Φ_T	Ref.	τ_{fl}/ns	Ref.	Φ_{fl}	Ref.	τ_{ph}/s	Ref.	Φ_{ph}	Ref.	E_T/kcal mol^{-1}	Ref.
	n					0.006	[6716]			0.14	[6716]	28214	[7021]
168	Pyrimidine, 2-amino												
	p					0.15	[6606]			0.22	[6606]		
169	4-Pyrimidone												
	p					0.02	[6606]			0.005	[6606]		
170	4-Pyrimidone, conjugate monoacid												
	p					0.01	[6606]			0.01	[6606]		
171	4-Pyrimidone ion (1−)												
	p					0.03	[6606]			0.008	[6606]		
172	Pyronine cation												
	p							1.6	[7720]	4×10^{-5}	[7720]	14900	[7720]
173	Quinazoline												
	n							0.56	[7823]			21925	[7823]
	p							0.68	[7823]			21900	[5905]
174	Quinoline												
	n	0.43	[7129]			$<1\times10^{-5}$	[7734]	1.04	[7734]			21590	[8025]
	p	0.25	[7326]			0.026	[8321]	1.3	[8321]	0.09	[8321]	21850	[8321]

Table 3d Low Temperature Data

No.	Solv.	Φ_T	Ref.	τ_{fl}/ns	Ref.	Φ_{fl}	Ref.	τ_{ph}/s	Ref.	Φ_{ph}	Ref.	E_T/ kcal mol^{-1}	Ref.
175	**Quinoline, 4-chloro-**												
	n							0.3	[8025]			21450	[8025]
	p					0.014	[8321]	0.4	[8025]	0.2	[8321]	21300	[8321]
176	**Quinoxaline**												
	n	0.27	[7129]					0.25	[6828]			21325	[8225]
	p	0.18	[7026]					0.27	[7225]	0.42	[7225]	21250	[5905]
177	**Reserpic acid**												
	p					0.95	[8336]	3.7	[8336]	0.14	[8336]		
178	**Reserpine**												
	p					0.33	[8336]	3.2	[8336]	0.05	[8336]		
179	**(all-E)-Retinoic acid**												
	n			1.7	[8034]	0.44	[8034]						
	p			7.5	[8034]	0.48	[8034]						
180	**Rhodamine, inner salt**												
	p					1	[6508]	2.2	[7720]	6×10^{-5}	[7720]	15930	[7720]
181	**Rhodamine, inner salt, N,N'-diethyl**												
	p	0.0015	[8301]					2.2	[7720]	6×10^{-5}	[7720]	15350	[7720]

Table 3d Low Temperature Data

No.	Solv.	Φ_T	Ref.	τ_{fl}/ns	Ref.	Φ_{fl}	Ref.	τ_{ph}/s	Ref.	Φ_{ph}	Ref.	E_T/kcal mol^{-1}	Ref.
182	**Rhodamine B, inner salt**												
	p	0.003	[8301]			1	[6508]	1.6	[7720]	5×10^{-5}	[7720]	14890	[7720]
183	**Rhodamine 6G cation**												
	p							1.7	[7527]	8×10^{-5}	[7720]	15100	[7720]
184	**Riboflavin**												
	p	0.7	[7735]			0.35	[7735]		[7735]	0.71	[7735]	17500	[6902]
185	**(E)-Stilbene**												
	n							0.1^x	[8535]			17200	[8015]
	p					0.9	[7921]	0.022^a	[7012]			17200	[8015]
186	**(E)-Stilbene, 4-iodo-**												
	n							0.0001	[8905]			16700	[8905]
	p					0.02	[8905]	8×10^{-5}	[8905]	<0.0001	[8905]	16900	[8015]
187	**(E)-Stilbene, 4-nitro-**												
	n					0.004	[8905]	0.009	[8905]	0.0003	[8905]	16300	[8905]
	p					0.003	[8905]	0.01	[8905]	0.0003	[8905]	17400	[4401]
188	**Styrene**												
	n			18.8	[7924]	0.46	[7924]			<0.001	[7924]	21600	[7723]

Table 3d Low Temperature Data

No.	Solv.	Φ_T	Ref.	τ_{fl}/ns	Ref.	Φ_{fl}	Ref.	τ_{ph}/s	Ref.	Φ_{ph}	Ref.	E_T/ kcal mol^{-1}	Ref.
189	*p*-Terphenyl												
	p			1.3	[6510]			2.6	[6414]			20400	[6701]
190	Thieno[2,3-*a*]carbazole, 2,3,5-trimethyl-												
	p					0.058		0.64	[0302]	0.154	[0302]	23256	[0302]
191	Thieno[2,3-*b*]carbazole, 2,3,4,10-tetramethyl-												
	p					0.081		0.45	[0302]	0.073	[0302]	20833	[0302]
192	Thiobenzophenone												
	n							2.5×10^{-5}	[7519]	0.021	[7519]	13760	[7519]
	p							4×10^{-5}	[7215]			14200	[7215]
193	Thiofluorenone												
	n							1.5×10^{-5}	[7519]	0.035	[7519]	13300	[7519]
194	Thioxanthene												
	n							0.025	[7724]	0.46	[7724]		
195	Thioxanthen-9-one												
	n							8.8×10^{-6}	[7528]	0.03	[7528]	22100	[8620]
196	Thymidine												
	n			1	[8027]	0.003	[8027]			0.04	[8027]		
	p					0.033	[9001]	<0.5	[9001]	0.006	[9001]		

Table 3d Low Temperature Data

No.	Solv.	Φ_T	Ref.	τ_{fl}/ns	Ref.	Φ_{fl}	Ref.	τ_{ph}/s	Ref.	Φ_{ph}	Ref.	E_T/ kcal mol^{-1}	Ref.
197	**Thymidine 5′-monophosphate**												
	p					0.14	[9001]	<0.4	[9001]	<0.01	[9001]		
198	**Thymidine 5′-monophosphate,conjugate acid**												
	p					0.11	[9001]			<0.02	[9001]		
199	**Thymidine 5′-monophosphateion (1−)**												
	p					0.1	[9001]			<0.02	[9001]		
200	**Thymine**												
	n			1	[8027]	0.005	[8027]	<0.075	[8027]	0.016	[8027]		
	p			<0.5	[8335]	0.012	[9001]	<0.6	[9001]	0.003	[9001]		
201	**Thymine, 1,3–dimethyl-**												
	n					0.032	[8027]						
	p					0.006	[9001]			0.001	[9001]		
202	**Thymine, 1-methyl**												
	p					0.009	[9001]			0.002	[9001]		
203	**Thymine, 1-methyl-, ion(1−)**												
	p					0.02	[6606]			0.002	[6606]		

Table 3d Low Temperature Data

No.	Solv.	Φ_T	Ref.	τ_{fl}/ns	Ref.	Φ_{fl}	Ref.	τ_{ph}/s	Ref.	Φ_{ph}	Ref.	E_T/ kcal mol^{-1}	Ref.
204	**Toluene**												
	n			53.3	[7128]	0.29	[8204]	7.73	[8204]	0.26	[8204]	28920	[6103]
	p					0.27	[7411]	7.7	[7411]	0.26	[7411]	29000	[6702]
205	**Triphenylene**												
	p	0.54	[6932]			0.07	[7225]	15.2	[7701]	0.41	[7225]	23400	[8436]
206	**Triphenylene-d_{12}**												
	p	0.88	[6932]			0.08	[7225]	21.4	[7225]	0.53	[7225]		
207	**Triphenylethylene**												
	n					0.9	[8213]					17500	[8213]
	p	<0.01	[8213]			0.9	[8213]						
208	**Triphenylmethane**												
	n			35.5	[7128]	0.15	[7128]	7.2	[7128]	0.53	[7128]		
	p							5.1	[6718]	0.011	[6718]		
209	**Uracil**												
	n					<0.0005	[8335]			0.0014	[8027]		
	p					0.0002	[9001]			0.001	[9001]		
210	**Uracil, 1,3-dimethyl-**												
	n					0.0001	[8027]						

Table 3d Low Temperature Data

No.	Solv.	Φ_T	Ref.	τ_{fl}/ns	Ref.	Φ_{fl}	Ref.	τ_{ph}/s	Ref.	Φ_{ph}	Ref.	E_T/kcal mol^{-1}	Ref.
	p					0.0003	[9001]			0.002	[9001]		
211	Uracil, 1,3-dimethyl-5-styryl-, (E)												
	n					0.43	[8714]	0.41	[8714]				
	p					0.67	[8714]	0.38	[8714]				
212	Uracil, 1-methyl-												
	p					0.0002	[9001]			0.001	[9001]		
213	Uracil, 1-methyl-, ion(1−)												
	p					0.01	[6606]			0.002	[6606]		
214	Uracil, 3-methyl-												
	p					0.0002	[9001]			0.001	[9001]		
215	Uracil, 6-methyl-												
	p					0.0002	[9001]			0.001	[9001]		
216	Uracil ion(1−)												
	p					0.01	[6606]			0.0006	[6606]		
217	Uridine												
	p					0.0005	[9001]	<0.5	[9001]	0.002	[9001]		

Table 3d Low Temperature Data

No.	Solv.	Φ_T	Ref.	τ_{fl}/ns	Ref.	Φ_{fl}	Ref.	τ_{ph}/s	Ref.	Φ_{ph}	Ref.	E_T/kcal mol^{-1}	Ref.
218	Uridine ion(1−)												
	p					0.002	[6606]			0.001	[6606]		
219	Uridine 5'-monophosphate												
	p	<0.003	[8245]			<0.002	[9001]	<0.6	[9001]	0.006	[9001]	27400	[7320]
220	Uridine 5'-monophosphate, conjugate acid												
	p					<0.002	[9001]			<0.006	[9001]		
221	Uridine 5'-monophosphate ion(1−)												
	p					<0.002	[9001]			0.006	[9001]		
222	9-Xanthione												
	n							4.3×10^{-5}	[7412]	0.11	[7412]	15143	[8128]
	p							4.8×10^{-5}	[7412]	0.15	[7412]		
223	Xanthone												
	n											25906	[7118]
	p							0.02	[7006]	0.44	[7006]	25905	[7118]
224	*m*-Xylene												
	n					0.28	[7208]			0.25	[7208]	28325	[6103]
	p							8.1	[5102]			28120	[6004]

Table 3d Low Temperature Data

No.	Solv.	Φ_T	Ref.	τ_{fl}/ns	Ref.	Φ_{fl}	Ref.	τ_{ph}/s	Ref.	Φ_{ph}	Ref.	E_T/kcal mol^{-1}	Ref.
225	**o-Xylene**												
	n			62.3	[8204]	0.33		7.46	[8204]	0.45	[8204]	28705	[6103]
	p							5.2	[5103]			28760	[6004]
226	**p-Xylene**												
	n					0.45		4.6	[7208]	0.49	[7208]	28135	[6103]
	p							6.2	[5103]			28145	[6004]
227	**Zinc(II) chlorophyll a**												
	p					0.23		0.001	[7910]	4×10^{-5}	[7910]	11000	[7910]
228	**Zinc(II) chlorophyll b**												
	p					0.08		0.0026	[7910]	0.0002	[7910]	11700	[7910]

[a] Decay of triplet-triplet absorption; [b] Aromatic solvent, benzene-like; [x] Crystalline medium.

3e ABSORPTION AND LUMINESCENCE SPECTRA: A SELECTION

In this section, electronic absorption and emission spectra of common sensitizers and/or quenchers, widely used dyes, and reference compounds for fluorescence and phosphorescence quantum yields are reported. The absorption spectra are presented as *molar absorption coefficients vs* UV, vis or IR wavelength (ε *vs* λ). When opportune, the spectra are on semilog plots to better show the wide range of ε. The emission spectra are in intensity (arbitrary units) *vs* wavelength.

The molar absorption coefficient at a given wavelength is given by

$$\varepsilon(\lambda) = \frac{A(\lambda)}{c \times l} \tag{3-7}$$

where c is the concentration of the absorbing species in mol L^{-1}, l is the optical path in cm, and A is the *absorbance* expressed by

$$A(\lambda) = \log_{10} \frac{P^0(\lambda)}{P(\lambda)} \tag{3-8}$$

where P^0 is the *spectral radiant power* ($P^0 = \int_\lambda I_\lambda d\lambda$) of the incident light beam, essentially monochromatic, and P is the radiant power of the same beam emerging from the sample of pathlength l. In practice, the absorbance is measured by the decadic logarithm of the ratio of the spectral radiant power trasmitted through the reference sample to that trasmitted by the solution of the compound under examination, both observed in identical cells. This definition supposes that all the incident light is only absorbed or trasmitted, with negligible reflection or scattering.

Most of the spectra are reproduced from the literature. The absorption spectra labeled with the asterisk have been performed by Serena Silvi, Dr. Paolo Passaniti and Dr. Alberto Di Fabio in the Photochemistry Laboratory of the Department of Chemistry of the University of Bologna (Italy). The compounds used were of highest purity commercially available or samples deriving from previous researches. For relevant photophysical data, see Sections 3a-d.

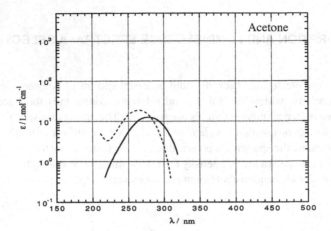

Fig. 3e-1. Absorption spectrum of acetone in water (----) and isooctane (——). Adapted from [8202].

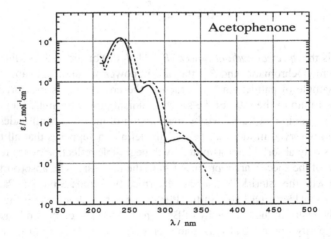

Fig. 3e-2. Absorption spectrum of acetophenone in ethanol (----) and hexane (——). Adapted from [6108].

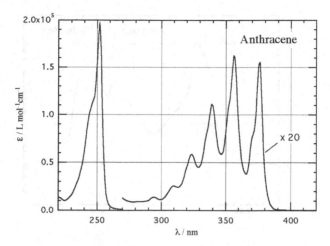

Fig. 3e-3. Absorption spectrum* of anthracene in ethanol.

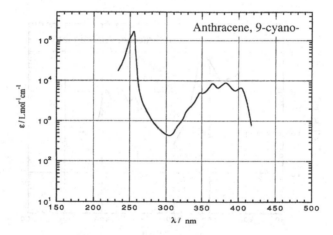

Fig. 3e-4. Absorption spectrum 9-cyano-anthracene in ethanol. Adapted from [5101].

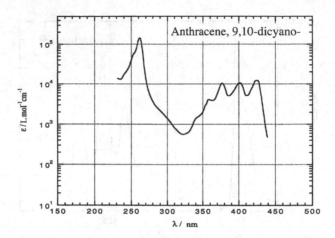

Fig. 3e-5. Absorption spectrum 9,10-dicyano anthracene in 2-
methyltetrahydrofuran.Adapted from [7105].

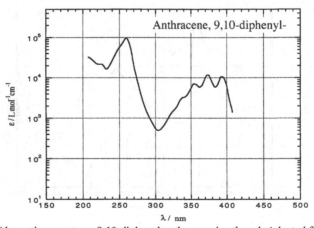

Fig. 3e-6. Absorption spectrum 9,10-diphenyl-anthracene in ethanol. Adapted from [7105].

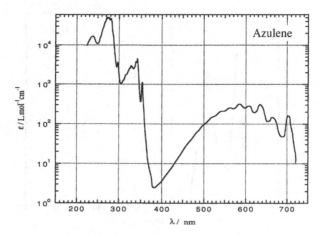

Fig. 3e-7. Absorption spectrum of azulene in heptane. Adapted from [7105].

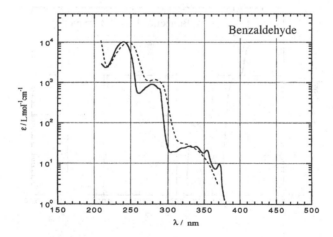

Fig. 3e-8. Absorption spectrum of benzaldehyde in ethanol (----) and hexane (——). Adapted from [6108].

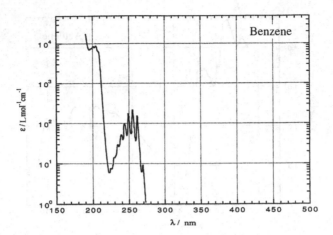

Fig. 3e-9. Absorption spectrum of benzene in hexane. Adapted from [7105].

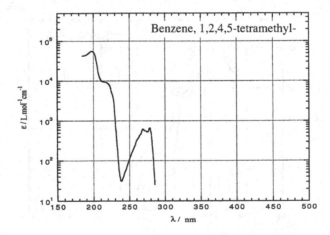

Fig. 3e-10. Absorption spectrum of 1,2,4,5-tetramethylbenzene (durene) in heptane.
Adapted from [7105].

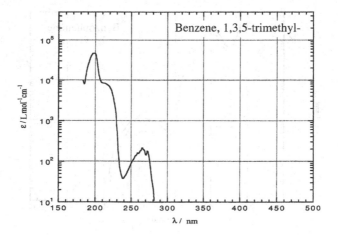

Fig. 3e-11. Absorption spectrum of 1,3,5-trimethyl-benzene (mesitylene) in ethanol. Adapted from [7105].

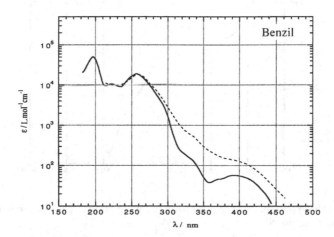

Fig. 3e-12. Absorption spectrum of benzil in ethanol (----) and heptane (——). Adapted from [6108] and [7105], respectively.

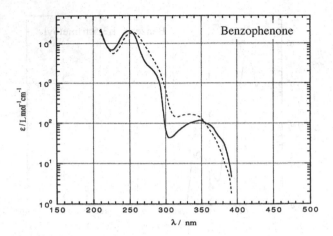

Fig. 3e-13. Absorption spectrum benzophenone in ethanol (----) and hexane (——). Adapted from [6108].

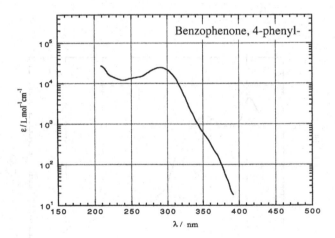

Fig. 3e-14. Absorption spectrum of 4-benzoylbiphenyl (4-phenyl-benzophenone) in methanol. Adapted from [6108].

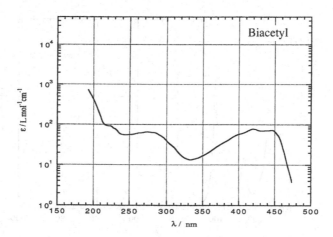

Fig. 3e-15. Absorption spectrum of biacetyl in hexane. Adapted from [7105].

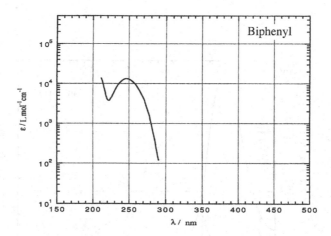

Fig. 3e-16. Absorption spectrum of biphenyl in ethanol. Adapted from [6108].

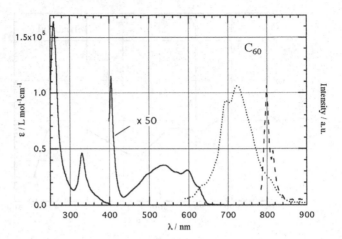

Fig. 3e-17. Absorption (——) and fluorescence (·····) spectra of fullerene C_{60} in toluene, RT. The phosphorescence (----) spectrum at 77 K is in iodoethane. Spectra performed by Gianluca Accorsi in ISOF-CNR laboratory in Bologna (Italy).

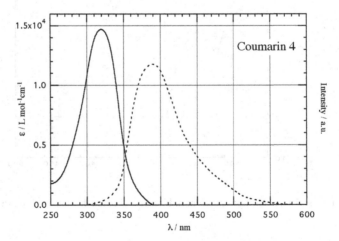

Fig. 3e-18. Absorption (——) and fluorescence (----) spectra of coumarine 4 (7-hydroxy-4-methyl-coumarin) in methanol. Adapted from Kodak optical products catalogue.

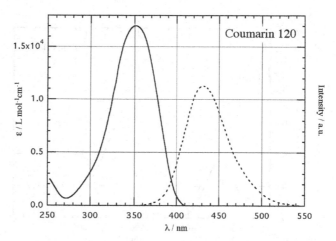

Fig. 3e-19. Absorption (——) and fluorescence (----) spectra of coumarine 120 (7-amino-4-methyl-coumarin) in methanol. Adapted from Kodak optical products catalogue.

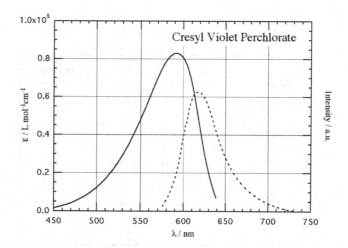

Fig. 3e-20. Absorption (——) and fluorescence (----) spectra of cresyl violet perchlorate in methanol. Adapted from Kodak optical products catalogue.

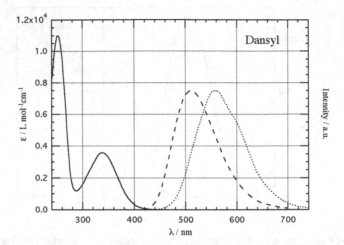

Fig. 3e-21. Absorption (——) and fluorescence (----) spectra* of dansyl-alanine ((1-dimethylamino-naphthalene-5-sulfonyl)-alanine in ethanol. Fluorescence spectrum* (·····) in water is also reported.

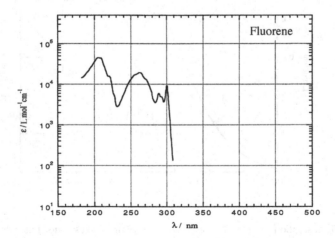

Fig. 3e-22. Absorption spectrum of fluorene in heptane. Adapted from [7105].

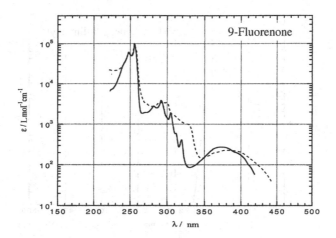

Fig. 3e-23. Absorption spectrum of 9-fluoreone in in ethanol (----) and cyclohexane (——).
Adapted from [7105] and [5101], respectively.

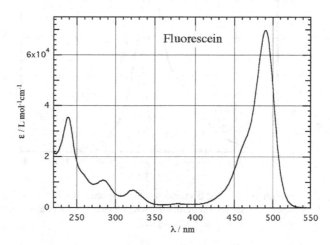

Fig. 3e-24. Absorption spectrum* of fluorescein in 0.1 M NaOH.

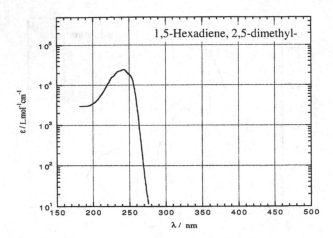

Fig. 3e-25. Absorption spectrum of 2,5-dimethyl-1,5-hexadiene in heptane. Adapted from [7105].

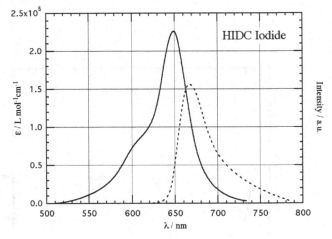

Fig. 3e-26. Absorption (——) and fluorescence (----) spectra of HIDC (1,1',3,3,3',3'-hexamethylindoledicarbocyanine) iodide in DMSO. Adapted from Kodak optical products catalogue.

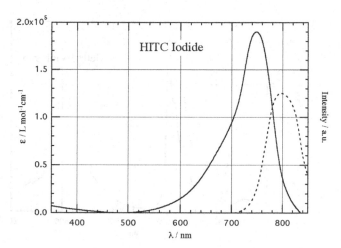

Fig. 3e-27. Absorption (——) and fluorescence (----) spectra of HITC (1,1′,3,3,3′,3′-hexamethylindoletricarbocyanine) iodide in DMSO. Adapted from Kodak optical products catalogue.

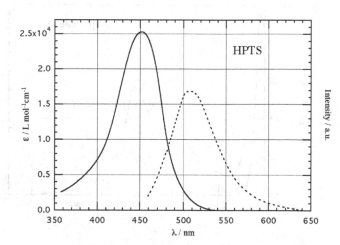

Fig. 3e-28. Absorption (——) and fluorescence (----) spectra of HPTS (1,3,6-pyrenetrisulfonic acid) in 0.05 M NaOH . Adapted from Kodak Optical Products catalogue.

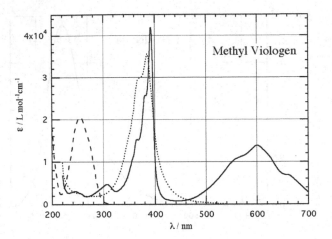

Fig. 3e-29. Absorption spectrum of MV^{2+} (----), MV^+ (——) and MV^0 (······) in water, RT.
[8452]

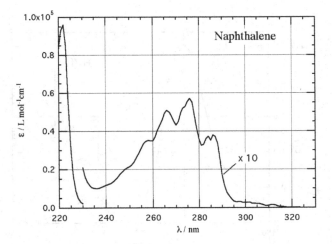

Fig. 3e-30. Absorption spectrum of naphthalene in heptane. Adapted from [7105].

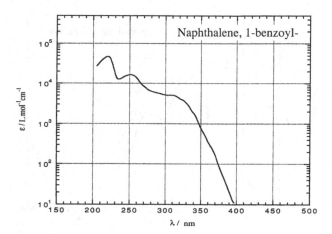

Fig. 3e-31. Absorption spectrum of 1-naphthyl phenyl ketone in methanol. Adapted from [7105].

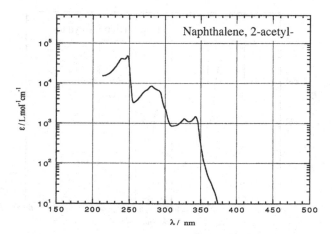

Fig. 3e-32. Absorption spectrum of 2′-acetonaphthone in cyclohexane. Adapted from [7105].

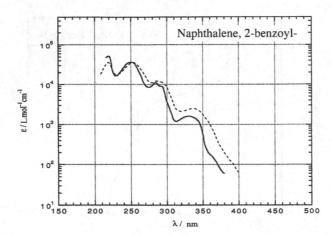

Fig. 3e-33. Absorption spectrum of 2-naphthyl phenyl ketone in ethanol (---) and cyclohexane. Adapted from [7105].

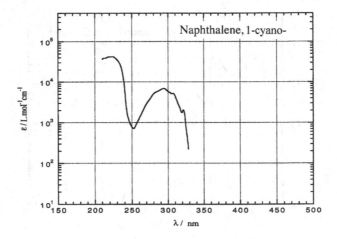

Fig. 3e-34. Absorption spectrum of 1-cyanonaphthalene in ethanol. Adapted from [7105].

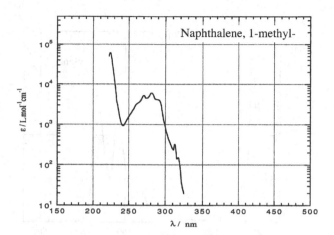

Fig. 3e-35. Absorption spectrum of 1-methylnaphthalene in ethanol. Adapted from [7105].

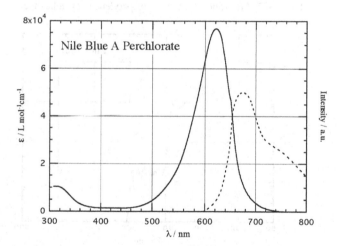

Fig. 3e-36. Absorption (——) and fluorescence (----) spectra of nile blue A perchlorate in methanol. Adapted from Kodak optical products catalogue.

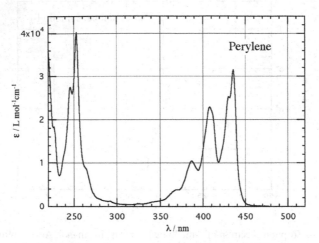

Fig. 3e-37. Absorption spectrum* of perylene in cyclohexane.

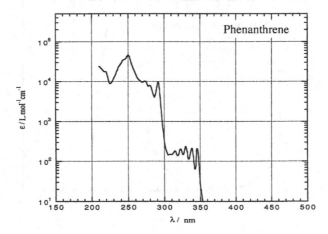

Fig. 3e-38. Absorption spectrum of phenanthrene in isooctane. Adapted from [7105].

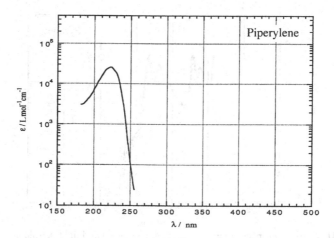

Fig. 3e-39. Absorption spectrum of 1,3-pentadiene in heptane. Adapted from [7105].

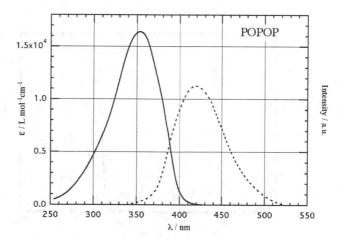

Fig. 3e-40. Absorption (——) and fluorescence (----) spectra of POPOP (1,4-bis(5-phenyl-2-oxazolyl)benzene) in cyclohexane. Adapted from Kodak optical products catalogue.

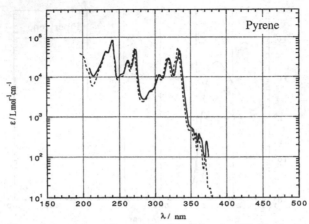

Fig. 3e-41. Absorption spectrum of pyrene in ethanol (-----) and light petroleum. Adapted from [7105].

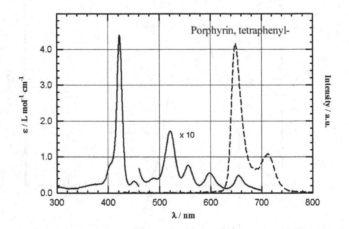

Fig. 3e-42. Absorption* (——) and fluorescence* (----) spectra of tetraphenylporphyrin in dichloromethane.

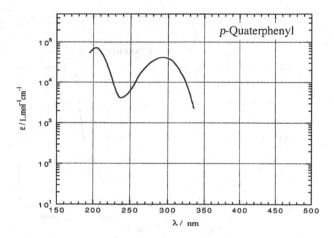

Fig. 3e-43. Absorption spectrum of *p*-quaterphenyl in hexane. Adapted from [7105].

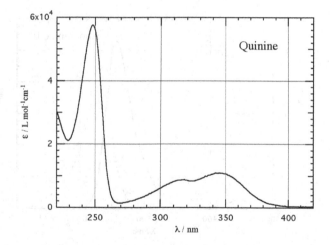

Fig. 3e-44. Absorption spectrum of quinine in heptane. Adapted from [7105].

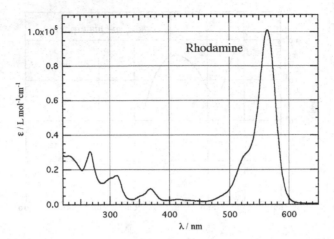

Fig. 3e-45. Absorption spectrum* of Rhodamine 101 in ethanol.

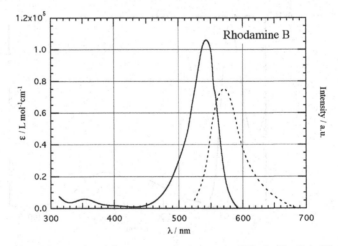

Fig. 3e-46. Absorption (——) and fluorescence (----) spectra of Rhodamine B iodide in methanol. Adapted from Kodak optical products catalogue.

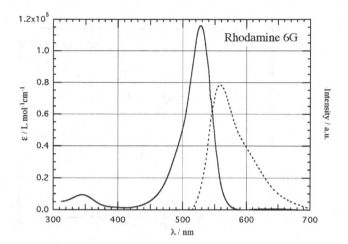

Fig. 3e-47. Absorption (——) and fluorescence (----) spectra of Rhodamine 6G in methanol. Adapted from Kodak optical products catalogue.

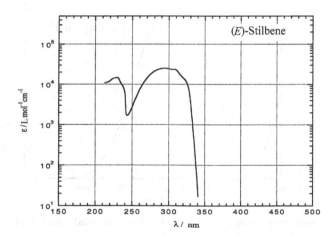

Fig. 3e-48. Absorption spectrum of (E)-stilbene in cyclohexane. Adapted from [7105].

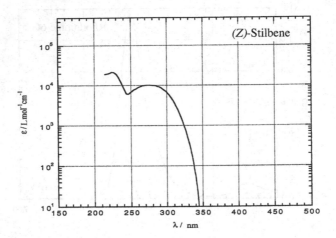

Fig. 3e-49. Absorption spectrum of (Z)-stilbene in cyclohexane. Adapted from [7105].

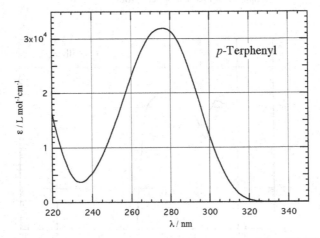

Fig. 3e-50. Absorption spectrum of *p*-terphenyl in ethanol. Adapted from [7105].

REFERENCES

[3901] Henri, V.; Pickett, L. W. *J. Chem. Phys.* **1939**, *7*, 439-440.

[4401] Lewis, G. N.; Kasha, M. *J. Am. Chem. Soc.* **1944**, *66*, 2100-2116.

[4701] Johnson, W. S.; Woroch, E.; Mathews, F. J. *J. Am. Chem. Soc.* **1947**, *69*, 566-571.

[4901] Kiss, A.; Molnar, J.; Sandorfy, C. *Bull. Soc. Chim. Fr.* **1949**, 275-280.

[4902] McClure, D. S. *J. Chem. Phys.* **1949**, *17*, 905-913.

[5001] Clar, E. *Spectrochim. Acta* **1950**, *4*, 116-121.

[5101] Friedel, R. A.; Orchin, M. *Ultraviolet Spectra of Aromatic Compounds*; Wiley; New York, 1951, 702p.

[5102] Dikun, P. P.; Petrov, A. A.; Sveshnikov, B. Ya. *Zh. Eksp. Teor. Fiz.* **1951**, *21*, 150-163.

[5103] Pyatnitskii, B. A. *Izv. Akad. Nauk SSSR, Ser. Fiz.* **1951**, *15*, 597-604.

[5401] Moodie, M. M.; Reid, C. *J. Chem. Phys.* **1954**, *22*, 252-254.

[5501] Sidman, J. W.; McClure, D. S. *J. Am. Chem. Soc.* **1955**, *77*, 6461-6470.

[5601] Beer, M. *J. Chem. Phys.* **1956**, *25*, 745-750.

[5602] Cherkasov, A. S.; Molchanov, V. A.; Vember, T. M.; Voldaikina, K. G. *Sot. Phys. Dokl.* **1956**, *1*, 427-429.

[5603] Padhye. M. R.; McGlynn, S. P.; Kasha, M. *J. Chem. Phys.* **1956**, *24*, 588-594.

[5604] Clar, E.; Zander. M. *Chem. Ber.* **1956**, *89*, 749-762.

[5605] Foster, R.; Hammick, D. L.; Hood, G. M.; Sanders, A. C. E. *J. Chem. Soc.* **1956**, 4865-4868.

[5701] Evans, D. F. *J. Chem. Soc.* **1957**, 1351-1357.

[5702] Weber, G.; Teale, F. W. J. *Trans. Faraday Soc.* **1957**, *53*, 646-655.

[5703] Brody, S. S. *Rev. Sci. Instrum.* **1957**, *28*, 1021-1026.

[5704] Dmitrievskii, O. D.; Ermolaev, V. L.; Terenin, A. N. *Dokl. Akad. Nauk SSSR* **1957**, *114*, 751-753.

[5801] Heckman, R. C. *J. Mol. Spectrosc.* **1958**, *2*, 27-41.

[5802] Linschitz, H.; Sarkanen, K. *J. Am. Chem. Soc.* **1958**, *80*, 4826-4832.

[5803] Goodman, L.; Kasha, M. *J. Mol. Spectrosc.* **1958**, *2*, 58-65.

[5804] Porter, G.; Windsor, M. W. *Proc. R. Soc. London, Ser.* **1958**, *245*, 238-258.

[5805] Livingston, R. *J. Am. Chem. Soc.* **1955**, *77*, 2179-2182.

[5901] Evans, D. F. *J. Chem. Soc.* **1959**, 2753-2757.

[5902] Clar, E.; Ironside, C. T.; Zander, M. *Tetrahedron* **1959**, *6*, 358-363.

[5903] Kanda, Y.; Shimada, R. *Spectrochim.Acta* **1959**, 211-224.

[5904] Claesson, S.; Lindqvist, L.; Holmstroem, B. *Nature (London)* **1959**, *183*, 661-662.

[5905] Mueller, R.; Doerr, F. *Z. Elektrochem.* **1959**, *63*, 1150-1156.

[6001] Ermolaev, V. L.; Terenin, A. N. *Sov. Phys. Usp.* **1960**, *3*, 423-426.

[6002] Backstrom, H. L. J.; Sandros, K. *Acta Chem. Scand.* **1960**, *14*, 48-62.

[6003] Lindqvist, L. *Ark. Kemi* **1960**, *16*, 79-138.

[6004] Blackwell, L. A.; Kanda, Y.; Sponer, H. *J. Chem. Phys.* **1960**, *32*, 1465-1476.

[6005] Evans, D. F. *J. Chem. Soc.* **1960**, 1735-1745.

[6006] Lang, K. F.; Zander, M.; Theiling, E. A. *Chem. Ber.* **1960**, *93*, 321-323.

[6101] Melhuish, W. H. *J. Phys. Chem.* **1961**, *65*, 229-235.

[6102] Ermolaev, V. L. *Opt. Spectrosc. (USSR)* **1961**, *11*, 266-269.

[6103] Kanda, Y.; Shimada, R. *Spectrochim. Acta* **1961**, *17*, 279-285.

[6104] Parker, C. A.; Hatchard, C. G. *Trans. Faraday Soc.* **1961**, *57*, 1894-1904.

[6105] Porter, G.; Wilkinson, F. *Proc. R. Soc. London, Ser. A* **1961**, *264*, 1-18.
[6106] Lippert, E.; Lueder, W.; Moll, F.; Naegele, W.; Boos, H.; Prigge, H.; Seibold-Blankenstein, I. *Angew. Chem.* **1961**, *73*, 695-706.
[6107] Paris, J. P.; Hirt, R. C.; Schmitt, R. G. *J. Chem. Phys.* **1961**, *34*, 1851-1852.
[6108] Lang, L. *Absorption spectra in the Ultraviolet and Visible Region.* Academic Press: New York (NY), **1961-1982**.
[6201] Ware, W. R. *J. Phys. Chem.* **1962**, *66*, 455-458.
[6202] Hoffman, M. Z.; Porter, G. *Proc. R. Soc. London, Ser. A* **1962**, *268*, 46-56.
[6203] Takei, K.; Kanda, Y. *Spectrochim. Acta* **1962**, *18*, 1201-1216.
[6204] Rhodes, W.; EI-Sayed, M. F. A. *J. Mol. Spectrosc.* **1962**, *9*, 42-49.
[6205] Parker, C. A. *Anal. Chem.* **1962**, *34*, 502-505.
[6206] Dawson, W.; Abrahamson, E. W. *J. Phys. Chem.* **1962**, *66*, 2542-2547.
[6207] Dyck, R. H.; McClure, D. S. *J. Chem.Phys.* **1962**, *36*, 2326-2345.
[6208] Wetlaufer, D. B. *Adv. Protein Chem.* **1962**, *17*, 303-390.
[6209] Muel, B.; Hubert-Habart, M. *Adv. Mol. Spectrosc.* **1962**, *2*, 647-657.
[6210] Parker, C. A.; Hatchard, C. G. *Analyst (London)* **1962**, *87*, 664-676.
[6211] Ermolaev, V. L. *Opt. Spectrosc. (USSR)* **1962**, *13*, 49-52.
[6301] Schuett, H. -U.; Zimmermann, H. *Ber. Bunsenges. Phys. Chem.* **1963**, *67*, 54-62.
[6302] Kanda, Y.; Shimada, R.; Takenoshita,Y. *Spectrochim. Acta* **1963**, *19*, 1249-1260.
[6303] Ermolaev, V. L. *Sov. Phys. Usp.* **1963**, *6*, 333-358.
[6304] Gropper, H.; Doerr, F. *Ber. Bunsenges. Phys. Chem.* **1963**, *67*, 46-54.
[6305] Bergmann, K.; O'Konski, C.T. *J. Phys. Chem.* **1963**, *67*, 2169-2177.
[6306] Teplyakov, P. A. *Opt. Spectrosc. (USSR)* **1963**, *15*, 350-352.
[6307] Becker, R. S.; Allison, J. B. *J. Phys. Chem.* **1963**, *67*, 2669-2675.
[6308] Doerr, F.; Gropper, H. *Ber. Bunsenges.Phys. Chem.* **1963**, *67*, 193-201.
[6309] Dubois, J. T.; Wilkinson, F. *J. Chem. Phys.* **1963**, *39*, 899-901.
[6310] Foerster, G. v., Z. Z. *Naturforsch., Teil A* **1963**, *18*, 620-626.
[6401] Trusov, V. V.; Teplyakov, P. A. *Opt. Spectrosc. (USSR)* **1964**, *16*, 27-30.
[6402] Kanda, Y.; Kaseda, H.; Matumura, T. *Spectrochim. Acta* **1964**, *20*, 1387-1396.
[6403] Ware, W. R.; Baldwin, B. A. *J. Chem. Phys.* **1964**, *40*, 1703-1705.
[6404] Herkstroeter, W. G.; Lamola, A. A.; Hammond, G. S. *J. Am. Chem. Soc.* **1964**, *86*, 4537-4540.
[6405] Birks, J. B.; Dyson, D. J.; King, T. A. *Proc. R. Soc. London, Ser. A* **1964**, *277*, 270-278.
[6406] Labhart, H. *Helv. Chim. Acta* **1964**, *47*, 2279-2288.
[6407] Parker, C. A.; Hatchard, C. G.; Joyce, T. A. *J. Mol. Spectrosc.* **1964**, *14*, 311-319.
[6408] Sponer, H.; Kanda, Y. *J. Chem. Phys.* **1964**, *40*, 778-787.
[6409] Clar, E. *Polycyclic Hydrocarbons*; Academic Press; New York, 1964, Vol. 2, 487.
[6410] Clar, E. *Polycyclic Hydrocarbons*; Academic Press; New York, 1964, Vol. 1, 487.
[6411] Buettner, A. V. *J. Phys. Chem.* **1964**, *68*, 3253-3259.
[6412] McGlynn, S. P.; Azumi, T.; Kasha, M. *J. Chem. Phys.* **1964**, *40*, 507-515.
[6413] Zander, M. *Chem. Ber.* **1964**, *97*, 2695-2699.
[6414] Kellogg, R. E.; Schwenker, R. P. *J. Chem. Phys.* **1964**, *41*, 2860-2863.
[6501] Ware, W. R.; Baldwin, B. A. *J. Chem. Phys.* **1965**, *43*, 1194-1197.
[6502] Medinger. T.; Wilkinson, F. *Trans. Faraday Soc.* **1965**, *61*, 620-630.
[6503] Lamola, A. A.; Hammond, G. S. *J. Chem. Phys.* **1965**, *43*, 2129-2134.
[6504] Kellogg, R. E.; Simpson, W. T. *J. Am. Chem. Soc.* **1965**, *87*, 4230-4234.
[6505] Aladekomo, J. B.; Birks, J. B. *Proc. R. Soc. London, Ser. A* **1965**, *284*, 551-565.
[6506] Shakhverdov, P. A.; Terenin, A. N. *Dokl. Phys. Chem.* **1965**, *160*, 163-165.

[6507] Cohen, B. J.; Baba, H.; Goodman, L. *J. Chem. Phys.* **1965**, *43*, 2902-2903.

[6508] Viktorova, E. N.; Gofman, I. A. *Russ. J. Phys. Chem.* **1965**, *39*, 1416-1419.

[6509] Berlman, I. B. *Handbook of Fluorescence Spectra of Aromatic Molecules*; Academic Press; New York, 1965, 258.

[6510] Laposa, J. D.; Lim, E. C.; Kellogg, R. E. *J. Chem. Phys.* **1965**, *42*, 3025-3026.

[6601] Parker, C. A.; Joyce, T. A. *Trans. Faraday Soc.* **1966**, *62*, 2785-2792.

[6602] Borkman, R. F.; Kearns, D. R. *J. Chem. Phys.* **1966**, *44*, 945-949.

[6603] Borkman, R. F.; Keams, D. R. *J. Am. Chem. Soc.* **1966**, *88*, 3467-3475.

[6604] Keams, D. R.; Case, W. A. *J. Am. Chem. Soc.* **1966**, *88*, 5087-5097.

[6605] Bennett, R. G.; McCartin, P. J. *J. Chem. Phys.* **1966**, *44*, 1969-1972.

[6606] Longworth, J. W.; Rahn, R. O.; Shulman, R. G. *Chem. Phys.* **1966**, *45*, 2930-2939.

[6607] Chessin, M.; Livingston, R.; Truscott, T. G. *Trans. Faraday Soc.* **1966**, *62*, 1519-1524.

[6608] Yang, N. C.; Murov, S. *J. Chem. Phvs.* **1966**, *45*, 4358.

[6609] Herkstroeter. W. G.; Hammond, G. S. *J. Am. Chem. Soc.* **1966**, *88*, 4769-4777.

[6610] Birks, J. B.; King, T. A. *Proc. R. Soc. London, Ser. A* **1966**, *291*, 244-256.

[6611] Selinger, B. K. *Aust. J. Chem.* **1966**, *19*, 825-834.

[6612] Parker, C. A.; Joyce, T. A. *Chem. Commun.* **1966**, 108-109.

[6613] Brines, J. S.; Koren, J. G.; Hodgson, W.G. *J. Chem. Phys.* **1966**, *44*, 3095-3099.

[6701] Marchetti, A. P.; Kearns, D. R. *J. Am. Chem. Soc.* **1967**, *89*, 768-777.

[6702] Murov, S. L. *Ph.D., Thesis*; Univ. Chicago; Chicago, IL, 1967, 227.

[6703] Yang, N. C.; McClure, D. S.; Murov, S. L.; Houser, J. J.; Dusenbery, R. *J. Am. Chem. Soc.* **1967**, *89*, 5466-5468.

[6704] Henson, R. C.; Jones, J. L. W.; Owen, E. D. *J. Chem. Soc. A* **1967**, 116-122.

[6705] Kellmann, A.; Lindqvist, L. *The Triplet State: Proceedings of an International Symposium*; Beirut, Lebanon, 1967, 43945.

[6706] Gueron, M.; Eisinger, J.; Shulman, R. G. *J. Chem. Phvs.* **1967**, *47*, 4077-4091.

[6707] Lim, E. C.; Chakrabarti, S. K. *Chem. Phys. Lett.* **1967**, *1*, 28-31.

[6708] Parker, C. A.; Joyce, T. A. *Chem. Commun.* **1967**, 744-745.

[6709] Horrocks, A. R.; Medinger, T.; Wilkinson, F. *Photochem. Photobiol.* **1967**, *6*, 21-28.

[6710] Zander, M. *Ber. Bunsenges. Phys. Chem* **1967**, *71*, 424-429.

[6711] Almgren, M. *Photochem. Photobiol.* **1967**, *6*, 829-840.

[6712] Kuboyama, A.; Yabe, S. *Bull. Chem. Soc. Jpn.* **1967**, *40*, 2475-2479.

[6713] Kearns, D.R.; Hollins, R. A.; Khan, A. U.; Chambers, R.W.; Radlick, P. *J. Am. Chem. Soc.* **1967**, *89*, 5455-5456.

[6714] Evans, T. R.; Leermakers, P. A. *J. Am. Chem. Soc.* **1967**, *89*, 4380-4382.

[6715] Bowers, P. G.; Porter, G. *Proc. R. Soc. London, Ser. A* **1967**, *299*, 348-353.

[6716] Cohen, B. J.; Goodman, L. *J. Chem .Phys.* **1967**, *46*, 713-721.

[6717] Hochstrasser, R. M.; Marzzacco, C. *J. Chem. Phys.* **1967**, *46*, 4155-4156.

[6718] Lewis, H. G.; Owen, E. D. *J. Chem. Soc. B* **1967**, 422-425.

[6719] Searle, R.; Williams, J. L. R.; DeMeyer, D. E.; Doty, J. C. *Chem. Commun.* **1967**, 1165.

[6720] Zander, M. *Fresenius' Z. Anal. Chem.* **1967**, *226*, 251-259.

[6721] Lam, E. Y. Y.; Valentine, D.; Hammond, G. S. *J. Am. Chem. Soc.* **1967**, *89*, 3482-3487.

[6722] Iwata, S.; Tanaka, J.; Nagakura, S. *J. Chem. Phys.* **1967**, *47*, 2203-2209.

[6723] Zimmerman, H. E.; Binkley, R. W.; McCullough, J. J.; Zimmerman, G. A. *J. Am. Chem. Soc.* **1967**, *89*, 6589-6595.

[6724] Chibisov, A. K.; Skvortsov, B. V.; Karyakin, A. V.; Shvindt, N. N. *Khim. Vys. Energ.* **1967**, *1*, 529-535.

[6725] Lamola, A. A.; Gueron, M.; Yamane, T.; Eisinger, J.; Shulman, R. G. *J. Chem. Phys.* **1967**, *47*, 2210-2217.

[6801] Parker, C. A. *Photoluminescence of Solutions. With Applications to Photochemistry and Analytical Chemistry*; Amsterdam; The Netherlands, 1968, 544.

[6802] Horrocks, A. R.; Wilkinson, F. *Proc. R. Soc. London, Ser. A* **1968**, *306*, 257-273.

[6803] Dawson, W. R.; Windsor, M. W. *J. Phys. Chem.* **1968**, *72*, 3251-3260.

[6804] Stevens, B.; Algar, B. E. *J. Phys. Chem.* **1968**, *72*, 3468-3474.

[6805] Brinen, J. S.; Koren, J. G. *Chem. Phys. Lett.* **1968**, *2*, 671-672.

[6806] Hurley, R.; Testa, A. C. *J. Am. Chem. Soc.* **1968**, *90*, 1949-1952.

[6807] Binet, D. J.; Goldberg, E. L.; Forster, L. S. *J. Phys. Chem.* **1968**, *72*, 3017-3020.

[6808] Dawson, W. R. *J. Opt. Soc. Am.* **1968**, *58*, 222-227.

[6809] Suppan, P. *Ber. Bunsenges. Phys. Chem.* **1968**, *72*, 321-326.

[6810] Arnold, D. R. *Adv. Photochem.* **1968**, *6*, 301-423.

[6811] Knowles, A.; Roe, E. M. F. *Photochem. Photobiol.* **1968**, *7*, 421-436.

[6812] Zander, M. *Phosphorimetry*; Academic Press; New York, 1968, 206.

[6813] Gollnick, K. *Adv. Photochem.* **1968**, *6*, 1-122.

[6814] Vander Donckt, E.; Nasielski, J.; Greenleaf, J. R.; Birks, J. B. *Chem. Phys. Lett.* **1968**, *2*, 409-410.

[6815] Stevens, B.; Thomaz, M. F. *Chem. Phys. Lett.* **1968**, *1*, 549-550.

[6816] Wilairat, P.; Selinger, B. *Aust. J. Chem.* **1968**, *21*, 733-746.

[6818] Birks, l. B.; Georghiou, S. *J. Phys. B* **1968**, *1*, 958-965.

[6819] Stevens, B.; Algar, B. E. *J. Phys. Chem.* **1968**, *72*, 2582-2587.

[6820] Yildiz, A.; Kissinger, P. T.; Reilley, C.N. *J. Chem. Phys.* **1968**, *49*, 1403-1406.

[6821] Dainton, F. S.; Peng, C. T.; Salmon, G.A. *J. Phys. Chem.* **1968**, *72*, 3801-3807.

[6822] Warwick, D. A.; Wells, C. H. J. *Spectrochim. Acta., Part A* **1968**, *24A*, 589-593.

[6823] Rosenberg, H. M.; Carson, S. D. *J. Phys. Chem.* **1968**, *72*, 3531-3534.

[6824] Marsh, G.; Kearns, D. R.; Schaffner, K. *Helv. Chim. Acta* **1968**, *51*, 1890-1899.

[6825] Pitts, J. N.; Burley, D. R.; Mani, J. C.; Broadbent, A. D. *J. Am. Chem. Soc.* **1968**, *90*, 5902-5903.

[6826] Yang, N. C.; Dusenbery, R. L. *J. Am. Chem. Soc.* **1968**, *90*, 5899-5900.

[6827] Windsor, M. W.; Dawson, W. R. *Mol. Cryst.* **1968**, *4*, 253-258.

[6828] Lim, E. C.; Yu, J. M. H. *J. Chem. Phis.* **1968**, *49*, 3878-3884.

[6901] Sandros, K. *Acta Chem. Scand.* **1969**, *23*, 2815-2829.

[6902] Chambers, R. W.; Kearns, D. R. *Photochem. Photohiol.* **1969**, *10*, 215-219.

[6903] McGlynn, S. P.; Azumi, T.; Kinoshita, M. *Molecular Spectroscopy of the Triplet State*; Prentice Hall; Englewood Cliffs, NJ, 1969, 434.

[6904] Rotkiewicz, K.; Grabowski, Z. R. *Trans. Faraday Soc.* **1969**, *65*, 3263-3278.

[6905] Heinzelmann, W.; Labhart, H. *Chem. Phys. Lett.* **1969**, *4*, 20-24.

[6906] Bera, S. Ch.; Mukherjee. R.; Chowdhury, M. *J. Chem. Phys.* **1969**, *51*, 754-761.

[6907] Dawson, W. R.; Kropp, J. L. *J. Phys. Chem.* **1969**, *73*, 1752-1758.

[6908] Yang, N. C.; Dusenbery, R. L. *Mol. Photochem.* **1969**, *1*, 159-171.

[6909] Shcheglova, N. A.; Shigorin, D. N.; Yakobson, G. G.;Tushishishvili, L. Sh. *Russ. J. Phys. Chem.* **1969**, *43*, 1112-1117.

[6910] Kemp, D. R.; Porter, G. *J. Chem. Soc. D* **1969**, 1029-1030.

[6911] Turro, N. J.; Engel, R. *Mol. Photochem.* **1969**, *1*, 235-238.

[6912] Dawson, W. R.; Kropp, J. L. *J. Phys. Chem.* **1969**, *73*, 693-699.

[6913] Stevens, B.; Algar, B. E. *J. Phys. Chem.* **1969**, *73*, 1711-1715.

[6914] Singer, L. A. *Tetrahedron Lett.* **1969**, 923-926.
[6915] Nemoto, M.; Kokubun, H.; Koizumi, M. *Bull. Chem. Soc. Jpn.* **1969**, *42*, 1223-1230.
[6916] Walker, M. S.; Bednar, T. W.; Lumry, R. *Molecular Luminescence*; Lim, E.C. (ed.); W. A. Benjamin, Inc.; New York, 1969, 135-52.
[6917] Eisinger, J.; Navon, G. *J. Chem. Phys.* **1969**, *50*, 2069-2077.
[6918] Clarke, R. H.; Hochstrasser, R. M. *J. Mol. Spectrosc.* **1969**, *32*, 309-319.
[6919] Pavlopoulos, T. G. *J. Chem. Phys.* **1969**, *51*, 2936-2940.
[6920] Seybold, P. G.; Gouterman, M. *J. Mol. Spectrosc.* **1969**, *31*, 1-13.
[6921] Thomson, A. J. *J. Chem. Phys.* **1969**, *51*, 4106-4116.
[6922] Bylina, A.; Grabowski, Z. R. *Trans. Faraday Soc.* **1969**, *65*, 458-463.
[6923] Kramer, H. E. A. *Z. Phys. Chem. (Frankfurt Am Main)* **1969**, *66*, 73-85.
[6924] Cundall, R. B.; Voss, A. J. R. *J. Chem. Soc. D* **1969**, 116.
[6925] Lim, E. C.; Li, R.; Li, Y. H. *J. Chem. Phys.* **1969**, *50*, 4925-4933.
[6926] Parkanyi, C.; Baum, E. J.; Wyatt, J.; Pitts, J. N., Jr. *J. Phys. Chem.* **1969**, *73*, 1132-1138.
[6927] Zander, M. *Z. Naturforsch., Teil A* **1969**, *24A*, 1387-1390.
[6928] Hoover, R. J.; Kasha, M. *J. Am. Chem. Soc.* **1969**, *91*, 6508-6510.
[6929] Scheiner, P. *Tetrahedron Lett.* **1969**, 4863-4866.
[6930] Cadogan, K. D.; Albrecht, A. C. *J. Phys. Chem.* **1969**, *73*, 1868-1877.
[6931] Kemp, T. J.; Roberts, J. P. *Trans. Faraday Soc.* **1969**, *65*, 725-731.
[6932] Hadley, S. G.; Keller, R. A. *J. Phys. Chem.* **1969**, *73*, 4356-4359.
[6933] Kropp, J. L.; Dawson, W. R.; Windsor, M. W. *J. Phys. Chem.* **1969**, *73*, 1747-1752.
[7001] Birks, J. B. *Photophysics of Aromatic Molecules*; Wiley-Interscience; New York, 1970, 704.
[7002] Yang, N. C.; Feit, E. D.; Hui, M. H.; Turro, N. J.; Dalton, J. C. *J. Am. Chem. Soc.* **1970**, *92*, 6974-6976.
[7003] Dalton, J. C.; Turro, N. J. *Annu. Rev. Phys. Chem.* **1970**, *21*, 499-560.
[7004] Sveshnikova, E. B.; Snegov, M. I. *Opt. Spectrosc. (USSR)* **1970**, *19*, 265-268.
[7005] Kearvell, A.; Wilkinson, F. *20th Reunion de la Societe de Chimie Physique*; Paris, May 27-30, 1969, 125-132.
[7006] Hunter, T. F. *Trans. Faraday Soc.* **1970**, *66*, 300-309.
[7007] Rehm, D.; Weller, A. *Isr. J. Chem.* **1970**, *8*, 259-271.
[7008] Hayashi, H.; Nagakura, S. *Mol. Phys.* **1970**, *19*, 45-53.
[7009] Loeff, I.; Lutz, H.; Lindqvist, L. *Isr. J. Chem.* **1970**, *8*, 141-146.
[7010] Slifkin, M. A.; Walmsley, R. H. *J. Phys. E* **1970**, *3*, 160-162.
[7011] Porter, G.; Topp, M. R. *Proc. R. Soc. London, Ser. A* **1970**, *315*, 163-184.
[7012] Heinrich, G.; Holzer, G.; Blume, H.; Schulte-Frohlinde, D. *Z. Naturforsch., Teil B* **1970**, *25*, 496-499.
[7013] Gallivan, J. B. *Mol. Photochem.* **1970**, *2*, 191-211.
[7014] Kropp, J. L.; Lou, J. J. *J. Phys. Chem.* **1970**, *74*, 3953-3959.
[7015] Bonnier, J. -M.; Jardon, P. *J. Chim. Phys. Phys.-Chim. Biol.* **1970**, *67*, 577-579.
[7016] El-Bayoumi, M. A.; Dalle, J. -P.; O'Dwyer, M. F. *J. Lumin.* **1970**, *1-2*, 716-725.
[7017] Wettack, F. S.; Renkes, G. D.; Rockley, M. G.; Turro, N. J.; Dalton, J. C. *J. Am. Chem. Soc.* **1970**, *92*, 1793-1794.
[7018] Gegiou, D.; Huber, J. R.; Weiss, K. *J. Am. Chem. Soc.* **1970**, *92*, 5058-5062.
[7019] Zanker, V.; Rudolph, E.; Prell, G. *Z. Naturforsch., Teil B* **1970**, *25B*, 1137-1143.
[7020] Gradyushko, A. T.; Sevchenko, A. N.; Solovyov, K. N.; Tsvirko, M. P. *Photochem. Photobiol.* **1970**, *11*, 387-400.
[7021] Nishi, N.; Shimada, R.; Kanda, Y. *Bull. Chem. Soc. Jpn.* **1970**, *43*, 41-46.

[7022] Eastwood, D.; Gouterman, M. *J. Mol. Spectrosc.* **1970**, *35*, 359-375.

[7023] Maria, H. J.; McGlynn, S. P. *J. Chem. Phys.* **1970**, *52*, 3399-3402.

[7024] Hadley, S. G. *J. Phys. Chem. 74* **1970**, *74*, 3551-3552.

[7025] Lim, E. C.; Li, Y. H.; Li, R. *J. Chem. Phys.* **1970**, *53*, 2443-2448.

[7026] Hadley, S. G. *Chem. Phys. Lett.* **1970**, *6*, 549-550.

[7027] Jones, P. F.; Calloway, A. R. *20th Reunion de la Societe de Chimie Physique*; Paris, May 27-30, 1969, 110-115.

[7101] Berlman, I. B. *Handbook of Fluorescence Spectra of Aromatic Molecules*; Second ed., 1971, 473.

[7102] Halpern, A. M.; Ware, W. R. *J. Chem. Phys.* **1971**, *54*, 1271-1276.

[7103] Porter, G.; Yip, R. W.; Dunston, J. M.; Cessna, A. J.; Sugamori, S. E. *Trans. Faraday Soc.* **1971**, *67*, 3149-3154.

[7104] Lutz, H.; Lindqvist, L., *J. Chem. Soc. D* **1971**, 493-494.

[7105] Perkampus, H. -H.; Sandeman, I.; Timmons, C. J. *DMS UV Atlas of Organic Compounds*; Verlag Chemie, Weinheim; Butterworths, London, England, 1966-1971, Vol. I-V, .

[7106] Clark, W. D. K.; Steel, C. *J. Am. Chem. Soc.* **1971**, *93*, 6347-6355.

[7108] Slifkin, M. A.; Walmsley, R. H. *Photochem. Photohiol.* **1971**, *13*, 57-65.

[7110] Rusakowicz, R.; Testa, A. C. *Spectrochim. Acta, Part A* **1971**, *27A*, 787-792.

[7111] Geacintov, N. E.; Burgos, J.; Pope, M.; Strom, C. *Chem. Phys. Lett.* **1971**, *11*, 504-508.

[7112] Japar, S. M.; Abrahamson, E. W. *J. Am. Chem. Soc.* **1971**, *93*, 4140-4144.

[7113] Vincett, P. S.; Voigt, E. M.; Rieckhoff, K. E. *J. Chem. Phys.* **1971**, *55*, 4131-4140.

[7114] Gradyushko, A. T.; Tsvirko, M. P. *Opt. Spectrosc. (USSR)* **1971**, *31*, 291-295.

[7115] Becker, R. S.; Inuzuka, K.; Balke, D. E. *J. Am. Chem. Soc.* **1971**, *93*, 38-42.

[7116] Sykes, A.; Truscott, T. G. *Trans. Faraday Soc.* **1971**, *67*, 679-686.

[7117] Kearvell, A.; Wilkinson, F. *Chem. Phys. Lett.* **1971**, *11*, 472-473.

[7118] Pownall, H. J.; Huber, J. R. *J. Am. Chem. Soc.* **1971**, *93*, 6429-6436.

[7119] Emeis, C. A.; Oosterhoff, L. J. *J. Chem. Phys.* **1971**, *54*, 4809-4819.

[7120] Lancelot, G.; Helene, C. *Chem. Phys.Lett.* **1971**, *9*, 327-331.

[7121] Yamamuro, T.; Tanaka, I.; Hata, N. *Bull. Chem. Soc. Jpn.* **1971**, *44*, 667-671.

[7122] Van Thielen, J.; Van Thien, T.; De Schryver, F. C. *Tetrahedron Lett.* **1971**, 3031-3034.

[7123] Gallivan, J. B.; Brinen, J. S. *Chem. Phys. Lett.* **1971**, *10*, 455-459.

[7124] Marsh, G.; Kearns, D. R.; Schaffner, K. *J. Am. Chem. Soc.* **1971**, *93*, 3129-3137.

[7125] Semeluk, G. P.; Stevens, R. D. S. *Can. J. Chem.* **1971**, *49*, 2452-2455.

[7126] Bensasson, R.; Land, E. J. *Trans. Faraday Soc.* **1971**, *67*, 1904-1915.

[7127] Davis, G. A.; Gresser, J. D.; Carapellucci, P. A. *J. Am. Chem. Soc.* **1971**, *93*, 2179-2182.

[7128] Watson, F. H., Jr.; El-Bayoumi, M. A. *J.Chem. Phys.* **1971**, *55*, 5464-5470.

[7129] Hadley, S. G. *J. Phys. Chem.* **1971**, *75*, 2083-2086.

[7201] Soep, B.; Kellmann, A.; Martin, M.; Lindqvist, L., *Chem. Phys. Lett.* **1972**, *13*, 241-244.

[7202] Hulme, B. E.; Land, E. J.; Phillips, G. O. *J. Chem. Soc., Faraday Trans. 1* **1972**, *68*, 2003-2012.

[7203] Birks, J. B. *Chem. Phys. Lett.* **1972**, *17*, 370-372.

[7204] Vander Donckt, E.; Vogels, C. *Spectrochim. Acta, Part A* **1972**, *28A*, 1969-1975.

[7205] Vander Donckt, E.; Lietaer, D. *J. Chem. Soc., Faraday Trans. 1* **1972**, *68*, 112-120.

[7206] Cundall, R. B.; Pereira, L. C. *J. Chem. Soc., Faraday Trans. 2* **1972**, *68*, 1152-1163.

[7207] Harrigan, E. T.; Wong, T. C.; Hirota, N. *Chem. Phys. Lett.* **1972**, *14*, 549-554.

[7208] Froehlich, P. M.; Morrison, H. A. *J. Phys. Chem.* **1972**, *76*, 3566-3570.

[7209] Almgren, M. *Mol. Photochem.* **1972**, *4*, 327-334.

[7210] Evans, D. F.; Tucker, J. N. *J. Chem. Soc,. Faraday Trans. 2* **1972**, *68*, 174-176.

[7211] Hudson, B. S.; Kohler, B. E. *Chem. Phys. Lett.* **1972**, *14*, 299-304.

[7212] Dempster. D. N.; Morrow, T.; Rankin, R.; Thompson, G. F. *J. Chem. Soc., Faraday Trans. 2* **1972**, *68*, 1479-1496.

[7213] Jackson, A. W.; Yarwood, A. J. *Can. J. Chem.* **1972**, *50*, 1331-1337.

[7214] Huber, J. R.; Mantulin, W. W. *J. Am. Chem. Soc.* **1972**, *94*, 3755-3760.

[7215] Blackwell, D. S. L.; Liao, C. C.; Loutfy, R. O.; de Mayo, P.; Paszyc, S. *Mol. Photochem.* **1972**, *4*, 171-188.

[7216] Kulberg, L. P.; Nurmukhametov, R. N.; Gorelik, M. V. *Opt. Spectrosc. (USSR)* **1972**, *32*, 476-478.

[7217] Dreeskamp, H.; Hutzinger, O.; Zander, M. *Z. Naturforsch.. Teil A* **1972**, *27A*, 756-759.

[7218] Nishimura, A. M.; Tinti, D. S. *Chem. Phys. Lett.* **1972**, *13*, 278-283.

[7219] Arnold, D. R.; Bolton, J. R.; Pedersen, J. A. *J. Am. Chem. Soc.* **1972**, *94*, 2872-2874.

[7220] Gallivan, J. B. *Can. J. Chem.* **1972**, *50*, 3601-3606.

[7221] Kawai, K.; Shirota, Y.; Tsubomura, H.; Mikawa, H., *Bull. Chem. Soc. Jpn.* **1972**, *45*, 77-81.

[7222] Tetreau, C.; Lavalette, D.; Land, E. J.; Peradejordi, F. *Chem. Phys. Lett.* **1972**, *17*, 245-247.

[7223] Hellner, C.; Lindqvist, L.; Roberge, P. C. *J. Chem. Soc., Faraday Trans. 2* **1972**, *68*, 1928-1937.

[7224] Dainton, F. S.; Robinson, E. A.; Salmon, G. A. *J. Phys. Chem.* **1972**, *76*, 3897-3904.

[7225] Li, R.; Lim, E. C. *J. Chem. Phys.* **1972**, *57*, 605-612.

[7226] Pantke, E. R.; Labhart, H. *Chem. Phys. Lett.* **1972**, *16*, 255-259.

[7227] Takemura, T.; Yamamoto, K.; Yamazaki, I.; Baba, H. *Bull. Chem. Soc. Jpn.* **1972**, *45*, 1639-1642.

[7228] Alvarez, V. L.; Hadley, S. G. *J. Phys. Chem.* **1972**, *76*, 3937-3940.

[7301] Merkel, P. B.; Kearns, D. R. *J. Chem. Phys.* **1973**, *58*, 398-400.

[7302] Adams, J. E.; Mantulin, W. W.; Huber, J. R. *J. Am. Chem. Soc.* **1973**, *95*, 5477-5481.

[7303] Mantulin, W. W.; Song, P. -S. *J. Am. Chem. Soc.* **1973**, *95*, 5122-5129.

[7304] Vander Donckt, E.; Barthels, M. R.; Delestinne, A. *J. Photochem.* **1973**, *1*, 429-432.

[7305] Cundall, R. B.; Pereira. L. C. *Chem. Phys. Lett.* **1973**, *18*, 371-374.

[7306] Cundall, R. B.; Ogilvie, S. McD.; Robinson, D. A. *J. Photochem.* **1973**, *1*, 417-422.

[7307] Murov, S. L. *Handbook of Photochemistry*; Dekker; New York, 1973, 272.

[7308] Dekker, R. H.; Srinivasan, B. N.; Huber, J. R.; Weiss, K. *Photochem. Photobiol.* **1973**, *18*, 457-466.

[7309] Olszowski, A.; Romanowski, H.; Ruziewicz, Z. *Bull. Acad. Pol. Sci., Ser. Sci., Math, Astron. Phys.* **1973**, *21*, 381-387.

[7310] Taylor, H. V.; Allred, A. L.; Hoffman, B. M. *J. Am. Chem. Soc.* **1973**, *95*, 3215-3219.

[7311] Dempster, D. N.; Morrow, T.; Rankin, R.; Thompson, G. F. *Chem. Phys. Lett.* **1973**, *18*, 488-492.

[7312] Stikeleather, J. A. *Chem. Phys. Lett.* **1973**, *21*, 326-329.

[7313] Burke, F. P.; Small, G. J.; Braun, J. R.; Lin, T. -S. *Chem. Phys. Lett.* **1973**, *19*, 574-579.

[7314] Berlman, l. B. *Energy Transfer Parameters of Aromatic Compounds*; Academic Press; New York, 1973, 379.

[7315] Kikuchi, K.; Watarai, H.; Koizumi, M. *Bull. Chem. Soc. Jpn.* **1973**, *46*, 749-754.

[7316] Fischer, G. *Chem. Phys. Lett.* **1973**, *21*, 305-308.

[7317] Tsvirko, M. P.; Sapunov, V. V.; Solovev, K. N. *Opt. Spectrosc. (USSR)* **1973**, *34*, 635-638.

[7318] Yip, R. W.; Szabo, A. G.; Tolg, P. K. *J.Am. Chem. Soc.* **1973**, *95*, 4471-4472.

[7319] Lakowicz, J. R.; Weber, G. *Biochemistry* **1973**, *12*, 4171-4179.

[7320] Lamola, A. A. *Pure Appl. Chem.* **1973**, *34*, 281-303.

[7321] Scharf, H. -D.; Leismann, H. *Z. Naturforsch., Teil B* **1973**, *28B*, 662-681.

[7322] Arnold, D. R.; Birtwell, R. J. *J. Am. Chem. Soc.* **1973**, *95*, 4599-4606.

[7323] Hotchandani, S.; Testa, A. C. *J. Chem. Phys.* **1973**, *59*, 596-600.

[7324] Lutz, H.; Breheret, E.; Lindqvist, L. *J. Phys. Chem.* **1973**, *77*, 1758-1762.

[7325] Truscott, T. G.; Land, E. J.; Sykes, A. *Photachem. Photobiol.* **1973**, *17*, 43-51.

[7326] Bagdasaryan, Kh. S.; Kiryukhin, Yu. l.; Sinitsina, Z. A. *J. Photochem.* **1973**, *1*, 225-240.

[7327] Lim, E. C.; Kedzierski, M. *Chem. Phys. Lett.* **1973**, *20*, 242-245.

[7401] Vigny, P.; Duquesne, M. *Excited States of Biological Molecules* **1976**, 167-177.

[7402] Dempster, D. N.; Morrow, T.; Quinn, M. F. *J. Photochem.* **1974**, *2*, 329-341.

[7403] Heinrich, G.; Schoof, S.; Gusten, H. *J. Photochem.* **1974**, *3*, 315-320.

[7404] Hochstrasser, R. M.; Lutz, H.; Scott, G.W. *Chem. Phys. Lett.* **1974**, *24*, 162-167.

[7405] Anderson, R. W. Jr.; Hochstrasser, R. M.; Lutz, H.; Scott, G. W. *Chem. Phys. Lett.* **1974**, *28*, 153-157.

[7406] Gouterman, M.; Khalil, G. -E. *J. Mol. Spectrosc.* **1974**, *53*, 88-100.

[7407] Smagowicz, J.; Wierzchowski, K. L. *J. Lumin.* **1974**, *8*, 210-232.

[7408] Dempster, D. N.; Morrow, T.; Quinn, M. F. *J. Photochem.* **1974**, *2*, 343-359.

[7409] Bent, D. V.; Schulte-Frohlinde, D. *J. Phys. Chem.* **1974**, *78*, 446-450.

[7410] Dalton, J. C.; Montgomery, F. C. *J. Am. Chem. Soc.* **1974**, *96*, 6230-6232.

[7411] Tournon, J.; Abu-Elgheit, M.; Avouris, P.; EI-Bayoumi, M. A. *Chem. Phys. Lett.* **1974**, *28*, 430-432.

[7412] Capitanio, D. A.; Pownall, H. J.; Huber, J. R. *J. Photochem.* **1974**, *3*, 225-236.

[7413] Blanchi, J. -P.; Watkins. A. R. *Mol. Photochem.* **1974**, *6*, 133-142.

[7414] Souto, M. A.; Wagner, P. J.; El-Sayed, M. A. *Chem. Phys.* **1974**, *6*, 193-204.

[7415] Metcalfe, J.; Rockley, M. G.; Phillips, D. *J. Chem. Soc., Faraday Trans. 2* **1974**, *70*, 1660-1666.

[7416] Brocklehurst, B.; Tawn, D. N. *Spectrochim. Acta, Part A* **1974**, *30A*, 1807-1815.

[7417] Kanamaru, N.; Nagakura, S.; Kimura, K. *Bull. Chem. Soc. Jpn.* **1974**, *47*, 745-746.

[7418] Kanamaru, N.; Bhattacharjee, H. R.; Lim, E. C. *Chem. Phys. Lett.* **1974**, *26*, 174-179.

[7419] Waddell, W. H.; Renner, C. A.; Turro, N. J. *Mol. Photochem.* **1974**, *6*, 321-324.

[7420] Kanamaru, N.; Long, M. E.; Lim, E. C. *Chem. Phys. Lett.* **1974**, *26*, 1-9.

[7501] Amand, B.; Bensasson, R. *Chem. Phys. Lett.* **1975**, *34*, 44-48.

[7502] DeToma, R. P.; Cowan, D. O. *J. Am. Chem. Soc.* **1975**, *97*, 3283-3291.

[7503] Wilkinson, F. *Organic Molecular Photophysics*; John Wiley; New York, 1975, Vol. 2, 95-158.

[7504] Herkstroeter, W. G. *J. Am. Chem. Soc.* **1975**, *97*, 4161-4167.

[7505] Bent, D. V.; Hayon, E. *J. Am. Chem. Soc.* **1975**, *97*, 2599-2606.

[7506] Harrigan, E. T.; Hirota, N. *J. Am. Chem. Soc.* **1975**, *97*, 6647-6652.

[7507] Lui, Y. H.; McGlynn, S. P. *J. Lumin.* **1975**, *10*, 113-121.

[7508] Lui, Y. H.; McGlynn, S. P. *J. Mol. Spectrosc.* **1975**, *55*, 163-174.

[7509] Bulska, H.; Chodkowska, A.; Grabowska, A.; Pakula, B.; Slanina, Z. *J. Lumin.* **1975**, *10*, 39-57.

[7510] Schreiner, S.; Steiner, U.; Kramer, H. E. A. *Photochem. Photobiol.* **1975**, *21*, 81-84.

[7511] Bent, D. V.; Hayon, E. *J. Am. Chem. Soc.* **1975**, *97*, 2612-2619.

[7512] Kikuchi, K.; Kokubun, H.; Kikuchi, M. *Bull. Chem. Soc. Jpn.* **1975**, *48*, 1378-1381.

[7513] Alkaitis, S. A.; Graetzel, M.; Henglein, A. *Ber. Bunsenges. Phys. Chem.* **1975**, *79*, 541-546.

[7514] Bent, D. V.; Hayon, E.; Moorthy, P. N. *J. Am. Chem. Soc.* **1975**, *97*, 5065-5071.

[7515] Tsvirko, M. P.; Solovev, K. N.; Gradyushko, A. T.; Dvomikov, S. S. *Opt. Spectrosc. (USSR)* **1975**, *38*, 400-404.

[7516] Quimby, D. J.; Longo, F. R. *J. Am. Chem.Soc.* **1975**, *97*, 5111-5117.

[7517] Wierzchowski, K. L.; Berens, K.; Szabo, A. G. *J. Lumin.* **1975**, *10*, 331-343.

[7518] Bensasson, R.; Land, E. J.; Truscott, T. G. *Photochem. Photobiol.* **1975**, *21*, 419-421.

[7519] Serafimov, O.; Bruehlmann, U.; Huber, J. R. *Ber. Bunsenges. Phys. Chem.* **1975**, *79*, 202-205.

[7520] Salet, C.; Bensasson, R. *Photochem. Photobiol.* **1975**, *22*, 231-235.

[7521] Gisin, M.; Wirz, J. *Helv. Chim. Acta* **1975**, *58*, 1768-1771.

[7522] Ferris, J. P.; Prabhu, K. V.; Strong, R. L. *J. Am. Chem. Soc.* **1975**, *97*, 2835-2839.

[7523] Lui, Y. H.; McGlynn, S. P. *J. Lumin.* **1975**, *9*, 449-458.

[7524] Bent, D. V.; Hayon, E. *Chem. Phys. Lett.* **1975**, *31*, 325-327.

[7525] Arce, R.; Ramirez, L. *Photochem. Photobiol.* **1975**, *21*, 13-19.

[7526] Acuna, A. U.; Ceballos, A.; Garcia Dominguez, J. A.; Molera, M. J. *An. Quim.* **1975**, *71*, 22-27.

[7527] Yamashita, M.; Ikeda, H.; Kashiwagi, H. *J. Chem. Phys.* **1975**, *63*, 1127-1731.

[7528] Mahaney, M.; Hubert, J.R. *Chem. Phys.* **1975**, *9*, 371-378.

[7529] Li, Y. –H; Chan, L.-M.; Tyer, L.; Moody, R. T.; Himel, C. M.; Hercules, D. M. *J. Am. Chem. Soc.* **1975**, *97*, 3118-3126.

[7601] Fushimi, K.; Kikuchi, K.; Kokubun, H. *J. Photochem.* **1976**, *5*, 457-468.

[7602] Ware, W. R ; Rothman, W. *Chem. Phys. Lett.* **1976**, *39*, 449-453.

[7603] Morris, J. V.; Mahaney, M. A.; Huber, J. R. *J. Phys. Chem.* **1976**, *80*, 969-974.

[7604] Arnold, D. R.; Maroulis, A.J. *J. Am. Chem. Soc.* **1976**, *98*, 5931-5937.

[7605] Alkaitis, S. A.; Graetzel, M. *J. Am. Chem. Soc.* **1976**, *98*, 3549-3554.

[7606] Wagner, P. J.; Thomas, M. J.; Harris, E. *J. Am. Chem. Soc.* **1976**, *98*, 7675-7679.

[7607] Wallace, W. L.; Van Duyne, R. P.; Lewis, F. D. *J. Am. Chem. Soc.* **1976**, *98*, 5319-5326.

[7608] Amouyal, E.; Bensasson, R. *J. Chem. Soc., Faraday Trans. 1* **1976**, *72*, 1274-1287.

[7609] Haink, H. J.; Huber, J. R. *Chem. Phys. Lett.* **1976**, *44*, 117-120.

[7610] Zander, M. *Z. Naturforsch., Teil A* **1976**, *31A*, 677-678.

[7611] Thulstrup, E. W.; Nepras, M.; Dvorak, V.; Michl, J. *J. Mol. Spectrosc.* **1976**, *59*, 265-285.

[7612] Bensasson, R.; Land, E. J.; Lafferty, J.; Sinclair, R.S.; Truscott, T. G. *Chem. Phys. Lett.* **1976**, *41*, 333-335.

[7613] Huber, J. R.; Mahaney, M.; Morris, J. V. *Chem. Phys.* **1976**, *16*, 329-335.

[7614] Fratev, F.; Polansky, O.E.; Zander, M. *Z. Naturforsch., Teil A* **1976**, *31A*, 987-989.

[7615] Ohno, T.; Kato, S. *Chem. Lett.* **1976**, 263-266.

[7616] Capellos, C.; Suryanarayanan, K. *Int. J. Chem. Kinet.* **1976**, *8*, 529-539.

[7617] Hirata, Y.; Tanaka, I. *Chem. Phys. Lett.* **1976**, *43*, 568-570.

[7618] Vogelmann, E.; Kramer, H. E. A. *Photochem. Photobiol.* **1976**, *24*, 595-597.

[7619] Vogelmann, E.; Kramer, H. E. A. *Photochem. Photobiol.* **1976**, *23*, 383-390.

[7620] Bensasson, R.; Land, E. J.; Maudinas, B. *Photochem. Photobiol.* **1976**, *23*, 189-193.

[7621] Pereira, L. C.; Ferreira, I. C.; Thomaz, M. P. F. *Chem. Phys. Lett.* **1976**, *43*, 157-161.

[7622] Garner, A.; Wilkinson, F. *J. Chem. Soc., Faraday Trans. 2* **1976**, *72*, 1010-1020.

[7623] Clar, E.; Schmidt, W. *Tetrahedron* **1976**, *32*, 2563-2566.

[7624] Metcalfe, J.; Chervinsky, S.; Oref, I. *Chem. Phys. Lett.* **1976**, *42*, 190-192.

[7625] Lancelot, G. *Mol. Phys.* **1976**, *31*, 241-254.

[7626] Schaffner, K. *Tetrahedron* **1976**, *32*, 641-653.

[7627] Hotchandani, S.; Testa, A.C. *Spectrochim. Acta, Part A* **1976**, *32A*, 1659-1663.

[7628] Kobayashi, T.; Nagakura, S. *Chem. Phys. Lett.* **1976**, *43*, 429-434.

[7629] Wilkinson, F.; Farmilo, A. *J. Chem. Soc., Faraday Trans. 2* **1976**, *72*, 604-618.

[7630] Barwise, A. J. G.; Gorman, A. A.; Rodgers, M. A. J. *Chem. Phys. Lett.* **1976**, *38*, 313-316.

[7631] Yang, N. C.; Shold, D. M.; Neywick, C. V. *J. Chem. Soc., Chem. Commun.* **1976**, 727-728.

[7632] Philen, D. L.; Hedges, R. M. *Chem. Phys. Lett.* **1976**, *43*, 358-362.

[7633] Friedrich, J.; Vogel, J.; Windhager, W.; Doerr, F. *Z. Naturforsch., Teil A* **1976**, *31*, 61-70.

[7701] Morgan, D. D.; Warshawsky, D.; Atkinson, T. *Photochem. Photobiol.* **1977**, *25*, 31-38.

[7702] Kellmann, A. *J. Phys. Chem.* **1977**, *81*, 1195-1198.

[7703] Wolf, M. W.; Brown, R. E.; Singer, L. A. *J. Am. Chem. Soc.* **1977**, *99*, 526-531.

[7704] Schuster, D. I.; Goldstein, M. D.; Bane, P. *J. Am. Chem. Soc.* **1977**, *99*, 187-193.

[7705] Brown, R. G.; Porter, G. *J. Chem.Soc., Faraday Trans. 1* **1977**, *73*, 1569-1573.

[7706] Grodowski, M. S.; Veyret, B.; Weiss, K. *Photochem. Photobiol.* **1977**, *26*, 341-352.

[7707] Irie, M.; Yorozu, T.; Yoshida, K.; Hayashi, K. *J. Phys. Chem.* **1977**, *81*, 973-976.

[7708] Zander, M. *Z. Naturforsch., Teil A* **1977**, *32A*, 339-340.

[7709] Yamamoto, S.; Kikuchi, K.; Kokubun, H. *Chem. Lett* **1977**, 1173-1176.

[7710] Bensasson, R.; Dawe, E. A.; Long, D. A.; Land, E. J. *J. Chem. Soc., Faraday Trans. 1* **1977**, *73*, 1319-1325.

[7711] Sheng, S. J.; El-Sayed, M. A. *Chem. Phys.* **1977**, *20*, 61-69.

[7712] Fleming, G. R.; Knight, A. W. E.; Morris, J. M.; Morrison, R. J. S.; Robinson, G. W. *J. Am. Chem. Soc.* **1977**, *99*, 4306-4311.

[7713] Pernot, C.; Lindqvist, L. *J. Photochem.* **1976/77**, *6*, 215-220.

[7714] Wilkinson, F.; Garner, A. *J. Chem. Soc., Faraday Trans. 2* **1977**, *73*, 222-233.

[7715] Aloisi, G. G.; Mazzucato, U.; Birks, J. B.; Minuti, L. *J. Am. Chem. Soc.* **1977**, *99*, 6340-6347.

[7716] Fouassier, J. -P.; Lougnot, D. -J.; Wieder, F.; Faure, J. *J. Photochem.* **1977**, *7*, 17-28.

[7717] Mardelli, M.; Olmsted, J. III. *J. Photochem.* **1977**, *7*, 277-285.

[7718] Bandyopadhyay, B. N.; Harriman, A. *J. Chem. Soc., Faraday Trans. 1* **1977**, *73*, 663-674.

[7719] Chantrell, S. J.; McAuliffe, C. A.; Munn, R. W.; Pratt, A. C.; Land, E.J. *J. Chem. Soc., Faraday Trans. 1* **1977**, 858-865.

[7720] Kunavin, N. I.; Nurmukhametov, R. N.; Khachaturova, G. T. *J. Appl. Spectrosc.* **1977**, *26*, 1023-1027.

[7721] Dunne, A.; Quinn, M. F. *J. Chem. Soc.. Faraday Trans. 1* **1977**, *73*, 1104-1110.

[7722] Korobov, V. E.; Shubin, V. V.; Chibisov, A. K. *Chem. Phys. Lett.* **1977**, *45*, 498-501.

[7723] Crosby, P. M.; Dyke, J. M.; Metcalfe, J.; Rest, A. J.; Salisbury, K.; Sodeau, J. R. *J. Chem. Soc., Perkin Trans. 2* **1977**, 182-185.

[7724] Jardon, P. *J. Chim. Phys. Phys.-Chim.Biol.* **1977**, *74*, 1177-1184.

[7725] Archer, M. D.; Ferreira, M. I. C.; Porter, G.; Tredwell, C. J. *Nouv. J. Chim.* **1977**, *1*, 9-12.

[7726] Ferreira, M. I. C.; Harriman, A. *J. Chem. Soc.. Faraday Trans. 1* **1977**, *73*, 1085-1092.

[7727] Volkert, W. A.; Kuntz, R. R.; Ghiron, C. A.; Evans, R. F.; Santus, R.; Bazin, M. *Photochem. Photobiol.* **1977**, *26*, 3-9.

[7728] Capellos, C.; Suryanarayanan, K. *Int. J. Chem. Kinet.* **1977**, *9*, 399-407.

[7729] Gonzenbach, H. -U.; Tegmo-Larsson, I. M.; Grosclaude, J. P.; Schaffner, K. *Helv. Chim. Acta* **1977**, *60*, 1091-1123.

[7730] Abdul-Rasoul, F.; Catherall, C. L. R.; Hargreaves, J. S.; Mellor, J. M.; Phillips, D. *Eur. Polym. J.* **1977**, *13*, 1019-1023.

[7731] Acuna, A. U.; Ceballos, A.; Molera, M. J. *J. Phys. Chem.* **1977**, *81*, 1090-1093.

[7732] Hotchandani, S.; Testa, A. C. *J. Chem. Phys.* **1977**, *67*, 5201-5206.

[7733] Ivanov, V. L.; Martynov, I. Yu.; Uzhinov, B. M.; Kuz'min, M. G. *High Energy Chem.* **1977**, *11*, 327-331.

[7734] Anton, M. F.; Moomaw, W. R. *J. Chem. Phys.* **1977**, *66*, 1808-1818.

[7735] Moore, W. M.; McDaniels, J. C.; Hen, J. A. *Photochem. Photobiol.* **1977**, *25*, 505-512.

[7801] Noe, L. J.; Degenkolb, E. O.; Rentzepis, P. M. *J. Chem. Phys.* **1978**, *68*, 4435-4438.

[7802] Romashov, L. V.; Borovkova, V. A.; Kiryukhin, Yu. I.; Bagdasar'yan, Kh. S. *High Energy Chem.* **1978**, *12*, 132-134.

[7803] Siegmund, M.; Bendig, J. *Ber. Bunsenges. Phys. Chem.* **1978**, *82*, 1061-1068.

[7804] Val'kova, G. A.; Shcherbo, S. N.; Shigorin, D. N. *Dokl. Phys. Chem.* **1978**, *240*, 491-493.

[7805] Bensasson, R. V.; Land, E. J.; Salet, C. *Photochem. Photobiol.* **1978**, *27*, 273-280.

[7806] Schoof, S.; Guesten, H.; von Sonntag, C. *Ber. Bunsenges. Phys. Chem.* **1978**, *82*, 1068-1073.

[7807] Soboleva, I. V.; Sadovskii, N. A.; Kuz'min, M. G. *Dokl. Phys. Chem.* **1978**, *238*, 70-73.

[7808] Zander, M. *Z. Naturforsch., Teil A* **1978**, *33A*, 998-1000.

[7809] Niizuma, S.; Kwan, L.; Hirota, N. *Mol. Phys.* **1978**, *35*, 1029-1046.

[7810] Braeuchle, C.; Kabza, H.; Voitlaender, J.; Clar, E. *Chem. Phys.* **1978**, *32*, 63-73.

[7811] Damschen, D. E.; Merritt, C. D.; Perry, D. L.; Scott, G. W.; Talley, L. D. *J. Phys. Chem.* **1978**, *82*, 2268-2272.

[7812] Bauer, H.; Reske. G. *J. Photochem.* **1978**, *9*, 43-54.

[7813] Das, P. K.; Becker, R. S. *J. Phys. Chem.* **1978**, *82*, 2093-2105.

[7814] Andrews, L. J.; Deroulede, A.; Linschitz, H. *J. Phys. Chem.* **1978**, *82*, 2304-2309.

[7815] Huggenberger, C.; Labhart, H. *Helv. Chim. Acta* **1978**, *61*, 250-257.

[7816] Das, P. K.; Becker, R. S. *J. Phys. Chem.* **1978**, *82*, 2081-2093.

[7817] Fang, H. L.-B.; Thrash, R. J.; Leroi, G. E. *Chem. Phys. Lett.* **1978**, *57*, 59-63.

[7818] Becker, R. S.; Bensasson, R. V.; Lafferty, J.; Truscott, T. G.; Land, E. J. *J. Chem. Soc., Faraday Trans. 2* **1978**, *74*, 2246-2255.

[7819] Latas, K. J.; Nishimura, A. M. *J. Phys. Chem.* **1978**, *82*, 491-495.

[7820] McVie, J.; Sinclair, R. S.; Truscott, T. G. *J. Chem. Soc., Faraday Trans. 2* **1978**, *74*, 1870-1879.

[7821] Gradyushko, A. T.; Knyukshto, V. N.; Solovev, K. N.; Shulga, A. M. *Opt. Spectrosc. (USSR)* **1978**, *44*, 268-272.

[7822] Madej, S. L.; Gillispie, G. D.; Lim, E. C. *Chem. Phys.* **1978**, *32*, 1-10.

[7823] Basara, H.; Ruziewicz, Z. *Acta Phys. Pol., A* **1978**, *A54*, 689-694.

[7824] Bensasson, R.; Land, E. J. *Nouv. J. Chim.* **1978**, *2*, 503-507.

[7825] Grellmann, K. H.; Hentzschel, P. *Chem. Phys. Lett.* **1978**, *53*, 545-551.

[7826] Kuz'min, V. A.; Tatikolov, A. S.; Borisevich, Yu. E. *Chem. Phys. Lett.* **1978**, *53*, 52-55.

[7827] Croteau, R.; Leblanc, R. M. *Photochem. Photobiol.* **1978**, *28*, 33-38.

[7828] Raemme, G. *J. Photochem.* **1978**, *9*, 439-447.

[7829] Pownall, H. J.: Schaffer, A. M.; Becker, R. S.; Mantulin, W. W. *Photochem. Photobiol.* **1978**, *27*, 625-628.

[7830] Basara, H.; Ruziewicz, Z.; Zawadzka, H. *J. Lumin.* **1978**, *17*, 283-290.

[7831] Coyle, J. D.; Newport, G.L.; Harriman, A. *J. Chem. Soc., Perkin Trans. 2* **1978**, 133-137.

[7832] Flicker, W. M.; Mosher, O. A.; Kuppermann, A. *J. Chem. Phys.* **1978**, *69*, 3311-3320.

[7833] Korobov, V. E.; Chibisov, A. K. *J. Photochem.* **1978**, *9*, 411-424.

[7834] Lui, Y. H.; McGlynn, S. P. *Spectrosc. Lett.* **1978**, *11*, 47-58.

[7835] Morris, J. M.; Yoshihara, K. *Mol. Phys.* **1978**, *36*, 993-1003.

[7901] Vogelmann, E.; Rauscher, W.; Kramer. H. E. A. *Photochem. Photobiol.* **1979**, *29*, 771-776.

[7902] Olmsted, J. III *J. Phys. Chem.* **1979**, *83*, 2581-2584.

[7903] Bendig, J; Siegmund, M. *J. Prakt. Chem.* **1979**, *321*, 587-600.

[7904] Dreeskamp, H.; Pabst, J. *J. Chem. Phys.Lett.* **1979**, *61*, 262-265.

[7905] Simons, W. W. *Sadtler Handbook of Ultraviolet Spectra*; Sadder Res. Lab.; Philadelphia, PA, 1979, 1016.

[7906] Carsey, T. P.; Findley, G. L.; McGlynn, S. P. *J. Am. Chem. Soc.* **1979**, *101*, 4502-4510.

[7907] Groenen, E. J. J.; Koelman, W. N. *J. Chem. Soc., Faraday Trans. 2* **1979**, *75*, 69-78.

[7908] Gschwind, R.; Haselbach, E. *Helv.Chim. Acta* **1979**, *62*, 941-955.

[7909] Das, P. K.; Becker, R. S. *J. Am. Chem. Soc.* **1979**, *101*, 6348-6353.

[7910] Dvomikov, S. S.; Knyukshto, V. N.; Solovev, K. N.; Tsvirko, M. P. *Opt. Spectrosc. (USSR)* **1979**, *46*, 385-388.

[7911] Land, E. J.; Truscott, T. G. *Photochem. Photobiol.* **1979**, *29*, 861-866.

[7912] Salet, C.; Bensasson, R.; Becker, R. S. *Photochem. Photobiol.* **1979**, *30*, 325-329.

[7913] George, M. V.; Kumar, Ch. V.; Scaiano, J. C. *J. Phys. Chem.* **1979**, *83*, 2452-2455.

[7914] Grellmann, K. -H.; Hentzschel, P.; Wismontski-Knittel, T.; Fischer, E. *J. Photochem.* **1979**, *11*, 197-213.

[7915] Bluemer, G. -P.; Zander, M. *Z. Naturforsch., Ted A* **1979**, *34A*, 909-910.

[7916] Dinse, K. P.; Winscom, C. J. *J. Lumin.* **1979**, *18-19*, 500-504.

[7917] Sa E Melo, M. T.; Averbeck, D.; Bensasson, R. V.; Land, E. J.; Salet, C. *Photochem. Photobiol.* **1979**, *30*, 645-651.

[7918] Beaumont, P. C.; Parsons, B. J.; Phillips, G. O.; Allen, J. C. *Biochim. Biophys. Acta* **1979**, *562*, 214-221.

[7919] Delouis, J. F.; Delaire, J. A.; Ivanoff, N. *Chem. Phys. Lett.* **1979**, *61*, 343-346.

[7920] Heisel, F.; Miehe, J. A.; Sipp, B. *Chem. Phys. Lett.* **1979**, *61*, 115-118.

[7921] Garner, H.; Schulte-Frohlinde, D. *J. Phys. Chem.* **1979**, *83*, 3107-3118.

[7922] Goemer, H.; Schulte-Frohlinde, D. *J. Am. Chem. Soc.* **1979**, *101*, 4388-4390.

[7923] Schulte-Frohlinde, D.; Goemer, H. *Pure Appl. Chem.* **1979**, *51*, 279-297.
[7924] Condirston, D. A.; Laposa, J. D. *Chem. Phys. Lett.* **1979**, *63*, 313-317.
[7925] Goerner, H.; Schulte-Frohlinde, D. *Chem. Phys. Lett.* **1979**, *66*, 363-369.
[7926] Greene, B. I.; Hochstrasser, R. M.; Weisman, R. B. *J. Chem. Phys.* **1979**, *70*, 1247-1259.
[7927] Eweg, J. K.; Mueller, F.; Visser, A. J. W. G.; Veeger, C.; Bebelaar, D.; van Voorst, J.D.W. *Photochem. Photobiol.* **1979**, *30*, 463-471.
[7928] Braeuchle, Chr.; Kabza, H.; Voitlaender, J. *Z. Naturforsch., Teil A* **1979**, *34A*, 6-12.
[7929] Alder, L.; Gloyna, D.; Wegener, W.; Pragst, F.; Henning, H. -G. *Chem. Phys. Lett.* **1979**, *64*, 503-506.
[7930] Pileni, M. P.; Giraud, M.; Santus, R. *Photochem. Photobiol.* **1979**, *30*, 251-256.
[7931] Latas, K. J.; Power, R. K.; Nishimura, A. M. *Chem. Phys. Lett.* **1979**, *65*, 272-277.
[7932] Encinas, M. V.; Scaiano, J. C. *J. Am. Chem. Soc.* **1979**, *101*, 7740-7741.
[7933] Arvis. M.; Mialoeq, J. -C. *J. Chem. Soc., Faraday Trans. 2* **1979**, *75*, 415-421.
[7934] Baugher, J.; Hindman, J. C.; Katz, J. J. *Chem. Phys. Lett.* **1979**, *63*, 159-162.
[7935] Werner, T. *J. Phys. Chem.* **1979**, *83*, 320-325.
[7936] Heinrich, G.; Guesten, H. *Z. Phys. Chem. (Wiesbaden)* **1979**, *118*, 31-41.
[8001] Nishida, Y.; Kikuchi, K.; Kokubun, H. *J. Photochem.* **1980**, *13*, 75-81.
[8002] Kalyanasundaram, K.; Dung, D. *J. Phys. Chem.* **1980**, *84*, 2251-2256.
[8003] Siegmund, M.; Bendig, J. *Z. Naturforsch., Teil A* **1980**, *35*, 1076-1086.
[8004] Perichet, G.; Chapelon, R.; Pouyet, B. *J. Photochem.* **1980**, *13*, 67-74.
[8005] Gupta, A. K.; Basu, S.; RohatgiMukherjee, K. K. *Can. J. Chem.* **1980**, *58*, 1046-1050.
[8006] Visser, R. J.; Varma, C. A. G. O. *J. Chem. Soc., Faraday Trans. 2* **1980**, *76*, 453-471.
[8007] Leigh, W. J.; Arnold, D. R. *J. Chem. Soc., Chem. Commun.* **1980**, 406-408.
[8008] Ronfard-Haret, J. -C.; Bensasson, R. V.; Amouyal, E. *J. Chem. Soc., Faraday Trans. 1* **1980**, *76*, 2432-2436.
[8009] Turro, N. J.; Shima, K.; Chung, C. J.; Tanielian, C.; Kanfer, S. *Tetrahedron Lett.* **1980**, *21*, 2775-2778.
[8010] Grieser, F.; Thomas, J. K. *J. Chem. Phys.* **1980**, *73*, 2115-2119.
[8011] Bennett, J. A.; Birge, R. R. *J. Chem. Phys.* **1980**, *73*, 4234-4246.
[8012] Jones, G. II; Jackson, W. R.; Halpern, A. M. *Chem. Phys. Lett.* **1980**, *72*, 391-395.
[8013] Toth, M. *Chem. Phys.* **1980**, *46*, 437-443.
[8014] Lee, E. K. C.; Loper, G. L. *Radiationless Transitions*; Academic Press; New York, 1980, 1-80.
[8015] Saltiel, J.; Khalil, G. -E.; Schanze, K. *Chem. Phys. Lett.* **1980**, *70*, 233-235.
[8016] Szabo, A. G.; Rayner, D. M. *J. Am. Chem. Soc.* **1980**, *102*, 554-563.
[8017] Davidson, R. S.; Bonneau, R.; JoussotDubien, J.; Trethewey, K. R. *Chem. Phys. Lett.* **1980**, *74*, 318-320.
[8018] Takahashi, T.; Kikuchi, K.; Kokubun, H. *J. Photochem.* **1980**, *14*, 67-76.
[8019] Val'kova, G. A.; Sakhno, T. V.; Shcherbo, S. N.; Shigorin, D. N.; Andrievskii, A. M.; Poplavskii, A. N.; Dyumaev, K. M. *Russ. J. Phys. Chem.* **1980**, *54*, 1382-1383.
[8020] Harriman, A. *J. Chem. Soc., Faraday Trans. 1* **1980**, *76*, 1978-1985.
[8021] Bonnets, R.; Charalambides, A. A.; Land, E. J.; Sinclair, R. S.; Tait, D.; Truscott, T. G. *J. Chem. Soc., Faraday Trans. 1* **1980**, *76*, 852-859.
[8022] Sinclair, R. S.; Tait, D.; Truscott, T. G. *J. Chem. Soc., Faraday Trans. 1* **1980**, *76*, 417-425.

[8023] Matthews, J. l.; Braslavsky, S. E.; Camilleri, P. *Photochem. Photobiol.* **1980**, *32*, 733-738.
[8024] Scott, G. W.; Talley, L. D.; Anderson, R. W. Jr. *J. Chem. Phys.* **1980**, *72*, 5002-5013.
[8025] Najbar, J.; Trzcinska, B. M.; Urbanek, Z. H.; Proniewicz, L. M. *Acta Phys. Pol., A* **1980**, *A58*, 331-344.
[8026] Bartocci, G.; Mazzucato, U.; Masetti, F.; Galiazzo, G. *J. Phys. Chem.* **1980**, *84*, 847-851.
[8027] Becker, R. S; Kogan, G. *Photochem. Photobiol.* **1980**, *31*, 5-13.
[8028] van der Velden, G. P. M.; de Boer, E.; Veeman, W. S. *J. Phys. Chem.* **1980**, *84*, 2634-2641.
[8029] Braeuchle, C.; Deeg, F. W.; Voitlaender, J. *J. Chem. Phys.* **1980**, *53*, 373-381.
[8030] Pileni, M.-P.; Graetzel, M. *J. Phys. Chem.* **1980**, *84*, 1822-1825.
[8031] Losev, A. P.; Zen'kevich, E. I.; Sagun, E. I. *Bull. Acad. Sci. USSR, Phys. Ser.* **1980**, *44*, 84-88.
[8032] Saucin, M.; Van de Vorst, A. *Radiat. Environ. Biophys.* **1980**, *17*, 159-168.
[8033] Arce, R.; Jimenez, L. A.; Rivera, V.; Torres, C. *Photochem. Photobiol.* **1980**, *32*, 91-95.
[8034] Takemura, T.; Chihara, K.; Becker, R. S.; Das, P. K.; Hug, G. L. *J. Am. Chem. Soc.* **1980**, *102*, 2604-2609.
[8101] Ghoshal, S. K.; Sarkar, S. K.; Kastha, G. S. *Bull. Chem. Soc. Jpn.* **1981**, *54*, 3556-3561.
[8102] Kasama, K.; Kikuchi, K.; Yamamoto, S.; Uji-ie, K.; Nishida, Y.; Kokubun, H. *J. Phys. Chem.* **1981**, *85*, 1291-1296.
[8103] Kasama, K.; Kikuchi, K.; Nishida, Y.; Kokubun, H. *J. Phys. Chem.* **1981**, *85*, 4148-4153.
[8104] Hirayama, S. *J. Am. Chem. Soc.* **1981**, *103*, 2934-2938.
[8105] Harriman, A.; Mills, A. *Photochem. Photobiol.* **1981**, *33*, 619-625.
[8106] Goemer, H.; Schulte-Frohlinde, D. *J. Photochem.* **1981**, *16*, 169-177.
[8107] Lewanowicz, A.; Lipinski, J.; Ruziewicz, Z. *J. Lumin.* **1981**, *26*, 159-175.
[8108] Mordzinski, A.; Grabowska, A. *J. Lumin.* **1981**, *23*, 393-404.
[8109] van der Velden, G. P. M.; de Boer, E.; Veeman, W. S. *Chem. Phys.* **1981**, *56*, 181-188.
[8110] Bendig, J.; Henkel, B.; Kreysig, D. *Ber. Bunsenges. Phys. Chem.* **1981**, *85*, 38-44.
[8111] Davydov, S. N.; Rodionov, A. N.; Shigorin, D. N.; Syutkina, O. P.; Krasnova, T. L. *Russ. J. Phys. Chem.* **1981**, *55*, 444-445.
[8112] Poletti, A.; Murgia, S. M.; Cannistraro, S. *Photobiochem. Photobiophys.* **1981**, *2*, 167-172.
[8113] Asano, M.; Koningstein, J. A. *Chem. Phys.* **1981**, *57*, 1-10.
[8114] Gorman, A. A.; Gould, I. R.; Hamblett, I. *J. Am. Chem. Soc.* **1981**, *103*, 4553-4558.
[8115] Klein, R.; Tatischeff, I.; Bazin, M.; Santus, R. *J. Phys. Chem.* **1981**, *85*, 670-677.
[8116] Capellos, C. *J. Photochem.* **1981**, *17*, 213-225.
[8117] Jordan, A. D.; Fischer, G.; Ross, I. G. *J. Mol. Spectrosc.* **1981**, *87*, 345-356.
[8118] Anderson, R. W.; Knox, W. *J. Lumin.* **1981**, *24-25*, 647-650.
[8119] Jacques, P.; Braun, A. M. *Helv. Chim. Acta* **1981**, *64*, 1800-1806.
[8120] Harriman, A.; Hosie, R. J. *J. Chem. Soc., Faraday Trans. 2* **1981**, *77*, 1695-1702.
[8121] Harriman, A. *J. Chem. Soc., Faraday Trans. 2* **1981**, *77*, 1281-1291.
[8122] Davidson, R. S.; Goodwin, D.; Fornier de Violet, Ph. *Chem. Phys. Lett.* **1981**, *78*, 471-474.

[8123] Damranyan. A. P.; Kuz'min, V. A. *Dokl. Phys. Chem.* **1981**, *260*, 938-941.

[8124] Herkstroeter, W. G.; Merkel, P. B. *J. Photochem.* **1981**, *16*, 331-341.

[8125] Goerner, H.; Schulte-Frohlinde, D. *J. Phys. Chem.* **1981**, *85*, 1835-1841.

[8126] Amirzadeh, G.; Schnabel, W. *Makromol. Chem.* **1981**, *182*, 2821-2835.

[8127] Becker, R. S.; Bensasson, R. V.; Salet, C. *Photochem. Photobiol.* **1981**, *33*, 115-116.

[8128] Mahaney, M.; Huber, J. R. *J. Mol. Spectrosc.* **1981**, *87*, 438-448.

[8129] Monti, S.; Gardini, E.; Bortolus, P.; Amouyal, E. *Chem. Phys. Lett.* **1981**, *77*, 115-119.

[8130] Palewska, K. *Chem. Phys.* **1981**, *58*, 21-28.

[8131] Jain, K. M.; Misra, T. N. *Spectrosc. Lett.* **1981**, *14*, 157-162.

[8132] Kemp, T. J.; Martins, L. J. A. *J. Chem. Soc., Faraday Trans. 1* **1981**, *77*, 1425-1435.

[8133] Leigh, W. J.; Arnold, D. R.; Baines, K. M. *Tetrahedron Lett.* **1981**, *22*, 909-912.

[8134] Wagner, P. J.; Siebert, E. J. *J. Am. Chem. Soc.* **1981**, *103*, 7329-7335.

[8135] Taherian, M. -R.; Maki, A. H. *Chem. Phys.* **1981**, *55*, 85-96.

[8136] Motten, A. G.; Kwiram, A. L. *J. Chem. Phys.* **1981**, *75*, 2608-2615.

[8137] Kuboyama, A.; Matsuzaki, S. Y. *Bull. Chem. Soc. Jpn.* **1981**, *54*, 3635-3638.

[8138] Sveshnikova, E. B.; Kondakova, V. P. *Opt. Spectrosc. (USSR)* **1981**, *50*, 477-479.

[8201] Lessing, H. E.; Richardt, D.; von Jena, A. *J. Mol. Struct.* **1982**, *84*, 281-292.

[8202] American Petroleum Institute Project-44 *Selected Ultraviolet Spectral Data*; Thermodyanamics Research Center Hydrocarbon Project; College Station, TX, 1945-1982, Vol. I-IV, .

[8203] Huppert D.; Rand, S. D.; Reynolds, A. H.; Rentzepis, P. M. *J. Chem. Phys.* **1982**, *77*, 1214-1224.

[8204] Shizuka, H.; Ueki, Y.; Iizuka, T.; Kanamaru, N. *J. Phys. Chem.* **1982**, *86*, 3327-3333.

[8205] Hirayama, S. *J. Chem. Soc., Faraday Trans. 1* **1982**, *178*, 2411-2421.

[8206] Inoue, H.; Hida, M.; Nakashima, N.; Yoshihara, K. *J. Phys. Chem.* **1982**, *86*, 3184-3188.

[8207] Koehler, G.; Kittel, G.; Getoff, N. *J. Photochem.* **1982**, *18*, 19-27.

[8208] Tway, P. C.; Love, L. J. C. *J. Phys. Chem.* **1982**, *86*, 5223-5226.

[8209] Guesten, H.; Heinrich, G. *J. Photochem.* **1982**, *18*, 9-17.

[8210] Shizuka, H.; Obuchi, H. *J. Phys. Chem.* **1982**, *86*, 1297-1302.

[8211] Chattopadhyay, S. K.; Das, P. K.; Hug, G. L. *J. Am. Chem. Soc.* **1982**, *104*, 4507-4514.

[8212] Velsko, S. P.; Fleming, G. R. *J. Chem.Phys.* **1982**, *76*, 3553-3562.

[8213] Goemer, H. *J. Phys. Chem.* **1982**, *86*, 2028-2035.

[8214] Boldridge, D. W.; Scott, G. W.; Spiglanin, T. A. *J. Phys. Chem.* **1982**, *86*, 1976-1979.

[8215] Velsko, S. P.; Fleming, G. R. *Chem.Phys.* **1982**, *65*, 59-70.

[8216] Yokoyama, K. *Chem. Phys. Lett.* **1982**, *92*, 93-96.

[8217] Yamauchi, S.; Ueno, T.; Hirota, N. *Mol. Phys.* **1982**, *47*, 1333-1348.

[8218] Venediktov, E. A.; Krasnovsky, A. A. Jr. *Zh. Prikl. Spektrosk.* **1982**, *36*, 152-154.

[8219] Kalyanasundaram, K.; Neumann Spallart, M. *J. Phys. Chem.* **1982**, *86*, 5163-5169.

[8220] Graetzel, C. K.; Graetzel, M. *J. Phys. Chem.* **1982**, *86*, 2710-2714.

[8221] Darwent, J. R.; Douglas, P.; Harriman, A.; Porter, G.; Richoux, M. -C. *Coord. Chem. Rev.* **1982**, *44*, 83-126.

[8222] Ronfard-Haret, J. C.; Averbeck, D.; Bensasson, R. V.; Bisagni, E.; Land, EJ. *Photochem. Photobiol.* **1982**, *35*, 479-489.

[8223] Hirano, H.; Azumi, T. *Chem. Phys. Lett.* **1982**, *86*, 109-112.

[8224] Wolleben, J.; Testa, A. C. *J. Photochem.* **1982**, *19*, 267-269.

[8225] Suga, K.; Kinoshita, M. *Bull. Chem. Soc. Jpn.* **1982**, *55*, 1695-1704.

[8226] Das, P. K.; Hug, G. L. *Photochem. Photobiol.* **1982**, *36*, 455-461.

[8227] Mazzucato, U. *Pure Appl. Chem.* **1982**, *54*, 1705-1721.

[8228] Bonneau, R. *J. Am. Chem. Soc.* **1982**, *104*, 2921-2923.

[8229] Bartocci, G.; Mazzucato, U. *J. Lumin.* **1982**, *27*, 163-175.

[8230] Taherian, M. -R.; Maki, A. H. *Chem.Phys.* **1982**, *68*, 179-189.

[8231] Specht, D. P.; Manic, P. A.; Farid, S. *Tetrahedron* **1982**, *38*, 1203-1211.

[8232] Mirbach, M. F.; Mirbach, M. J.; Saus, A. *Chem. Rev.* **1982**, *82*, 5976.

[8233] Verheijdt, P. L.; Cerfontain, H. *J. Chem. Soc., Perkin Trans 2* **1982**, 1541-1547.

[8234] Cundall, R. B.; Grant, D. J. W.; Shulman, N. H. *J. Chem. Soc., Faraday Trans. 2* **1982**, *78*, 737-750.

[8235] Deeg, F. W.; Braeuchle, Chr.; Voitlaender, J. *J. Chem. Phys.* **1982**, *64*, 427-436.

[8236] Tetreau, C ; Lavalette, D.; Cabaret, D.; Geraghty, N.; Welvart, Z. *Nouv. J. Chim.* **1982**, *6*, 461-465.

[8237] Wong, P. C. *Can. J. Chem.* **1982**, *60*, 339-441.

[8238] Martins, L. J. A.; Kemp, T. J. *J. Chem. Soc., Faraday Trans. 1* **1982**, *78*, 519-531.

[8239] Clarke, R. H.; Mitra, P.; Vinodgopal, K. *J. Chem. Phys.* **1982**, *77*, 5288-5297.

[8240] Kokai, F.; Azumi, T. *J. Chem. Phys.* **1982**, *77*, 2757-2762.

[8241] Kobashi, H.; Ikawa, H.; Kondo, R.; Morita. T. *Bull. Chem. Soc. Jpn.* **1982**, *55*, 3013-3018.

[8242] Bolotko, L. M.; Gruzinskii, V. V.; Danilova, V. I.; Kopylova, T. N. *Opt. Spectrosc. (USSR)* **1982**, *52*, 379-381.

[8243] Teply, J.; Mehnert, R.; Brede, O.; Fojtik, A. *Radiochem. Radioanal. Lett.* **1982**, *53*, 141-151.

[8244] Gurinovich, G. P.; Zenkevich, E. I.; Sagun, E. I. *J. Lumin.* **1982**, *26*, 297-317.

[8245] Maki, A. H. *Triplet State ODMR Spectroscopy: Techniques and Applications to Biophysical Systems*; Wiley; New York, 1982, 479-557.

[8246] Zander, M. *Z. Naturforsch., Teil A* **1982**, *37A*, 1348-1352.

[8247] Arce, R.; Rivera, M. *Photochem. Photobiol.* **1982**, *35*, 737-740.

[8301] Korobov, V. E.; Chibisov, A. K. *Russ. Chem. Rev.* **1983**, *52*, 27-42.

[8302] Meech, S. R.; Phillips, D. *J. Photochem.* **1983**, *23*, 193-217.

[8303] Chattopadhyay, S. K.; Kumar, Ch. V.; Das, P. K. *Chem. Phys. Lett* **1983**, *98*, 250-254.

[8304] Bunce, N. J.; Hayes, P. J.; Lemke, M. E. *Can. J. Chem.* **1983**, *61*, 1103-1104.

[8305] Leismann, H.; Scharf, H. -D.; Strassburger, W.; Wollmer, A. *J. Photochem.* **1983**, *21*, 275-280.

[8306] Wermuth, G. *Z. Naturforsch., Teil A* **1983**, *38A*, 368-377.

[8307] Visser, R. J.; Varma, C. A. G. O.; Konijnenberg, J.; Weisenborn, P. C. M. *J. Mol. Struct.* **1984**, *114*, 105-112.

[8308] Brenner, K.; Lipinski, J.; Ruziewicz, Z. *J. Lumin.* **1983**, *28*, 13-26.

[8309] Das, P. K.; Hug, G. L. *J. Phys. Chem.* **1983**, *87*, 49-54.

[8310] Darmanyan, A. P. *Chem. Phys. Lett.* **1983**, *96*, 383-389.

[8311] Pepmiller, C.; Bedwell, E.; Kuntz, R. R.; Ghiron, C. A. *Photochem. Photobiol.* **1983**, *38*, 273-280.

[8312] Meech, S. R.; O'Connor, D. V.; Phillips, D.; Lee, A. G. *J. Chem. Soc., Faraday Trans. 2* **1983**, *79*, 1563-1584.

[8313] Basara, H. *J. Lumin.* **1983**, *28*, 73-86.

[8314] Petrushenko, K. B.; Vokin, A. I.; Turchaninov, V. K.; Frolov, Yu. L. *Bull. Acad. Sci. USSR, Div. Chem. Sci.* **1983**, *32*, 2151-2152.

[8315] Kalyanasundaram, K. *J. Chem. Soc., Faraday Trans. 2* **1983**, *79*, 1365-1374.

[8316] Harriman, A.; Porter, G.; Wilowska, A. *J. Chem. Soc., Faraday Trans. 2* **1983**, *79*, 807-816.

[8317] Craw, M.; Bensasson, R. V.; RonfardHaret, J. C.; Sa E Melo, M. T.; Truscott, T. G. *Photochem. Photobiol.* **1983**, *37*, 611-615.

[8318] Lee, J.; Li, F.; Bernstein, E. R. *J. Phys. Chem.* **1983**, *87*, 260-265.

[8319] Kumar, C. V.; Chattopadhyay, S. K.; Das, P. K. *Photochem. Photobiol.* **1983**, *38*, 141-152.

[8320] Hamai, S.; Hirayama, F. *J. Phys. Chem.* **1983**, *87*, 83-89.

[8321] Najbar, J.; Jarzeba, W.; Urbanek, Z. H. *Chem. Phys.* **1983**, *79*, 245-253.

[8322] Safarzadeh-Amiri, A.; Verrall, R. E.; Steer, R. P. *Can. J. Chem.* **1983**, *61*, 894-900.

[8323] Gorman, A. A.; Hamblett, I. *Chem. Phys. Lett* **1983**, *97*, 422-426.

[8324] Baiardo, J.; Vala, M.; Trabjerg, I. *Chem. Phys.* **1983**, *80*, 305-315.

[8325] Taherian, M. R.; Maki, A. H. *Chem. Phys. Lett.* **1983**, *96*, 541-546.

[8326] Sarphatie, L. A.; Verheijdt, P. L.; Cerfontain, H. *Recl. Trav. Chim. Pays-Bas* **1983**, *102*, 9-13.

[8327] Nishimoto, S.; Izukawa, T.; Kagiya. T. *J. Chem. Soc., Perkin Trans. 2* **1983**, 1147-1152.

[8328] Swaminathan, M.; Dogra, S. K. *Spectrochim. Acta, Part A* **1984**, *39A*, 973-977.

[8329] Ito, Y.; Nishimura, H.; Umehara, Y.; Yamada, Y.; Tone, M.; Matssura, T. *J. Am. Chem. Soc.* **1983**, *105*, 1590-1597.

[8330] Hilburn, S. G.; Power, R. K.; Martin, K. A.; Nishimura, A. M. *Chem. Phys. Lett.* **1983**, *100*, 429-435.

[8331] Sarkar, S. K.; Ghoshal, S. K.; Kastha, G. S. *Proc. - Indian Acad. Sci., Chem. Sci.* **1983**, *92*, 47-58.

[8332] Jardon, P.; Azarnouche, B.; Corval, A.; Gautron, R. *J. Chim. Phys. Phys.-Chim. Biol.* **1983**, *80*, 603-608.

[8333] Slama, H; Braeuchle, C.; Voitlaender, J. *Chem. Phys. Lett.* **1983**, *102*, 307-311.

[8334] Hurley, J. K.; Sinai, N.; Linschitz, H. *Photochem. Photobiol.* **1983**, *38*, 9-14.

[8335] Murray, D.; Becker, R. S. *J. Phys. Chem.* **1983**, *87*, 625-628.

[8336] Savory, B.; Tumbull, J. H. *J. Photochem.* **1983**, *23*, 171-181.

[8401] Naito, I.; Schnabel, W. *Bull. Chem. Soc. Jpn.* **1984**, *57*, 771-775.

[8402] Diverdi, L. A.; Topp, M. R. *J. Phys. Chem.* **1984**, *88*, 3447-3451.

[8403] Usacheva, M. N.; Osipov, V. V.; Drozdenko, I. V.; Dilung, I.I. *Russ. J. Phys. Chem.* **1984**, *58*, 1550-1553.

[8404] Craw, M.; Truscott, T. G.; Dall'Acqua, F.; Guiotto, A.; Vedaldi, D.; Land, E. J. *Photobiochem. Photobiophys.* **1984**, *7*, 359-365.

[8405] Hamanoue, K.; Tai, S.; Hidaka, T.; Nakayama, T.; Kimoto, M.; Teranishi, H. *J. Phys. Chem.* **1984**, *88*, 4380-4384.

[8406] Darmanyan, A. P. *Chem. Phys. Lett.* **1984**, *110*, 89-94.

[8407] Bubekov, Yu. J.; Kabelka, V.; Lysak, N. A.; Milyauskas, A.; Tolstorozhev, G. B. *Bull. Acad. Sci. USSR, Phys. Ser.* **1984**, *48*, 137-141.

[8408] Gorman, A. A.; Hamblett, I.; Harrison, R. J. *J. Am. Chem. Soc.* **1984**, *106*, 6952-6955.

[8409] Previtali, C. M.; Ebbesen, T. W. *J. Photochem.* **1984**, *27*, 9-15.

[8410] Yip, R. W.; Sharma, D. K.; Giasson, R.; Gravel, D. *J. Phys. Chem.* **1984**, *88*, 5770-5772.

[8411] Hashimoto, S.; Thomas, J. K. *J. Phys. Chem.* **1984**, *88*, 4044-4049.
[8412] Das, P. K.; Muller, A. J.; Griffin, G. W.; Gould, I. R.; Tung, C. -H.; Turro, N. J. *Photochem. Photobiol.* **1984**, *39*, 281-285.
[8413] Chattopadhyay, S. K.; Kumar, C. V.; Das, P. K. *J. Photochem.* **1984**, *26*, 39-47.
[8414] Kumar, C. V.; Chattopadhyay, S. K.; Das, P. K. *Chem. Phys. Lett.* **1984**, *106*, 431-436.
[8415] Jones, G. II; Bergmark, W. R.; Jackson, W. R. *Opt. Commun.* **1984**, *50*, 320-324.
[8416] Grajcar, L.: Ivanoff, N.; Delouis, J. F.; Faure, J. *J. Chim. Phys. Phys.-Chim. Biol.* **1984**, *81*, 33-38.
[8417] Chattopadhyay, S. K.; Kumar, C. V.; Das, P. K. *J. Chem. Soc., Faraday Trans. 1* **1984**, *80*, 1151-1161.
[8418] Tine, A.; Aaron, J. -J. *Can. J. Spectrosc.* **1984**, *29*, 121-130.
[8419] Eftink, M. R.; Ghiron, C. A. *Biochemistry* **1984**, *23*, 3891-3899.
[8420] Pac, C.; Fukunaga, T.; Ohtsuki, T.; Sakurai, H. *Chem. Lett* **1984**, 1847-1850.
[8421] Davis, H. F.; Chattopadhyay, S. K.; Das, P. K. *J. Phys. Chem.* **1984**, *88*, 2798-2803.
[8422] Das, P. K.; Muller, A. J.; Griffin, G. W. *J. Org. Chem.* **1984**, *49*, 1977-1985.
[8423] Wismontski-Knittel, T.; Das. P. K. *J. Phys. Chem.* **1984**, *88*, 2803-2808.
[8424] Bensasson, R. V.; Land, E. J.; Liu, R. S. H.; Lo. K. K. N.; Truscott, T. G. *Photochem. Photobiol.* **1984**, *39*, 263-265.
[8425] Biczok, L.; Berces, T.; Forgeteg, S.; Marta, F. *J. Photochem.* **1984**, *27*, 41-48.
[8426] Laszlo, B.; Forgeteg, S.; Berces, T.; Marta, F. *J. Photochem.* **1984**, *27*, 49-59.
[8427] Sarkar, S. K.; Maiti, A.; Kastha, G. S. *Chem. Phys. Lett.* **1984**, *105*, 355-358.
[8428] Nakamura, S.; Kanamaru, N.; Nohara, S.; Nakamura, H.; Saito, Y.; Tanaka, J.; Sumitani, M.; Nakashima, N.; Yoshihara, K. *Bull. Chem. Soc. Jpn.* **1984**, *57*, 145-150.
[8429] Kalyanasundaram, K. *Inorg. Chem.* **1984**, *23*, 2453-2459.
[8430] Lee, W. A.; Graelzel, M.: Kalyanasundaram, K. *Chem. Phys. Lett.* **1984**, *107*, 308-313.
[8431] Bensasson, R. V. *NATO ASI Ser., Ser. A* **1985**, *85*, 241-254.
[8432] Malkin, Y. N.; Dvomikov, A. S.; Kuz'min, V. A. *J. Photochem.* **1984**, *27*, 343-354.
[8433] Goerner, H. *Ber. Bunsenges. Phys. Chem.* **1984**, *88*, 1199-1208.
[8434] Kumar, C. V.; Qin, L.; Das, P. K. *J. Chem. Soc.. Faraday Trans. 2* **1984**, *80*, 783-793.
[8435] Maeda, Y.; Okada, T.; Malaga, N.; Irie, M. *J. Phys. Chem.* **1984**, *88*, 1117-1119.
[8436] Zander, M. *Z. Naturforsch., Teil A* **1984**, *39A*, 1145-1146.
[8437] Paone, S.; Moule, D.C.; Bruno, A. E.; Steer, R. P. *J. Mol. Spectrosc.* **1984**, *107*, 1-11.
[8438] Takuma, K.; Kirmura, T.; Sonoda, T.;Kobayashi, H. *Chem. Lett.* **1984**, 881-4.
[8439] Savory, B.; Turnbull, J. H. *J. Photochem.* **1984**, *24*, 355-371.
[8440] Iwasaki, N.; Misra, T. N.; Kinoshita, M. *J. Lumin.* **1984**, *29*, 83-92.
[8441] Yamauchi, S.; Miyake, K.; Hirota, N. *Mol. Phys.* **1984**, *53*, 479-491.
[8442] Shizuka, H.; Sato, Y.; Ueki, Y.; Ishikawa, M.; Kumada, M. *J. Chem. Soc., Faraday Trans. 1* **1984**, *80*, 341-357.
[8443] Ghoshal, S. K.; Maiti, A. K.; Kastha, G. S. *J. Lumin.* **1984**, *31-32*, 541-445.
[8444] Kumar, C. V.; Davis, H. F.; Das, P. K. *Chem. Phys. Lett.* **1984**, *109*, 184-189.
[8445] Baral-Tosh, S.; Chattopadhyay, S. K.; Das, P. K. *J. Phys. Chem.* **1984**, *88*, 1404-1408.
[8446] Malkin, Ya. N.; Pirogov, N. O.; Kuz'min, V. A. *J. Photochem.* **1984**, *26*, 193-202.
[8447] Mehnert, R.; Brede, O.; Cserep, G. *Radiat. Phys. Chem.* **1984**, *24*, 455-457.
[8448] Urruti, E. H.; Kilp, T. *Macromolecules* **1984**, *17*, 50-54.

[8449] Menzel, R.; Rapp, W. *Chem. Phys.* **1984**, *89*, 445-455.

[8450] Maiti, A.; Sarkar, S. K.; Kastha, G. S. *Proc. - Indian Acad. Sci., Chem. Sci.* **1984**, *93*, 1-11.

[8451] Shizuka, H.; Obuchi, H.; Ishikawa, M.; Kumada, M. *J. Chem. Soc., Faraday Trans. 1* **1984**, *80*, 383-401.

[8452] Venturi, M.; Mulazzani, Q. G.; Hoffman, M. Z. *Radiat. Phys. Chem.* **1984**, *23*, 229-236.

[8501] Chattopadhyay, S. K.; Kumar, C. V.; Das, P. K. *J. Photochem.* **1985**, *30*, 81-91.

[8502] Komorowski, S. J.; Grabowski. Z. R.; Zielenkiewicz, W. *J. Photochem.* **1985**, *30*, 141-151.

[8503] Bryukhanov, V. V.; Levshin, L. V.; Smagulov, Zh. K.; Muldakhmetov, Z. M. *Opt. Spectrosc. (USSR)* **1985**, *59*, 540-542.

[8504] Hirayama, S.; Lampert, R. A.; Phillips, D. *J. Chem. Soc., Faraday Trans. 2* **1985**, *81*, 371-382.

[8505] Jardon, P.; Gautron, R. *J. Chim. Phys. Phys.-Chim. Biol.* **1985**, *82*, 353-360.

[8506] Abdullah, K. A.; Kemp, T. *J. Photochem.* **1985**, *28*, 61-69.

[8507] Rohatgi-Mukherjee, K. K.; Bhattacharyya, K.; Das, P. K. *J. Chem. Soc.. Faraday Trans. 2* **1985**, *81*, 1331-1344.

[8508] Zalesskaya, G. A.; Blinov, S. I. *Sov. Phys. Dokl.* **1985**, *30*, 297-299.

[8509] Mordzinski, A.; Komorowski, S. J. *Chem. Phys. Lett.* **1985**, *114*, 172-177.

[8510] Tinnemans, A. H. A.; den Ouden, B.; Bos, H. J. T.; Mackor, A. *Neth. Chem. Soc.* **1985**, *104*, 109-116.

[8511] Previtali, C. M.; Ebbesen, T. W. *J. Photochem.* **1985**, *30*, 259-267.

[8512] Alfassi, Z. B.; Previtali, C. M. *J. Photochem.* **1985**, *30*, 127-132.

[8513] Baumann, H.; Becker, H. G. O.; Kronfeld, K. P.; Pfeifer, D.; Timpe. H. -J. *J. Photochem.* **1985**, *28*, 393-403.

[8514] Visser, R. J.; Weisenborn, P. C. M.; Varma, C. A. G. O. *Chem. Phys. Lett.* **1985**, *113*, 330-336.

[8515] Suter, G. W.; Wild, U. P.; Brenner, K.; Ruziewicz, Z. *Chem. Phys.* **1985**, *98*, 455-463.

[8516] Zander, M. *Z. Naturforsch., A. Phys., Phys. Chem., Kosmophys.* **1985**, *40A*, 497-502.

[8517] Palmer, T. F.; Parmar, S. S.. *J. Photochem.* **1985**, *31*, 273-288.

[8518] Jones, G. II; Jackson, W. R.; Choi, C.; Bergmark, W. R. *J. Phys. Chem.* **1985**, *89*, 294-300.

[8519] Jinguji, M.; Ashizawa, M.; Nakazawa, T.; Tobita, S.; Hikida, T.; Mori, Y. *Chem. Phys. Lett.* **1985**, *121*, 400-404.

[8520] Mau, A. W. -H.; Johansen, O.; Sasse, W. H. F. *Photochem. Photobiol.* **1985**, *41*, 503-509.

[8521] Baba, M. *J. Chem. Phys.* **1985**, *83*, 3318-3326.

[8522] Chattopadhyay, S. K.; Kumar, C. V.; Das, P. K. *Photochem. Photobiol.* **1985**, *42*, 17-24.

[8523] Kanemoto, A.; Kikuchi, K.; Kokubun, H. *J. Phys. Chem.* **1985**, *89*, 3567-3570.

[8524] Bhattacharyya, K.; Das, P. K. *Chem. Phys. Lett.* **1985**, *116*, 326-332.

[8525] Gorman, A. A.; Hamblett, I.; Irvine, M.; Raby, P.; Standen, M. C.; Yeates, S. *J. Am. Chem. Soc.* **1985**, *107*, 4404-4411.

[8526] Bhattacharyya, K.; Kumar, C. V.; Das, P. K.; Jayasree, B.; Ramamurthy, V. *J. Chem. Soc., Faraday Trans. 2* **1985**, *81*, 1383-1393.

[8527] Zimmerman, H. E.; Lynch. D. C. *J. Am. Chem. Soc.* **1985**, *107*, 7745-7756.

[8528] Arnold, B.; Donald, L.; Jurgens, A.; Pincock, J.A. *Can. J. Chem.* **1985**, *63*, 3140-3146.

[8529] Zimmerman, H. E.; Caufield. C. E.; King. R. K. *J. Am. Chem. Soc.* **1985**, *107*, 7732-7744.

[8530] Val'kova, G. A.; Sakhno, T. V.; Shigorin, D. N.; Davydov, S. N.; Andrievskii, A. N.; Eremenko, L. V. *Russ. J. Phys. Chem.* **1985**, *59*, 1050-1053.

[8531] Maiti, A. K.; Sarkar, S. K.; Kastha, G. S. *Proc. - Indian Acad. Sci., Chem. Sci.* **1985**, *95*, 409-419.

[8532] Slama, H.; Braeuchle, C.; Voitlaender, J. *Chem. Phys* **1985**, *92*, 91-96.

[8533] Previtali, C. M. *J. Photochem.* **1985**, *31*, 2338.

[8534] Shizuka, H.; Hagiwara, H.; Fukushima, M. *J. Am. Chem. Soc.* **1985**, *107*, 7816-7823.

[8535] Ikeyama, T.; Azumi, T. *J. Phys. Chem.* **1985**, *89*, 5332-5333.

[8601] Kramer, H. E. A. *Chimia* **1986**, *40*, 160-169.

[8602] Maiti, A.; Kastha, G. S. *Indian J. Phys., B* **1986**, *60B*, 336-346.

[8603] Khasawneh, I. M.; Winefordner, J. D. *Anal. Chim. Acta* **1986**, *184*, 307-310.

[8604] Hoshi, M.; Kikuchi, K.; Kokubun, H.; Yamamoto, S. -A. *J. Photochem.* **1986**, *34*, 63-71.

[8605] Wasielewski, M. R.; Kispert, L. D. *Chem. Phys. Lett.* **1986**, *128*, 238-243.

[8606] Jabben, M.; Garcia, N. A.; Braslavsky, S. E.; Schaffner, K. *Photochem. Photobiol.* **1986**, *43*, 127-131.

[8607] Boldridge, D. W.; Scott, G. W. *J. Chem. Phys.* **1986**, *84*, 6790-6798.

[8608] Matsushima, R.; Sakai, K. *J. Chem. Soc., Perkin Trans. 2* **1986**, 1217-1222.

[8609] Bhattacharyya, K.; Ramaiah, D.; Das, P. K.; George, M. V. *J. Phys. Chem.* **1986**, *90*, 5984-5989.

[8610] Surer, G. W.; Wild, U. P.; Holzwarth, A. R. *Chem. Phys.* **1986**, *102*, 205-214.

[8611] Waluk, J.; Komorowski, S. J. *J. Mol. Struct.* **1986**, *142*, 159-162.

[8612] Kikuchi, K.; Takahashi, T.; Koike, T.; Kokubun, H. *J. Photochem.* **1986**, *32*, 341-391.

[8613] Kim, D. *Bull. Korean Chem. Soc.* **1986**, *7*, 416-421.

[8614] Feitelson, J.; Barboy, N. *J. Phys. Chem.* **1986**, *90*, 271-274.

[8615] Aramendia, P. F.; Redmond, R. W.; Nonell, S.; Schuster, W.; Braslavsky, S. E.; Schaffner, K.; Vogel, E. *Photochem. Photobiol.* **1986**, *44*, 555-559.

[8616] Lopez-Arbeloa, I.; Rohatgi-Mukherjee, K. K. *Chem. Phys. Lett.* **1986**, *129*, 607-614.

[8617] Lewitzka, F.; Loehmannsroeben, H. -G. *Z. Phys. Chem. (Munich)* **1986**, *150*, 69-86.

[8618] Bhattacharyya, K.; Das, P. K.; Ramamurthy, V.; Rao, V. P. *J. Chem. Soc., Faraday Trans. 2* **1986**, *82*, 135-147.

[8619] Evans, C.; Weir, D.; Scaiano, J. C.; Mac Eachern, A.; Arnason, J. T.; Morand, P.; Hollebone, B.; Leitch, L. C.; Philogene, B. J. R. *Photochem. Photobiol.* **1986**, *44*, 441-451.

[8620] Meier, K.; Zweifel, H. *J. Photochem.* **1986**, *35*, 353-366.

[8621] Bhattacharyya, K.; Ramamurthy, V.; Das, P. K.; Sharat, S. *J. Photochem.* **1986**, *35*, 299-309.

[8622] Herkstroeter, W. G.; Farid, S. *J. Photochem.* **1986**, *35*, 71-85.

[8623] Wang, G. -C.; Winnik, M. A.; Schaefer, H. J.; Schmidt, W. *J. Photochem.* **1986**, *33*, 291-296.

[8624] Wagner, P. J.; Truman, R. J.; Puchalski, A. E.; Wake, R. *J. Am. Chem. Soc.* **1986**, *108*, 7727-7738.

[8625] Palit, D. K.; Mukherjee, T.; Mittal, J. P. *J. Indian Chem. Soc.* **1986**, *63*, 35-42.

[8626] Wagner, P. J.; Thomas, M. J.; Puchalski, A. E. *J. Am. Chem. Soc.* **1986**, *108*, 7739-7744.

[8627] Ndiaye, S. A.; Aaron, J. J.; Gamier, F. *J. Photochem.* **1986**, *35*, 389-394.

[8628] Arce, R.; Rodriguez, G. *J. Photochem.* **1986**, *33*, 89-97.

[8701] Malkin, Ya. N.; Ruziev, Sh.; Pirogov, N. O.; Kuz'min, V. A. *Bull. Acad. Sci. USSR, Div. Chem. Sci.* **1987**, *36*, 51-56.

[8702] Rotkiewicz, K.; Koehler, G. *J. Lumin.* **1987**, *37*, 219-225.

[8703] Duguid, R.; Maxwell, B. D.; Munozsola, Y.; Muthuramu, K.; Rasbury, V.; Singh, T. -V.; Morrison, H.; Das, P. K.; Hug, G. L. *Chem. Phys. Lett.* **1987**, *139*, 475-478.

[8704] Shim, S. C.; Kang, H. K.; Park, S. K.; Shin, E. J. *J. Photochem.* **1987**, *37*, 125-137.

[8705] Terazima, M.; Azumi, T. *Chem. Phys. Lett.* **1987**, *141*, 237-240.

[8706] Palewska, K.; Ruziewicz, Z.; Chojnacki, H. *J. Lumin.* **1987**, *39*, 75-85.

[8707] Kemp, T. J.; Parker, A. W.; Wardman, P. *Photochem. Photobiol.* **1987**, *45*, 663-666.

[8708] Ono, L; Hata, N. *Bull. Chem. Soc. Jpn.* **1987**, *60*, 2891-2897.

[8709] Rudenko, N. A.; Val'kova, G. A.; Shigorin, D. N.; Poplavskii, A. N.; Andrievskii, A.M. *J. Phys. Chem.* **1987**, *61*, 871-873.

[8710] Olba, A.; Tomas, F.; Zabala, J.; Medina, P. *J. Photochem.* **1987**, *39*, 263-272.

[8711] Arai, T.; Oguchi, T.; Wakabayashi, T.; Tsuchiya, M.; Nishimura, Y.; Oishi, S.; Sakuragi, H.; Tokumaru, K. *Bull. Chem. Soc. Jpn.* **1987**, *60*, 2937-2943.

[8712] Carmichael, I.; Helman, W. P.; Hug, G. L. *J. Phys. Chem. Ref. Data* **1987**, *16*, 239-260.

[8713] Hamanoue, K.; Nakayama, T.; Kajiwara, Y.; Yamaguchi, T.; Teranishi, H. *J. Chem. Phys.* **1987**, *86*, 6654-6659.

[8714] Shim, S. C.; Shin, E. J.; Kang, H. K.; Park, S. K. *J. Photochem.* **1987**, *36*, 163-175.

[8801] Ryzhikov, M. B.; Rodionov, A. N.; Nesterova, O. V.; Shigorin, D. N.; *Russ. J. Phys. Chem.* **1988**, *62*, 552-554.

[8802] Thompson, R. B.; Gratton, E. *Anal. Chem.* **1988**, *60*, 670-674.

[8803] Bright, F. V. *Appl. Spectrosc.* **1988**, *42*, 1531-1537.

[8804] Rtishchev, N. I.; Lebedeva, G. K.; Kvitko, I. Ya.; El'tsov, A. V. *J. Gen. Chem., USSR* **1988**, *58*, 1914-1927.

[8805] Druzhinin, S. I.; Rodchenkov, G. M.; Uzhinov, B.M. *Chem. Phys.* **1988**, *128*, 383-394.

[8806] Hedstrom, J.; Sedarous, S.; Prendergast, F. G. *Biochemistry* **1988**, *27*, 6203-6208.

[8807] Sa e Melo, T.; Macanita, A.; Prieto. M.; Bazin, M. *Photochem. Photobiol.* **1988**, *48*, 429-437.

[8808] Kunjappu, J. T.; Rao, K. N.; *Indian J. Chem., Sect. A* **1988**, *27A*, 1-3.

[8809] Terazima, M.; Azumi, T. *Chem. Phys. Lett.* **1988**, *145*, 286-288.

[8901] Timpe, H. -J.; Kronfeld, K. -P. *J. Photochem. Photobiol., A* **1989**, *46*, 253-267.

[8902] Kikuchi, K. *Triplet-Triplet Absorption Spectra*; Bunshin Publishing Co., Tokyo, Japan, 1989, 189.

[8903] Schoof, S.; Guesten, H. *Ber. Bunsenges. Phys. Chem.* **1989**, *93*, 864-870.

[8904] Tanigaki, K.; Taguchi, N.; Yagi, M.; Higuchi, J. *Bull. Chem. Soc. Jpn.* **1989**, *62*, 668-673.

[8905] Goerner, H. *J. Phys. Chem.* **1989**, *93*, 1826-1832.

[8906] Gopidas, K. R.; Kamat, P. V. *J. Photochem. Photobiol., A* **1989**, *48*, 291-301.

[8907] Kirstein, S.; Moehwald, H.; Shimomura, M. *Chem. Phys. Lett.* **1989**, *154*, 303-308.

[8908] Bruce, J. M.; Gorman, A. A.; Hamblett, I.; Kerr, C. W.; Lambert, C.; McNeeney, S. P. *Photochem. Photobiol.* **1989**, *49*, 439-445.

[8909] Maiti, A. K.; Kastha, G. S. *J. Lumin.* **1989**, *43*, 383-385.

[8910] Zander, M.; Kirsch, G. Z. Naturforsch., A, Phys. Sci. 1989, 44A, 205-209.

[8911] Boate, D. R.; Johnston, L. J.; Scaiano, J. C. Can. J. Chem. 1989, 67, 927-932.

[8912] Ito, Y.; Uozu, Y.; Arai, H.; Matsuura, T. J. Org. Chem. 1989, 54, 506-509.

[8913] Guerin, B.; Johnston, L. J. Can. J. Chem. 1989, 67, 473-480.

[8914] Netto-Ferreira, J. C.; Weir, D.; Scaiano, J. C. J. Photochem. Photobiol., A 1989, 48, 345-352.

[8915] Schafer, O.; Allan, M.; Haselbach, E.; Davidson, R. S. Photochem. Photobiol. 1989, 50, 717-719.

[9001] Goerner, H. J. Photochem. Photobiol., B 1990, 5, 359-377.

[9101] Arbogast, J. W.; Darmanyan, A. P.; Foote, C. S.; Rubin, Y.; Diederich, F. N.; Alvarez, M. M.; Anz, S. J.; Whetten, R. L. J. Phys. Chem. 1991, 95, 11-12.

[9102] Ebbesen, T. W.; Tanigaki, K.; Kuroshima, S. Chem. Phys. Lett. 1991, 181, 501-504.

[9103] Hung, R. R.; Grabowski, J. J. J. Phys. Chem. 1991, 95, 6073-6075.

[9104] Arbogast, J. W.; Foote, C. S. J. Am. Chem. Soc. 1991, 113, 8886-8889.

[9105] Tanigaki, K.; Ebbesen, T. W.; Kuroshima, S. Chem. Phys. Lett. 1991, 185, 189-192.

[9106] Wasielewski, M. R.; O'Neil, M. P.; Lykke, K. R.; Pellin, M. J.; Gruen, D. M. J. Am. Chem. Soc. 1991, 113, 2774-2776.

[9107] Brun, A. M.; Harriman, A. J. Am. Chem. Soc. 1991, 113, 8153-8159.

[9108] Bonneau, R.; Carmichael, I.; Hug, G. L. Pure Appl. Chem. 1991, 63, 289-299.

[9201] Palit, D. K.; Sapre, A. V.; Mittal, J. P.; Rao, C. N. R. Chem. Phys. Lett. 1992, 195, 1-6.

[9202] Dimitrijevic, N. M.; Kamat, P. V. J. Phys. Chem. 1992, 96, 4811-4814.

[9203] Hung, R. R.; Grabowski, J. J. Chem. Phys. Lett. 1992, 192, 249-253.

[9204] Haley, J. L.; Fitch, A. N.; Goyal, R.; Lambert, C.; Truscott, T. G.; Chacon, J. N.; Stirling. D.; Schalch, W. J. Chem. Soc., Chem. Commun. 1992, 1175-1176.

[9205] Biczok, L.; Berces, T.; Marta, F. J. Phys. Chem. 1993, 97, 8895-8899.

[9206] Schael, F.; Loehmannsroeben, H. -G. J. Photochem. Photobiol., A: Chemistry 1992, 69, 27-32.

[9207] Terazima, M.; Hirota, N. Chem. Phys. Lett. 1992, 189, 560-564.

[9208] Kamat, P. V.; Das, S.; Thomas, K. G.; George, M. V. J. Phys. Chem. 1992, 96, 195-199.

[9209] Das, S.; Thomas, K. G.; Ramanathan, R.; George, M. V.; Kamat, P. V. J. Phys. Chem. 1993, 97, 13625-13628.

[9210] Szymanski, M.; Maciejewski, A.; Steer, R. P. J. Phys. Chem. 1992, 96, 7857-7863.

[9211] Schuster, D. I.; Woning, J.; Kaprinidis, N. A.; Pan, Y.; Cai, B.; Barra, M.; Rhodes, C.A. J. Am. Chem. Soc. 1992, 114, 7029-7034.

[9301] Becker, R. S.; Natarajan, L. V. J. Phys. Chem. 1993, 97, 344-349.

[9401] Jones, G. II; Feng, Z.; Bergmark, W. R. J. Phys. Chem. 1994, 98, 4511-4516.

[9402] Anderson, L. J.; An Y-Z.;Rubin, Y.; Foote, C. S. J. Am. Chem. Soc. 1994, 116, 9763-9764.

[9403] Adam, W.; Fragale, G.; Klapstein, D.; Nau, W. M.; Wirz, J. J. Am. Chem. Soc. 1995, 117, 12578-12592.

[9404] Wintgens, V.; Valat, P.; Kossanyi, J.; Biczok, L.; Demeter, A.; Berces, T. J. Chem. Soc., Faraday Trans. 1994, 90, 411-421.

[9405] Demeter, A.; Berces, T.; Biczok, L.; Wintgens, V.; Valat, P.; Kossanyi, J. J. Chem. Soc., Faraday Trans. 1994, 90, 2635-2641.

[9406] Jenks, W. S.; Lee, W.; Shutters, D. J. Phys. Chem. 1994, 98, 2282-2289.

[9501] Sander, T.; Loehmannsroeben, H. -G.; Langhals, H. J. Photochem. Photobiol., A 1995, 86, 103-108.

[9502] Lin, S.-K.; Shiu, L.-L.; Chien, K.-M; Luh T.-Y.; Lin T.-I. *J. Phys. Chem.* **1995**, *99*, 105-111.

[9601] Elidei, F.; Aloisi, G. G.; Lattarini, C.; Latterini, L.; Dall'Acqua, F.; Guiotto, A. *Photochem. Photobiol.* **1996**, *64*, 67-74.

[9602] Maciejewski, A.; Szymanski, M.; Steer, R. P. *J. Photochem. Photobiol., A* **1996**, *100*, 43-52.

[9603] Becker, R. S.; de Melo, J. S.; Macanita, A. L.; Elisei, F. *J. Phys. Chem.* **1996**, *100*, 18683-18695.

[9604] Wintgens, V.; Valat, P.; Kossanyi, J.; Demeter, A.; Biczok, L.; Berces, T. *J. Photochem. Photobiol., A* **1996**, *93*, 109-117.

[9605] Pal, P.; Zeng, H.; Durocher, G.; Girard, D.; Li, T.; Gupta, A. K.; Giasson, R.; Blanchard, L.; Gaboury, L.; et al. *Photochem. Photobiol.* **1996**, *63*, 161-168.

[9701] Manoj, N.; Ajit Kumar, R.; Gopidas, K. R. *J. Photochem. Photobiol., A* **1997**, *109*, 109-118.

[9702] Jovanovic, S. V.; Morris, D. G.; Pliva, C. N.; Scaiano, J. C. *J. Photochem. Photobiol., A* **1997**, *107*, 153-158.

[9801] Srividya, N.; Ramamurthy, P.; Ramakrishnan, V. T. *Spectrochim. Acta, Part A* **1998**, *52A*, 245-253.

[9802] Grabner, G.; Rechthaler, K.; Koehler, G. *J. Phys. Chem. A* **1998**, *102*, 689-696.

[9803] Palit, D. K.; Mohan, H.; Mittal, J. P. *J. Phys. Chem. A* **1998**, *102*, 4456-4461.

[9804] Fujitsuka, M.; Sato, T.; Sezaki, F.; Tanaka, K.; Watanabe, A.; Ito, O. *J. Chem. Soc., Faraday Trans.* **1998**, *94*, 3331-3337.

[9805] Thomas, K. G.; Biju V.; George, M. V.; Guldi, D. M.; Kamat, P. V. *J. Phys. Chem. A* **1998**, *102*, 5341-5348.

[9806] Nemes, P.; Demeter, A.; Biczok, L.; Berces, T.; Wintgens, V.; Valat, P.; Kossanyi, J. *J. Photochem. Photobiol., A* **1998**, *113*, 225-231.

[9807] Romani, A.; Ortica, F.; Favaro, G. *Chem. Phys.* **1998**, *237*, 413-424.

[9808] Credi, A. *Ph.D., Thesis*; Univ. Bologna; Bologna (Italy), **1998**, 53.

[9901] Bartocci, G.; Spalletti, A.; Becker, R.S.; Elisei, F.; Floridi, S.; Mazzucato, U. *J. Am. Chem. Soc.* **1999**, *121*, 1065-1075.

[9902] Zhao, N.; Strehmel, B.; Gorman, A.A.; Hamblett, I.; Neckers, D. C. *J. Phys. Chem. A* **1999**, *103*, 7757-7765.

[9903] Kircher, T.; Lohmannsroben, H. -G. *Phys. Chem. Chem. Phys.* **1999**, *1*, 3987-3992.

[9904] Lopez Arbeloa, T.; Lopez Arbeloa, F.; Lopez Arbeloa, I. *Phys. Chem. Chem. Phys.* **1999**, *1*, 791-795.

[9905] Manoj, N.; Gopidas, K. R. *J. Photochem. Photobiol., A* **1999**, *127*, 31-37.

[9906] Lewis, F. D.; Kalgutkar, R. S.; Yang, J. -S. *J. Am. Chem. Soc.* **1999**, *121*, 12045-12053.

[9907] Benedetto, A. F.; Bachilo, S. M.; Weisman, R. B.; Nossal, J. R.; Billups, W. E. *J. Phys. Chem. A* **1999**, *103*, 10842-10845.

[9908] Griesbeck, A. G.; Goerner, H. *J. Photochem. Photobiol., A* **1999**, *129*, 111-119.

[0001] Liu, B.; Barkley, M. D.; Morales, G. A.; McLaughlin, M. L.; Callis, P. R. *J. Phys. Chem. B* **2000**, *104*, 1837-1843.

[0002] Singh, A. K.; Bhasikuttan, A. C.; Palit, D. K.; Mittal, J. P. *J. Phys. Chem. A* **2000**, *104*, 7002-7009.

[0003] de Melo, S. J.; Elisei, F.; Gartner, C.; Aloisi, G. G.; Becker, R. S. *J. Phys. Chem. A* **2000**, *104*, 6907-6911.

[0004] Khopde, S. M.; Priyadarsini, K. I.; Palit, D. K.; Mukherjee, T. *Photochem. Photobiol.* **2000**, *72*, 625-631.

[0005] Elisei, F.; Lima, J.C.; Ortica, F.; Aloisi, G. G.; Costa, M.; Leitao, E.; Abreu, I.; Dias, A.; Bonifacio, V.; Medeiros, J.; Macanita, A.L.; Becker, R. S. *J. Phys. Chem. A* **2000**, *104*, 6095-6102.

[0006] Biczók, L., Bérces, T., Yatsuhashi, T., Tachibana, H.; Inoue, H. *Phys. Chem. Chem. Phys.* **2001**, *3*, 980-985.

[0007] El-Daly, S.A.; Fayed, T.A. *J. Photochem. Photobiol.*, *A* **2000**, *137*, 15-19.

[0008] Romani, A.; Ortica, F.; Favaro, G. *J. Photochem. Photobiol.*, *A* **2000**, *135*, 127-134.

[0009] Lewis, F. D.; Kalgutkar, R.S. *J. Phys. Chem. A* **2001**, *105*, 285-291.

[0010] Aloisi, G. G.; Elisei, F.; Moro, S.; Miolo, G.; Dall'Acqua, F. *Photochem. Photobiol.* **2000**, *71*, 506-513.

[0011] Cardona, C. M.; Alvarez, J.; Kaifer, A. E.; McCarley, T. D.; Pandey, S.; Baker, J. A.; Bonzagni, N. J.; Bright, F. V. *J. Am. Chem. Soc.* **2000**, *122*, 6140-6144.

[0101] Yoshihara, T.; Shimada, H.; Shizuka, H.; Tobita, S. *Phys. Chem. Chem. Phys.* **2001**, *3*, 4972-4978.

[0102] Lee, W.; Jenks, W. S. *J. Org. Chem.* **2001**, *66*, 474-480.

[0103] Biczok, L.; Cser, A.; Nagy, K. *J. Photochem. Photobiol.*, *A* **2001**, *146*, 59-62.

[0104] Birckner, E.; Grummt, U. -W.; Goeller, A.H.; Pautzsch, T.; Egbe, D. A. M.; Al-Higari, M.; Klemm, E. *J. Phys. Chem. A* **2001**, *105*, 10307-10315.

[0105] Fujitsuka, M.; Takahashi, H.; Kudo, T.; Tohji, K.; Kasuya, A.; Ito, O. *J. Phys. Chem. A* **2001**, *105*, 675-680.

[0106] Fujitsuka, M.; Fujiwara, K.; Murata, Y.; Uemura, S.; Kunitake, M.; Ito, O.; Komatsu, K. *Chem. Lett.* **2001**, *30*, 384-385.

[0107] Demeter, A.; Berces, T.; Zachariasse, K. A. *J. Phys. Chem. A* **2001**, *105*, 4611-4621.

[0108] Lewis, F. D.; Weigel, W.; Zuo, X. *J. Phys. Chem. A* **2001**, *105*, 4691-4696.

[0109] Yamaji, M.; Takehira, K.; Itoh, T.; Shizuka, H.; Tobita, S. *Phys. Chem. Chem. Phys.* **2001**, *3*, 5470-5474.

[0110] Peon, J.; Tan, X.; Hoerner, D.; Xia, C.; Luk, Y. F.; Kohler, B. *J. Phys. Chem. A* **2001**, *105*, 5768-5777.

[0201] Shin, E. J.; Stackow, R.; Foote, C. S. *Phys. Chem. Chem. Phys.* **2002**, *4*, 5088-5095.

[0202] Dosche, C.; Loehmannsroeben, H. -G.; Bieser, A.; Dosa, P .I.; Han, S.; Iwamoto, M.; Schleifenbaum, A.; Vollhardt, K. P. C. *Phys. Chem. Chem. Phys.* **2002**, *4*, 2156-2161.

[0203] Uchiyama, S.; Takehira, K.; Kohtani, S.; Santa, T.; Nakagaki, R.; Tobita, S.; Imai, K. *Phys. Chem. Chem. Phys.* **2002**, *4*, 4514-4522.

[0204] Bonesi, S. M.; Erra-Balsells, R. *J. Lumin.* **2002**, *97*, 83-101.

[0205] Tirapattur, S.; Belletete, M.; Drolet, N.; Bouchard, J.; Ranger, M.; Leclerc, M.; Durocher, G. *J. Phys. Chem. B* **2002**, *106*, 8959-8966.

[0206] Singh, A. K.; Palit, D. K. *Chem. Phys. Lett.* **2002**, *357*, 173-180.

[0207] Balon, M.; Angulo, G.; Carmona, C.; Munoz, M. A.; Guardado, P.; Galan, M. *Chem. Phys.* **2002**, *276*, 155-165.

[0208] Nagy, K.; Biczok, L.; Demeter, A.; Kover, P.; Riedl, Z. *J. Photochem. Photobiol.*, *A* **2002**, *153*, 83-88.

[0301] Jones, G. II; Kumar, S.; Klueva, O.; Pacheco, D. *J. Phys. Chem. A* **2003**, *107*, 8429-8434.

[0302] de Melo,S. J.; Rodrigues, L. M.; Serpa, C.; Arnaut, L. G.; Ferreira, I. C. F. R.; Queiroz, M. -J. R. P. *Photochem. Photobiol.* **2003**, *77*, 121-128.

[0401] Sikorska, E.; Khmelinskii, I. V.; Prukala, W.; Williams, S. L.; Patel, M.; Worrall, D. R.; Bourdelande, J. L.; Koput, J.; Sikorski, M. *J. Phys. Chem. A* **2004**, *108*, 1501-1508.

[0402] Sikorska, E.; Khemlinskii, I. V.; Worrall, D. R.; Williams, S. L.; Gonzalez-Moreno, R.; Bourdelande, J. L.; Koput, J.; Sikorski, M. *J. Photochem. Photobiol., A* **2004**, *162*, 193-201.

4

EPR and ODMR Parameters of the Triplet State

For many organic molecules the g-factor of the excited triplet state is almost isotropic and close to that of the free electron. However a large splitting of the first-order ($\Delta M_S = \pm 1$) EPR spectrum is common. This doublet, due to dipolar interactions between the unpaired electron spins, is strongly anisotropic leading to very broad EPR transitions in nonoriented samples. The splitting is present even in the absence of an applied external magnetic field.

The traceless (spin) dipole-dipole interaction tensor is often described by two terms which are linear combinations of the principal axis components, (D_q ; $q = x,y,z$) as

$$D = {}^3/_2\, D_z \quad \text{and} \quad E = {}^1/_2\, (D_x - D_y)$$

Determination of the absolute signs of the so-called zero-field splitting parameters, D and E, is difficult and often only magnitudes are reported. These magnitudes depend on the choice of axis system used to characterize the molecular environment. Care should thus be taken to compare measurements between several laboratories, and it is recommended that the original references be consulted in cases of conflict.

Second order ($\Delta M_S = \pm 2$) transitions are also detected as a single line, for a given orientation, at roughly half the field (or frequency) value of the $\Delta M_S = \pm 1$ transition. Though intrinsically weaker in intensity, these half-field resonances generally show much less anisotropy and have proven extremely useful for investigations of randomly oriented triplets. Often only the signal derived from the transitions occuring at minimum magnetic fields are measured and a combination of the magnitudes of D and E is reported as

$$D^* = (D^2 + 3E^2)^{1/2}$$

Many modern techniques couple optical detection schemes, for example phosphorescence, with microwave irradiation to conveniently characterize the excited state parameters.

Table 4 **EPR Parameters of Triplet Excited States**

No.	Host	D (cm^{-1})	E (cm^{-1})	D^* (cm^{-1})	Ref.
1	**Acenaphthene**				
	EtOH (77 K)	0.0966	−0.1040		[6801]
2	**Acetaldehyde**				
	MCH (77 K)	0.148	0.028		[9001]
3	**Acetone**				
	MCH (77)	0.142	0.058		[9001]
4	**Acetophenone**				
	1,4-Dibromobenzene	−0.3350	0.0492		[7401]
5	**Acetophenone, 4′-chloro-**				
	1,4-Dibromobenzene	−0.1981	0.0617		[7401]
6	**Acetophenone, 4′-methoxy-**				
	1,4-Dimethoxybenzene	−0.1051	0.0611		[7901]
7	**Acetophenone, 2′-methyl-**				
	Toluene/EtOH (77 K)	±0.060	±0.0025		[8701]
8	**Acetophenone, 4′-phenyl-**				
	EtOH (77 K)	0.0954	−0.0082		[7701]
9	**Acetylene, diphenyl-**				
	EPA			0.152	[7201]
10	**Acridine**				
	MeOH/H$_2$O (90 K)			0.069	[7402]
11	**Acridine, 3-amino-**				
	MeOH/H$_2$O (90 K)			0.062	[7402]
12	**Acridine, 9-amino-**				
	MeOH/H$_2$O (90 K)			0.065	[7402]
13	**Acridine, 3,6-bis(diethylamino)-**				
	MeOH/H$_2$O (90 K)			0.069	[7402]
14	**Acridine, 3,6-diamino-**				
	MeOH/H$_2$O (90 K)			0.076	[7402]
15	**Acridine, 3,9-diamino-**				
	MeOH/H$_2$O (90 K)			0.069	[7402]
16	**Acridine Orange, free base**				
	EtOH (77 K)	0.061	0.015	0.069	[7501]

Table 4 EPR Parameters of Triplet Excited States

No.	Host	D (cm^{-1})	E (cm^{-1})	D^* (cm^{-1})	Ref.
17	**Acridine Red cation**				
	EtOH (77 K)			0.068	[7501]
18	**Acriflavine cation**				
	MeOH/H$_2$O (90 K)			0.076	[7402]
19	**Adenosine 5′-monophosphate**				
	PEG/H$_2$O (77 K)	0.119	0.027	0.128	[6601]
20	**Aniline**				
	EPA (77 K)			0.1317	[6201]
	1,4-Dibromobenzene (1.8 K)	0.1160	−0.0597		[8601]
21	**Aniline, *N,N*-diphenyl-**				
	Lucite (77 K)			0.0801	[6401]
22	**Aniline, 4-methyl-**				
	1,4-Dibromobenzene (1.8 K)	0.1119	−0.0533		[8601]
23	**Aniline, *N*-phenyl-**				
	EPA (77 K)			0.0994	[6201]
24	**Anthracene**				
	EPA	0.072	0.007		[6402]
	Toluene (50 K)	0.0712	−0.0080		[9901]
25	**Anthracene, 9-bromo-**				
	Toluene (50 K)	0.0678	−0.0045		[9901]
26	**Anthracene, 9,10-dibromo-**				
	Toluene (50 K)	0.0610	±0.0010		[9901]
27	**Anthracene, 9,10-dichloro-**				
	Toluene (50 K)	0.0680	−0.0070		[9901]
28	**9,10-Anthraquinone, 1,8-dichloro-**				
	EtOH (10 K)	−0.018	0.0003		[8702]
29	**9,10-Anthraquinone**				
	EPA (77 K)	−0.351	0.005		[8401]
	Octane (77K)	−0.2894	0.0041		[8401]
30	**7-Azaindole**				
	1-PrOH (77 K)	0.0919	0.0465	0.123	[8001]
	3-MP (77 K)			0.123	[8001]

Table 4 **EPR Parameters of Triplet Excited States**

No.	Host	D (cm^{-1})	E (cm^{-1})	D^* (cm^{-1})	Ref.
31	**Bacteriochlorophyll** *a*				
	MTHF (100 K)	0.0232	0.0058		[7801]
32	**Bacteriopheophytin** *a*				
	MTHF (100 K)	0.0256	0.0054		[7801]
33	**Benzaldehyde**				
	1,4-Dibromobenzene	−0.3350	0.0492		[7401]
34	**Benzaldehyde, 4-chloro-**				
	1,4-Dibromobenzene	−0.1944	0.0587		[7401]
35	**Benzaldehyde, 4-(dimethylamino)-**				
	Durene	−0.0796	0.0460		[7401]
36	**Benzaldehyde, 2-hydroxy-**				
	MCH (77 K)	0.0697	0.014		[8801]
37	**Benzaldehyde, 4-methoxy-**				
	1,4-Dimethoxybenzene	−0.188	0.0525		[7403]
38	**Benzaldehyde, 4-methyl-**				
	1,4-Dibromobenzene	−0.1547	0.0649		[7401]
39	**Benz[*a*]anthracene**				
	2-MTHF	0.079	±0.014	0.083	[6602]
40	**Benzene**				
	EtOH (77 K)	0.1581	0.0044	0.1583	[7001]
	Cyclohexane (1.2 K)	0.1593	±0.0093		[7503]
41	**Benzene, 1,4-dibromo-**				
	Xylene	0.268	±0.0597		[8002]
42	**Benzene, 1,4-dichloro-**				
	Xylene	0.144	−0.0228		[7301]
43	**Benzene, 1,4-dicyano-**				
	1,4-Dibromobenzene (77 K)	0.128	0.018		[7202]
44	**Benzene, 1,4-dihydroxy-**				
	EPA (77 K)			0.1321	[6201]
45	**Benzene, 1,4-dimethoxy-**				
	1,4-Dibromobenzene (77 K)	0.131	0.027		[7202]
46	**Benzene, 1,2,4,5-tetrachloro-**				
	Durene (4.2 K)	0.1518	0.02908		[7203]

Table 4 EPR Parameters of Triplet Excited States

No.	Host	D (cm^{-1})	E (cm^{-1})	D^* (cm^{-1})	Ref.
47	**Benzene, 1,2,4,5-tetracyano-**				
	Hexane (1.25 K)	0.1243	0.0136	0.1265	[7802]
48	**Benzene, 1,3,5-triphenyl-**				
	3-MP (77 K)	0.111	<<0.001	0.113	[6603]
49	**Benzil**				
	Benzil (77 K)	±0.119	±0.0464		[7502]
50	**Benzimidazole**				
	Benzoic acid (4.2 K)	0.1123	±0.0270		[8402]
	1,4-Dibromobenzene	0.1133	−0.0409		[7504]
51	**Benzofuran**				
	1,4-Dibromobenzene	0.1076	−0.0530		[7504]
52	**Benzoic acid**				
	EPA (77 K)			0.1385	[6201]
53	**Benzoic acid, methyl ester**				
	EtOH/MeOH (4:1) (77 K)	0.133	0.011	0.134	[7202]
54	**Benzonitrile**				
	1,4-Dibromobenzene (77 K)	0.1370	0.0072		[7205]
	EtOH (90 K)	0.1363	0.0062	0.1389	[7601]
55	**Benzonitrile, 4-bromo-**				
	Xylene	0.1681	±0.0129		[7803]
56	**Benzonitrile, 4-chloro-**				
	Xylene (1.8 K)	0.1349	±0.0045		[7803]
57	**Benzonitrile, 4-fluoro-**				
	EtOH (90 K)	0.1412	0.0032	0.1423	[7601]
58	**Benzonitrile, 4-methoxy-**				
	EtOH (77 K)	±0.1260	±0.0216	0.1329	[9101]
59	**Benzonitrile, 4-methyl-**				
	1,4-Dibromobenzene (77 K)	0.136	0.005		[7202]
60	**Benzophenone**				
	Benzophenone	±0.152	∓0.021		[6901]
61	**Benzophenone, 4-phenyl-**				
	EtOH (77 K)	0.0946	−0.0087		[7701]

Table 4 **EPR Parameters of Triplet Excited States**

No.	Host	D (cm^{-1})	E (cm^{-1})	D^* (cm^{-1})	Ref.
62	**Benzo[*a*]pyrene**				
	EPA (77 K)			0.0758	[6201]
63	**Benzo[*e*]pyrene**				
	2-MTHF (77 K)	0.090	±0.023	0.098	[6602]
64	**Benzo[*h*]quinoline**				
	EtOH (77 K)	0.1011	−0.0468		[6801]
65	**1,4-Benzoquinone**				
	EPA (77 K)	−0.330	0.019		[8501]
66	**Benzo[*b*]tryphenylene**				
	Nonane (2 K)	±0.0984	±0.0129		[7602]
67	**Biacetyl**				
	Cyclohexane (77 K)	0.208	−0.016		[8201]
	EtOH (77 K)	0.214	−0.019		[8201]
68	**Biphenyl**				
	3-MP (77 K)	0.110	±0.004	0.110	[6603]
	EtOH (77 K)	±0.1094	±0.0036		[6801]
69	**Biphenyl, 4,4′-difluoro-**				
	EtOH (77 K)	±0.1140	±0.0011	0.1140	[8901]
70	**Biphenyl, 4-hydroxy-**				
	EG/H$_2$O (77 K)	±0.1058	∓0.0036	0.1060	[8003]
71	**2,2′-Bipyridine**				
	Heptane (4.2 K)	0.1125	±0.0128		[8802]
	EtOH (77 K)	±0.1104	±0.0121	0.1124	[8004]
72	**4,4′-Bipyridine**				
	MeOH/H$_2$O (77 K)	0.1225	0.0044	0.1209	[8202]
73	**2,2′-Biquinoline**				
	EtOH (77 K)	±0.0992	±0.0368	0.1179	[8004]
74	**(*E*)-1,2-Bis(2-pyridyl)ethylene**				
	MeOH/EtOH (1:1) (77 K)	0.0912	−0.0443		[9401]
75	**(*E*)-1,2-Bis(4-pyridyl)ethylene**				
	EPA (77 K)	0.0947	−0.0464		[9401]
76	**2-Butanone**				
	MCH (77 K)	0.150	0.050		[9001]

Table 4 EPR Parameters of Triplet Excited States

No.	Host	D (cm^{-1})	E (cm^{-1})	D^* (cm^{-1})	Ref.
77	**Butyraldehyde**				
	MCH (77 K)	0.158	0.021		[9001]
78	**Carbazole**				
	Et$_2$O (77 K)	0.1022	0.0066	0.1043	[6604]
	EPA (77 K)			0.1044	[6201]
79	**2,2'-Carbocyanine, 1,1'-diethyl-**				
	MeOH/H$_2$O (77 K)	±0.0681	±0.0097	0.063	[8101]
80	**β-Carotene**				
	Micelle (160 K)	±0.0333	±0.0037		[8005]
81	**Chlorin, tetraphenyl-**				
	Toluene (115 K)	±0.0361	±0.0061		[9701]
82	**Chlorophyll a**				
	MTHF (4.2 K)	0.0288	0.0042		[7804]
	EtOH (100 K)	0.0272	0.0032		[7801]
83	**Chlorophyll b**				
	MTHF (4.2 K)	0.0294	0.0049		[7804]
	Octane (2 K)	±0.0320	±0.0041		[7702]
84	**Chrysene**				
	2-MTHF (77 K)	0.095	±0.025	0.104	[6602]
85	**(E)-Cinnamate anion**				
	MeOH/H$_2$O (4:1) (77 K)	0.1057	−0.0514		[9702]
86	**(E)-Cinnamic acid**				
	MeOH/H$_2$O (4:1) (77 K)	0.1041	−0.0480		[9702]
87	**Cinnoline**				
	EtOH (77 K)	±0.0936	±0.0140		[8703]
88	**Codeine**				
	Hexane (1.4 K)	0.0976	±0.0163		[8803]
89	**Coronene**				
	EPA (77 K)			0.0971	[6201]
90	**Coumarin**				
	1,4-Dibromobenzene	0.1001	±0.0425		[7603]
91	**2,2'-Cyanine, 1,1'-diethyl-**				
	MeOH/H$_2$O (77 K)	±0.0668	±0.0136	0.075	[8101]

Table 4 **EPR Parameters of Triplet Excited States**

No.	Host	D (cm^{-1})	E (cm^{-1})	D^* (cm^{-1})	Ref.
92	**1,4-Cyclohexanedione**				
	1,4-Cyclohexanedione (1.4K)	−0.1353	0.0371		[8704]
93	**Cyclohexanone**				
	MCH (77 K)	0.143	0.034		[9001]
94	**2-Cyclohexen-1-one**				
	CF$_3$CH$_2$OH (77 K)	−0.169	±0.00067		[8804]
95	**Cyclopentanone**				
	Hexane (4.2 K)	±0.1404	∓0.0271		[7206]
96	**2-Cyclopentenone**				
	CF$_3$CH$_2$OH (77 K)	-0.185	±0.0015		[8804]
97	**2′-Deoxyadenosine**				
	LiCl/H$_2$O Glass (77 K)			0.133	[8602]
98	**2′-Deoxyadenosine 5′-monophosphate**				
	LiCl/H$_2$O Glass (77 K)			0.134	[8602]
99	**9,10-Diazaphenanthrene**				
	Biphenyl (77 K)	−0.068	−0.111		[0301]
100	**Dibenz[*a,h*]antracene**				
	2-MTHF (77 K)	0.090	±0.025	0.100	[6602]
101	**Dibenzocycloheptadienylidene**				
	EtOH (20 K)	±0.201	±0.0080		[0302]
102	**Dibenzofuran**				
	Et$_2$O (77 K)	0.1071	0.0092	0.1092	[6604]
103	**Dibenzothiophene**				
	Et$_2$O (77 K)	0.1130	0.0021	0.1144	[6604]
104	**Dimethyl isophthalate**				
	EtOH/MeOH (4:1) (77 K)	0.136	0.036	0.149	[7204]
105	**Dimethyl phthalate**				
	EtOH/MeOH (4:1) (77 K)	0.121	0.034	0.140	[7204]
106	**Dimethyl terephthalate**				
	EtOH/MeOH (4:1) (77 K)	0.122	0.005	0.122	[7204]
107	**Diphenylmethylene**				
	EtOH (20 K)	±0.200	±0.0080		[0302]

Table 4 EPR Parameters of Triplet Excited States

No.	Host	D (cm^{-1})	E (cm^{-1})	D^* (cm^{-1})	Ref.
108	**Fluoranthene**				
	2-PrOH (77 K)			0.076	[6301]
109	**Fluorene**				
	Et$_2$O (77 K)	0.1075	0.0033	0.1092	[6604]
110	**Fluorescein, 2′,7′-dichloro-**				
	EtOH (77 K)			0.075	[7501]
111	**Fluorescein dianion**				
	EtOH (77 K)	0.065	0.019	0.075	[7501]
112	**Fluorescein dianion, 4′,5′-dibromo-2′,7′-dinitro-**				
	EtOH (77 K)			0.082	[7501]
113	**Fluorescein dianion, 2′,4′,5′,7′-tetrabromo-**				
	EtOH (77 K)			0.081	[7501]
114	**Formaldehyde**				
	EPA	0.42	0.04		[6403]
115	**Fullerene C$_{60}$**				
	Toluene (5 K)	−0.0114	0.00069		[9601]
	Toluene (77 K)	±0.0114	±0.00069		[9102]
	Toluene (233 K)	−0.0114	±0.0007		[9301]
	Liquid crystal (E-7) (8 K)	−0.0114	±0.0002		[9301]
	Liquid crystal (E-7) (298 K)	−0.0030	±0.0007		[9301]
116	**Fullerene C$_{60}$, (EtOOC)$_2$-methano-**				
	Toluene (4.2 K)	−0.0096	0.0002		[0201]
117	**Fullerene C$_{60}$, N-methylpyrrolidine-**				
	Toluene (4.2 K)	−0.00902	0.00141		[9601]
	Toluene (100 K)	−0.00807	0.00109		[9601]
	PMMA (5 K)	−0.0096	0.0014		[9602]
118	**Fullerene C$_{60}$ oxide**				
	Toluene (8 K)	−0.00944	0.0005		[9601]
119	**Fullerene C$_{70}$**				
	Toluene (8 K)	−0.0053	−0.0006		[9302]
	Toluene (77 K)	±0.0052	±0.00069		[9102]
	Toluene (213 K)	−0.00523	−0.00169		[9302]

Table 4 EPR Parameters of Triplet Excited States

No.	Host	$D\,(\mathrm{cm}^{-1})$	$E\,(\mathrm{cm}^{-1})$	$D^*\,(\mathrm{cm}^{-1})$	Ref.
120	**Imidazole**				
	Benzoic acid (4.2 K)	0.1077	±0.0293		[8402]
121	**1-Indanone**				
	Durene	−0.439	0.011		[7805]
122	**2-Indanone**				
	Durene (4.2 K)	±0.1303	±0.0359		[7902]
123	**Indazole**				
	1,4-Dibromobenzene	0.1003	−0.0382		[7504]
	Benzoic acid (4.2 K)	0.1077	0.0293		[8301]
124	**Indene**				
	1,4-Dibromobenzene	0.1079	−0.0472		[7504]
125	**Indole**				
	1,4-Dibromobenzene	0.0978	−0.0453		[7504]
	EtOH (77 K)	0.1011	0.0416	0.1241	[7001]
126	**Indole, 3-methyl-**				
	EtOH (77 K)	0.0965	0.0436	0.1225	[7001]
127	**Isoquinoline**				
	EtOH (77 K)	0.0999	−0.0113		[8805]
	Durene	±0.1004	∓0.0117		[6501]
128	**Methylene Blue cation**				
	MeOH/H_2O (98 K)			0.066	[7505]
129	**Morphine hydrochloride**				
	THF (1.4 K)	0.0994	±0.0163		[8803]
130	**2-Naphthaldehyde**				
	EPA	0.094	0.029		[6902]
131	**Naphthalene**				
	EtOH (77 K)	0.1008	−0.0154		[8805]
	Biphenyl (77 K)	±0.0992	∓0.01545		[6202]
132	**Naphthalene, 2-acetyl-**				
	EPA	0.096	0.027	0.105	[6902]
133	**Naphthalene, 1-amino-**				
	EtOH (77 K)			±0.0943	[8203]

Table 4 EPR Parameters of Triplet Excited States

No.	Host	$D\,(\mathrm{cm}^{-1})$	$E\,(\mathrm{cm}^{-1})$	$D^*\,(\mathrm{cm}^{-1})$	Ref.
134	**Naphthalene, 2-amino-**				
	EtOH (77 K)			±0.0953	[8203]
135	**Naphthalene, 1,4-dimethyl-**				
	EPA	0.0935	−0.0133	0.0959	[7101]
136	**Naphthalene, 1,8-dimethyl-**				
	EPA	0.0948	−0.0137	0.0978	[7101]
137	**Naphthalene, 1,4-dinitro-**				
	EPA (77 K)	0.0765	−0.0092		[8902]
138	**Naphthalene, 1,8-dinitro-**				
	EPA (77 K)	0.0921	−0.0109		[9103]
139	**Naphthalene, 1-hydroxy-**				
	EtOH (77 K)	0.0931	−0.0138		[6701]
140	**Naphthalene, 2-hydroxy-**				
	EtOH (77 K)	0.0942	−0.0146		[6701]
141	**Naphthalene, 1-methyl-**				
	EPA	0.0963	−0.0143	0.0995	[7101]
142	**Naphthalene, 2-methyl-**				
	EPA	0.0958	−0.0137	0.0990	[7101]
143	**Naphthalene, 1-nitro-**				
	EPA (77 K)	0.0855	−0.0051		[8902]
144	**Naphthalene, 2-nitro-**				
	EPA (77 K)	0.0870	−0.0247		[9103]
145	**Naphthalene, 1-phenyl-**				
	EtOH (77 K)	±0.0911	±0.0093	±0.0923	[8203]
146	**Naphthalene, 2-phenyl-**				
	EtOH (77 K)	±0.0963	±0.0274	±0.1007	[8203]
147	**1,2-Naphthoquinone**				
	EtOH (15 K)	0.114	0.033	0.1275	[8903]
148	**1,4-Naphthoquinone**				
	EPA (77 K)	−0.330	0.019		[8501]
149	**2,2′-Oxacarbocyanine, 3,3′-diethyl-**				
	MeOH/H$_2$O (77 K)	±0.1022	±0.0165	0.113	[8101]

Table 4 EPR Parameters of Triplet Excited States

No.	Host	D (cm^{-1})	E (cm^{-1})	D^* (cm^{-1})	Ref.
150	**Pentacene**				
	Naphthacene (1.2 K)	0.0463	−0.0014		[8006]
	Benzoic acid crystals (298 K)	0.046510	±0.001823		[0003]
151	**3-Pentanone**				
	MCH (77 K)	0.149	0.035		[9001]
152	**Perinaphthone**				
	EtOH (77 K)	±0.075	−0.014		[0001]
153	**Phenanthrene**				
	Biphenyl (78 K)	±0.10043	∓0.04658		[6404]
	EtOH/Et$_2$O/ 2,2-DMB/Pentane (80 K)	0.1053	−0.0467	0.1325	[8007]
154	**9,10-Phenanthrenequinone**				
	EtOH (15 K)	0.0935	0.023	0.1016	[8903]
	MTHF (15 K)	0.19	0.035	0.1911	[8903]
155	**Phenanthridine**				
	EtOH/Et$_2$O/ 2,2-DMB/Pentane (80 K)	0.1065	−0.0422	0.1292	[8007]
156	**1,10-Phenanthroline**				
	EtOH (77 K)	0.1038	−0.0485		[6801]
157	**Phenazine, 1,2,3,4-tetrahydro-**				
	MeOH/H$_2$O (103 K)	0.0942	−0.0180		[7703]
158	**Phenol**				
	EtOH (77 K)	0.1352	0.0451	0.1561	[7001]
159	**Phenol, 4-methyl-**				
	EtOH (77 K)	0.1251	0.0590	0.1615	[7001]
160	**Phenothiazine**				
	PrOH (77 K)			0.125	[0004]
161	**Phenothiazine, *N*-phenyl-**				
	PrOH (77 K)			0.125	[0004]
162	**Phenoxazinium, 3,7-bis(dimethylamino)-**				
	MeOH/H$_2$O (98 K)			0.058	[7504]
163	**Phenoxazinium, 3,7-diamino-**				
	MeOH/H$_2$O (98 K)			0.067	[7504]

Table 4 EPR Parameters of Triplet Excited States

No.	Host	D (cm^{-1})	E (cm^{-1})	D^* (cm^{-1})	Ref.
164	**Phenoxide ion**				
	EtOH/0.25M NaOH (77 K)	0.1133	0.0337	0.1275	[7001]
165	**Phenoxide ion, 4-methyl-**				
	EtOH/0.25M NaOH (77 K)	0.1071	0.0540	0.1422	[7001]
166	**Phenylalanine**				
	EtOH (77 K)	0.1475	0.0439	0.1517	[7001]
167	**Phenyl *t*-butyl ketone**				
	MCH (77 K)	−0.084	±0.017		[9905]
	EG/H$_2$O (77 K)	0.104	±0.003		[9905]
168	**4-Phenylphenoxide ion**				
	EG/H$_2$O (77 K)	±0.0985	∓0.0062	0.0991	[8003]
169	**4-Phenylpyridine**				
	MeOH/H$_2$O (77 K)	0.1125	0.0051	0.1127	[8202]
170	**Pheophytin *a***				
	MTHF (100 K)	0.0341	0.0033		[7801]
	EtOH (100 K)	0.0344	0.0026		[7801]
171	**Pheophytin *b***				
	Octane (2 K)	±0.0368	±0.0049		[7702]
172	**Phthalazine**				
	Biphenyl (4.2 K)	−0.012	−0.026	0.038	[8502]
	EtOH (77 K)	−0.073	0.022	0.052	[8502]
173	**Phthalocyanine, zinc(II)**				
	MTHF (4.2 K)	−0.012	−0.026	0.038	[8502]
	MTHF (80 K)	0.0247	±0.0042		[0101]
174	**Phthalocyanine, tetrakis(*t*-butyl)-**				
	Toluene/CHCl$_3$ (1:1) (20 K)	0.0253	±0.0023		[9902]
175	**Phthalocyanine, tetrakis(*t*-butyl)-, gallium(III)(OH)**				
	Toluene/CHCl$_3$ (1:1) (20 K)	0.0220	±0.0037		[0002]
176	**Phthalocyanine, tetrakis(*t*-butyl)-, germanium(IV)(OH)$_2$**				
	Toluene/CHCl$_3$ (1:1) (20 K)	0.0210	±0.0061		[0002]
177	**Phthalocyanine, tetrakis(*t*-butyl)-, magnesium(II)**				
	Toluene (20 K)	0.0238	±0.0056		[9903]

Table 4 EPR Parameters of Triplet Excited States

No.	Host	D (cm^{-1})	E (cm^{-1})	D^* (cm^{-1})	Ref.
178	**Phthalocyanine, tetrakis(t-butyl)-, silicon(IV)(OH)$_2$**				
	Toluene (20 K)	0.0208	±0.0053		[9904]
179	**Phthalocyanine, tetrakis(t-butyl)-, zinc(II)**				
	Toluene/CHCl$_3$ (1:1) (20 K)	0.0238	±0.0054		[0002]
180	**Picene**				
	p-Terphenyl (77 K)	±0.0937	±0.0365		[7903]
181	**Piperonal**				
	Durene	−0.1283	0.0310		[7401]
182	**Porphycene**				
	Liquid crystal (E-7) (108 K)	0.0295	0.0033		[8705]
183	**Porphyrin**				
	Octane (1.3 K)	0.0437	0.00664		[7404]
184	**Porphyrin, magnesium(II)**				
	Pyridine	±0.0321	±0.0100		[7904]
185	**Porphyrin, octaethyl-, magnesium(II)**				
	EtOH/Toluene (1:1) (20 K)	0.0377	±0.0035		[9603]
186	**Porphyrin, tetrakis(4-methylphenyl)-**				
	Octane (80 K)	±0.0369	±0.0076		[7405]
187	**Porphyrin, tetrakis(1-methylpyridinium-4-yl)-**				
	Sucrose/H$_2$O/HCl (4.2 K)	0.0418	0.0041		[8603]
188	**Porphyrin, tetrakis(4-sulfonatophenyl)-, zinc(II)**				
	CH$_3$Cl/MeOH (10 K)	0.0298	0.0099		[8604]
189	**Porphyrin, tetrakis(4-sulfonatophenyl)-**				
	H$_2$O/Glycerol (100 K)	0.0391	−0.0075		[8403]
190	**Porphyrin, tetrakis(4-trimethylammoniophenyl)-, zinc(II)**				
	H$_2$O/Glycerol (10 K)	0.0323	0.0095		[8503]
191	**Porphyrin, tetrakis(4-trimethylammoniophenyl)-**				
	H$_2$O/Glycerol (10 K)	0.0400	0.0075		[8503]
192	**Porphyrin, tetraphenyl-**				
	Octane (80 K)	±0.0359	±0.0079		[7405]
	EtOH/Et$_2$O (100 K)	±0.0369	∓0.0082		[7406]
	Liquid crystal (E-7) (100 K)	0.0383	±0.0080		[8606]

Table 4 EPR Parameters of Triplet Excited States

No.	Host	D (cm^{-1})	E (cm^{-1})	D* (cm^{-1})	Ref.
193	**Porphyrin, tetraphenyl-, gallium(III)(OH)**				
	Toluene/CHCl$_3$ (1:1) (20 K)	0.0325	±0.0105		[0002]
194	**Porphyrin, tetraphenyl-, germanium(IV)(Br)$_2$**				
	Toluene/CHCl$_3$ (1:1) (20 K)	−0.0437	±0.0053		[0002]
195	**Porphyrin, tetraphenyl-, germanium(IV)(OH)$_2$**				
	Toluene/CHCl$_3$ (1:1) (20 K)	0.0315	±0.0098		[0002]
196	**Porphyrin, tetraphenyl-, magnesium(II)**				
	EtOH/Toluene (1:1) (20 K)	0.0310	±0.0092		[9603]
	Phase V (120 K)	±0.0304	∓0.0084		[8008]
197	**Porphyrin, tetraphenyl-, zinc(II)**				
	Toluene/CHCl$_3$ (1:1) (20 K)	0.0307	±0.0097		[0002]
	Liquid crystal (E-7) (100 K)	±0.0298	±0.0098		[8505]
	Phase V (120 K)	±0.0304	∓0.0090		[8008]
198	**Propionaldehyde**				
	MCH (77 K)	0.163	0.023		[9001]
199	**Purine**				
	Benzoic acid (4.2 K)	0.1042	±0.0608		[8401]
	1,4-Dibromobenzene	0.1009	−0.0584		[7503]
200	**Pyrazine, tetrachloro-**				
	Durene (1.8 K)	±0.1398	±0.0197		[8302]
201	**Pyrazine, tetramethyl-**				
	Durene	±0.0973	±0.014		[8404]
202	**Pyrene**				
	Octane (1.3 K)	0.0858	0.0170		[8009]
	EPA (77 K)			0.0929	[6201]
203	**Pyridazine**				
	EtOH (3.0 K)	−0.138	0.116		[8504]
	EtOH (77 K)	−0.108	−0.127		[0301]
204	**Pyridazine, 3,6-dichloro-**				
	EtOH (77 K)	−0.098	−0.114		[0301]
205	**Pyridinium**				
	EG/H$_2$SO$_4$/HCl	0.134	±0.030		[7704]

Table 4 EPR Parameters of Triplet Excited States

No.	Host	D (cm^{-1})	E (cm^{-1})	D^* (cm^{-1})	Ref.
206	**Pyridinium, 4-phenyl-**				
	MeOH/H$_2$O (77 K)	0.1032	0.0094	0.1044	[8202]
207	**Pyridol[2,3-*b*]pyrazine**				
	Durene (10 K)	−0.1034			[8706]
208	**Quinoline**				
	EtOH (77 K)	0.1014	−0.0164		[8805]
	Durene (77 K)	±0.1030	∓0.0162		[6501]
209	**Quinoline, conjugate acid**				
	EtOH (77 K)	0.0921	−0.0149		[8805]
210	**Quinoxaline**				
	Durene (77 K)	±0.1007	∓0.0182		[6302]
	MeOH/H$_2$O (103 K)	0.0954	−0.0184		[7703]
211	**Quinoxaline, conjugate monoacid**				
	EtOH/1-PrOH/Sulfuric acid (103 K)	0.0834	−0.0172		[7703]
212	**(*all-E*)-Retinal**				
	Polyethylene film (1.2 K)	−0.7272	0.0634		[9201]
213	**Rh(bpy)$_3$(ClO$_4$)$_2$**				
	(1.4 K)	~±0.1	~±0.02		[9703]
214	**Rhodamine B, inner salt**				
	EtOH (77 K)	0.058	0.017	0.065	[7501]
215	**Rhodamine 6G cation**				
	EtOH (77 K)	0.054	0.019	0.063	[7501]
216	**Rhodamine S cation**				
	EtOH (77 K)	0.057	0.017	0.064	[7501]
217	**Salicylamide**				
	MCH (77 K)	0.1017	0.012		[8801]
	EtOH/Toluene (77 K)	0.0772	0.015		[8801]
218	**(*E*)-Stilbene**				
	3-MP (77 K)			0.1122	[8605]
	1-Pentanol (77 K)			0.1119	[8605]
219	***p*-Terphenyl**				
	EPA (77 K)			0.0961	[6201]

Table 4 EPR Parameters of Triplet Excited States

No.	Host	D (cm^{-1})	E (cm^{-1})	D^* (cm^{-1})	Ref.
220	**Tetraazabacteriochlorin, 2,3:12,13-bis(9,10-dihydro-2,6-di(t-butyl) anthracene-9,10-diyl)-**				
	Toluene (20 K)	0.0278	0.00910		[0202]
221	**Tetraazachlorin, 2,3-(9,10-dihydroanthracene-9,10-diyl)-**				
	Toluene (20 K)	0.0285	0.00817		[0202]
222	**Tetraazaisobacteriochlorin, 2,3:7,8-bis(9,10-dihydro-2,6-di(t-butyl) anthracene-9,10-diyl)-**				
	Toluene (20 K)	0.0460	0.0113		[0202]
223	**Tetraazaporphyrin, tetrabutyl-**				
	Toluene (20 K)	0.0330	0.0010		[0202]
224	**Tetracene**				
	Nonane (2 K)	0.0573	0.0043		[7602]
225	**7,7,8,8-Tetracyanoquinodimethane**				
	Phenazine (300 K)	±0.05975	±0.00760		[8904]
226	**α-Tetralone**				
	Durene	0.429	0.017		[7207]
227	**2,2′-Thiacarbocyanine, 3,3′-diethyl-**				
	MeOH/H$_2$O (77 K)	±0.1038	±0.0186	0.107	[8101]
228	**Thionine cation**				
	MeOH/H$_2$O (98 K)			0.069	[7505]
229	**Toluene**				
	Cyclohexane/Decalin			0.17	[7102]
	EtOH (77 K)	0.1454	0.0250	0.1517	[7001]
230	**1,3,5-Triazine, 2,4-diphenyl-**				
	3-MP (77 K)	0.120	<0.001	0.122	[6603]
231	**1,3,5-Triazine, 2-phenyl-**				
	3-MP (77 K)	0.119	±0.003	0.122	[6603]
232	**1,3,5-Triazine, 2,4,6-triphenyl-**				
	3-MP (77 K)	0.124	<<0.001	0.125	[6603]
233	**Triphenylene**				
	MeOH/H$_2$O (103 K)	0.1353	0.000		[7703]
234	**Tryptophan**				
	EtOH (77 K)	0.0984	0.0410	0.1213	[7001]

Table 4 EPR Parameters of Triplet Excited States

No.	Host	D (cm^{-1})	E (cm^{-1})	D^* (cm^{-1})	Ref.
235	**Tyrosine**				
	EtOH/0.1M HCl (77 K)	0.1301	0.0558	0.1621	[7001]
236	**9-Xanthione**				
	Hexane (4.2 K)	−15.9	0.06144		[8806]
237	**Xanthone**				
	EPA (77 K)	±0.171	±0.019		[8807]
	α-Cyclodextrin (77 K)	−0.150	0.019		[8905]

REFERENCES

[6201] Smaller, B. *J. Chem. Phys.* **1962**, *37*, 1578-1579.

[6202] Brandon, R.W.; Gerkin, R. E.; Hutchison, C. A., Jr. *J. Chem. Phys.* **1962**, *37*, 447-448.

[6301] Foerster, G. v. *Z. Naturforsch., Teil A* **1963**, *18*, 620-626.

[6302] Vincent, J. S.; Maki, A. H. *J. Chem. Phys.* **1963**, *39*, 3088-3096.

[6401] Thomson, C. *J. Chem. Phys.* **1964**, *41*, 1-6.

[6402] Van der Waals, J. H.; ter Maten, G. *Mol. Phys.* **1964**, *8*, 301-318.

[6403] Raynes, W. T. *J. Chem. Phys.* **1964**, *41*, 3020-3032.

[6404] Brandon, R. W.; Gerkin, R. E.; Hutchison, C. A., Jr. *J. Chem. Phys.* **1964**, *41*, 3717-3726.

[6501] Vincent, J. S.; Maki. A. H. *J. Chem. Phys.* **1965**, *42*, 865-868.

[6601] Rahn, R. O.; Yamane, T.; Eisinger, J.; Longworth, J. W.; Shulman, R. G. *J. Chem. Phys.* **1966**, *45*, 2947-2954.

[6602] Brinen, J. S.; Orloff, M. K. *J. Chem. Phys.* **1966**, *45*, 4747-4750.

[6603] Brinen, J. S.; Koren, J. G.; Hodgson, W. G. *J. Chem. Phys.* **1966**, *44*, 3095-3099.

[6604] Siegel, S.; Judeikis, H. S. *J. Phys. Chem.* **1966**, *70*, 2201-2204.

[6701] Grivet, J. -P.; Ptak, M. *C. R. Hebd. Seances Acad. Sci., Ser. B* **1967**, *265*, 972-975.

[6801] Gondo, Y.; Maki, A. H. *J. Phys. Chem.* **1968**, *72*, 3215-3222.

[6901] Sharnoff, M. *J. Chem. Phvs.* **1969**, *51*, 451-452.

[6902] Wells, C. H. J. *J. Chem. Soc., Chem. Commun.* **1969**, 393-394.

[7001] Zuclich, J. *J. Chem. Phys.* **1970**, *52*, 3586-3591.

[7101] Yamanashi, B. S.; Bowers, K. W. *J. Magn. Reson.* **1971**, *5*, 109-114.

[7102] Gueron, M. in *Creation and Detection of the Excited State*; Lamola, A. A. (ed.); Marcel Dekker: New York, 1971, volume 1, pp. 303-342.

[7201] Wade, C. G.; Webber, S. E. *J. Chem. Phys.* **1972**, *56*, 1619-1625.

[7202] Harrigan, E. T.; Wong, T. C.; Hirota, N. *Chem. Phys. Lett.* **1972**, *14*, 549-554.

[7203] Chen, C. R.; Mucha, J. A.; Pratt, D. W. *Chem. Phys. Lett.* **1972**, *15*, 73-78.

[7204] Arnold, D. R.; Bolton, J. R.; Pedersen, J. A. *J. Am. Chem. Soc.* **1972**, *94*, 2872-2874.

[7205] Mao, S. W.; Wong, T. C.; Hirota, N. *Chem. Phys. Lett.* **1972**, *13*, 199-204.

[7206] Shain, A. L.; Chiang, W.- T; Sharnoff, M. *Chem. Phys. Lett.* **1972**, *16*, 206-210.

[7207] Nishimura, A. M.; Vincent, J. S. *Chem. Phys. Lett.* **1972**, *13*, 89-92.

[7301] Kothandaraman, G.; Tinti, D. S. *Chem. Phys. Lett.* **1973**, *19*, 225-230.

[7401] Cheng, T. H.; Hirota, N. *Mol. Phys.* **1974**, *27*, 281-307.

[7402] Schmidt, H.; Zellhofer, R. *Z. Phys. Chem. (Frankfurt)* **1974**, *91*, 204-218.

[7403] Harrigan, E. T.; Hirota, N. *Chem. Phys. Lett.* **1974**, *27*, 405-410.

[7404] Van Dorp, W. G.; Soma, M.; Kooter, J. A.; Van der Waals, J. H. *Mol. Phys.* **1974**, *28*, 1551-1568.

[7405] Scherz, A.; Orbach, N.; Levanon, H. *Isr. J. Chem.* **1974**, *12*, 1037-1048.

[7406] Levanon, H.; Wolberg, A.; *Chem. Phys. Lett.* **1974**, *24*, 96-98

[7501] Yamashita, M.; Ikeda, H.; Kashiwagi, H. *J. Chem. Phys.* **1975**, *63*, 1127-1131.

[7502] Chan, I. Y.; Nelson, B. N. *J. Chem. Phys.* **1975**, *62*, 4080-4088.

[7503] Vergragt, P. J.; Van der Waals, J. H. *Chem. Phys. Lett.* **1975**, *36*, 283-289.

[7504] Harrigan, E. T.; Hirota, N. *J. Am. Chem. Soc.* **1975**, *97*, 6647-6652.

[7505] Vogelmann, E.; Schmidt, H.; Steiner, U.; Kramer, H. E. A. *Z. Phys. Chem. (Frankfurt)* **1975**, *94*, 101-106.

[7601] Wagner, P. J.; May, M. J. *Chem. Phys. Lett.* **1976**, *39*, 350-352.
[7602] Clarke, R. H.; Frank, H. A. *J. Chem. Phys.* **1976**, *65*, 39-47.
[7603] Harrigan, E. T.; Chakrabarti, A.; Hirota, N. *J. Am. Chem. Soc.* **1976**, *98*, 3460-3465.
[7701] Vyas, H. M.; Wan, J. K. S. *Can. J. Chem.* **1977**, *55*, 1175-1180.
[7702] Tria, J. J.; Johnsen, R. H., *J. Phys. Chem.* **1977**, *81*, 1274-1278.
[7703] Chodkowska, A.; Grabowska, A.; Herbich, J. *Chem. Phys. Lett.* **1977**, *51*, 365-369.
[7704] Motten, A. G.; Kwiram, A. L. *Chem. Phys. Lett.* **1977**, *45*, 217-220.
[7801] Kleibeuker, J. F.; Platenkamp, R. J.; Schaafsma, T. J. *Chem. Phys.* **1978**, *27*, 51-64.
[7802] Yagi, M.; Nishi, N.; Kinoshita, M.; Nagakura, S. *Mol. Phys.* **1978**, *35*, 1369-1379.
[7803] Niizuma, S.; Kwan, L.; Hirota, N. *Mol. Phys.* **1978**, *35*, 1029-1046.
[7804] Haegele, W.; Schmid, D.; Wolf, H. C. *Z. Naturforsch., Teil A* **1978**, *33A*, 83-93.
[7805] Niizuma, S.; Hirota, N. *J. Phys. Chem.* **1978**, *82*, 453-459.
[7901] Niizuma, S.; Hirota, N. *J. Phys. Chem.* **1979**, *83*, 706-713.
[7902] Hirota, N.; Baba, M.; Hirata, Y.; Nagaoka, S. *J. Phys. Chem.* **1979**, *83*, 3350-3354.
[7903] Kim, S. S. *Chem. Phys. Lett.* **1979**, *61*, 327-330.
[7904] Connors, R. E.; Comer, J. C.; Durand, R. R., Jr. *Chem. Phys. Lett.* **1979**, *61*, 270-274.
[8001] Bulska, H.; Chodkowska, A. *J. Am. Chem. Soc.* **1980**, *102*, 3259-3261.
[8002] Shinohara, H.; Hirota, N. *J. Chem. Phys.* **1980**, *72*, 4445-4457.
[8003] Yagi, M.; Higuchi, J. *Chem. Phys. Lett.* **1980**, *72*, 135-138.
[8004] Higuchi, J.; Yagi, M.; Iwaki, T.; Bunden, M.; Tanigaki, K.; Ito, T. *Bull. Chem. Soc. Jpn.* **1980**, *53*, 890-895.
[8005] Frank, H. A.; Bolt, J. D.; Costa, S. M. de B.; Sauer, K. *J. Am. Chem. Soc.* **1980**, *102*, 4893-4898.
[8006] Van Strien, A. J.; Schmidt, J. *Chem. Phys. Lett.* **1980**, *70*, 513-517.
[8007] Schaaf, R.; Perkampus, H. -H. *Chem. Phys. Lett.* **1980**, *71*, 467-470.
[8008] Grebel, V.; Levanon, H. *Chem. Phys. Lett.* **1980**, *72*, 218-224.
[8009] Braeuchle, C.; Kabza, H.; Voitlaender, J. *Chem. Phys.* **1980**, *48*, 369-385.
[8101] Schmidt, H.; Roedder, H. D.; Dietzel, U. *Photogr. Sci. Eng.* **1981**, *25*, 21-28.
[8201] Murai, H.; Imamura, T.; Obi, K. *J. Phys. Chem.* **1982**, *86*, 3279-3281.
[8202] Yagi, M.; Matsunaga, M.; Higuchi, J. *Chem. Phys. Lett.* **1982**, *86*, 219-222.
[8203] Stoesser, R.; Thurner, J. -U.; Hanke, T.; Sarodnick, G. *J. Prakt. Chem.* **1982**, *324*, 761-768.
[8301] Noda, M.; Hirota, N. *J. Am. Chem. Soc.* **1983**, *105*, 6790-6794.
[8302] von Borczyskowski, C.; Fallmer, E. *Chem. Phys. Lett.* **1983**, *102*, 433-437.
[8401] Murai, H.; Hayashi, T.; I'Haya, Y. J. *Chem. Phys. Lett.* **1984**, *106*, 139-142.
[8402] Noda, M.; Nagaoka, S.; Hirota, N. *Bull. Chem. Soc. Jpn.* **1984**, *57*, 2376-2382.
[8403] Chandrashekar, T. K.; van Willigen, H.; Ebersole, M. H. *J. Phys. Chem.* **1984**, *88*, 4326-4332.
[8404] Yamauchi, S.; Miyake, K.; Hirota, N. *Mol. Phys.* **1984**, *53*, 479-491.
[8501] Murai, H.; Minami, M.; Hayashi, T.; I'Haya, Y. J. *Chem. Phys.* **1985**, *93*, 333-338.
[8502] Terazima, M.; Yamauchi, S.; Hirota, N. *J. Chem. Phys.* **1985**, *83*, 3234-3243.
[8503] van Willigen, H.; Das, U.: Ojadi, E.; Linschitz, H. *J. Am. Chem. Soc.* **1985**, *107*, 7784-7785.
[8504] Terazima, M.; Yamauchi, S.; Hirota, N. *Chem. Phys. Lett.* **1985**, *120*, 321-326.
[8505] Gonen, O.; Levanon, H. *J. Phys. Chem.* **1985**, *89*, 1637-1643.
[8601] Nagaoka, S.; Harrigan. E. T.; Noda, M.; Hirota, N.; Higuchi, J. *Bull. Chem. Soc. Jpn.* **1986**, *59*, 355-361.
[8602] Arce, R.; Rodriguez, G. *J. Photochem.* **1986**, *33*, 89-97.

[8603] Hofstra, U.; Koehorst, R. B. M.; Schaafsma, T. J. *Chem. Phys. Lett.* **1986**, *130*, 555-559.

[8604] van Willigen, H.; Chandrashekar, T. K.; Das, U.; Ebersole, M. H. *ACS Symp. Ser.* **1986**, *321*, 140-153.

[8605] Yagi, M. *Chem. Phys. Lett.* **1986**, *124*, 459-462.

[8606] Gonen, O.; Levanon, H. *J. Chem. Phys.* **1986**, *84*, 4132-4136.

[8701] Akiyama, K.; Ikegami, Y.; Tero-Kubota, S. *J. Am. Chem. Soc.* **1987**, *109*, 2538-2539.

[8702] Yamauchi, S.; Hirota, N. *J. Chem. Phys.* **1987**, *86*, 5963-5970.

[8703] Terazima, M. *J. Chem. Phys.* **1987**, *87*, 3789-3795.

[8704] Tro, N. J.; Tro, J. J.; Marten, D. F.; Nishimura, A. M. *J. Photochem.* **1987**, *36*, 141-148.

[8705] Ofir, H.; Regev, A.; Levanon, H.; Vogel, E.; Koecher, M.; Balci, M. *J. Phys. Chem.* **1987**, *91*, 2686-2688.

[8706] Yamauchi, S.; Hirota, N. *J. Phys. Chem.* **1987**, *91*, 1754-1760.

[8801] Yamauchi, S.; Hirota, N. *J. Am. Chem. Soc.* **1988**, *110*, 1346-1351.

[8802] Suisalu, A. P.; Aslanov, L. A.; Kamyshnyi, A. L.; Zakharov, V. N.; Avarmaa, R. A. *Bull. Acad. Sci. USSR, Phys. Ser.* **1988**, *52*, 26-29.

[8803] Slama, H.; Basche, T.; Braeuchle, C.; Voitlaender, J. *Photochem. Photobiol.* **1988**, *47*, 661-667.

[8804] Yamauchi, S.; Hirota, N., Higuchi, J. *J. Phys. Chem.* **1988**, *92*, 2129-2133.

[8805] Yagi, M.; Komura, A.; Higuchi, J. *Chem. Phys. Lett.* **1988**, *148*, 37-40.

[8806] Petrin, M. J.; Ghosh, S.; Maki, A. H. *Chem. Phys.* **1988**, *120*, 299-309.

[8807] Murai, H.; Minami, M.; I'Haya, Y. J. *J. Phys. Chem.* **1988**, *92*, 2120-2124.

[8901] Tanigaki, K.; Yagi, M.; Higuchi, J. *J. Magn. Reson.* **1989**, *84*, 282-295.

[8902] Shioya, Y.; Yagi, M.; Higuchi, J. *Chem. Phys. Lett.* **1989**, *154*, 25-28.

[8903] Shimoishi, H.; Tero-Kubota, S.; Akiyama, K.; Ikegami, Y. *J. Phys. Chem.* **1989**, *93*, 5410-5414.

[8904] Gundel, D.; Frick, J.; Krzystek, J.; Sixl, H.; von Schuetz, J. U.; Wolf, H. C. *Chem. Phys.* **1989**, *132*, 363-372.

[8905] Murai, H.; I'Haya, Y. J. *Chem. Phys.* **1989**, *135*, 131-137.

[9001] Tominaga, K.; Yamauchi, S.; Hirota, N. *J. Phys. Chem.* **1990**, *94*, 4425-4431.

[9101] Wagner, P. J.; May, M. L. *J. Phys. Chem.* **1991**, *95*, 10317-10321.

[9102] Wasielewski, M. R.; O'Neil, M. P.; Lykke, K. R.; Pellin, M. J.; Gruen, D. M. *J. Am. Chem. Soc.* **1991**, *113*, 2774-2776.

[9103] Yagi, M.; Shioya, Y.; Higuchi, J. *J. Photochem. Photobiol., A* **1991**, *62*, 65-73.

[9201] Ros, M.; Hogenboom, M. A.; Kok, P.; Groenen, E. J. J. *J. Phys. Chem.* **1992**, *96*, 2975-2982.

[9301] Regev, A.; Gamliel, D.;Meiklyar, V.; Michaeli, S.; Levanon, H. *J. Phys. Chem.* **1993**, *97*, 3671-3679.

[9302] Levanon, H.; Meiklyar, V.; Michaeli, S.; Gamliel, D. *J. Am. Chem. Soc.* **1993**, *115*, 8722-8727.

[9401] Shioya, Y.; Mikuni, K.; Higuchi, J.; Yagi, M. *J. Phys. Chem.* **1994**, *98*, 12521-12525.

[9601] Agostini, G.; Corvaja, C.; Pasimeni, L. *Chem. Phys.* **1996**, *202*, 349-356.

[9602] Agostini, G.; Corvaja, C.; Maggini, G.; Pasimeni, L.; Prato, M. *J. Phys. Chem.* **1996**, *100*, 13416-13420.

[9603] Yamauchi, S.; Matsukawa, Y.; Ohba, Y.; Iwaizumi, M. *Inorg. Chem.* **1996**, *35*, 2910-2914.

[9701] Kay, C. W. M.; Di Valentin, M.; Möbius, K. *J. Chem. Soc., Perkin Trans. 2* **1997**, 2563-2568.

[9702] Kakuho, S.; Seki, K.; Yagi, M. *Chem. Phys. Lett.* **1997**, *277*, 326-330.

[9703] Yersin, H.; Humbs, W.; Strasser, J. *Coord. Chem. Rev.* **1997**, *159*, 325-358.

[9901] Kamata, Y.; Akiyama, K.; Tero-Kubota, S. *J. Phys. Chem. A* **1999**, *103*, 1714-1718.

[9902] Ishii, K.; Kobayashi, N.; Higashi, Y.; Osa, T.; LeLièvre, D.; Simon, J.; Yamauchi, S. *Chem. Commun.* **1999**, 969-970.

[9903] Ishii, K.; Ishizaki, T.; Kobayashi, N. *J. Phys. Chem. A* **1999**, *103*, 6060-6062.

[9904] Ishii, K.; Hirose, Y.; Kobayashi, N. *J. Phys. Chem. A* **1999**, *103*, 1986-1990.

[9905] Ikoma, T.; Akiyama, K.; Tero-Kubota, S. *Mol. Phys.* **1999**, *96*, 813-820.

[0001] Okutsu, T.; Noda, S.; Tanaka, S.; Kawai, A.; Obi, K.; Hiratsuka, H. *J. Photochem. Photobiol. A: Chem.* **2000**, *132*, 37-41.

[0002] Ishii, K.; Abiko, S.; Kobayashi, N. *Inorg. Chem.* **2000**, *39*, 468-472.

[0003] Yang, T.-C.; Sloop, D. J.; Weissman, S. I.; Lin, T.-S. *J. Chem. Phys.* **2000**, *113*, 11194-11201.

[0004] Borowicz, P.; Herbich, J.; Kapturkiewicz, A.; Anulewicz-Ostrowska, R.; Nowacki, J.; Grampp, G. *Phys. Chem. Chem. Phys.* **2000**, *2*, 4275-4280.

[0101] Barbon, A.; Brustolon, M.; van Faassen, E. E. *Phys. Chem. Chem. Phys.* **2001**, *3*, 5342-5347.

[0201] Corvaja, C. in *Developments in Fullerene Science*; Guldi, D. M.; Martín, N. (eds.); Kluwer: Dordrecht (The Netherlands), 2002, volume 4, pp. 213-236.

[0202] Miwa, H.; Makarova, E. A.; Ishii, K.; Luk'yanets, E. A.; Kobayashi, N. *Chem. Eur. J.* **2002**, *8*, 1082-1090.

[0301] Hirota, N.; Yamauchi, S. *J. Photochem. Photobiol. C: Photochem. Rev.* **2003**, *4*, 109-124.

[0302] Akiyama, K.; Suzuki, A.; Morikuni, H.; Tero-Kuroba, S. *J. Phys. Chem. A* **2003**, *107*, 1447-1451.

5

Photophysical Properties of Transition Metal Complexes

5a PHOTOPHYSICAL PARAMETERS IN SOLUTION

In this section are collected the photophysical data of a series of luminescent transition metal complexes. Among the wide number of transition metal complexes reported in the literature, in this table are gathered those that can be seen as archetypes for the different families, or those whose photophysical properties are of particular interest for practical applications. In this contest, a particular attention has been devoted to the family of Ru polypyridine complexes, that is by far the most studied since, thanks to their peculiar properties, the complexes of this family have been found wide applications in important fields such as medical diagnostics, analytical sciences, photocatalysis, and solar energy conversion. A detailed introduction to the world of Ru polypyridine complexes and a wider selection of data can be found in ref. [8801].

Although nowadays a large number of polynuclear complexes is reported in the literature, their photophysical data are not reported in this table. In fact one of the scope of this section is to give the essential elements for predicting the properties of new multi-component supramolecular systems from the data of the single subunits. For the same reason, some important complexes of non transition metals are instead included in the table.

The table lists the lowest energy absorption maximum (and its molar absorbivity, ε) at room temperature, the emission maximum, the emission lifetime, and the emission quantum yield both at room temperature and at 77K. In table the complexes are listed alphabetically according to the metal centre. For each element, the ligand are listed in the following order: polypyridine ligands (2,2'-bipyridine and its derivatives, 1,10-phenanthroline and its derivatives, 2,2':6',2''-

terpyridine and its derivatives), cyclometallating ligands, porphyrins and related macrocycles, and other ligands.

Table 5a Photophysical Parameters of Transition Metal Complexes

		Room temperature				77 K			
Complex	Solvent	λ_{abs} nm (ε) (L mol⁻¹ cm⁻¹)	λ_{em} nm	τ_{em}[a] μs	Φ_{em}[a]	λ_{em} nm	τ_{em} μs	Φ_{em}	Ref.
Al									
[Al(QO)₃]	DMF	388	540	0.0099	<1×10⁻⁴	475[b]	<0.020		[8601]
[Al(TPP)]	EtOH	595		0.0051	0.11	597[c], 770[d]	30000[d]		[8201]
[Al(TPP)(Ac)]	CHCl₃	590		0.0082	0.14	593[c], 780[d]		0.048[d]	[8301]
[Al(TPP)(Cl)]	CHCl₃	590		0.0056	0.10	593[c], 780[d]			[8301]
[AlPcTs]	DMF		686	0.00604	0.559				[9801]
[AlNPcTs]	H₂O/Py		775	0.0025	0.20				[9801]
[ClAl(salen)]	CH₃CN	355 (9300)	476	0.0066	0.18				[0301]
Cr									
[Cr(bpy)₃]³⁺	H₂O	458 (269)	694	77, 48[e]	8.9×10⁻⁴ᵉ	694	6500		[7801], [7901], [8001]
[Cr(phen)₃]³⁺	H₂O	454 (324)	699	250		699	5300		[8302], [7901]

Table 5a Photophysical Parameters of Transition Metal Complexes

Complex	Solvent	Room temperature				77 K			Ref.
		λ_{abs} (ε) nm (L mol⁻¹ cm⁻¹)	λ_{em} nm	τ_{em} [a] μs	Φ_{em} [a]	λ_{em} nm	τ_{em} μs	Φ_{em}	
trans-[Cr(en)₂(SCN)₂]⁺	H₂O		726	10.5[e]	8.6×10⁻⁴ᵉ				[8001]
[Cr(en)₃]³⁺	H₂O		670	1.85	6.2×10⁻⁵				[8001]
[Cr(NH₃)₆]³⁺	H₂O		667	2.2	5.5×10⁻⁵				[8001]
trans-[Cr(cyclam)(CN)₂]⁺	H₂O		724	335	≈ 2×10⁻³		355[f]		[8303]
Cd									
[Cd(TPP)]	MeCH			6.5×10⁻⁵ᶜ	4×10⁻⁴		2400[d]	0.04[d]	[8101]
Cu									
[Cu(OEP)]	CH₂Cl₂	560 (16900)	688[g]	0.11					[9701]
[Cu(TTP)]	CH₂Cl₂	538 (19200)	805[g]	0.040					[9701]
[Cu(dmp)₂]⁺	CH₂Cl₂	454 (7950)	750	0.090	2.7×10⁻⁴	738[h]	0.82		[8002], [8701], [0102]
[Cu(bcp)₂]⁺	CH₂Cl₂	479 (14200)	770	0.080	3.2×10⁻⁴				[0101], [8701]

Table 5a Photophysical Parameters of Transition Metal Complexes

Complex	Solvent	Room temperature				77 K			Ref.
		λ_{abs} (ε) nm (L mol⁻¹ cm⁻¹)	λ_{em} nm	τ_{em}[a] μs	Φ_{em}[a]	λ_{em} nm	τ_{em} μs	Φ_{em}	
[Cu(dnp)₂]⁺	CH₂Cl₂		715	0.250	1.2×10^{-3}				[8701]
[Cu(dpp)₂]⁺	CH₂Cl₂	441 (3800)	710	0.25	1.1×10^{-3}				[0101], [8701]
[Cu(tpp)₂]⁺	CH₂Cl₂	450 (6400)	745	0.23	1.48×10^{-3}				[0101], [8701]
[Cu(bfp)₂]⁺	CH₂Cl₂	462 (10900)	665	0.165	$\approx4\times10^{-3}$				[9802]
[Cu(dsbp)₂]⁺	CH₂Cl₂	455 (6600)	690	0.400	4.5×10^{-3}				[9703]
	CH₃CN	452	670	0.130					
[Cu(dbtmp)₂]⁺	CH₂Cl₂	453	670	0.920	6.3×10^{-3}				[9901]
[Cu(dap)₂]⁺	CH₂Cl₂	436 (2310)	710	0.260	8.3×10^{-4}				[8901]
Eu									
[Eu(2.2.1)]³⁺	H₂O	298(111)		220	3×10^{-3i}		340		[9702]
[Eu(bpy.bpy.bpy)]³⁺	H₂O	303(2800)		340	0.0		810		[9702]
[Eu(1,4,7-(bpy)₃-triazacyclononane)]³⁺	H₂O	311(24500)		500	0.05		870		[9101]
[Eu(Calixarene-(bpy)₂]³⁺	CH₃CN	305(22700)		1260	0.23				[0001]

Table 5a Photophysical Parameters of Transition Metal Complexes

Complex	Solvent	Room temperature				77 K			Ref.
		λ_{abs} nm (L mol⁻¹ cm⁻¹) (ε)	λ_{em} nm	τ_{em}^{a} µs	Φ_{em}^{a}	λ_{em} nm	τ_{em} µs	Φ_{em}	
Hf									
[Hf(OEP)(Oac)₂]	CH₂Cl₂	564 (29400)	570		0.020	565^c		0.020^c	[7503]
[Hf(OEP)(Oac)₂]	Toluene		633^c 718^d		1×10^{-3c} 8×10^{-5d}	704^d	65000^d	0.096^d	[0201]
Ir									
[Ir(bpy)₃]³⁺	H₂O	440 (8)		2.4		441	80^j		[7401], [7501]
[Ir(phen)₃]³⁺	MeOH		444	2.9		469			[7501]
[Ir(phen)₂Cl₂]⁺	H₂O	360 (3300)	490	0.31^k		490			[7502], [7701]
[Ir(Me₂phen)₂Cl₂]⁺	H₂O	390 (4400)	495	0.86^k					[7502], [7701]
[Ir(ppy)₂(bpy)]⁺	MeOH	465 (580)	606	0.34		532	5.2		[8802]
[Ir(ppy)₂Cl(py)]	CH₂Cl₂		500	0.14		483	4.3		[8401]
[Ir(ppy)₃]³⁺	Toluene	~380	530	2.0	0.4	495^j	5.0		[9201]

Table 5a Photophysical Parameters of Transition Metal Complexes

Complex	Solvent	Room temperature				77 K			Ref.
		λ_{abs} nm (ε) (L mol⁻¹ cm⁻¹)	λ_{em} nm	τ_{em}[a] μs	Φ_{em}[a]	λ_{em} nm	τ_{em} μs	Φ_{em}	
[Ir(bpy-C³)(bpy)₂]²⁺	CH₂Cl₂		460	10		460[j]	11.6		[9201]
[Ir(Hbpy-C³)(bpy)₂]³⁺	CH₂Cl₂		480	12.2		470[j]	14.7		[9201]
[Ir(ppy)₂(μ-Cl)]₂	CH₂Cl₂	484 (1100)	518	0.14		483[j]	4.8		[8401]
[Ir(bhq)₂(μ-Cl)]₂	CH₂Cl₂	480 (3100)	550	1.4		510[j]	30		[8401]
[Ir(ptpy)₂(μ-Cl)]₂	CH₂Cl₂	482 (1000)	510	0.15		490[j]	5.2		[8401]
[Ir(mppy)₂(μ-Cl)]₂	CH₂Cl₂	484 (1400)	520	0.06		497[j]	5.0		[8401]
[Ir(ptpy)₂(bpy)]⁺	CH₂Cl₂	468 (710)	596	0.25		526[j]	4.9		[8401]
[Ir(mppy)₂(bpy)]⁺	CH₂Cl₂	467 (870)	594	0.20		528[j]	4.4		[8401]
Ir(QO)₃	DMF	448	660	2.5	8.6×10⁻³	593	10		[8601]
Mg									
MgTPP	MeCH	565 (22000)[l]	610	0.0092	0.15		45000[d]	0.015[d]	[8201], [6901]
MgETIO		579[j]		0.0124	0.25				[8201]

Table 5a Photophysical Parameters of Transition Metal Complexes

Complex	Solvent	Room temperature				77 K			Ref.
		λ_{abs} (ε) nm (L mol⁻¹ cm⁻¹)	λ_{em} nm	τ_{em}[a] μs	Φ_{em}[a]	λ_{em} nm	τ_{em} μs	Φ_{em}	
MgPc		675 (87100)	688	0.0072	0.6				[8201]
Os									
[Os(bpy)$_3$]$^{2+}$	CH$_3$CN		743	0.060	0.005				[8602]
[Os(bpy)$_3$]$^{2+}$	H$_2$O	590 (3200)	715[g]	0.019			1.8[j]	0.038[j]	[8003], [8004]
[Os(bpy)(dppy)$_2$]$^{2+}$	CH$_3$CN		537	1.68	0.382				[8602]
[Os(bpy)$_2$(Me$_2$SO$_2$)$_2$]$^{2+}$	CH$_3$CN		575	1.5	0.300				[8602]
[Os(bpy)$_2$(diars)]$^{2+}$	CH$_3$CN		682	0.151	0.020				[8602]
[Os(bpy)(diars)$_2$]$^{2+}$	CH$_3$CN		591	1.64	0.21				[8602]
[Os(bpy)$_2$(CO)(py)]$^{2+}$	CH$_3$CN		599	1.5	0.1				[8602]
[Os(phen)$_3$]$^{2+}$	CH$_3$CN	650	720	0.262	0.021				[8602]
[Os(phen)$_3$]$^{2+}$	H$_2$O	560 (4400)	700[g]	0.084			2.8[j]	0.12[j]	[8003], [8004]
[Os(phen)$_2$(diars)]$^{2+}$	CH$_3$CN		665	0.608	0.061				[8602]

Table 5a Photophysical Parameters of Transition Metal Complexes

Complex	Solvent	Room temperature				77 K			Ref.
		λ_{abs} nm (ε) (L mol⁻¹ cm⁻¹)	λ_{em} nm	τ_{em}^{a} μs	Φ_{em}^{a}	λ_{em} nm	τ_{em} μs	Φ_{em}	
[Os(phen)(diars)₂]²⁺	CH₃CN		576	4.321	0.435				[8602]
[Os(phen)₂(dppy)]²⁺	CH₃CN		609	1.84	0.20				[8602]
[Os(phen)(dppy)₂]²⁺	CH₃CN		519	3.6	0.36				[8602]
[Os(phen)₂(DP)]²⁺	CH₂Cl₂			0.054	0.0042				[9902]
[Os(tpy)₂]²⁺	Alchools	657 (3650)	718	0.269	0.014	689	3.9	0.124	[9401]
[Os(tpy)₂]²⁺	CH₃CN	698	729	0.269 / <0.040ᵉ					[8507]
[Os(ttpy)₂]²⁺	Nitriles	667 (6600)	734	0.220	0.021	740ᵐ	0.54	0.049	[9401]
[Os(tphtpy)₂]²⁺	Alchools	692 (9200)	751	0.266	0.028				[9401]
[Os(ttpy)(dpb)]⁺	Nitriles	765 (2000)	824	<3×10⁻⁵	0.9×10⁻⁶	832			[9401]
[Os(OEP)(CO)(py)]	3MP	540				720	<6	6×10⁻⁴	[7802]
[Os(OEP)(NO)(OMe)]	3MP	568				688	116/35	3×10⁻³	[7802]

Table 5a Photophysical Parameters of Transition Metal Complexes

Complex	Solvent	Room temperature				77 K			Ref.
		λ_{abs} (ε) nm (L mol⁻¹ cm⁻¹)	λ_{em} nm	τ_{em}[a] μs	Φ_{em}[a]	λ_{em} nm	τ_{em} μs	Φ_{em}	
Pd									
[Pd(bpy)$_2$]$^{2+}$	CH$_3$CN	312 (22400)							[8803]
[Pd(phpy)$_2$]	PN	349 (8800) 346(9500)				458	480	<10^{-2}	[8803], [8804]
[Pd(thpy)$_2$]	BN	380 (8600)				536	280	1	[8804]
[Pd(bhq)$_2$]	BN	391 (5900)				472	2600	<10^{-2}	[8804]
[Pd(bph)(bpy)]	CH$_3$CN	349(2200)				473^{n}	250		[8803]
PdTPP	EPAF	551		2×10^{-5}	2.0×10^{-4}	688	2430	0.2	[7001], [8101]
PdEtio	EPAF	544				660	1930	0.5	[7001]
Pt									
[Pt(bpy)$_2$]$^{2+}$	CH$_3$CN	321 (23000)				455	24.0		[8803]
[Pt(bpy)(NH$_3$)$_2$]$^{2+}$	CH$_3$CN	319 (18000)	488	<0.01	2.0×10^{-4}	453	25		[8901]

Table 5a Photophysical Parameters of Transition Metal Complexes

Complex	Solvent	Room temperature				77 K			Ref.
		λ_{abs} (ε) nm (L mol^{-1} cm^{-1})	λ_{em} nm	τ_{em}[a] μs	Φ_{em}[a]	λ_{em} nm	τ_{em} μs	Φ_{em}	
[Pt(phen)(C≡CPh)$_2$]			580	0.907	0.090				[0002]
[Pt(5-Ph-phen)(C≡CPh)$_2$]			575	0.98	0.098				[0002]
[Pt(5-Ph-phen)(CN)$_2$]	CH$_2$Cl$_2$			223	0.097				[9402]
[Pt(hph-ph-phen)]	CH$_2$Cl$_2$	504 (7200)	586	5.3	0.6	555[j]	13[j]		[0302]
[Pt(phpy)$_2$]	Nitriles	402 (12800)				491	4		[9201]
[Pt(thpy)$_2$]	Nitriles	418 (10500)	578	4.8	0.36	570	12		[9201]
[Pt(bhq)$_2$]	Nitriles	421 (9200)				492	6.5		[9201]
[Pt(bph)(bpy)]	Nitriles	430 (5700)				581	1.1		[9201]
[Pt(H-bph)$_2$(bpy)]	Nitriles	442 (1000)				528	5.7		[9201]
[Pt(ppz)$_2$]	Nitriles	340 (13000)				431	14		[9201]
[Pt(ppz)(thpy)]	Nitriles	377 (8700)	565	15.0	0.40	555	19		[9201]
[Pt(phpy)(Cl)$_2$]$^+$	CH$_2$Cl$_2$	379 (4400)				498[n]	13.2		[9201]

Table 5a Photophysical Parameters of Transition Metal Complexes

Complex	Solvent	Room temperature				77 K			Ref.
		λ_{abs} (ε) nm (L mol^{-1} cm^{-1})	λ_{em} nm	τ_{em}[a] μs	Φ_{em}[a]	λ_{em} nm	τ_{em} μs	Φ_{em}	
[Pt(phpy)(H-ppy)Cl]	CH$_2$Cl$_2$	404 (800)	484	5.9		478[n]	14.2		[9201]
[Pt(bph)(CH$_3$CN)$_2$]	CH$_2$Cl$_2$	325	493	14	0.03				[9201]
[Pt(phpy)$_2$(CH$_2$Cl)Cl]	CH$_2$Cl$_2$	306 (15000)	447	150	0.15	444[n]	300		[9201]
[Pt(phpy)$_2$(CHCl$_2$)Cl]	CH$_2$Cl$_2$	307 (15000)	446	100	0.10	444[n]	360		[9201]
[Pt(thpy)$_2$(CH$_2$Cl)Cl]	CH$_2$Cl$_2$	344 (17000)[b]	513	270	0.05	507[n]	340		[9201]
[Pt(thpy)$_2$(CHCl$_2$)Cl]	CH$_2$Cl$_2$	344 (16000)[b]	513	200	0.05	507[n]	430		[9201]
[Pt(C^N^N-phbpy)(C≡CPh]	CH$_2$Cl$_2$	455 (4940)	582	0.4	0.04	540[j]	7.8[j]		[0401]
[Pt(C^N^N-dpp)(etpy]$^+$	Acetone			46	0.30				[9301]
[PtTPP]	EPAF	536				654	281	0.45	[7001]
[PtTPP]	CH$_2$Cl$_2$	539 (9900)	665	50	0.046				[0402]
[PtF$_{28}$TPP]	CH$_2$Cl$_2$	533 (13600)	650	5.8	0.043				[0402]
PtEtio	EPAF	534		65		641	121	0.9	[7001]

Table 5a Photophysical Parameters of Transition Metal Complexes

Complex	Solvent	λ_{abs} (ε) nm (L mol⁻¹ cm⁻¹)	Room temperature			77 K			Ref.
			λ_{em} nm	τ_{em}^a μs	Φ_{em}^a	λ_{em} nm	τ_{em} μs	Φ_{em}	
[Pt(QO)₂]	DMF	478	655	2.7	9.5×10⁻³	623	14		[8601]
[Pt(Shiffbase)]	CH₃CN		550	3.5	0.19				[0403]
Re									
[Re(bpy)(CO)₃Cl]	MeTHF	400 370 (2450)°	642	0.039	0.0031	540	3.12 ^{m}	0.0045 ^{m}	[9102]
[Re{4,4'-DEA-bpy}(CO)₃Cl]	MeTHF	367 (9790)°	575	0.412	0.033	501	12.5 ^{m}	0.125 ^{m}	[9102]
[Re{4,4'-da-bpy}(CO)₃Cl]	MeTHF	350 (6600)°	573	0.262	0.020	502	11.0 ^{m}	0.154 ^{m}	[9102]
[Re{4,4'-BAA-bpy}(CO)₃Cl]	MeTHF	364°	620	0.065	0.0073	535	4.6		[9102]
[Re{4,4'-(DMO-bpy}(CO)₃Cl]	MeTHF	356 (3880)°	630	0.026	0.0028	525	3.66		[9102]
[Re{4,4'-dm-bpy}(CO)₃Cl]	MeTHF	364 (3630)°	626	0.049	0.0057	530	3.45 ^{m}	0.0102 ^{m}	[9102]
[Re{4,4'-dph-bpy}(CO)₃Cl]	MeTHF	384°	647	0.056	0.0084	560	4.36		[9102]
[Re{4,4'-dCl-bpy}(CO)₃Cl]	MeTHF	390 (3630)°	700	0.009	0.0006	580	1.15		[9102]

389

Table 5a Photophysical Parameters of Transition Metal Complexes

Complex	Solvent	Room temperature				77 K			Ref.
		λ_{abs} (ε) nm (L mol⁻¹ cm⁻¹)	λ_{em} nm	τ_{em}[a] μs	Φ_{em}[a]	λ_{em} nm	τ_{em} μs	Φ_{em}	
[Re{4,4'-DCE-bpy}(CO)₃Cl]	MeTHF	412 (4670)°	715	0.015	0.0014	598	2.93[m]	0.0019[m]	[9102]
[Re{4,4'-dm-bpy}(CO)₃Cl]	MeTHF	448 (4510)°	780	<0.006	<0.0001	670	0.86		[9102]
[Re(phen)(CO)₃Cl]	CH₂Cl₂	377 (4000)	577	0.3	0.036	528[p]	9.6	0.33	[7402]
[Re(5-m-phen)(CO)₃Cl]	CH₂Cl₂	380 (4100)	588	≤0.65	0.03	531[p]	4.0	0.33	[7402]
[Re(5-Br-phen)(CO)₃Cl]	CH₂Cl₂	387 (3900)	584	≤0.65	0.02	535[p]	7.6	0.20	[7402]
Rh									
[Rh(bpy)₃]³⁺	MeOH/EtOH	321 (38300)				450	2210		[7101]
[Rh(phen)₃]³⁺	MeOH/EtOH	351 (3080)				450	48400		[7101]
[Rh(phpy)₂(bpy)]⁺	MeOH	367 (8000)	454	<0.05		454	170		[9201]
[Rh(thpy)₂(bpy)]⁺	MeOH	379 (9000)	526	1		521	500		[9201]
[Rh(phpy)₂(phen)]⁺	MeOH	355 (6900)				454	190		[9201]

Table 5a Photophysical Parameters of Transition Metal Complexes

Complex	Solvent	Room temperature				77 K			Ref.
		λ_{abs} nm (ε) (L·mol⁻¹ cm⁻¹)	λ_{em} nm	τ_{em}[a] μs	Φ_{em}[a]	λ_{em} nm	τ_{em} μs	Φ_{em}	
[Rh(thpy)₂(phen)]⁺	MeOH	380 (8500)				521	500		[9201]
[Rh(bhq)₂(phen)]⁺	MeOH	393 (5500)	490	<0.01		483	4250		[9201]
[Rh(phpy)₂(biq)]⁺	CH₂Cl₂	360 (27500)				544	80		[9201]
[Rh(thpy)₂(biq)]⁺	CH₂Cl₂	363 (29400)	526[j]			546	40		[9201]
[Rh(ppz)₂(bpy)]⁺	CH₂Cl₂	320 (8500)				437[n]	320		[9201]
[Rh(ppz)₂(biq)]⁺	CH₂Cl₂	366 (21500)	547			538[n]	100		[9201]
[Rh(4-NO₂-ppz)₂(biq)]⁺	CH₂Cl₂	354 (30500)	548	28		540[n]	150		[9201]
[Rh(phpy)₂(μ-Cl)]₂⁺	CH₂Cl₂	393 (7700)				461[j]	93		[9201]
[Rh(bhq)₂(μ-Cl)]₂⁺	CH₂Cl₂	410 (8200)				484[j]	2600		[9201]
[Rh(QO)₃]	EtOH	425				622	220		[8601]
[RhTMPyP(H₂O)₂]⁵⁺	H₂O	417(140000)	696	150	0.006	684[q]	260	0.01	[0302]
[RhTTMAPP(H₂O)₂]⁵⁺	H₂O		708	130	0.007	703[q]	240	0.012	[0302]

Table 5a Photophysical Parameters of Transition Metal Complexes

Complex	Solvent	Room temperature				77 K			Ref.
		λ_{abs} nm	(ε) (L mol^{-1} cm^{-1})	λ_{em} nm	τ_{em}[a] μs	Φ_{em}[a]	λ_{em} nm	τ_{em} μs	Φ_{em}	
[RhTSPP(H$_2$O$_2$)$_2$]$^{5+}$	H$_2$O			708	120	0.007	702^q	270	0.018	[0302]
[(dma)$_2$RhEtioCl]	MeTHF	546	(18300)^r	552		2×10^{-4}	552^c	730	0.23	[7301]
Ru										
[Ru(bpy)$_2$(CN)$_2$]	MeOH	458		636	0.39	0.038	578	3.9		[8501]
[Ru(bpy)$_2$(CN)$_2$]	MeOH/EtOH							3.96	0.269	[7102]
[Ru(bpy)$_2$(en)]$^{2+}$	MeOH	487			0.070		624			[8502]
[Ru(bpy)$_2$(en)]$^{2+}$	MeOH/EtOH							0.96	0.0222	[7102]
[Ru(bpy)$_2$(NH$_3$)$_2$]$^{2+}$	MeOH	488			0.033		734			[8502]
[Ru(bpy)$_3$]$^{2+}$	CH$_3$CN	452	(13000)	615	1.10					[8102]
[Ru(bpy)$_3$]$^{2+}$	CH$_3$CN	450		611	0.89	0.059				[8202]
[Ru(bpy)$_3$]$^{2+}$	BN						600	4.1		[9903]

Table 5a Photophysical Parameters of Transition Metal Complexes

Complex	Solvent	Room temperature				77 K			Ref.
		λ_{abs} nm (ε) (L mol⁻¹ cm⁻¹)	λ_{em} nm	τ_{em}^a µs	Φ_{em}^a	λ_{em} nm	τ_{em} µs	Φ_{em}	
[Ru(bpy)₃]²⁺	Benzo-nitrile	454	609	0.92	0.073				[8202]
[Ru(bpy)₃]²⁺	CH₃COOH	451	619	0.85	0.065				[8202]
[Ru(bpy)₃]²⁺	DMF	453	628	0.93	0.068				[8202]
[Ru(bpy)₃]²⁺	DMSO	454	630	0.96	0.077				[8202]
[Ru(bpy)₃]²⁺	Eglyc	452	611		0.070				[8202]
[Ru(bpy)₃]²⁺	EPA			0.926	0.270		3.75	0.746	[6902]
[Ru(bpy)₃]²⁺	EtOH	450	608	0.74	0.052				[8202]
[Ru(bpy)₃]²⁺	H₂O	452 (14600)	607	0.60					[7601]
[Ru(bpy)₃]²⁺	H₂O	452	628	0.65	0.042 0.028ᵉ				[8203]
[Ru(bpy)₃]²⁺	MeOH	449	609	0.72	0.045				[8202]
[Ru(bpy)₃]²⁺	MeOH/EtOH	450 (14300)	630	1.15	0.089	578	5.2	0.35	[7601], [8402]
[Ru(bpy)₃]²⁺	Py	454	619	0.92	0.087				[8202]

Table 5a Photophysical Parameters of Transition Metal Complexes

Complex	Solvent	Room temperature				77 K			Ref.
		λ_{abs} nm (ε) L mol⁻¹ cm⁻¹	λ_{em} nm	τ_{em}[a] μs	Φ_{em}[a]	λ_{em} nm	τ_{em} μs	Φ_{em}	
[Ru(bpy)$_2$(4-n-bpy)]$^{2+}$	CH$_3$CN	493 (11100)	625	0.78					[8204]
[Ru(bpy)$_2$(3,3'-dm-bpy)]$^{2+}$	CH$_3$CN	448 (11500)	620	0.74					[8102]
[Ru(bpy)$_2$(3,3'-dm-bpy)]$^{2+}$	H$_2$O	453 (12600)	630	0.41	0.00036 0.00030[d]				[8203]
[Ru(bpy)$_2$(3,3'-dm-bpy)]$^{2+}$	MeOH/EtOH					588	5.2	0.252	[8203]
[Ru(bpy)$_2$(4,4'-dm-bpy)]$^{2+}$	H$_2$O	454	620	0.470	0.025				[8503]
[Ru(bpy)$_2$(4,4'-dm-bpy)]$^{2+}$	MeOH/EtOH					586	5.2	0.35	[8103]
[Ru(bpy)$_2$(4,4'-dph-bpy)]$^{2+}$	MeOH/EtOH	457 (19700)	630	1.92	0.197 0.041[e]	593	5.6	0.46	[8103], [8402]
[Ru(bpy)$_2$(4,4'-dc-bpy)]$^{2+}$	H$_2$O[s]		686	0.299	0.006				[8403]
[Ru(bpy)$_2$(4,4'-dst-bpy)]$^{2+}$	CHCl$_3$	482	668	0.900	0.180				[7702]
[Ru(bpy)$_2$(4,4'-DCM-bpy)]$^{2+}$	H$_2$O	475	660	0.615	0.045				[7702]

Table 5a Photophysical Parameters of Transition Metal Complexes

Complex	Solvent	Room temperature				77 K			Ref.
		λ_{abs} nm (ε) (L mol^{-1} cm^{-1})	λ_{em} nm	τ_{em} [a] µs	Φ_{em} [a]	λ_{em} nm	τ_{em} µs	Φ_{em}	
[Ru(bpy)$_2$(4,4'-DCE-bpy)]$^{2+}$	EtOH			1.03	0.074				[8005]
[Ru(bpy)$_2$(4,4'-DCE-bpy)]$^{2+}$	H$_2$O			0.37	0.027				[8005]
[Ru(bpy)$_2$(phen)]$^{2+}$	CH$_2$Cl$_2$		601	0.31	0.033				[8304]
[Ru(bpy)$_2$(phen)]$^{2+}$	H$_2$O			0.684					[8305]
[Ru(bpy)$_2$(phen)]$^{2+}$	MeOH/EtOH					575	6.6	0.44	[7601]
[Ru(bpy)$_2$(py)$_2$]$^{2+}$	CH$_3$CN	457	606	0.500	2.3×10^{-4}				[8205]
[Ru(bpy)$_2$(DP)]$^{2+}$	EtOH	448 (15700)	610		0.027				[8504]
[Ru(bpy)$_2$(DP)]$^{2+}$	2-Propanol		622		0.210				[9001]
[Ru(bpy)$_2$(DPP)]$^{2+}$	H$_2$O	430 (12000)	675	0.135	0.049				[8404]
[Ru(bpy)$_2$(4,4'-dm-bpy)$_2$]$^{2+}$	H$_2$O	456	631	0.385	0.014				[8503]
[Ru(bpy)$_2$(4,4'-dph-bpy)$_2$]$^{2+}$	MeOH/EtOH	465 (14700)	635	2.12	0.098 0.025[e]				[8402]

Table 5a Photophysical Parameters of Transition Metal Complexes

Complex	Solvent	Room temperature				77 K			Ref.
		λ_{abs} (ε) nm (L mol^{-1} cm^{-1})	λ_{em} nm	τ_{em}[a] μs	Φ_{em}[a]	λ_{em} nm	τ_{em} μs	Φ_{em}	
[Ru(4-a-bpy)$_3$]$^{2+}$	MeOH/EtOH	472 (12100)	675	0.35	0.021 0.006[e]				[8402]
[Ru(4-dma-bpy)$_3$]$^{2+}$	MeOH/EtOH	475 (15000)	680	0.35	0.023 0.006[e]				[8402]
[Ru(4-mo-bpy)$_3$]$^{2+}$	MeOH/EtOH	461 (14100)	660	0.67	0.039 0.014[e]				[8402]
[Ru(4-n-bpy)$_3$]$^{2+}$	CH$_3$CN	480 (20400)	700	0.12		643[j]	3.8[j]		[8204]
[Ru(4-bo-bpy)$_3$]$^{2+}$	MeOH/EtOH	460 (14800)	650	0.60	0.053 0.017[e]				[8402]
[Ru(3,3'-dm-bpy)$_3$]$^{2+}$	CH$_3$CN	456 (10800)	625	0.21		601[j]	6.4[j]	0.113[j]	[8102], [8203]
[Ru(4,4'-dm-bpy)$_3$]$^{2+}$	CH$_2$Cl$_2$	459	618	0.931	0.12	593			[8603]
[Ru(4,4'-dm-bpy)$_3$]$^{2+}$	H$_2$O	459 (14900)	642		0.026 0.021[e]				[8203]
[Ru(4,4'-dm-bpy)$_3$]$^{2+}$	MeOH/EtOH	455 (17000)	640	0.95	0.086 0.027[e]	593	4.6	0.283	[8405], [8402], [8203]
[Ru(4,4'-dn-bpy)$_3$]$^{2+}$	MeOH/EtOH	473	700	0.25	0.002 0.001[e]				[8402]

Table 5a Photophysical Parameters of Transition Metal Complexes

Complex	Solvent	λ_{abs} nm (ε) (L mol^{-1} cm^{-1})	λ_{em} nm	τ_{em} [a] μs	Φ_{em} [a]	λ_{em} nm	τ_{em} μs	Φ_{em}	Ref.
			Room temperature			77 K			
$[Ru(4,4'-da-bpy)_3]^{2+}$	MeOH/ EtOH	504 (10500)	705	0.10	0.004 0.002[e]				[8405], [8402]
$[Ru(4,4'-DEA-bpy)_3]^{2+}$	MeOH/ EtOH	518 (14500)	700	0.13	0.010 0.005[e]				[8402]
$[Ru(4,4'-BAA-bpy)_3]^{2+}$	MeOH/ EtOH	479 (15000)	670	0.41	0.027 0.014[e]				[8402]
$[Ru(4,4'-DEO-bpy)_3]^{2+}$	MeOH/ EtOH	477 (13000)	675	0.30	0.020 0.009[e]				[8405], [8402]
$[Ru(4,4'-DPO-bpy)_3]^{2+}$	MeOH/ EtOH	479 (13400)	670	0.35	0.038 0.018[e]				[8402]
$[Ru(4,4'-DBO-bpy)_3]^{2+}$	MeOH/ EtOH	476 (13500)	670	0.31	0.032 0.015[e]				[8402]
$[Ru(4,4'-dph-bpy)_3]^{2+}$	MeOH/ EtOH	473 (28000)	635	1.95	0.306 0.058[e]				[8405], [8402]
$[Ru(4,4'-dbz-bpy)_3]^{2+}$	MeOH/ EtOH	460 (16100)	640	1.25	0.098 0.030[e]				[8402]
$[Ru(4,4'-dsty-bpy)_3]^{2+}$	MeOH/ EtOH	487 (33000)	680	0.72	0.030 0.009[e]				[8402]
$[Ru(4,4'-DCE-bpy)_3]^{2+}$	CH_2Cl_2	467	629	2.230	0.30	607			[8603]

397

Table 5a Photophysical Parameters of Transition Metal Complexes

Complex	Solvent	Room temperature				77 K			Ref.
		λ_{abs} (ε) nm (L mol⁻¹ cm⁻¹)	λ_{em} nm	τ_{em} a μs	Φ_{em} a	λ_{em} nm	τ_{em} μs	Φ_{em}	
[Ru(4,4'-DCE-bpy)₃]²⁺	MeOH/EtOH	464 (23300)	655	1.65	0.200 / 0.076^e				[8405], [8402]
[Ru(5,5'-dm-bpy)₃]²⁺	MeOH/EtOH	440 (14700)	620	0.35	0.037 / 0.015^e				[8402]
[Ru(5,5'-DCE-bpy)₃]²⁺	MeOH/EtOH	495 (9900)	720	0.23	0.004 / 0.003^e				[8405], [8402]
[Ru(5,5'-BAA-bpy)₃]²⁺	MeOH/EtOH	445 (14900)	630	2.40	0.126 / 0.021^e				[8402]
[Ru(phen)₂(CN)₂]	H₂O	421 (9500)	626	0.71					[7703]
[Ru(phen)₂(CN)₂]	MeOH	448 (10400)	638	1.58					[7703]
[Ru(phen)₃]²⁺	CH₃CN	442	604	0.46	0.028				[8406]
[Ru(phen)₃]²⁺	DMF			0.68	0.063				[8406]
[Ru(phen)₃]²⁺	H₂O	447 (18100)	604	0.962 / 0.459^e	0.058 / 0.032^e				[8203], [8407], [8505]
[Ru(phen)₃]²⁺	MeOH/EtOH	445 (20000)	595	0.45	0.019	567	9.8	0.58	[7601], [8103], [8405], [8506]

Table 5a Photophysical Parameters of Transition Metal Complexes

Complex	Solvent	Room temperature				77 K			Ref.
		λ_{abs} nm (ε) (L mol^{-1} cm^{-1})	λ_{em} nm	τ_{em} [a] μs	Φ_{em} [a]	λ_{em} nm	τ_{em} μs	Φ_{em}	
[Ru(phen)$_2$(4,7-dm-phen)]$^{2+}$	H$_2$O		611	1.389					[8407]
[Ru(phen)$_2$(4,7-Ph$_2$-phen)]$^{2+}$	H$_2$O			3.07					[8306]
[Ru(phen)$_2$(4,7-dsph-phen)]	H$_2$O			3.46, 0.920^e					[8306], [8505]
[Ru(phen)$_2$(DMCH)]$^{2+}$	MeOH/EtOH	522 (10600)^o	728°			710	2.0		[8408], [8409]
[Ru(4-m-phen)$_3$]$^{2+}$	MeOH/EtOH	445 (22000)	602	0.80	0.054				[8506]
[Ru(4-Ph-phen)$_3$]$^{2+}$	MeOH/EtOH	450 (22100)	605	4.00	0.109				[8506]
[Ru(4-tol-phen)$_3$]$^{2+}$	MeOH/EtOH	453 (23600)	605	3.50	0.117				[8506]
[Ru(4-bph-phen)$_3$]$^{2+}$	MeOH/EtOH	454 (28700)	612	4.9	0.184				[8506]
[Ru(4-MoPh-phen)$_3$]$^{2+}$	MeOH/EtOH	452 (25300)	615	3.25	0.130				[8506]
[Ru(5-m-phen)$_3$]$^{2+}$	MeOH/EtOH	445 (21000)	597	0.85	0.061				[8506]
[Ru(5-Ph-phen)$_3$]$^{2+}$	H$_2$O	448 (24600)	595	1.29					[7602]

Table 5a Photophysical Parameters of Transition Metal Complexes

Complex	Solvent	Room temperature				77 K			Ref.
		λ_{abs} (ε) nm (L mol⁻¹ cm⁻¹)	λ_{em} nm	τ_{em}[a] μs	Φ_{em}[a]	λ_{em} nm	τ_{em} μs	Φ_{em}	
[Ru(5-Ph-phen)₃]²⁺	MeOH/EtOH	448 (21700)	597	1.15	0.033				[8506]
[Ru(2,9-dm-phen)₃]²⁺	MeOH/EtOH					588	2.3	0.0479	[8006]
[Ru(4,7-dm-phen)₃]²⁺	MeOH/EtOH	448 (22700)	600	0.85	0.059				[8506]
[Ru(4,7-Ph₂-phen)₃]²⁺	MeOH/EtOH	463 (28600)	618	6.40	0.366		9.58	0.682	[8405], [8506], [7201]
[Ru(4,7-bmo-phen)₃]²⁺	MeOH/EtOH	468 (37900)	615	6.10	0.387				[8506]
[Ru(4,7-bph₂-phen)₃]²⁺	MeOH/EtOH	465 (36300)	620	7.20	0.360				[8506]
[Ru(dtol-phen)₃]²⁺	MeOH/EtOH	464 (27400)	618	6.43	0.280				[8506]
[Ru(4,7-bpbr-phen)₃]²⁺	MeOH/EtOH	465 (31000)	623	6.00	0.310				[8506]
[Ru(4,7-bpmo-phen)₃]²⁺	MeOH/EtOH	468 (34800)	620	6.05	0.305				[8506]
[Ru(4,7-bpph-phen)₃]²⁺	MeOH/EtOH	465 (33700)	618	6.10	0.330				[8506]
[Ru(4,7-brmo-phen)₃]²⁺	MeOH/EtOH	464 (25500)	625	5.95	0.237				[8506]

Table 5a Photophysical Parameters of Transition Metal Complexes

Complex	Solvent	Room temperature				77 K			Ref.
		λ_{abs} nm (ε) (L mol^{-1} cm^{-1})	λ_{em} nm	τ_{em}[a] μs	Φ_{em}[a]	λ_{em} nm	τ_{em} μs	Φ_{em}	
[Ru(5,6-dm-phen)$_3$]$^{2+}$	MeOH/EtOH	450 (19800)	598	2.50	0.143				[8506]
[Ru(tm1-phen)$_3$]$^{2+}$	MeOH/EtOH	439 (23300)	595	0.48	0.032				[8506]
[Ru(thpy)$_3$]$^{2+}$	H$_2$O	463	668	0.22	0.026				[8307]
[Ru(biq)$_2$(CN)$_2$]	DMF	635 (8900)	842	0.11	0.007	780	0.8		[8501]
[Ru(biq)$_2$(CN)$_2$]	MeOH	580	790	0.17	0.009		0.93		[8501]
[Ru(i-biq)$_2$(CN)$_2$]	DMF	437 (17600)	632	0.45	0.023	577	40		[8501]
[Ru(tpy)$_2$]$^{2+}$	CH$_3$CN	474 (14600)		2.5×10^{-5}					[8408], [9803]
[Ru(tpy)$_2$]$^{2+}$	H$_2$O	473 (16200)	610	<0.005					[7602]
[Ru(tpy)$_2$]$^{2+}$	H$_2$O	473	628	<0.005					[8003]
[Ru(tpy)$_2$]$^{2+}$	alcohols	476 (17700)		2.5×10^{-4}		598	11.0	0.48	[9401]
[Ru(Cl-tpy)$_2$]$^{2+}$	Nitriles	480 (16000)	630	1.0	$<10^{-5}$	615	8.6		[9401]
[Ru(tro)$_2$]$^{2+}$	MeOH/EtOH					629	12.3	0.45	[8104]

Table 5a Photophysical Parameters of Transition Metal Complexes

Complex	Solvent	Room temperature				77 K			Ref.
		λ_{abs} (ε) nm (L mol⁻¹ cm⁻¹)	λ_{em} nm	τ_{em} [a] μs	Φ_{em} [a]	λ_{em} nm	τ_{em} μs	Φ_{em}	
[Ru(Me₂N-tpy)₂]²⁺	Nitriles	490 (15400)				656	5.4		[9401]
[Ru(MeSO₂-tpy)₂]²⁺	Nitriles	485	650	0.025	4×10⁻⁴	632	10.5		[9401]
[Ru(ph-tpy)₂]²⁺	Nitriles	488 (30000)	670		2.5×10⁻⁵				[9401]
[Ru(MeO-phtpy)₂]²⁺	alcohols	495 (24400)	650	0.0048	3×10⁻⁵	632	13	0.4	[9401]
[Ru(tphtpy)₂]²⁺	alcohols	501 (38400)	650	0.0038	7.4×10⁻⁴	633	7.8	0.57	[8104], [9401]
[Ru(ttpy)₂]²⁺	Nitriles	490 (28000)	640	9.5×10⁻⁴	3.2×10⁻⁵	628	9.1		[9401]
[RuOEP(CO)]	CH₂Cl₂			55					[8805]
[RuOEP(CO)py]	3MP	549 (24550)ᶠ		26		653	405	6×10⁻²	[8805], [7802]
[RuTPP(CO)]	CH₂Cl₂			47					[8805]
Sc									
(OEP)ScOAc	BI	571 (16200)	577		0.14	571ᶜ / 700ᵈ	400000	0.16-0.21ᶜ / 0.046-0.056ᵈ	[7503]

Table 5a Photophysical Parameters of Transition Metal Complexes

Complex	Solvent	Room temperature				77 K			Ref.
		λ_{abs} (ε) nm (L·mol⁻¹·cm⁻¹)	λ_{em} nm	τ_{em}^{a} μs	Φ_{em}^{a}	λ_{em} nm	τ_{em} μs	Φ_{em}	
Tb									
[Tb(2.2.1)]³⁺	H₂O	368(0.3)		1300	0.3		1300		[9702]
[Tb(bpy.bpy.bpy)]³⁺	H₂O	304(29000)		330	0.03		1700		[9702]
[Tb(1,4,7-(bpy)₃-triazacyclononane)]³⁺	H₂O	311(20400)		1500	0.37		1400		[9101]
[Tb(Calixarene-(bpy)₂]³⁺	CH₃CN	305(23400)		1860	0.39				[0001]
Ti									
[Ti(OEP)(O)]	3MP		577		0.02	582ᶜ / 714ᵈ	175000ᵈ	0.06ᶜ	[7503]
Th									
[Th(OEP)(Oac)₂]	Toluene		651ᶜ / 766ᵈ		4×10⁻⁴ᶜ / 2×10⁻³ᵈ				[0201]
Zn									
[Zn(tpy)₂]²⁺	CH₃CN		353		0.65				[9803]

Table 5a Photophysical Parameters of Transition Metal Complexes

Complex	Solvent	Room temperature				77 K			Ref.
		λ_{abs} (ε) nm (L mol^{-1} cm^{-1})	λ_{em} nm	$\tau_{em}{}^{a}$ μs	$\Phi_{em}{}^{a}$	λ_{em} nm	τ_{em} μs	Φ_{em}	
[ZnOEP]				0.0022	0.04		57000^d		[8206]
[ZnTPP]	Benzene	586 (3680)	598	1.9	0.033	598			[7504]
[ZnTPP]	MeCH			2.7	0.04	781^d	26000^d	0.012^d	[8101]
[ZnOMeTPP]	Benzene	584 (2030)	595	1.5	0.027				[7504]
[ZnpClTPP]	Benzene	589 (3910)	600	1.05	0.020				[7504]
[ZnPcTs]		672 (282000)		3.8	0.3		1100^d		[8206]
[ZnPcTs]	DMF		686	0.00288	0.277				[9801]
[ZnNc]	DMF	759	786	0.0088					[0303]
Zr									
(OEP)Zr(Oac)$_2$	CH$_2$Cl$_2$	565 (31600)	576		0.016	570^c 704^d	41000^d	0.017^c 0.034^d	[7503]

[a] Lifetime and quantum yield values are for deareated solutions unless otherwise noted; [b] In EtOH; [c] Fluorescence; [d] Phosphorescence; [e] Aerated solution; [f] In DMSO; [g] Uncorrected spectra; [h] CH$_2$Cl$_2$/MeOH solution; [i] D$_2$O solution; [j] EtOH/MeOH solution; [k] DMF solution; [l] Benzene solution; [m] 155 – 160 K; [n] Nitriles as solvent; [o] In CH$_3$CN; [p] EPA (EtOH/Isopentane/Et$_2$O); [q] Ethylene glycol/H$_2$O; [r] In CH$_2$Cl$_2$; [s] pH-dependent.

ABBREVIATIONS

Ligands

3,3'-dm-bpy	3,3'-dimethyl-2,2'-bipyridine
4,4'-(DMO)-bpy	4,4'-dimethoxy-2,2'-bipyridine
4,4'-DCE-bpy	4,4'-dicarboxyethyl-2,2'-bipyridine
4,4'-dCl-bpy	4,4'-dichloro-2,2'-bipyridine
4,4'-DEA-bpy	4,4'-bis(diethylamino)-2,2'-bipyridine
4,4'-dm-bpy	4,4'-dimethyl-2,2'-bipyridine
4,4'-dn-bpy	4,4'-dinitro-2,2'-bipyridine
4,4'-dph-bpy	4,4'-diphenyl-2,2'-bipyridine
4,4'-BAA-bpy	4,4'-bis(acetoamido)-2,2'-bipyridine
4,4'-da-bpy	4,4'-diamino-2,2'-bipyridine
4,4'-dbz-bpy	4,4'-dibenzyl-2,2'-bipyridine
4,4'-dc-bpy	4,4'-dicarboxy-2,2'-bipyridine
4,4'-DCM-bpy	4,4'-dicarboxymethyl-2,2'-bipyridine
4,4'-DEO-bpy	4,4'-diethoxy-2,2'-bipyridine
4,4'-DPO-bpy	4,4'-diphenoxy-2,2'-bipyridine
4,4'-dst-bpy	4,4'-distearyl-2,2'-bipyridine
4,4'-dsty-bpy	4,4'-distyryl-2,2'-bipyridine
4,7-bmo-phen	4,7-bis(p-methoxyphenyl)-1,10-phenanthroline
4,7-bpbr-phen	4-p-biphenylyl-7-p-bromophenyl-1,10-phenanthroline
4,7-bph2-phen	4,7-bis(p-methoxyphenyl)-1,10-phenanthroline
4,7-bpmo-phen	4-p-biphenylyl-7-p-methoxyphenyl-1,10-phenanthroline
4,7-bpph-phen	4-p-biphenylyl-7-phenyl-1,10-phenanthroline
4,7-brmo-phen	4-p-bromophenyl-7-p-methoxyphenyl-1,10-phenanthroline
4,7-dm-phen	4,7-dimethyl-1,10-phenanthroline
4,7-dsph-phen	disulfonated 4,7-diphenyl-1,10-phenanthroline
4,7-dtol-phen	4,7-ditolyl-1,10-phenanthroline
4,7-Ph2-phen	4,7-diphenyl-1,10-phenanthroline
4-a-bpy	4-amino-2,2'-bipyridine
4-bo-bpy	4-benzyl-oxy-2,2'-bipyridine
4-bph-phen	4-p-biphenylyl-1,10-phenanthroline
4-dma-bpy	4-dimethylamino-2,2'-bipyridine
4-mo-bpy	4-methoxy-2,2'-bipyridine
4-MoPh-phen	4-p-methoxyphenyl-1,10-phenanthroline
4-m-phen	4-methyl-1,10-phenanthroline
4-n-bpy	4-nitro-2,2'-bipyridine
4-Ph-phen	4-phenyl-1,10-phenanthroline
4-tol-phen	4-tolyl-1,10-phenanthroline

5,5'-BAA-bpy	5,5'-bisacetoamido-2,2'-bipyridine
5,5'-dm-bpy	5,5'-dimethyl-2,2'-bipyridine
5,6-dm-phen	5,6-dimethyl-1,10-phenanthroline
5-Br-phen	5-bromo-1,10-phenanthroline
5-m-phen	5-methyl-1,10-phenanthroline
5-m-phen	5-methyl-1,10-phenanthroline
5-Ph-phen	5-phenyl-1,10-phenanthroline
bcp	2,9-dimethyl-4,7-diphenyl-1,10-phenanthroline
bfp	2,9-bis-(trifluoromethyl)-1,10-phenantroline
bhq	benzo[h]quinoline
biq	2,2'-biquinoline
bph2-	1,1'-biphenyl-2,2'-diyl dianion
bpy	2,2'-bipyridine
calixarene-(bpy)$_2$	dimethoxy-p-tert-butylcalix[4]arene)-26,28-[2,10-bis(2,2'-bipyridine-6-methyl)oxy-methyl]crown-4
dbtmp	2,9-dibutyl-3,4,7,8tetramethyl-1,10-phenantroline
diars	1,2-bis(dimethyl-arsino)benzene
dma	dimethylamine
DMCH	5,6-dihydro-4,7-dimethyldibenzo[3,2-*b*:2',3'-*j*][1,10]phenanthroline
dmp	2,9-dimethyl-1,10-phenantroline
dnp	(2,9-bis(1-naphthyl)-1,10-phenantroline
DP	dipyrido[3,2-*a*:2',3'-*c*]phenazine
dpp	2,9-diphenyl-1,10-phenantroline
DPP	2,3-bis(2-pyridyl)-pyrazine
dppb	1,2-bis(diphenyl-phospino)ethane
dsbp	2,9-bis-(*sec*-butyl)-1,10-phenantroline
etpy	2-ethyl pyridine
F$_{28}$TPP	2,3,7,8,12,13,17,18-octafluoro-5,10,15,20-tetrakis(pentafluorophenyl)porphyirin
hph-ph-phen	2,9-Bis(2'-hydroxyphenyl)-4,7-diphenyl-1,10-phenanthroline, dianion
i-biq	3,3'-biisoquinoline
mtpy	2-(*m*-Tolyl)pyridine
NpcTs	naphthalocyanine tetrasulphonate
OEP	octaethyl-porphyrin
PcTs	phthalocyanine tetrasulphonate
phbpy	6-phenyl-2,2'-bipyridine
phen	1,10-phenantroline
ppy	2-phenylpyridine
ppz	1-phenylpyrazole
ptpy	2-(*p*-Tolyl)pyridine
py	pyridine
QOH	8-hydroxyquinoline

Shiffbase	*N, N'*-bis(salycilidene)-1,2-ethylenediamine
thpy	2-(2-thienyl)pyridine
tml-phen	3,4,7,8-tetramethyl-1,10-phenanthroline
TMPyP	*meso*-tetrakys(4-N-methyl Pyridyl) porphyrin
tphtpy	4,4',4"-triphenyl-2,2':6',2"-terpyridine
tpp	2,4,7,9-tetraphenyl-1,10-phenanthroline
TPP	tetraphenyl-porphyrin
tpy	2,2':6',2"-terpyridine
tro	4'-phenyl-2,2',2"-tripyridine
TSPP	*meso*-tetrakis(4-sulfonate phenyl) porphirin
TTMAPP	*meso*-tetrakis(4-*N,N,N*-trimethyl aminophenyl) porphirin
ttpy	4'-(*p*-tolyl)-2,2':6',2"-terpyridine

Solvents

3MP	3-methyl pentane
BI	butanol-isopentane
BN	butyronitrile
Eglyc	ethylene glycol
EPAF	diethylether, isopentane, dimethylformamide, ethanol, 12:10:6:1
MeCH	methyl cyclohexane
PN	propionitrile
Py	pyridine

5b ABSORPTION AND LUMINESCENCE SPECTRA: A SELECTION

In this section absorption and emission spectra (at room temperature) of some transition metal complexes are displayed. The selection was made in order to present the spectra of widely studied complexes of the most representative metals of the transition series. The preference has been given to complexes showing luminescence at room temperature. At the end of the section, aqueous absorption spectra of the most common anions (in all cases, sodium as counterion) are also reported. The $CH_3SO_3^-$, $CH_3SO_3^-$, ClO_4^-, and PF_6^- anions show very weak absorption only in the far UV region (at 190 nm $\varepsilon < 25$ l mol^{-1} cm^{-1}). The absorption spectra (full lines) are presented as molar absorption coefficients vs wavelength (ε vs λ). The emission spectra (dashed lines) are presented as intensity (arbitrary units) vs wavelength. For other relevant photophysical properties , see Table 5a.

Part of the spectra are reproduced from the literature. The spectra labeled with the asterisk have been performed by Serena Silvi, Dr. Paolo Passaniti and Dr. Alberto Di Fabio in the photochemistry laboratory of the department of chemistry of the University of Bologna (Italy). The compounds used were of highest purity commercially available or samples deriving from previous researches.

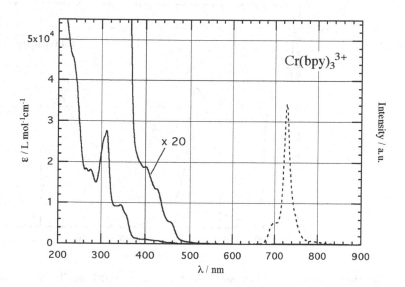

Fig. 5b-1. Absorption* and emission* spectra
of $[Cr(bpy)_3](ClO_4)_3$ in 0.1 M H_2SO_4.

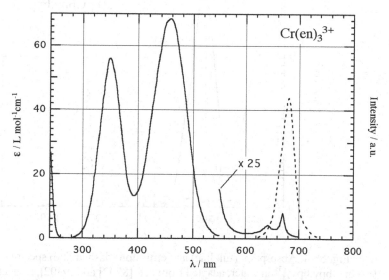

Fig. 5b-2. Absorption and emission spectra of $[Cr(en)_3]^{3+}$
in 3×10^{-3} M $HClO_4$, adapted from ref. [7103]

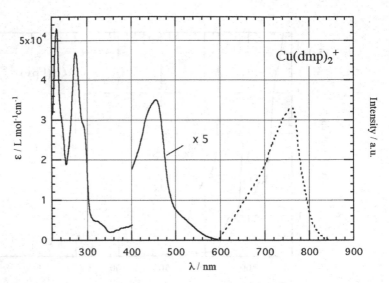

Fig. 5b-3. Absorption and emission spectra of [Cu(dmp)₂]⁺ in CH₂Cl₂, adapted from refs. [0101] and [8002], respectively.

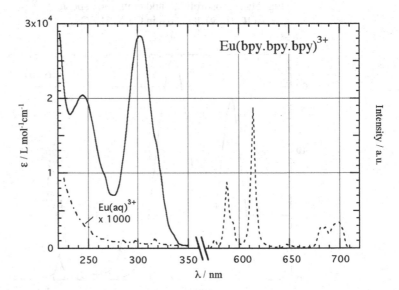

Fig. 5b-4. Absorption (full line) and emission (dashed line) spectra of [Eu(bpy.bpy.bpy)]³⁺ in water, adapted from ref. [8702] and [9302], respectively.

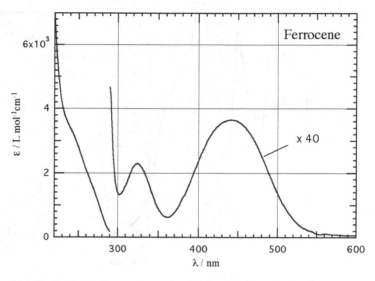

Fig. 5b-5. Absorption spectrum* of ferrocene in acetonitrile.

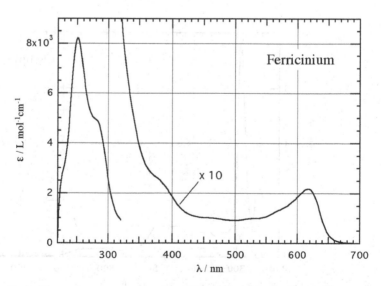

Fig. 5b-6. Absorption spectrum* of ferrocinium tetrafluoborate in water.

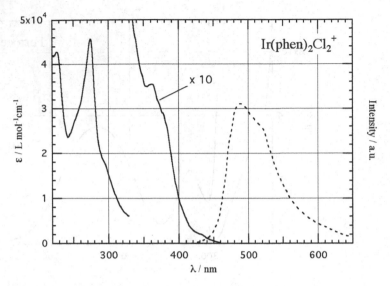

Fig. 5b-7. Absorption and emission spectra of [Ir(phen)₂Cl₂)]⁺
in water, adapted from ref. [7701].

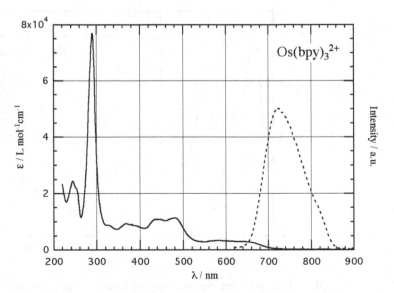

Fig. 5b-8. Absorption* and emission* spectra
of [Os(bpy)₃](PF₆)₂ in acetonitrile.

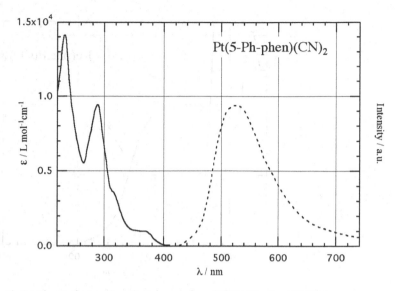

Fig. 5b-9. Absorption and emission spectra of [Pt(5-Ph-phen)(CN)$_2$] in CH$_2$Cl$_2$, adapted from ref. [9402].

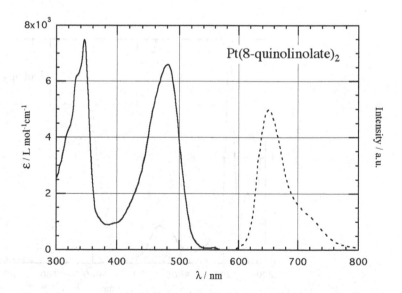

Fig. 5b-10. Absorption and emission spectra of [Pt(8-quinolinate)$_2$] in DMF, adapted from ref. [7803].

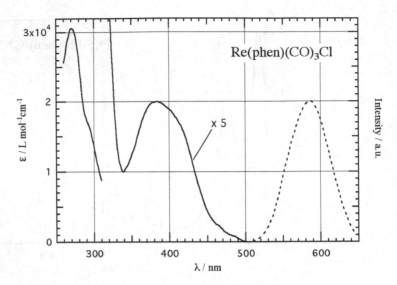

Fig. 5b-11. Absorption and emission spectra of [Re(phen)(CO)₃Cl]
in EPA, adapted from ref. [7402].

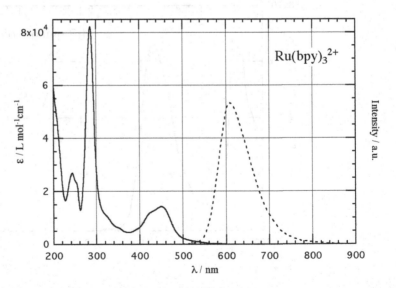

Fig. 5b-12. Absorption* and emission* spectra
of [Ru(bpy)₃](PF₆)₂ in acetonitrile.

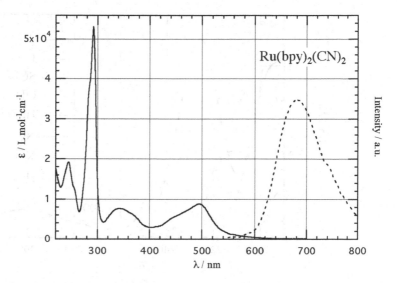

Fig. 5b-13. Absorption and emission spectra of [Ru(bpy)$_2$CN$_2$)]
in acetonitrile, adapted from ref. [8902].

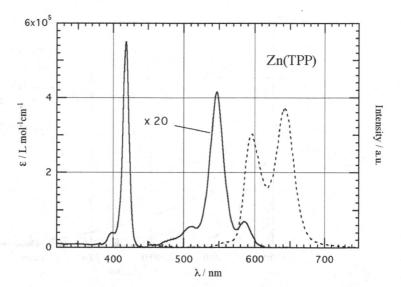

Fig. 5b-14. Absorption and emission spectra of Zn(TPP)
in CH$_2$Cl$_2$, adapted from ref. [7504].

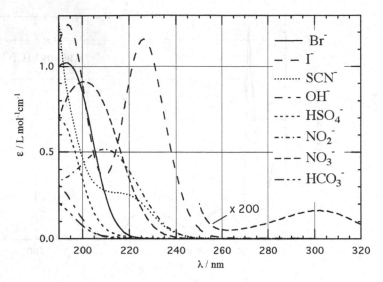

Fig. 5b-15. Absorption spectra* of some anions in water.

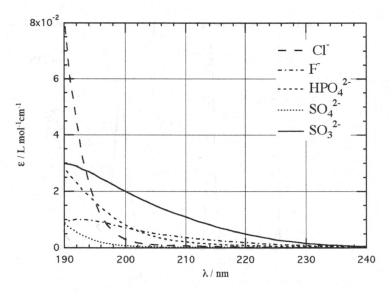

Fig. 5b-16. Absorption spectra* of some anions in water.

REFERENCES

[6901] Seybold, P. G., Gouterman, M. *J. Mol. Spectrosc.* **1969**, *31*, 1-13.

[6902] F.E. Lytle F. E.; Hercules, D. M. *J. Am. Chem. Soc.* **1969**, *91*, 253-257.

[7001] Eastwood, D.; Gouterman, M. *J. Mol. Spectrosc.* **1970**, *35*, 359-375.

[7101] DeArmond, M. K.; Hillis, J. E. *J. Chem. Phys.* **1971**, *54*, 2247-2253.

[7102] Demas J. N.; Crosby, G. A. *J. Am. Chem. Soc.* **1971**, *93*, 2841-2847.

[7103] Balzani, V.; Ballardini, R.; Gandolfi, M. T.; Moggi, L. *J. Am. Chem. Soc.* **1971**, *93*, 339-345.

[7201] Watts, R. J.; Crosby, G. A. *J. Am. Chem. Soc.* **1972**, *94*, 2606-2614.

[7301] Hanson, L. K.; Gouterman, M.; Hanson, J. C. *J. Am. Chem. Soc.* **1973**, *95*, 4822-4829.

[7401] Flynn, C. M. Jr.; Demas, J. N. *J. Am. Chem. Soc.* **1974**, *96*, 1959-1960.

[7402] Wrighton, M.; Morse, D. L. *J. Am. Chem. Soc.* **1974**, *96*, 998-1003.

[7501] Demas, J. N.; Harris, E. W.; Flynn, C. M.; Demiente, D. *J. Am. Chem. Soc.* **1975**, *97*, 3838-3839.

[7502] Ballardini, R.; Varani, G.; Moggi, L.; Balzani, V.; Olson, K. R.; Scandola, F.; Hoffman, M. Z. *J. Am. Chem. Soc.* **1975**, *97*, 728-736.

[7503] Gouterman, M.; Hanson, L. K.; Khalil, G.-E.; Buchler, J. W.; Rohbock, K.; Dolphin, D. *J. Am. Chem. Soc.* **1975**, *97*, 3142-3149.

[7504] Quimby, D. J.; Longo, F. R. *J. Am. Chem. Soc.* **1975**, *94*, 5111-5117.

[7601] Crosby G. A.; Elfring Jr., W. H. *J. Phys. Chem.* **1976**, *80*, 2206-2211.

[7602] Lin, C. T.; Boettcher, W.; Chou, M.; Creutz, C.; Sutin, N. *J. Am. Chem. Soc.* **1976**, *98*, 6536-6544.

[7701] Ballardini, R.; Varani, G.; Moggi, L.; Balzani, V. *J. Am. Chem. Soc.* **1977**, *99*, 6881-6884.

[7702] Harriman, A. *Chem. Commun.* **1977**, 777-778.

[7703] Demas, J. N.; Addington, J. W.; Peterson, S. H.; Harris, E. W. *J. Phys. Chem.* **1977**, *81*, 1039-1043.

[7801] Maestri, M.; Bolletta, F.; Moggi, L.; Balzani, V.; Henry, M. S.; Hoffman, M. Z. *J. Am. Chem. Soc.* **1978**, *100*, 2694-2701.

[7802] Antipas, A., Buchler, J. W.; Gouterman, M.; Smith, P. D. *J. Am. Chem. Soc.* **1978**, *100*, 3015-3024.

[7803] Ballardini, R.; Indelli, M. T.; Varani, G.; Bignozzi, C. A., Standola F. *Inorg. Chim. Acta*, **1978**, *31*, L423-L424.

[7901] Serpone, N.; Jamieson, M. A.; Henry, M. S.; Hoffman, M. Z.; Bolletta, F.; Maestri, M. *J. Am. Chem. Soc.* **1979**, *101*, 2907-2916.

[8001] Kirk, A. D.; Porter, G. B. *J. Phys. Chem.* **1980**, *84*, 887-891.

[8002] Blaskie, M. W.; McMillin, D. R. *Inorg. Chem.* **1980**, *19*, 3519-3522.

[8003] Creutz, C.; Chou, M.; Netzel, T. L.; Okumura, M.; Sutin, N. *J. Am. Chem. Soc.* **1980**, *102*, 1309-1319.

[8004] Lacky, D. E.; Pankuch, B. J.; Crosby, G. A. *J. Phys. Chem.* **1980**, *84*, 2068-2074.

[8005] Johansen, O.; Launikonis, A.; Mau, A. W. H.; Sasse, W. H. F. *Aust. J. Chem.* **1980**, *33*, 1643-1648.

[8006] Fabian, R. H.; Klassen, D. M.; Sonntag, R. W. *Inorg. Chem.* **1980**, *19*, 1977-1982.

[8101] Harriman, A. *J. Chem. Soc., Faraday Trans. II* **1981**, *77*, 1281-1291.

[8102] Juris, A.; Balzani, V.; Belser, P.; von Zelewsky, A. *Helv. Chim. Acta* **1981**, *64*, 2175-2182.

[8103] Elfring Jr., W. H.; Crosby, G. A. *J. Am. Chem. Soc.* **1981**, *103*, 2683-2687.

[8104] Stone, M. L.; Crosby, G. A. *Chem. Phys. Lett.* **1981**, *79*, 169-173.

[8201] Harriman, A. *J. Chem. Soc., Faraday Trans. I* **1982**, *78*, 2727-2734.

[8202] Nakamaru, K. *Bull. Chem. Soc. Jpn.* **1982**, *55*, 1639-1640.

[8203] Nakamaru, K. *Bull. Chem. Soc. Jpn.* **1982**, *55*, 2697-2705.

[8204] Juris, A.; Barigelletti, F.; Balzani, V.; Belser, P.; von Zelewsky, A. *Isr. J. Chem.* **1982**, *22* , 89-90.

[8205] Calvert, J. M.; Caspar, J. V.; Binstead, R. A.; Westmoreland, T. D.; Meyer, T. J. *J. Am. Chem. Soc.* **1982**, *104*, 6620-6627.

[8206] Darwent, J. R.; Douglas, P.; Harriman, A.; Porter, G.; Richoux, M. C. *Coord. Chem. Rev.* **1982**, *44*, 83-126.

[8301] Harriman, A.; Osborne, A. *J. Chem. Soc., Faraday Trans. I* **1983**, *79*, 765-772.

[8302] Bolletta, F.; Maestri, M.; Moggi, L.; Jamieson, M. A.; Serpone, N.; Henry, M. S.; Hoffman, M. Z. *Inorg. Chem.* **1983**, *22*, 2502-2509.

[8303] Kane-Maguire, N. A. P.; Crippen, W. S.; Miller, P. K. *Inorg. Chem.* **1983**, *22*, 696-698.

[8304] Caspar, J. V.; Meyer, T. J. *Inorg. Chem.* **1983**, *22*, 2444-2453.

[8305] Baggott, J. E.; Gregory, G. K.; Pilling, M. J.; Anderson, S.; Seddon, K. R.; Turp, J. E. *J. Chem. Soc., Faraday Trans. II* **1983**, *79*, 195-210.

[8306] Hauenstein Jr., B. L.; Dressick., W. J.; Buell, S. L.; Demas, J. N.; DeGraff, B. A. *J. Am. Chem. Soc.* **1983**, *105*, 4251-4255.

[8307] Fitzpatrick, L. J.; Goodwin, H. A.; Launikonis, A.; Mau, A. W. H.; Sasse, W. H. F. *Aust. J. Chem.* **1983**, *36*, 2169-2173.

[8401] Sprouse, S.; King, K. A., Spellane, P. J.; Watts, R. J. *J. Am. Chem. Soc.* **1984**, *106*, 6647-6653.

[8402] Cook, M. J.; Lewis, A. P.; McAuliffe, G. S. G.; Skarda, V.; Thomson, A. J.; Glasper, J. L.: Robbins, D. J. *J. Chem. Soc., Perkin Trans. 2* **1984**, 1293-1301.

[8403] Cherry, W. R.; Henderson Jr., L. J. *Inorg. Chem.* **1984**, *23*, 983-986.

[8404] Brauenstein, C. H.; Baker, A. D.; Strekas, T. C.; Gafney, H. D. *Inorg. Chem.* **1984**, *23*, 857-864.

[8405] Cook, M.J.; Thomson, A. J. *Chem. Brit.* **1984**, 914-917.

[8406] Kawanishi, Y.; Kitamura, N.; Kim, Y.; Tazuke, S. *Riken Q.* **1984**, *78*, 212-219.

[8407] Dressick, W. J.; Hauenstein Jr., B. L.; Gilbert, T. B.; Demas, J. N.; DeGraff, B. A. *J. Phys. Chem.* **1984**, *88*, 3337-3340.

[8408] Balzani, V.; Juris, A.; Barigelletti, F.; Belser, P.; von Zelewsky, A. *Riken Q.* **1984**, *78*, 78-85.

[8409] Belser, P.; von Zelewsky, A.; Juris, A.; Barigelletti, F.; Balzani, V. *Chem. Phys. Lett.* **1984**, *104*, 100-104.

[8501] Belser, P.; von Zelewsky, A.; Juris, A.; Barigelletti, F.; Balzani, V. *Gazz. Chim. Ital.,* **1985**, *115*, 723-729.

[8502] Kobayashi H.; Kaizu, Y. *Coord. Chem. Rev.* **1985**, *64*, 53-64.

[8503] Mc Clanahan, S. F.; Dallinger, R. F.; Holler, F. J.; Kincaid, J. R. *J. Am. Chem. Soc.* **1985**, *107*, 4853-4860.

[8504] Chambron, J.-C.; Sauvage, J.-P.; Amouyal, E.; Koffi, P. *Nouv. J. Chim.* **1985**, *9*, 527-529.

[8505] Cline J. I., III; Dressick, W .J.; Demas, J. N.; DeGraff, B. A. *J. Phys. Chem.* **1985**, *89*, 94-97.

[8506] Alford, P. C.; Cook, M. J.; Lewis, A. P.; McAuliffe, G. S.; Skarda, G. V.; Thomson, A. J.; Glasper, J. L.; Robbins, D. J. *J. Chem. Soc., Perkin Trans. II* **1985**, , 705-709.

[8507] Kober, E. M.; Marschall, J. L.; Dressick, W. J.; Sullivanm B. P.; Caspar, J. V.; Meyer, T. J. *Inorg. Chem.* **1985**, *24*, 2755-2763.

[8601] Ballardini, R.; Varani, G.; Indelli, M. T.; Scandola, F. *Inorg. Chem.* **1986**, *25*, 3858-3865.

[8602] Kober, E. M.; Caspar, J. V.; Lumpkin, R. S.; Meyer, T. J. *J. Phys. Chem.* **1986**, *90*, 3722-3734.

[8603] Wacholtz, W. F.; Auerbach, R. A.; Schmehl, R. H. *Inorg. Chem.* **1986**, *25*, 227-234.

[8701] Ichinaga, A. K.; Kirchhoff, J. R.; McMillin, D. R.; Dietrich-Buchecker, C. O.; Marnot, P. A., Sauvage, J.-P. *Inorg. Chem.* **1987**, *26*, 4290-4292.

[8702] Sabbatini, N.; Perathoner, S.; Balzani, V.; Alpha, B.; Lehn, J.-M. in Supramolecular Photochemistry, Ed: Balzani, V.; Reidel, Dordrecht (NL), 1987, 187-206.

[8801] Juris, A.; Balzani, V.; Barigelletti, F.; Campagna, S.; Belser, P.; Von Zelewsky, A. *Coord. Chem. Rev.* **1988**, *84*, 85-277.

[8802] Garges, F. O.; King, K. A.; Watts, R. J. *Inorg. Chem.* **1988**, *27*, 3464-3471.

[8803] Maestri, M.; Sandrini, D.; Balzani, V.; von Zelewsky, A.; Deuschel-Cornioley, A.; Jolliet, P. *Helv. Chim. Acta* **1988**, *71*, 1053-1059.

[8804] Maestri, M.; Sandrini, D.; Balzani, V.; von Zelewsky, A.; Jolliet, P. *Helv. Chim. Acta* **1988**, *71*, 134-139.

[8805] Leanna, M. A.; Holten, D. *J. Phys. Chem.* **1988**, *92*, 714-720.

[8901] Gushurst, A. K. I.; D. R. McMillin, D. R.; Dietrich-Bucheckert, C. O.; Sauvage, J.-P. *Inorg. Chem.* **1989**, *28*, 4070-4072.

[8902] Bignozzi, C. A.; Roffia, S.; Chiorboli, C.; Davila, J.; Indelli, M. T.; Scandola, F. *Inorg. Chem.* **1989**, *28*, 4350-4358.

[9001] Friedman, A. E.; Chambron, J.-C.; Sauvage, J.-P.; Turro, N. J.; Barton, J. K. *J. Am. Chem. Soc.* **1990**, *112*, 4960-4962.

[9101] Prodi, L.; Maetsri, M.; Ziessel, R.; Balzani, V. *Inorg. Chem.* **1991**, *30*, 3788-3802.

[9102] Worl, L. A.; Duesing, R.; Chen. P; Della Ciana, L.; Meyer, T. J. *J. Chem. Soc., Dalton Trans.* **1991**, , 849-858.

[9201] Maestri, M.; Balzani, V.; Deuschel-Cornioley, C.; von Zelewsky, A. *Advances in Photochemistry*; Eds: Volman, D.; Hammond, G.; Neckers, D. Wiley & Sons, 1992, 17, 1-68.

[9301] Chan, C.-W.; Lai, T.-F.; Che, C.-M.; Peng, S.-M. *J. Am. Chem. Soc.* **1993**, *115*, 11245-11253.

[9302] Guardagli, M. Ph. D. Thesis, Dept of Chemistry, University of Bologna, 1993.

[9303] Sacksteder, L.; Lee, M. Demas, J. N.; DeGraff, B. A. *J. Am. Chem. Soc.* **1993**, *115*, 8230-8238.

[9401] Sauvage, J.-P.; Collin, J.-P.; Chambron, J.-C.; Guillerz, S.; Coudret, C.; Balzani, V.; Barigelletti, F.; De Cola, L.; Flamigni, L. *Chem. Rev.* **1994**, *94*, 993-1019.

[9402] Chan, C.-W.; Cheng, L.-K.; Che, C.-M. *Coord. Chem. Rev.* **1994**, *132*, 87-97.

[9701] Cunningham, K. L.; McNett, K. M.; Pierce, R. A.; Davis, K. A.; Hariis, H. H.; Falk, D. M.; McMillin, D. R. *Inorg. Chem.* **1997**, *36*, 608-613.

[9702] Sabbatini, N.; Guardigli, M.; Manet, I. *Advances in Photochemistry*; Eds: Neckers, D.C; Volman, D. H.; von Bunau, G., J. Wiley & Sons, 1997, 23, 213-278.

[9703] Eggleston, M. K.; McMillin, D. R. *Inorg. Chem.* **1997**, *36*, 172-176.

[9801] Owens, J. W.; Smith, R.; Robinson, R.; Robins, M. *Inorg. Chim. Acta* **1998**, *279*, 226-231.

[9802] Miller, M. T.; Gantzel, P. T.; Karpishin, T. B. *Angew. Chem. Int. Ed.* **1998**, *37*, 1556-1558.

[9803] Albano, G.; Balzani, V.; Constable, E. C.; Maestri, M.; Smith, D. R. *Inorg. Chim. Acta* **1998**, *277*, 225-231.

[9901] Cunningham, C. T.; Kuratan, L. H.; Cunningham, K. L.; Michaelec, J. F.; McMillin, D. R. *Inorg. Chem.* **1999**, *38*, 4388-4392.

[9902] Holmin, R. E.; Yao, J. A.; Barton, J. K. *Inorg. Chem.* **1999**, *38*, 174-189.

[9903] Albano, G.; Belser, P.; De Cola, L.; Gandolfi, M. T. *Chem. Commun.* **1999**, 1171-1172.

[0001] Fisher, C.; Sarti, G.; Casnati, A.; Carrettoni, B.; Manet, I.; Shuurman, R.; Guardigli, M.; Sabbatini, N.; Ungaro, R. *Chem. Eur. J.* **2000**, *6*, 1026-1064.

[0002] Hisler, M.; McGarrah, J. E.; Connick, W. B.; Geiger, D. K.; Cummings, S. D.; Eisenberg, R. *Coord. Chem. Rev.* **2000**, *208*, 115-137.

[0101] Armaroli, N. *Chem. Soc. Rev.* **2001**, *30*, 113-124.

[0102] Felder, D.; Nierengarten, J.-F.; Barigelletti, F.; Ventura, B.; Armaroli, N. *J. Am. Chem. Soc.* **2001**, *123*, 6291-6299.

[0201] Knor, G.; Strasser, A. *Inorg. Chem. Commun.* **2002**, *5*, 993-995.

[0301] Cozzi, P.G.; Dolci, L. S.; Garelli, A.; Montalti, M.; Prodi, L.; Zaccheroni, N. *New J. Chem.* **2003**, *27*, 692-697.

[0302] Lin, Y.-Y.; Chan, S.-C.; Chan, M. C. W.; Hou, Y.-J.; Zhu, N., Che, C.-M.;, Liu, Y.; Wang Y. *Chem. Eur. J.* **2003**, *9*, 1263-1272.

[0401] Lu, W.; Mi, B.-X.; Chan, M. C. W.; Hui, Z.; Che, C.-M.; Zhu, N.; Lee, S.-T.; *J. Am. Chem. Soc.* **2004**, *126*, 4958-1971.

[0402] Lai, S.-W; Hou, Y.-J.; Che, C.-M.; Pang, H.-L.; Wong, K.-Y.; Chang, C. K.; Zhu, N. *Inorg. Chem.* **2004**, *43*, 3724-3732.

[0403] Che, C.-M.; Chan, S.-C.; Xiang, H.-F.; Chan, M. C. W.; Liu, Y.; Wang, Y. *Chem. Comun.* **2004**, 1484-1485.

6

Rate Constants of Excited-State Quenching

A substrate A can be promoted from the electronic ground state to an electronically excited state (*A) by the absorption of a quantum of light (process 6-I).

$$A + h\nu \longrightarrow *A \tag{6-I}$$

Subsequently, the excites deactivates along routes that typically, according to the Kasha's rule, lead to the formation of the lowest excited singlet state (S_1) or the lowest excited triplet state (T_1) of the molecule. In the most general sense, *A, when unperturbed, can decay via radiative (6-II), nonradiative (6-III), and reactive intramolecular (6-IV) processes:

$$*A \xrightarrow{k_r} A + h\nu' \tag{6-II}$$

$$*A \xrightarrow{k_{nr}} A + heat \tag{6-III}$$

$$*A \xrightarrow{k_{rx}} products \tag{6-IV}$$

The probabilities of the various unimolecular processes (6-II)-(6-IV) have the dimensions of a first-order rate constant. The lifetime of unperturbed *A, τ_0, is defined as $\tau_0 = 1/k_0$, where $k_0 = k_r + k_{nr} + k_{rx}$.

Irrespective of the nature of the specific system and the multiplicity of the excited state, it should be reminded that *A and A are different species: *A is more energetic, is a stronger reductant and oxidant, and, because of its unique electronic configuration, has the possibility of exhibiting greatly different patterns of reactivity (acid-base, substitution, etc.). Because of its peculiar reactivity, the excited state *A can also deactivate because of the interaction with other molecules. This decay mode, named quenching process, can occur when the lifetime of the excited state is long enough for the excited state substrate to encounter a suitable species Q (quencher). The subject of this section is exactly the second order constant k_q for the bimolecular reaction between an electronically excited species *A and a quencher Q which contributes to the deactivation of *A. The phenomenological approach [8901] to quenching kinetic involves the idea of a collision complex *A–Q from which the final products or excites states are formed, according to reaction 6-V:

$$*A + Q \xrightleftharpoons[k_{-d}]{k_d} *A\text{–}Q \xrightarrow{k_{de}} \text{products} \tag{6-V}$$

In this scheme, the collision complex *A–Q is held within the solvent cage for a short period of time (sub-ns); the deactivation of the collision complex to yields products with a rate constant k_{de} must compete with the break up of the solvent cage and the release of *A and Q into the bulk solvent.

Following the quenching scheme described in reaction 6-V and using the steady-state approximation for the concentration of *A–Q, , the observed rate constant k_q can be expressed by Eq. 6-1:

$$k_q = \frac{k_d\, k_{de}}{k_{-d} + k_{de}} \tag{6-1}$$

In this mechanism, k_d is the rate constant for the bimolecular diffusional encounter of *A and Q, which can be computed from Smoluchowski's theory of diffusion controlled reaction, and k_{-d} is the rate constant for the unimolecular dissociation of *A–Q into the original species. In the case of outer-sphere encounters, k_{-d} can be calculated by means of the Eigen equation [5401]. In the limit that product formation k_{de} is much faster than the break-up of the collision process (k_{-d}) into the original species, k_q in Eq. 6-1 reduces to the diffusion rate constant k_d. Diffusion-controlled rate constants for some common solvents calculated at 20 and 25 °C from Smoluchowski's theory using viscosity data [8616, 0304] are gathered in Table 6a.

Quenching can occur via one or more of the following primary mechanisms simultaneously:
1) Transfer of electronic energy from *A to Q, with the formation of an electronically excited state of Q, *Q (ET, Energy transfer mechanism);
2) Transfer of one electron from Q to *A (RT, reductive electron transfer mechanism);
3) Transfer of one electron from *A to Q (OT, oxidative electron transfer mechanism);
4) Hydrogen abstraction between *A and Q (HA, Hydrogen abstraction mechanism; this mechanisms in Tables 6b and 6c has been subsumed under the reductive transfer mechanism);

5) Formation of an exciplex between *A and Q (EX, exciplex fomation mechanism), possibly with charge-transfer character (CT);

6) Transfer of one proton from *A to Q or vice versa (PT, proton transfer mechanism)

The knowledge of the occurring mechanism can be of great importance for practical applications, since the dependence of the rate constants on the solvent, the *A–Q distance, the temperature, and the nature of the reactants are closely related to the process involved. For this reason, the quenching mechanism has been included, when known, in the tables of this section.

The experimental determination of k_q is based on the measurement of a property that is proportional to the concentration of one or more of the species involved in the overall photochemical process. The most common properties are the intensity of emission from *A under continuous steady state excitation, and the decay of emission from *A, or absorption by *A, after pulsed excitation. By assuming that the deactivation of *A is represented by processes (6-II)-(6-V), the Eq. 6-2 (Stern-Volmer equation, SV) can be obtained, in which k_q is related to the intensity of the emission from *A in the absence (I^0) and presence (I) of Q.

$$I^0/I = 1 + (k_q/k_0)[Q] \qquad (6\text{-}2)$$

Similarly, an equivalent equation (6-3) can also be derived:

$$\tau_0/\tau = 1 + (k_q/k_0)[Q] \qquad (6\text{-}3)$$

The Stern-Volmer constant (K_{SV}) is then defined as:

$$K_{SV} = k_q/k_0 = k_q\tau_0 \qquad (6\text{-}4)$$

Stern-Volmer plots of I^0/I or τ_0/τ vs [Q] are then predicted to be linear, with slopes equal to K_{sv} provided that , for plots derived from Eq. 6-2, Q does not absorb the exciting and emitting light, or that appropriate corrections have been made (see Section 10a).

In Tables 6b, 6c, and 6d are gathered representative examples of available rate constants related to quenching processes in which *A is (i) an organic molecule in its lowest excited singlet state (S_1),(ii) an organic molecule in its lowest excited triplet state (T_1), and (iii) a transition metal complex, respectively. It has to be noted that for the latter two classes, a wider set of data has been reported in the literature, because of the usually longer lifetime of the represented species, due to the spin restrictions governing their conversion back to the ground state.

Table 6a Diffusion-Controlled Rate Constants

No.	Solvent	η (20°C) (10^{-3} Pa s)	k_{diff} (20°C) (L mol^{-1} s^{-1})	η (25°C) (10^{-3} Pa s)	k_{diff} (25°C) (L mol^{-1} s^{-1})
1	Isopentane	0.225	2.9×10^{10}	0.215	3.1×10^{10}
2	Diethyl ether	0.242	2.7×10^{10}	0.224	3.0×10^{10}
3	Pentane	0.235	2.8×10^{10}	0.225	2.9×10^{10}
4	Hexane	0.3126	2.1×10^{10}	0.2942	2.2×10^{10}
5	Acetone	0.322	2.0×10^{10}	0.307	2.1×10^{10}
6	Acetonitrile			0.341	1.9×10^{10}
7	Heptane	0.4181	1.6×10^{10}	0.3967	1.7×10^{10}
8	Dichloromethane	0.434	1.5×10^{10}	0.414	1.6×10^{10}
9	Tetrahydrofuran	0.575	1.3×10^{10}	0.460	1.4×10^{10}
10	Isooctane	0.504	1.3×10^{10}		
11	Octane	0.5466	1.2×10^{10}	0.5151	1.3×10^{10}
12	Chloroform	0.564	1.2×10^{10}	0.5357	1.2×10^{10}
13	Methanol	0.5929	1.1×10^{10}	0.5513	1.2×10^{10}
14	Toluene	0.5859	1.1×10^{10}	0.5525	1.2×10^{10}
15	Benzene	0.6487	1.0×10^{10}	0.6028	1.1×10^{10}
16	Nitromethane			0.630	1.0×10^{10}
17	Methylcyclohexane	0.734	8.8×10^{9}	0.685	9.6×10^{9}
18	1,2-Dichloroethane			0.779	8.5×10^{9}
18	*N,N*-Dimethylformamide	0.86	7.6×10^{9}	0.794	8.3×10^{9}
19	Decane	0.9284	7.0×10^{9}	0.8614	7.6×10^{9}
20	Pyridine	0.952	6.8×10^{8}	0.884	7.5×10^{9}
21	Carbon tetrachloride	0.9785	6.6×10^{9}	0.9004	7.3×10^{9}
22	Water	1.0019	6.5×10^{9}	0.89025	7.4×10^{9}
23	Cyclohexane	0.9751	6.7×10^{9}	0.898	7.4×10^{9}
24	Ethanol	1.21	5.4×10^{9}	1.0826	6.1×10^{9}
25	Dimethyl sulfoxide	2.2159	2.9×10^{9}	1.991	3.3×10^{9}
26	2-Propanol	2.55	2.5×10^{9}	2.0436	3.2×10^{9}
27	Glycerol	1412	4.6×10^{6}	945	7.0×10^{6}

Table 6b Singlet-State Quenching of Organic Molecules

No.	Quencher	Mechanism	Solvent	K_q (L mol^{-1} s^{-1})	Ref.
1	**Acenaphthene**				
	O_2		Cyclohexane	2.6×10^{10}	[8806]
2	**Acetone**				
	Dibutyldiazene	ET	MeCN	7.10×10^9	[7507]
	1,4-Dioxene	ET	MeCN	1.4×10^9	[7505]
	2-Methoxypropene	ET	MeCN	2.2×10^8	[7505]
	1,3-Pentadiene	ET	Hexane	9×10^7	[7008]
3	**2-Adamantanone**				
	Dibutyldiazene	ET	MeCN	2.9×10^9	[7507]
	Ethanol	RT	Hexane	1.1×10^6	[7411]
	2-Propanol	RT	Hexane	1.9×10^6	[7411]
	Tributylstannane	RT	Hexane	4.8×10^8	[7411]
4	**Alloxazine, 7,8-dimethyl-**				
	Aniline	RT	MeOH	1.4×10^{10}	[0202]
	Diphenylamine	RT	MeOH	1.06×10^{10}	[0202]
	TMPD	RT	MeOH	1.6×10^{10}	[0202]
5	**Anthracene**				
	O_2		Benzene	3.1×10^{10}	[6201]
	O_2		Cyclohexane	2.5×10^{10}	[7810]
	O_2		EtOH	2.5×10^{10}	[6201]
	Triphenylphosphine	EX	Benzene	2.2×10^9	[7506]
	N,N-Diethylaniline	RT	MeCN	2.1×10^{10}	[6803]
	Indole	RT	MeCN	5.6×10^9	[9803]
6	**Anthracene, 9-bromo-**				
	O_2		Cyclohexane	2.9×10^{10}	[7810]
	Triethylamine	EX	MeCN	2.3×10^{10}	[8415]
	Triphenylphosphine	RT	Benzene	3.7×10^9	[7506]
7	**Anthracene, 9-chloro-**				
	O_2		Cyclohexane	3.1×10^{10}	[7810]
	Triphenylphosphine	RT	Benzene	1.9×10^9	[7506]

Table 6b Singlet-State Quenching of Organic Molecules

No.	Quencher	Mechanism	Solvent	K_q (L mol^{-1} s^{-1})	Ref.
8	**Anthracene, 9-cyano-**				
	O_2		Cyclohexane	6.7×10^9	[7810]
	MV^{2+}	OT	MeCN/H$_2$O (9:1)	$\sim2\times10^{10}$	[8107]
	Triphenylphosphine	RT	Benzene	5.1×10^9	[7506]
9	**Anthracene, 9-cyano-10-phenyl-**				
	O_2		Cyclohexane	6.9×10^9	[7810]
10	**Anthracene, 9,10-diacetoxy-**				
	O_2		Cyclohexane	2.5×10^{10}	[7810]
11	**Anthracene, 9,10-dibromo-**				
	O_2		Cyclohexane	2.4×10^{10}	[7810]
	Biphenyl	EX	MeCN	1.2×10^8	[8426]
	Naphthalene	EX	MeCN	1.8×10^9	[8426]
	Naphthalene	EX	Toluene	5.2×10^8	[8426]
	Pyrene	EX	MeCN	2.4×10^{10}	[8426]
	Triethylamine	EX	MeCN	1.8×10^{10}	[8415]
	Triphenylphosphine	RT	Benzene	3.9×10^9	[7506]
12	**Anthracene, 9,10-dichloro-**				
	O_2		Benzene	2.4×10^{10}	[6201]
	O_2		EtOH	1.7×10^{10}	[6201]
	Naphthalene	EX	Toluene	1.5×10^7	[8426]
	Pyrene	EX	Toluene	1.6×10^9	[8426]
	Triphenylphosphine	RT	Benzene	2.5×10^9	[7506]
13	**Anthracene, 9,10-dicyano-**				
	O_2		Cyclohexane	4.7×10^9	[7810]
	Biphenyl	EX	Toluene	2.5×10^7	[8426]
	Naphthalene	EX	Toluene	2.1×10^9	[8426]
	Pyrene	EX	Toluene	1.7×10^{10}	[8426]
	Indole	RT	MeCN	2.5×10^{10}	[9803]
	Triphenylphosphine	RT	Benzene	1.1×10^{10}	[7506]

Table 6b Singlet-State Quenching of Organic Molecules

No.	Quencher	Mechanism	Solvent	K_q (L mol^{-1} s^{-1})	Ref.
14	**Anthracene, 9,10-dimethoxy-**				
	O_2		Cyclohexane	2.3×10^{10}	[7810]
	Triphenylphosphine	EX	Benzene	6.0×10^{7}	[7506]
15	**Anthracene, 9,10-dimethyl-**				
	O_2		Cyclohexane	2×10^{10}	[7810]
	Benzophenone	ET	Benzene	4.5×10^{9}	[8131]
	Benzonitrile	OT	MeCN	4.6×10^{7}	[7005]
	1,4-Dicyanobenzene	OT	MeCN	1.6×10^{10}	[7005]
	Indole	RT	MeCN	4×10^{7}	[9803]
16	**Anthracene, 9,10-diphenyl-**				
	O_2		Benzene	3.6×10^{10}	[6802]
	O_2		Cyclohexane	1.7×10^{10}	[7810]
	O_2		EtOH	2.3×10^{10}	[6201]
17	**Anthracene, 9-methoxy-**				
	O_2		Cyclohexane	2.7×10^{10}	[7810]
	Triphenylphosphine	EX	Benzene	3.6×10^{8}	[7506]
18	**Anthracene, 9-methyl-**				
	O_2		Cyclohexane	2.8×10^{10}	[7810]
	Triphenylphosphine	EX	Benzene	4.7×10^{8}	[7506]
	Indole	RT	MeCN	8×10^{8}	[9803]
19	**Anthracene, 9-phenyl-**				
	O_2		Cyclohexane	1.9×10^{10}	[7810]
20	**1,5-Anthracenedisulfonate ion**				
	Di-*tert*-butylnitroxide		H_2O	1.3×10^{10}	[8518]
	O_2		H_2O	1.0×10^{10}	[8518]
21	**1-Anthracenedisulfonate ion**				
	Di-*tert*-butylnitroxide		H_2O	7.5×10^{9}	[8518]
	O_2		H_2O	1.2×10^{10}	[8518]

Table 6b Singlet-State Quenching of Organic Molecules

No.	Quencher	Mechanism	Solvent	K_q (L mol^{-1} s^{-1})	Ref.
22	**2-Anthracenedisulfonate ion**				
	Di-*tert*-butylnitroxide		H_2O	1.4×10^{10}	[8518]
	O_2		H_2O	2.0×10^{10}	[8518]
23	**9-Anthronate ion**				
	MV^{2+}	OT	H_2O (pH 5.0)	9.0×10^{10}	[8308]
24	**Benz[*a*]aceanthrylene**				
	O_2		Heptane	2.4×10^{10}	[8215]
25	**Benz[*c*]acridine**				
	1,4-Dimethoxybenzene	RT	MeCN	6.7×10^9	[7005]
	N,N-Diethylaniline	RT	MeCN	1.7×10^{10}	[7005]
	Phenol	RT	MeCN	1.1×10^9	[7005]
	TMPD	RT	MeCN	1.8×10^{10}	[7005]
26	**Benz[*a*]antracene**				
	O_2		Benzene	2.6×10^{10}	[8806]
	O_2		Cyclohexane	3.0×10^{10}	[7004]
	Chlorobenzene	EX	Cyclohexane	7.0×10^5	[9801]
	Benzonitrile	OT	MeCN	3×10^6	[7005]
	Tetracyanoethylene	OT	MeCN	2.4×10^{10}	[7005]
	1,4-Dimethoxybenzene	RT	MeCN	1.2×10^8	[7005]
	N,N-Diethylaniline	RT	MeCN	1.4×10^{10}	[7005]
	TMPD	RT	MeCN	1.5×10^{10}	[7005]
27	**Benz[*a*]antracene, 9,10-dimethyl-**				
	O_2		Benzene	2.8×10^{10}	[6802]
	Azulene	ET	Benzene	8.3×10^9	[7003]
28	**Benzene**				
	O_2		Cyclohexane	2.3×10^{10}	[8806]
29	**Benzene, 1,4-dimethoxy-**				
	O_2		Cyclohexane	3.1×10^{10}	[8806]
30	**Benzene, methoxy-**				
	O_2		Cyclohexane	3.1×10^{10}	[8806]

Table 6b Singlet-State Quenching of Organic Molecules

No. Quencher	Mechanism	Solvent	K_q (L mol^{-1} s^{-1})	Ref.
31 Benzene, 1,2,4-trimethyl-				
O$_2$		Cyclohexane	3.1×10^{10}	[8806]
32 Benzene, 1,3,5-trimethyl-				
O$_2$		Cyclohexane	2.6×10^{10}	[8806]
33 Benzidine, *N,N,N',N'*-tetramethyl-				
O$_2$		MeCN	4.0×10^{10}	[8414]
Dibenzyl sulfone	EX	MeCN	1.0×10^{10}	[8433]
1,4-Dicyanonaphthalene	OT	MeCN	2.2×10^{10}	[8414]
34 Benzo[*b*]fluoranthene				
O$_2$		Heptane	5.1×10^{9}	[8215]
35 Benzo[*j*]fluoranthene				
O$_2$		Heptane	1.8×10^{10}	[8215]
36 Benzo[*k*]fluoranthene				
O$_2$		Heptane	2.3×10^{10}	[8215]
37 Benzo[*rst*]pentaphene				
O$_2$		Cyclohexane	2.7×10^{10}	[7004]
38 Benzo[*ghi*]perylene				
1,4-Dicyanobenzene	OT	MeCN	8.7×10^{9}	[7005]
Tetracyanoethylene	OT	MeCN	2.1×10^{10}	[7005]
39 Benzo[*g*]pteridine-2,4-dione,3,7,8,10-tetramethyl-				
Triethylamine	EX	MeCN	1.1×10^{10}	[8228]
Triethylamine	EX	Dioxane	3.7×10^{9}	[8228]
40 Benzo[*g*]pteridine-2,4-dione,7,8,10-trimethyl-				
1,3-Dimethoxybenzene	RT	MeOH	1.1×10^{10}	[8119]
EDTA	RT	H$_2$O (pH 7)	5.8×10^{9}	[8513]
Imidazole	RT	H$_2$O (pH 7)	5.8×10^{9}	[8513]
Naphthalene	RT	MeOH	8.9×10^{9}	[8119]
41 Benzo[*a*]pyrene				
O$_2$		Cyclohexane	2.9×10^{10}	[7004]

Table 6b Singlet-State Quenching of Organic Molecules

No.	Quencher	Mecha-nism	Solvent	K_q (L mol^{-1} s^{-1})	Ref.
42	**Benzo[*h*]quinoline**				
	Phenol	RT	Cyclohexane	1.2×10^{10}	[8609]
	Phenol	RT	MeCN	2.4×10^9	[7005]
43	**Benzo[*b*]triphenylene**				
	O_2		Cyclohexane	2.6×10^{10}	[7004]
44	**Biacetyl**				
	1,3-Dioxole	EX	MeCN	1.0×10^9	[8701]
	SCN$^-$	EX	H_2O	4.2×10^9	[7509]
	N,N-Diethylaniline	RT	Benzene	1.1×10^{10}	[6903]
	N,N-Diethylaniline	RT	EtOH	1.6×10^{10}	[6903]
	N,N-Diethylaniline	RT	MeCN	1.6×10^{10}	[6903]
	Phenol	RT	Benzene	2.1×10^9	[6901]
	Phenol	RT	EtOH	1×10^8	[6903]
	Phenol	RT	MeCN	1×10^8	[6903]
	Triphenylamine	RT	Benzene	4.9×10^9	[6906]
	Triphenylamine	RT	EtOH	1.3×10^{10}	[6903]
	Triphenylamine	RT	MeCN	1.3×10^{10}	[6906]
45	**Bicyclo[2.2.1]heptane-2-one, 7,7-dimethyl-**				
	Dibutyldiazene	ET	MeCN	5.2×10^9	[7507]
46	**1,1′-Binaphthyl**				
	O_2		Cyclohexane	3.3×10^{10}	[8806]
47	**2,2′-Binaphthyl**				
	O_2		Cyclohexane	2.6×10^{10}	[8806]
48	**Biphenyl**				
	O_2		Cyclohexane	2.8×10^{10}	[8806]
	SCN$^-$	RT	MeCN	8.6×10^9	[7306]
49	**Biphenyl, 4-benzyl-**				
	O_2		Cyclohexane	3.0×10^{10}	[8806]
50	**Biphenyl, 4-methoxy-**				
	O_2		Cyclohexane	3.3×10^{10}	[8806]

Table 6b Singlet-State Quenching of Organic Molecules

No. Quencher	Mecha-nism	Solvent	K_q (L mol^{-1} s^{-1})	Ref.
51 Biphenyl, 4-phenoxy-				
O$_2$		Cyclohexane	3.2×10^{10}	[8806]
52 1,3-Butadiene, 1-4-diphenyl-				
EtI		Cyclohexane	8.3×10^9	[8221]
O$_2$		Cyclohexane	6.0×10^{10}	[8221]
53 Carbazole				
p-Dicyanobenzene		MeCN	3.4×10^{10}	[0402]
Pyridine	RT	Cyclohexane	1.0×10^{10}	[8431]
54 Carbazole, *N*-methyl-				
p-Dicyanobenzene		MeCN	2.52×10^{10}	[0402]
55 Chrysene				
O$_2$		Cyclohexane	2.9×10^{10}	[7004]
Benzophenone	ET	Benzene	1.1×10^{10}	[8131]
56 Coronene				
Pb^{2+}	EX	DMF	3.3×10^8	[8429]
Eu^{3+}	OT	MeCN	4.0×10^{10}	[8806]
N,N-Diethylaniline	RT	MeCN	1.9×10^{10}	[6803]
57 Decacyclene				
O$_2$		Benzene	1.7×10^{10}	[7505]
58 Dibenzo[*a,h*]anthracene				
O$_2$		Cyclohexane	2.9×10^{10}	[7008]
59 Dibenzo[*def,mno*]chrysene				
O$_2$		Benzene	3.5×10^{10}	[6802]
60 Dibenzofuran				
O$_2$		Cyclohexane	2.9×10^{10}	[8806]
61 Diphenylmethane				
O$_2$		Cyclohexane	2.9×10^{10}	[8806]
62 Fluoranthene				
O$_2$		Cyclohexane	4.4×10^9	[8806]
Diphenyl disulfide	ET	Benzene	3×10^8	[7610]

Table 6b Singlet-State Quenching of Organic Molecules

No.	Quencher	Mecha-nism	Solvent	K_q (L mol^{-1} s^{-1})	Ref.
62	**Fluoranthene** (continued)				
	Br$^-$	RT	MeCN	1.1×10^9	[7404]
	NO$_3^-$	RT	MeCN	4×10^6	[7404]
63	**Fluorene**				
	Br$^-$	RT	MeCN	1.1×10^9	[7404]
	Fluorene	RT	MeCN	2.2×10^9	[7404]
64	**9-Fluorenone**				
	Triethylamine	RT	Benzene	7×10^9	[6908]
65	**Fluorescein dianion, 2′,4′,5′,7′-tetrabromo-**				
	Phenoxide ion	RT	H$_2$O	1.5×10^{10}	[6101]
66	**Fullerene C$_{60}$, 1,2,5-triphenylpyrrolidino-**				
	N,N-Dimethylaniline	RT	Toluene	1.14×10^{10}	[0101]
	Ferrocene	RT	Toluene	2.76×10^{10}	[0101]
67	**1,3,5-Hexatriene, 1,6-diphenyl-**				
	EtI		Cyclohexane	1.7×10^8	[8221]
	O$_2$		Cyclohexane	2.10×10^{10}	[8221]
68	**Indole**				
	Benzonitrile		MeCN	1.38×10^{10}	[0302]
	Chlorobenzene		MeCN	2.57×10^9	[0302]
	O$_2$		H$_2$O	6.5×10^9	[7305]
	Succinimide		H$_2$O	4.8×10^9	[8424]
	Anthracene	ET	Cyclohexane	7×10^{10}	[8120]
69	**Indole, 5-methoxy-**				
	Succinimide		H$_2$O	4.6×10^9	[8424]
70	**Indole, 1-methyl-**				
	Benzonitrile		MeCN	7.28×10^9	[0302]
	Chlorobenzene		MeCN	5.0×10^8	[0302]
	Succinimide		H$_2$O	5.2×10^9	[8424]

Table 6b Singlet-State Quenching of Organic Molecules

No. Quencher	Mecha-nism	Solvent	K_q $(L\ mol^{-1}\ s^{-1})$	Ref.
71 Indole, 3-methyl-				
Benzonitrile		MeCN	2.02×10^{10}	[0302]
Chlorobenzene		MeCN	1.2×10^9	[0302]
Succinimide		H_2O	4.6×10^9	[8424]
72 Methylene Blue cation, conjugate monoacid				
Fe^{II}	RT	MeCN/H_2O (1:1) (pH 2)	3.2×10^9	[8007]
73 Naphthalene				
O_2		MeCN	3.36×10^{10}	[9301]
O_2		Cyclohexane	2.7×10^{10}	[8806]
Succinimide		H_2O	4.0×10^8	[8424]
Biacetyl	ET	Cyclohexane	1.0×10^{10}	[7009]
Chlorobenzene	EX	Cyclohexane	5.4×10^6	[9801]
Hexachlorobenzene	EX	Cyclohexane	6.8×10^9	[9801]
Benzonitrile	OT	MeCN	4×10^6	[7005]
1,4-Dicyanobenzene	OT	MeCN	1.8×10^{10}	[7005]
Br^-	RT	MeCN	1.10×10^9	[7404]
ClO_4^-	RT	MeCN	5×10^6	[7404]
74 Naphthalene, 1-cyano-				
MV^{2+}	OT	MeCN/H_2O (9:1)	$\sim 2 \times 10^{10}$	[8107]
75 Naphthalene, 1,4-dicyano-				
O_2		MeCN	1.3×10^{10}	[8414]
1,4-Dimethoxybenzene	EX	Heptane	4.3×10^{10}	[8421]
2,5-Dimethyl-2,4-hexadiene	EX	MeCN	1.9×10^{10}	[8414]
Mesitylene	EX	Heptane	1.8×10^{10}	[8421]
Triethylamine	RT	MeCN	1.6×10^{10}	[8414]
76 Naphthalene, 2,3-dimethyl-				
O_2		Cyclohexane	2.9×10^{10}	[8806]
77 Naphthalene, 2,6-dimethyl-				
O_2		Cyclohexane	2.7×10^{10}	[8806]

Table 6b Singlet-State Quenching of Organic Molecules

No. Quencher	Mechanism	Solvent	K_q (L mol^{-1} s^{-1})	Ref.
78 Naphthalene, 1,4-diphenyl-				
O$_2$		Cyclohexane	2.7×10^{10}	[8806]
79 Naphthalene, 1,5-diphenyl-				
O$_2$		Cyclohexane	2.9×10^{10}	[8806]
80 Naphthalene, 2-hydroxy-				
Pyridine	OT	MeCN	2.6×10^9	[7005]
81 Naphthalene, 1-methyl-				
O$_2$		Cyclohexane	3.2×10^{10}	[8806]
82 Naphthalene, 2-methyl-				
O$_2$		Cyclohexane	2.5×10^{10}	[8806]
Chlorobenzene	EX	Cyclohexane	3.0×10^6	[9801]
Hexachlorobenzene	EX	Cyclohexane	6.8×10^9	[9801]
83 Naphthalene, 1-phenyl-				
O$_2$		Cyclohexane	2.4×10^{10}	[8806]
84 Naphthalene, 1-styryl-, (*E*)-				
Diethylamine	EX	Hexane	1.8×10^{10}	[7710]
85 Naphthalene, 2-styryl-, (*E*)-				
Diethylamine	EX	Hexane	9.0×10^9	[7710]
86 Norcamphor				
1,4-Dioxene	EX	MeCN	8.5×10^8	[7505]
87 1,3,5,7-Octatetraene, 1,8-diphenyl-				
EtI		Cyclohexane	6.0×10^7	[8221]
O$_2$		Cyclohexane	1.9×10^{10}	[8221]
O$_2$		MeOH	2.6×10^{10}	[8221]
88 Oxazole, 2,5-diphenyl-				
CCl$_4$		Cyclohexane	2.0×10^9	[8016]
Fe^{2+}		H$_2$O/MeOH (3:2)	2.7×10^{10}	[9704]
Co^{2+}		H$_2$O/MeOH (3:2)	2.9×10^9	[9704]
Ni^{2+}		H$_2$O/MeOH (3:2)	9.8×10^9	[9704]
Cu^{2+}		H$_2$O/MeOH (3:2)	8.8×10^9	[9704]

Table 6b Singlet-State Quenching of Organic Molecules

No.	Quencher	Mechanism	Solvent	K_q (L mol^{-1} s^{-1})	Ref.
88	**Oxazole, 2,5-diphenyl-** (continued)				
	Zn^{2+}		H_2O/MeOH (3:2)	3.5×10^9	[9704]
	Hg^{2+}		H_2O/MeOH (3:2)	1.1×10^{10}	[9704]
89	**Pentahelicene**				
	O_2		MCH	1.6×10^{10}	[7908]
90	**2-Pentanone**				
	1,3-Pentadiene	ET	Hexane	4×10^7	[7008]
91	**Perylene**				
	I^-		MeCN	1.1×10^{10}	[7404]
	O_2		Benzene	2.7×10^{10}	[6802]
	O_2		Cyclohexane	2.2×10^{10}	[8806]
	O_2		EtOH	2.7×10^{10}	[6201]
	N,N-Diethylaniline	RT	MeCN	2.0×10^{10}	[6803]
92	**Phenanthrene**				
	I^-		MeCN	4.7×10^9	[7404]
	O_2		Cyclohexane	$2.3 \times x 10^{10}$	[8806]
	Di-*tert*-butyl disulfide	ET	MeCN	2.2×10^8	[7610]
	Chlorobenzene	EX	Cyclohexane	1.3×10^6	[9801]
	Hexachlorobenzene	EX	Cyclohexane	5.7×10^8	[9801]
93	**Phenol**				
	Succinimide		H_2O	5.5×10^9	[8424]
94	**Phenoxazinium, 3,7-diamino-**				
	EDTA	RT	H_2O (pH 8.0)	6.2×10^9	[7712]
95	**Phenyl ether**				
	O_2		Cyclohexane	3.1×10^{10}	[8806]
96	**Pinacolone**				
	1,3-Pentadiene	ET	Benzene	2×10^7	[8806]
97	**Porphyrin, tetrakis(4-sulfonatophenyl)-**				
	1,4-Benzoquinone	OT	H_2O (pH 7)	2.7×10^{10}	[8121]
	Nitrobenzene	OT	H_2O (pH 7)	1.9×10^{10}	[8121]

Table 6b Singlet-State Quenching of Organic Molecules

No.	Quencher	Mechanism	Solvent	K_q (L mol^{-1} s^{-1})	Ref.
98	**Porphyrin-2,18-dipropanoic acid, 7,12-diethenyl-3,8,13,17-tetramethyl-, dimethyl ester**				
	O$_2$		Benzene	1.8×10^{10}	[7706]
99	**Pterin**				
	CH$_3$COO$^-$		H$_2$O (pH 6.2)	2.1×10^9	[0401]
	HPO$_4^{2-}$		H$_2$O (pH 6.2)	1.8×10^8	[0401]
100	**Pyrene**				
	I$^-$		MeCN	4.5×10^9	[7404]
	O$_2$		Cyclohexane	2.5×10^{10}	[7004]
	O$_2$		Toluene	3.42×10^{10}	[9301]
	Benzophenone	ET	Benzene	8×10^9	[8131]
	DABCO	ET	Cyclohexane	3.2×10^9	[7915]
	N,N-Diethylaniline	ET	Toluene	1.0×10^{10}	[7309]
	Chlorobenzene	EX	Cyclohexane	2.9×10^5	[9801]
	Hexachlorobenzene	EX	Cyclohexane	7.7×10^8	[9801]
	Benzonitrile	OT	MeCN	4×10^6	[7005]
	1,4-Dicyanobenzene	OT	MeCN	2.1×10^{10}	[7005]
	SCN$^-$	RT	MeCN	4.4×10^7	[7404]
101	**Pyrene, 1-hydroxy-**				
	Pyridine	OT	MeCN	2.0×10^9	[7005]
102	**1-Pyrenecarboxylic acid**				
	1,4-Dimethoxybenzene	RT	MeCN	1.1×10^{10}	[7005]
	N,N-Diethylaniline	RT	MeCN	1.8×10^{10}	[7005]
	TMPD	RT	MeCN	1.6×10^{10}	[7005]
103	**2-Pyrenecarboxylic acid**				
	1,4-Dimethoxybenzene	RT	MeCN	7.0×10^7	[7005]
	N,N-Diethylaniline	RT	MeCN	1.7×10^{10}	[7005]
	TMPD	RT	MeCN	1.8×10^{10}	[7005]
104	**Pyrimido[4,5-*b*]quinoline-2,4-dione, 3,7,8,10-tetramethyl-**				
	1,2,3-Trimethoxybenzene	RT	MeOH	6.3×10^9	[8119]

Table 6b Singlet-State Quenching of Organic Molecules

No.	Quencher	Mechanism	Solvent	K_q (L mol^{-1} s^{-1})	Ref.
105	**Riboflavin**				
	Aniline	RT	MeOH	1.17×10^{10}	[0303]
	DABCO	RT	MeOH	4.7×10^{9}	[0303]
	p-Dimethoxybenzene	RT	MeOH	4.4×10^{9}	[0303]
	TMPD	RT	MeOH	1.23×10^{10}	[0303]
	Triethylamine	RT	MeOH	2.6×10^{9}	[0303]
106	**Rubrene**				
	O_2		Benzene	1.2×10^{10}	[6802]
107	***p*-Terphenyl**				
	I^-		MeCN	2.5×10^{10}	[7404]
	Br^-	RT	MeCN	1.8×10^{10}	[7404]
	ClO_4^-	RT	MeCN	1.1×10^{9}	[7404]
108	**Tetracene**				
	O_2		Benzene	2.4×10^{10}	[6802]
	Eu^{3+}	OT	MeCN	4.5×10^{10}	[7806]
	N,N-Diethylaniline	RT	MeCN	1.2×10^{10}	[6803]
109	**Thionine cation, conjugate monoacid**				
	Fe^{2+}	RT	H_2SO_4 (pH 2.5)	3.5×10^{9}	[7707]
110	**Thioxanthen-9-one**				
	Br^-	EX	MeCN/H_2O (3:2)	2.5×10^{9}	[8516]
111	**Toluene**				
	O_2		Cyclohexane	2.8×10^{10}	[8806]
112	**Triphenylene**				
	O_2		Cyclohexane	2.0×10^{10}	[8806]
	Dibutyldiazene	ET	Benzene	1.1×10^{10}	[7507]
	Dibutyl disulfide	ET	Benzene	5.7×10^{8}	[7610]
113	**Tryptophan**				
	I^-		H_2O	1.9×10^{9}	[7305]
	O_2		H_2O	5.9×10^{9}	[7305]

Table 6b Singlet-State Quenching of Organic Molecules

No.	Quencher	Mecha-nism	Solvent	K_q (L mol^{-1} s^{-1})	Ref.
114	L-Tryptophan				
	Succinimide		H_2O	3.5×10^9	[8424]
115	L-Tryptophan, *N*-acetyl-				
	Succinimide		H_2O	4.4×10^9	[8424]
116	L-Tryptophanamide, *N*-acetyl-				
	Succinimide		H_2O	3.9×10^9	[8424]
117	*m*-Xylene				
	O_2		Cyclohexane	2.6×10^{10}	[8806]
118	*o*-Xylene				
	O_2		Cyclohexane	2.6×10^{10}	[8806]
119	*p*-Xylene				
	O_2		Cyclohexane	2.9×10^{10}	[8806]

Table 6c Triplet-State Quenching of Organic Molecules

No.	Quencher	Mechanism	Solvent	K_q (L mol^{-1} s^{-1})	Ref.
1	**Acenaphthene**				
	O_2		MeCN	6.5×10^9	[0201]
2	**Acenaphthylene**				
	MV^{2+}	OT	MeCN/H$_2$O (9:1)	2.0×10^9	[8310]
3	**Acetone**				
	O_2		Cyclohexane	8.4×10^9	[8017]
	O_2		MeCN	5.4×10^{10}	[8413]
	Dibutyldiazene	ET	MeCN	3.1×10^9	[7507]
	I^-	ET	H$_2$O	7.1×10^9	[7615]
	1,3-Pentadiene	ET	MeCN	4.4×10^9	[8019]
	Tetrachloroethylene	OT	MeCN	6.9×10^8	[7709]
	Benzene	RT	MeCN	1.7×10^6	[7312]
	Ethanol	RT	MeCN	5.4×10^5	[6902]
	Triethylamine	RT	H$_2$O	2.7×10^8	[7204]
4	**Acetophenone**				
	Hexamethyldisilane		Benzene	1.95×10^5	[9703]
	O_2		Benzene	4.5×10^9	[8510]
	O_2		MeCN	3.7×10^9	[8510]
	Naphthalene	ET	Benzene	7.7×10^9	[6901]
	MV^{2+}	OT	MeCN/H$_2$O (9:1)	5.2×10^9	[8412]
	I^-	RT	H$_2$O/MeCN (4:1)	2.3×10^9	[8208]
	2-Propanol	RT	MeCN	2.1×10^6	[7808]
	Triethylamine	RT	Benzene	1.9×10^9	[7812]
5	**Acetophenone, 4'-amino-**				
	O_2		Benzene	9.4×10^9	[8510]
6	**Acetophenone, 4'-cyano-**				
	Hexamethyldisilane		Benzene	3.53×10^5	[9703]
	O_2		Benzene	5.2×10^8	[8510]
	2,5-Dimethyl-2,4-hexadiene	ET	Benzene	5.8×10^9	[8510]
	MV^{2+}	OT	MeCN/H$_2$O (9:1)	1.5×10^9	[8108]

Table 6c Triplet-State Quenching of Organic Molecules

No. Quencher	Mechanism	Solvent	K_q (L mol^{-1} s^{-1})	Ref.
6 Acetophenone, 4′-cyano- (continued)				
Phenoxide ion	RT	MeCN/H$_2$O (1:1)	9.5×10^9	[8110]
7 Acetophenone, 4′-fluoro-				
O$_2$		Benzene	3.8×10^9	[8510]
MV^{2+}	OT	MeCN/H$_2$O (9:1)	3.1×10^9	[8412]
8 Acetophenone, 3′-methoxy-				
O$_2$		H$_2$O	3.3×10^9	[0002]
Phenol	RT	H$_2$O (pH 8)	5.1×10^8	[0002]
L-Tyrosine	RT	H$_2$O (pH 7)	6.6×10^9	[0002]
9 Acetophenone, 4′-methoxy-				
Di-*tert*-butylnitroxide		Benzene	1.8×10^9	[8514]
O$_2$		Benzene	6.0×10^9	[8432]
Azulene	ET	Benzene	7.5×10^9	[8514]
MV^{2+}	OT	MeCN/H$_2$O (9:1)	8.0×10^9	[8412]
Phenoxide ion	RT	MeCN/H$_2$O (1:1)	6.8×10^9	[8110]
Triethylamine	RT	Benzene	1.0×10^8	[8432]
10 Acetophenone, 4′-methyl-				
O$_2$		Benzene	4.0×10^9	[8510]
β-Ionone	ET	Toluene	3.2×10^9	[8515]
MV^{2+}	OT	MeCN/H$_2$O (9:1)	4.1×10^9	[8412]
2-Propanol	RT	Benzene	1.3×10^5	[7311]
11 Acetophenone, 2,2,2-trifluoro-				
Anisole	CT	MeCN	1.9×10^8	[8018]
MV^{2+}	OT	MeCN/H$_2$O (9:1)	2.6×10^9	[8108]
1,4-Dimethoxybenzene	RT	Benzene	4.1×10^9	[8129]
12 Acetophenone, 4′-(trifluoromethyl)-				
O$_2$		Benzene	2.5×10^9	[8510]
p-Xylene	CT	MeCN	4.7×10^6	[8614]
Cyclohexane	RT	MeCN	2.0×10^6	[7808]
2-Propanol	RT	MeCN	8.8×10^6	[7808]

Table 6c Triplet-State Quenching of Organic Molecules

No.	Quencher	Mechanism	Solvent	K_q (L mol^{-1} s^{-1})	Ref.
13	**Acridine**				
	O_2		Benzene	1.8×10^9	[7816]
	(E)-Azobenzene	ET	Benzene	5.1×10^9	[8118]
	Cr(acac)$_3$	ET	Benzene	1.3×10^9	[8126]
	Fullerene C$_{60}$	ET	Benzene	9.3×10^9	[9101]
	Fullerene C$_{70}$	ET	Benzene	8.1×10^9	[9102]
	2-Naphthol	RT	Cyclohexane	2.9×10^9	[7310]
14	**Acridine Orange, conjugate monoacid**				
	Triphenylamine	RT	MeOH	2.5×10^8	[7916]
15	**Acridine Orange, free base**				
	TMPD	RT	MeOH	5.0×10^9	[7916]
16	**9-Acridinethione, 10-methyl-**				
	O_2		Cyclohexane	4.8×10^9	[8425]
17	**Acridinium, 3,6-bis(dimethylamino)-10-methyl-**				
	N,N-Dimethylaniline	RT	MeOH	1.3×10^9	[7916]
18	**Alloxazine, 7,8-dimethyl-**				
	Aniline	RT	MeOH	1.7×10^9	[0202]
	Diphenylamine	RT	MeOH	3.7×10^9	[0202]
	TMPD	RT	MeOH	4.1×10^9	[0202]
19	**Aniline, N,N-dimethyl-**				
	O_2		Cyclohexane	$\sim 1 \times 10^{10}$	[7607]
20	**Aniline, N,N-diphenyl-**				
	O_2		MCH	1.5×10^{10}	[7206]
21	**Anthracene**				
	Di-*tert*-butylnitroxide		Benzene	8.8×10^6	[8423]
	O_2		Cyclohexane	3.9×10^9	[8214]
	O_2		Toluene	3.96×10^9	[9301]
	Azulene	ET	Benzene	6.7×10^9	[8423]
	β-Carotene	ET	Hexane	1.1×10^{10}	[6601]
	Cr(acac)$_3$	ET	Benzene	1.3×10^9	[8126]

Table 6c Triplet-State Quenching of Organic Molecules

No.	Quencher	Mecha-nism	Solvent	K_q (L mol^{-1} s^{-1})	Ref.
21	**Anthracene** (continued)				
	Ferrocene	ET	Benzene	4.2×10^9	[8423]
	Fullerene C_{60}	ET	Benzene	6.1×10^9	[9101]
	Fullerene C_{70}	ET	Benzene	6.2×10^9	[9102]
	(*all-E*)-Retinal	ET	Benzene	3.0×10^{10}	[7715]
	1,1'-Dibenzyl-4,4'-bipyridinium	OT	MeOH	4.1×10^9	[8310]
	MV^{2+}	OT	MeOH	5.1×10^9	[8310]
22	**Anthracene, 9-bromo-**				
	Azulene	ET	Benzene	4.5×10^9	[8130]
23	**Anthracene, 9,10-dibromo-**				
	(*E*)-Azobenzene	ET	Benzene	2.4×10^8	[8118]
24	**Anthracene, 9,10-dimethyl-**				
	O_2		Benzene	4.0×10^9	[7003]
25	**Anthracene, 9,10-diphenyl-**				
	Di-*tert*-butylnitroxide		Benzene	5.1×10^6	[8423]
26	**Anthracene, 9,10-diphenyl-**				
	O_2		Benzene	1.9×10^9	[8423]
	Anthracene	ET	Toluene	3.0×10^8	[8316]
	MV^{2+}	OT	MeCN/H$_2$O (9:1)	4.3×10^9	[8310]
27	**1,8-Anthracenedisulfonate ion**				
	I^-		H$_2$O	1.1×10^5	[8518]
	O_2		MeOH	1.5×10^9	[8518]
28	**1-Anthracenesulfonate ion**				
	I^-		H$_2$O	4.4×10^4	[8518]
	O_2		MeOH	2.4×10^9	[8518]
29	**2-Anthracenesulfonate ion**				
	I^-		H$_2$O	4.0×10^4	[8518]
	O_2		MeOH	2.4×10^9	[8518]

Table 6c Triplet-State Quenching of Organic Molecules

No.	Quencher	Mechanism	Solvent	K_q (L mol^{-1} s^{-1})	Ref.
30	**9,10-Anthraquinone**				
	O_2		Benzene	1.4×10^9	[7704]
	Anthracene	ET	Benzene	2.0×10^{10}	[6902]
	Biacetyl	ET	Benzene	1.2×10^9	[6902]
	2-Propanol	RT	Benzene	2.1×10^7	[6902]
31	**9,10-Anthraquinone-2-sulfonate ion**				
	I^-	CT	H_2O	4.2×10^9	[8311]
32	**9-Anthroate ion**				
	MV^{2+}	OT	H_2O (pH 5.0)	2.2×10^9	[8308]
33	**Anthrone**				
	O_2		Benzene	2×10^9	[8213]
	$Cu(acac)_2$	ET	Benzene	5.0×10^9	[8213]
	1,3-Octadiene	ET	Benzene	6.8×10^9	[8213]
34	**1-Azaxanthone**				
	Toluene	CT	MeCN	1.6×10^8	[0001]
	p-Xylene	CT	MeCN	1.7×10^9	[0001]
35	**Azulene**				
	O_2		Benzene	6×10^9	[8130]
	O_2		MeCN	8×10^9	[8130]
	Anthracene	ET	Benzene	7.8×10^7	[8427]
	Perylene	ET	Benzene	9.5×10^9	[8427]
36	**Benzaldehyde**				
	Benzene	RT	MeCN	2.2×10^7	[7402]
	SCN^-	RT	H_2O/MeCN (4:1)	1.7×10^{10}	[8208]
37	**Benz[*a*]anthracene**				
	Di-*tert*-butylnitroxide		Benzene	2.5×10^7	[8422]
	O_2		Cyclohexane	1.9×10^9	[7004]
	Azulene	ET	Benzene	8.3×10^9	[8422]
	(*all-E*)-Retinal	ET	Hexane	5×10^9	[6907]
	MV^{2+}	OT	MeCN/H_2O (9:1)	5.2×10^9	[8310]

Table 6c Triplet-State Quenching of Organic Molecules

No.	Quencher	Mechanism	Solvent	K_q (L mol^{-1} s^{-1})	Ref.
38	**Benz[*a*]anthracene, 9,10-dimethyl-**				
	O_2		Benzene	3.0×10^9	[7003]
39	**Benzanthrone**				
	Cu(acac)$_2$	ET	Benzene	2.3×10^9	[6602]
	Ferrocene	ET	Benzene	5.7×10^9	[6602]
40	**Benzene**				
	Benzophenone	ET	Benzene	1×10^{11}	[7504]
41	**Benzene, methoxy-**				
	O_2		H$_2$O (pH 8.5)	6.3×10^9	[7511]
42	**Benzene, nitro-**				
	(*E*)-Piperylene	ET	THF	1×10^9	[8428]
	Triethylamine	RT	Benzene	1.2×10^9	[8435]
43	**Benzene, 1,3,5-trimethyl-**				
	O_2		Hexane	7×10^8	[8220]
44	**Benzene, 1,3,5-triphenyl-**				
	O_2		Benzene	1.1×10^9	[7307]
45	**Benzidine, *N,N,N',N'*-tetramethyl-**				
	O_2		MeCN	2.2×10^{10}	[8414]
	Naphthalene	ET	Cyclohexane	2×10^9	[7613]
	Duroquinone	OT	MeOH	2.3×10^{10}	[7613]
	MV^{2+}	OT	MeCN	8.4×10^9	[8414]
46	**Benzil**				
	Cu(acac)$_2$	CT	MeOH	9.5×10^8	[8321]
	1,3-Cyclohexadiene	CT	Benzene	3.0×10^8	[6602]
	β-Ionone	CT	Toluene	3.5×10^8	[8515]
	Phenoxide ion	RT	MeCN/H$_2$O (1:1)	3.7×10^9	[8110]
	Tryethylamine	RT	Cyclohexane	8.8×10^8	[7812]
	1,3,5-Trimethoxybenzene	RT	MeCN	8.0×10^5	[8129]

Table 6c Triplet-State Quenching of Organic Molecules

No. Quencher	Mechanism	Solvent	K_q (L mol^{-1} s^{-1})	Ref.
47 Benzo[*rst*]pentaphene				
O$_2$		Cyclohexane	2.9×10^9	[7004]
Rubrene	ET	Benzene	1.7×10^9	[8125]
48 Benzo[*ghi*]perylene				
TEMPO		MeCN	1.2×10^7	[8014]
1,4-Benzoquinone	CT	Benzene	8.2×10^9	[7906]
49 Benzo[*c*]phenanthrene				
O$_2$		Cyclohexane	1.1×10^9	[7004]
50 Benzophenone				
Hexamethyldisilane		Benzene	1.72×10^5	[9703]
O$_2$		Benzene	2.3×10^9	[8430]
O$_2$		H$_2$O	2.6×10^9	[0002]
O$_2$		MeCN	2.3×10^9	[8510]
Azulene	ET	Benzene	1.2×10^{10}	[8427]
Cu(acac)$_2$	ET	MeOH	3.6×10^9	[8321]
Ferrocene	ET	MeCN	1.8×10^{10}	[8130]
Naphthalene	ET	Benzene	4.7×10^9	[7007]
(*E*)-Stilbene	ET	Benzene	6×10^9	[8123]
Fumaronitrile	OT	MeCN	1.2×10^9	[7709]
MV^{2+}	OT	H$_2$O	3.7×10^9	[8512]
Anisole	RT	Benzene	2.1×10^6	[9804]
Cyclohexene	RT	MeCN	9.6×10^7	[8114]
DABCO	RT	Benzene	3.1×10^9	[8406]
p-Dimethoxybenzene	RT	Benzene	5.1×10^8	[9804]
Hexamethylbenzene	RT	Benzene	1.1×10^9	[9804]
Phenol	RT	Benzene	1.3×10^9	[8113]
Phenol	RT	H$_2$O (pH 8)	3.9×10^9	[0002]
2-Propanol	RT	Benzene	1.1×10^6	[7203]
Toluene	RT	Benzene	3.1×10^5	[9804]
Triethylamine	RT	Benzene	4.1×10^9	[8106]

Table 6c Triplet-State Quenching of Organic Molecules

No. Quencher	Mechanism	Solvent	K_{q} (L mol^{-1} s^{-1})	Ref.
50 Benzophenone (continued)				
L-Tyrosine	RT	H$_2$O (pH 7)	2.6×10^9	[0002]
51 Benzophenone, 4,4′-bis(dimethylamino)-				
O$_2$		Benzene	1.3×10^{10}	[8510]
O$_2$		MeCN	1.1×10^{10}	[8510]
β-Ionone	ET	Toluene	4.6×10^9	[8515]
Naphthalene	ET	Cyclohexane	9.9×10^9	[7718]
52 Benzophenone, 4-(carboxymethyl)-, ion(1−)				
Fumaronitrile	OT	CCl$_4$	6.2×10^9	[8227]
53 Benzophenone, 4-chloro-				
O$_2$		MeCN	1.9×10^9	[8510]
MV^{2+}	OT	MeCN/H$_2$O (9:1)	2.8×10^9	[8412]
Triethylamine	RT	MeCN/H$_2$O (9:1)	2.6×10^9	[8611]
54 Benzophenone, 4-cyano-				
Quadricyclane	RT	MeCN	6.5×10^9	[8705]
55 Benzophenone, 4,4′-dichloro-				
O$_2$		MeCN	1.4×10^9	[8510]
MV^{2+}	OT	MeCN/H$_2$O (9:1)	2.1×10^9	[8412]
Triethylamine	RT	MeCN/H$_2$O (9:1)	3.3×10^9	[8611]
56 Benzophenone, 4,4′-dimethoxy-				
O$_2$		MeCN	5.6×10^9	[8510]
β-Ionone	ET	Toluene	4.2×10^9	[8515]
Triethylamine	RT	MeCN/H$_2$O (9:1)	5.5×10^8	[8611]
57 Benzophenone, 4-fluoro-				
O$_2$		MeCN	2.3×10^9	[8510]
MV^{2+}	OT	MeCN/H$_2$O (9:1)	1.3×10^9	[8412]
Triethylamine	RT	MeCN/H$_2$O (9:1)	2.7×10^9	[8611]
58 Benzophenone, 4-methoxy-				
O$_2$		Benzene	3.7×10^9	[8510]
β-Ionone	ET	Toluene	3.0×10^9	[8515]

Table 6c Triplet-State Quenching of Organic Molecules

No.	Quencher	Mechanism	Solvent	K_q (L mol^{-1} s^{-1})	Ref.
58	**Benzophenone, 4-methoxy-** (continued)				
	MV^{2+}	OT	MeCN/H$_2$O (9:1)	5.6×10^9	[8412]
	Triethylamine	RT	MeCN/H$_2$O (9:1)	1.2×10^9	[8611]
59	**Benzophenone, 4-methyl-**				
	Di-*tert*-butyl peroxide		Benzene	4.1×10^9	[8432]
	O$_2$		Benzene	2.3×10^9	[8510]
	2,5-Dimethyl-2,4-hexadiene	CT	Benzene	6.2×10^9	[8432]
	MV^{2+}	OT	MeCN/H$_2$O (9:1)	3.3×10^9	[8412]
	Triethylamine	RT	MeCN/H$_2$O (9:1)	1.5×10^9	[8611]
60	**Benzophenone, 4-phenyl-**				
	2-Nitrophene	ET	MeCN	2.3×10^9	[8209]
	2,3-Dimethyl-2-butene	RT	MeCN	5.2×10^6	[7709]
61	**Benzophenone, 4-(trifluoromethyl)-**				
	O$_2$		MeCN	1.6×10^9	[8510]
	MV^{2+}	OT	MeCN/H$_2$O (9:1)	1.7×10^9	[8412]
	Triethylamine	RT	MeCN/H$_2$O (9:1)	4.2×10^9	[8611]
62	**Benzo[*g*]pteridine-2,4-dione**				
	O$_2$		H$_2$O	1.7×10^9	[7719]
63	**Benzo[*g*]pteridine-2,4-dione, 3,7,8,10-tetramethyl-**				
	Allylthiourea	RT	H$_2$O (pH 7)	2.8×10^9	[8434]
	Triethylamine	RT	MeCN	5.8×10^9	[8228]
64	**Benzo[*g*]pteridine-2,4-dione, 7,8,10-trimethyl-**				
	O$_2$		H$_2$O (pH 7)	1.2×10^9	[8801]
	Adenine	RT	H$_2$O (pH 7.2)	1.6×10^9	[8509]
	EDTA	RT	H$_2$O (pH 7)	3.1×10^8	[8513]
	Indole	RT	H$_2$O (pH 7.2)	3.7×10^9	[8801]
	Tryptophan	RT	H$_2$O (pH 7.2)	3.7×10^9	[8509]

Table 6c Triplet-State Quenching of Organic Molecules

No.	Quencher	Mechanism	Solvent	K_q (L mol^{-1} s^{-1})	Ref.
65	**Benzo[*g*]pteridine-2,4-dione, 7,8,10-trimethyl-, conjugate monoacid**				
	EDTA	RT	H_2O (pH 3)	1.4×10^8	[8513]
	Naphthalene	RT	MeOH	7.9×10^7	[8119]
66	**Benzo[*g*]pteridine-2,4-dione, 7,8,10-trimethyl-, ion(1–)**				
	EDTA	RT	H_2O (pH 14)	1.5×10^8	[8513]
67	**Benzo[*a*]pyrene**				
	Di-*tert*-butylnitroxide		Benzene	8.0×10^6	[8423]
	O_2		Benzene	3.2×10^9	[8423]
	O_2		Cyclohexane	2.6×10^9	[7004]
	Azulene	ET	Benzene	5.6×10^9	[8423]
	Biacetyl	ET	Benzene	4.9×10^7	[6402]
	Ferrocene	ET	Benzene	4.5×10^9	[8423]
	MV^{2+}	OT	MeCN/H_2O (9:1)	5.4×10^9	[8310]
68	**Benzo[*e*]pyrene**				
	Methyl cinnamate	ET	Benzene	7.3×10^7	[8612]
69	**1,4-Benzoquinone**				
	O_2		Heptane	3×10^8	[8212]
	O_2		MeCN	3×10^8	[8212]
70	**1,4-Benzoquinone, 2,5-bis(4-chlorophenyl)-**				
	O_2		Heptane	3.6×10^9	[8212]
	O_2		MeCN	1.1×10^9	[8212]
71	**1,4-Benzoquinone, 4-chlorophenyl-**				
	O_2		MeCN	9×10^8	[8212]
72	**1,4-Benzoquinone, 2,6-diphenyl-**				
	Triphenylamine	CT	Toluene	5.0×10^9	[7913]
73	**1,4-Benzoquinone, tetrachloro-**				
	Naphthalene	CT	Benzene	1.4×10^8	[8803]
	Indole	RT	MeCN	1.3×10^{10}	[8511]

Table 6c Triplet-State Quenching of Organic Molecules

No. Quencher	Mechanism	Solvent	K_{q} (L mol^{-1} s^{-1})	Ref.
74 1,4-Benzoquinone, tetramethyl-				
O$_2$		Cyclohexane	1.3×10^9	[7101]
O$_2$		MeCN	7×10^8	[8212]
Anthracene	ET	Benzene	6.8×10^9	[7101]
Cu(acac)$_2$	ET	MeOH	3.8×10^9	[8321]
I$^-$	ET	H$_2$O (pH 7)	9×10^9	[8013]
I$^-$	RT	MeCN	9.9×10^9	[8210]
Triphenylamine	RT	MeCN	1.0×10^{10}	[7717]
75 Benzo[*b*]triphenylene				
O$_2$		Benzene	1.8×10^9	[7307]
O$_2$		Cyclohexane	1.5×10^9	[7004]
Cr(acac)$_3$	ET	Benzene	1.2×10^9	[8126]
β-Ionone	ET	Toluene	5.0×10^8	[8515]
Methyl cinnamate	ET	Benzene	2.4×10^6	[8612]
MV^{2+}	OT	MeCN/H$_2$O (9:1)	6.1×10^9	[8310]
76 Biacetyl				
Ferrocene		Benzene	1.7×10^{10}	[9601]
Acridine	ET	Benzene	9×10^9	[6906]
Anthracene	ET	Dichloromethane	7.0×10^9	[9601]
Benzil	ET	Cyclohexane	1.5×10^9	[7314]
2,2′-Binaphthyl	ET	Benzene	2.9×10^9	[9601]
1,8-Dinitronaphthalene	ET	Benzene	2.6×10^9	[6402]
I$^-$	ET	H$_2$O	5.5×10^9	[7509]
Naphthalene	ET	Benzene	2.6×10^6	[9601]
Pyrene	ET	Dichloromethane	5.0×10^9	[9601]
Piperylene	ET	CCl$_4$	2.5×10^8	[7401]
Aniline	RT	Benzene	2.9×10^8	[7002]
Benzidine	RT	Dichloromethane	1.0×10^{10}	[9601]
DABCO	RT	Benzene	1.5×10^7	[9601]
Diphenylamine	RT	Dichloromethane	4.0×10^9	[9601]

Table 6c Triplet-State Quenching of Organic Molecules

No. Quencher	Mechanism	Solvent	K_q (L mol^{-1} s^{-1})	Ref.
76 Biacetyl (continued)				
1-Naphthylamine	RT	Dichloromethane	1.2×10^5	[9601]
2-Propanol	RT	MeCN	1.3×10^4	[6903]
TMPD	RT	Dichloromethane	5.0×10^9	[9601]
Triethylamine	RT	Benzene	5.0×10^7	[6906]
77 9,9′-Bicarbazole				
Naphthalene	ET	Cyclohexane	6.2×10^9	[7813]
78 Bilirubin				
O_2		Benzene	8.2×10^8	[7609]
79 2,2′-Binaphthyl				
Biacetyl	ET	Benzene	1.3×10^9	[6402]
80 Biphenyl				
I$^-$		MeCN	6.2×10^3	[7403]
O_2		Benzene	1.51×10^9	[9902]
O_2		Cyclohexane	7.8×10^8	[9902]
O_2		MeCN	2.85×10^9	[9902]
β-Carotene	ET	Cyclohexane	8.4×10^9	[8313]
(all-*E*)-Retinal	ET	Hexane	2.0×10^{10}	[7711]
MV^{2+}	OT	MeCN/H$_2$O (9:1)	9.3×10^9	[8310]
Diethyl sulfide	RT	Cyclohexane	4.8×10^6	[8410]
Diethyl sulfide	RT	MeCN	1.0×10^7	[8410]
81 Biphenyl, 4,4′-dibromo-				
O_2		Benzene	1.07×10^9	[9902]
O_2		MeCN	1.46×10^9	[9902]
82 Biphenyl, 4,4′-dichloro-				
O_2		Benzene	1.00×10^9	[9902]
O_2		MeCN	1.77×10^9	[9902]
83 Biphenyl, 4,4′-dimethoxy-				
O_2		Benzene	9.15×10^9	[9902]
O_2		MeCN	1.26×10^{10}	[9902]

Table 6c Triplet-State Quenching of Organic Molecules

No.	Quencher	Mechanism	Solvent	K_q (L mol^{-1} s^{-1})	Ref.
84	**Biphenyl, 4,4'-dimethyl-**				
	O_2		Benzene	3.71×10^9	[9902]
	O_2		MeCN	5.93×10^9	[9902]
85	**Biphenyl, 4-methoxy-**				
	O_2		Benzene	5.94×10^9	[9902]
	O_2		MeCN	8.56×10^9	[9902]
86	**Biphenyl, 4-methyl-**				
	O_2		Benzene	2.46×10^9	[9902]
	O_2		MeCN	4.36×10^9	[9902]
87	**2,2'-Bipyridine**				
	O_2		Cyclohexane	5.2×10^8	[8802]
88	**1,3-Butadiene, 1,4-diphenyl-**				
	O_2		Benzene	5.0×10^9	[8423]
	O_2		Cyclohexane	5.3×10^9	[8221]
	O_2		MeCN	4.0×10^9	[8222]
	Azulene	ET	Benzene	2.9×10^9	[8423]
	Ferrocene	ET	Benzene	1.5×10^9	[8222]
89	**1,3-Butadiene, 1,1,4,4-tetraphenyl-**				
	β-Carotene	ET	Toluene	4.3×10^9	[8417]
90	**2-Butanone**				
	O_2		Hexane	3.4×10^{10}	[8413]
	O_2		MeCN	4.3×10^{10}	[8413]
91	**Butyraldehyde**				
	2,5-Dimethyl-2,4-hexadiene	ET	Pentane	1.5×10^{10}	[8127]
92	**Carbazole**				
	O_2		Benzene	5.7×10^9	[7704]
	Ferrocene	ET	Cyclohexane	6×10^9	[7813]
	Pyridine	RT	Cyclohexane	4.9×10^7	[8431]

Table 6c Triplet-State Quenching of Organic Molecules

No. Quencher	Mechanism	Solvent	K_q (L mol^{-1} s^{-1})	Ref.
93 β-*apo*-14′-Carotenal				
Di-*tert*-butylnitroxide		MCH	1.0×10^9	[8418]
O_2		Cyclohexane	4.7×10^9	[8430]
94 β-Carotene				
Di-*tert*-butylnitroxide		MCH	6.8×10^8	[8418]
O_2		Benzene	3.6×10^9	[7313]
I_2	ET	Benzene	2.5×10^9	[8408]
I_2	ET	MeCN	1.5×10^9	[8408]
95 Chlorophyll *a*				
O_2		Benzene	1.1×10^9	[5701]
β-Carotene	ET	Benzene	1.2×10^9	[7313]
Tetracene	ET	Toluene	6×10^8	[6905]
1,4-Benzoquinone	OT	Benzene	2.4×10^9	[5701]
α-Tocopherylquinone	OT	EtOH	7×10^8	[7010]
TMPD	RT	EtOH	2.7×10^6	[7817]
96 Chlorophyll *b*				
1,6-Diphenyl-1,3,5-hexatriene	ET	Hexane	2×10^5	[7608]
1,4-Benzoquinone	OT	Benzene	1.5×10^9	[5701]
Phenylhydrazine	RT	Toluene	5.0×10^4	[6905]
97 Chrysene				
Di-*tert*-butylnitroxide		Hexane	9.6×10^8	[7308]
O_2		Benzene	1.4×10^9	[7307]
O_2		Cyclohexane	1.0×10^9	[7004]
Azulene	ET	Benzene	8×10^9	[8123]
Azulene	ET	MeCN	1.3×10^{10}	[8123]
Cr(acac)$_3$	ET	Benzene	1.5×10^9	[8126]
MV^{2+}	OT	MeCN/H$_2$O (9:1)	7.4×10^9	[8310]
98 Coronene				
Di-*tert*-butylnitroxide		Hexane	3.3×10^8	[7308]
O_2		Benzene	4.2×10^8	[7307]

Table 6c Triplet-State Quenching of Organic Molecules

No.	Quencher	Mechanism	Solvent	K_q (L mol^{-1} s^{-1})	Ref.
98	**Coronene** (continued)				
	Cr(acac)$_3$	ET	Benzene	1.6×10^9	[8126]
	Methyl cinnamate	ET	Benzene	3.1×10^8	[8612]
	MV^{2+}	OT	MeCN/H$_2$O (9:1)	5.4×10^9	[8310]
99	**Coumarin, 3-benzoyl-**				
	Methyl cinnamate	ET	Benzene	3.7×10^9	[8612]
100	**Coumarin, 3,3'-carbonylbis-**				
	Methyl cinnamate	ET	Benzene	2.8×10^9	[8612]
101	**Coumarin, 3,3'-carbonylbis(5,7-diethoxy-**				
	Methyl cinnamate	ET	Benzene	1.5×10^9	[8612]
102	**Coumarin, 3,3'-carbonylbis(7-diethamino-**				
	Methyl cinnamate	ET	Benzene	7.7×10^6	[8612]
103	**Coumarin, 3,3'-carbonylbis(5,7-dimethoxy-**				
	Methyl cinnamate	ET	Benzene	1.6×10^9	[8612]
104	**Coumarin, 7-(diethylamino)-4-methyl-**				
	Anthracene	ET	EtOH	6.9×10^9	[7406]
105	**Crocetin**				
	O$_2$		H$_2$O (pH 8)	1.4×10^9	[8312]
106	**Cyclobutene, 1,2-diphenyl-**				
	O$_2$		Benzene	3.4×10^9	[8123]
107	**1,3-Cycloheptadiene**				
	β-Carotene	ET	Toluene	9.0×10^9	[8417]
	Azulene	ET	Benzene	1.0×10^{10}	[8427]
	MV^{2+}	OT	MeOH	2.6×10^9	[8224]
108	**Cyclohexanone**				
	9,10-Dibromoanthracene	ET	Cyclohexane	4.5×10^9	[8419]
109	**Cyclohexene, 1-phenyl-**				
	MV^{2+}	OT	MeOH	1.4×10^9	[8225]
110	**2-Cyclohexenethione, 3,5,5-trimethyl-**				
	O$_2$		Benzene	3.7×10^9	[8610]

Table 6c Triplet-State Quenching of Organic Molecules

No.	Quencher	Mechanism	Solvent	K_q (L mol^{-1} s^{-1})	Ref.
111	**2-Cyclohexen-1-one**				
	O$_2$		Cyclohexane	5×10^9	[8012]
	Triethylamine	RT	Cyclohexane	9.2×10^7	[8804]
	Triethylamine	RT	MeCN	9.0×10^7	[8804]
112	**2-Cyclohexen-1-one, 4,4-dimethyl-**				
	Triethylamine	RT	MeCN	3.7×10^7	[8804]
113	**2-Cyclohexen-1-one, 4,4,6,6-tetramethyl-**				
	Triethylamine	RT	Cyclohexane	1.3×10^8	[8804]
114	**Cyclopentadiene**				
	Anthracene	ET	Benzene	9.8×10^9	[8124]
	Naphthalene	ET	Benzene	2.0×10^7	[8124]
115	**Cyclopentene, 1-phenyl-**				
	MV^{2+}	OT	MeOH	4.5×10^9	[8225]
116	**2-Cyclopentenone**				
	O$_2$		Cyclohexane	5×10^9	[8012]
117	**Dibenz[*a,h*]anthracene**				
	O$_2$		Benzene	1.6×10^9	[7307]
	O$_2$		Cyclohexane	1.3×10^9	[7004]
	Ferrocene	ET	EtOH	5.0×10^9	[7408]
	Rubrene	ET	Benzene	2.2×10^9	[8125]
	MV^{2+}	OT	MeCN/H$_2$O (9:1)	7.5×10^9	[8310]
118	**Dibenzo[*b,def*]chrysene**				
	Di-*tert*-butylnitroxide		Hexane	1.9×10^7	[7308]
	O$_2$		Benzene	2.8×10^9	[7307]
	(*E*)-Azobenzene	ET	Benzene	1.0×10^7	[8118]
119	**Dibenzo[*def,mno*]chrysene**				
	Ferrocene	ET	Benzene	5.8×10^8	[7508]
	Rubrene	ET	Benzene	1.6×10^9	[8125]
120	**Dibenzo[*g,p*]chrysene**				
	Methyl cinnamate	ET	Benzene	3.4×10^5	[8612]

Table 6c Triplet-State Quenching of Organic Molecules

No.	Quencher	Mechanism	Solvent	K_q (L mol^{-1} s^{-1})	Ref.
121	**2,6-Dimethylquinoline**				
	Methyl cinnamate	ET	Benzene	4.0×10^9	[8612]
122	**1,4-Dioxin, 2,3,5,6-tetraphenyl-**				
	O_2		Benzene	1.7×10^9	[7909]
123	**Dodecapreno-β-carotene**				
	O_2		Benzene	5.6×10^9	[7313]
124	**Ethene, tetraphenyl**				
	O_2		Benzene	2.4×10^9	[8218]
	O_2		MeCN	2.5×10^9	[8218]
125	**Flavine mononucleotide**				
	Glycyl-L-tyrosine	RT	H_2O (pH 7.0)	1.0×10^9	[7910]
	L-Methionine	RT	H_2O (pH 7.0)	1.7×10^7	[7910]
126	**Fluoranthene**				
	TEMPO		MeCN	4.3×10^8	[8014]
	Biacetyl	ET	Benzene	2.1×10^7	[6402]
	MV^{2+}	OT	MeCN/H_2O (9:1)	1.0×10^{10}	[8310]
127	**Fluorene**				
	O_2		Benzene	4.9×10^9	[7704]
	Dibutyldiazene	ET	Benzene	1.8×10^8	[7507]
	Ruthenocene	ET	Benzene	6.4×10^8	[7716]
	MV^{2+}	OT	MeCN/H_2O (9:1)	8.2×10^9	[8310]
128	**9-Fluorenone**				
	Norbornadiene		Benzene	4.6×10^5	[7815]
	Azulene	ET	Benzene	8×10^9	[8130]
	Cu(acac)$_2$	ET	Benzene	1.0×10^9	[6602]
	Ferrocene	ET	Benzene	5.1×10^9	[6602]
	Triethylamine	RT	Cyclohexane	1.0×10^7	[7812]
	Triethylamine	RT	MeCN	4.7×10^8	[7812]

Table 6c Triplet-State Quenching of Organic Molecules

No.	Quencher	Mechanism	Solvent	K_q (L mol^{-1} s^{-1})	Ref.
129	**Fluorescein, conjugate monoacid**				
	O_2		H_2O (pH ~4)	2×10^8	[6401]
	Fe^{3+}	OT	H_2O (pH 1.6)	1×10^9	[6001]
130	**Fluorescein dianion**				
	O_2		H_2O (pH 12)	1.7×10^9	[6401]
	Allylthiourea	RT	H_2O (pH 12)	2×10^5	[6001]
	p-Phenylenediamine	RT	H_2O (pH 12)	5×10^9	[6001]
131	**Fluorescein dianion, 2′,4′,5′,7′-tetrabromo-**				
	O_2		H_2O	$\sim 3 \times 10^8$	[6701]
	Ferrocene	ET	EtOH	3.5×10^9	[7408]
	MV^{2+}	OT	EtOH/H_2O (1:1)	3.2×10^9	[8704]
	Allylthiourea	RT	H_2O (pH 7.1)	4.9×10^5	[7410]
	Aniline	RT	H_2O (pH 10.4)	1.4×10^9	[6801]
	2-Naphthol	RT	H_2O (pH 10.5)	1.5×10^9	[6801]
	L-Tryptophan	RT	H_2O (pH 7.1)	8.4×10^8	[7410]
132	**Fluorescein dianion, 3,4,5,6-tetrachloro-2′,4′,5′,7′-tetraiodo-**				
	O_2		H_2O (pH 7.2)	9.8×10^8	[8801]
	EDTA	RT	H_2O (pH 5.0)	1.5×10^6	[8508]
133	**Fluorescein monoanion**				
	O_2		H_2O (pH 4.5)	1.2×10^9	[6401]
	EDTA	RT	H_2O (pH 5.0)	3.9×10^6	[8508]
134	**Fluorescein monoanion, 2′,4′,5′,7′-tetrabromo-**				
	I^-	OT	H_2O (pH 6.5)	1.5×10^7	[8112]
	$Fe(CN)_6^{4-}$	RT	H_2O (pH 6.5)	5.7×10^8	[8112]
135	**Fullerene C$_{60}$**				
	β-Carotene	ET	Benzene	7.8×10^9	[9701]
	O_2	ET	Benzene	1.9×10^9	[9101]
	(*all-E*)-Retinoic acid	ET	Benzonitrile	2.0×10^9	[9901]
	(*all-E*)-Retinol	ET	Benzonitrile	1.8×10^9	[9901]

Table 6c Triplet-State Quenching of Organic Molecules

No.	Quencher	Mechanism	Solvent	K_q (L mol^{-1} s^{-1})	Ref.
135	**Fullerene C$_{60}$** (continued)				
	TEMPO	CT	Toluene	3.3×10^9	[9201]
	Ferrocene	RT	Toluene	3.3×10^9	[9802]
136	**Fullerene C$_{60}$, 1,2,5-triphenylpyrrolidino-**				
	O$_2$	ET	Toluene	2.14×10^9	[9802]
	TEMPO	CT	Toluene	6.6×10^8	[9802]
	N,N-Dimethylaniline	RT	Toluene	5.6×10^8	[0101]
	Ferrocene	RT	Toluene	3.0×10^9	[0101]
	N-Methylphenothiazine	RT	Toluene	3.3×10^9	[0101]
137	**Fullerene C$_{70}$**				
	O$_2$	ET	Benzene	9.4×10^8	[9102]
	Tetracene	ET	Benzene	2.3×10^9	[9102]
138	**Fullerene C$_{76}$**				
	β-Carotene	ET	Toluene	2.6×10^9	[9702]
	TMPD	RT	Benzonitrile	7.8×10^8	[9702]
139	**2-Furoic acid, 5-nitro-**				
	O$_2$		Acetone	1.7×10^9	[8111]
	Azulene	ET	Acetone	1.2×10^{10}	[8111]
	I$^-$	RT	MeCN	6.6×10^9	[8111]
	TMPD	RT	Acetone	6.0×10^9	[8111]
140	**D-Glucose phenylosazone**				
	O$_2$		EtOH	1×10^9	[8314]
141	**2-Heptanone**				
	Styrene		Benzene	4.0×10^9	[7920]
142	**(*E,E,E*)-2,4,6-Heptarienal, 5-methyl-7-(2,6,6-trimethyl-1-cyclohexen-1-yl)-**				
	O$_2$		Cyclohexane	3.3×10^9	[8430]
	O$_2$		MeOH	4.0×10^9	[8430]
	Azulene	ET	Toluene	3.1×10^7	[8418]
	Tetracene	ET	Toluene	3.3×10^9	[8418]

Table 6c Triplet-State Quenching of Organic Molecules

No. Quencher	Mechanism	Solvent	K_q (L mol^{-1} s^{-1})	Ref.
143 (E,E)-2,4-Hexadiene				
MV^{2+}	OT	MeOH	1.0×10^9	[8224]
144 2,4-Hexadiene, 2,5-dimethyl-				
Anthracene	ET	Benzene	2.8×10^9	[8124]
MV^{2+}	OT	MeOH	2.0×10^9	[8224]
145 1,3,5-Hexatriene, 1,6-diphenyl-				
EtI		Cyclohexane	7.0×10^4	[8221]
EtI		MeOH	5.2×10^4	[8221]
O$_2$		Benzene	5.6×10^9	[8222]
O$_2$		Cyclohexane	4.4×10^9	[8221]
O$_2$		MeCN	5.8×10^9	[8222]
146 Indene, 2-phenyl-				
O$_2$		Benzene	3.8×10^9	[8123]
147 Indeno[2,1-a]indene				
O$_2$		Benzene	3.4×10^9	[8123]
Azulene	ET	MeCN	1.5×10^{10}	[8009]
148 Indole				
O$_2$		Cyclohexane	1.6×10^{10}	[8805]
O$_2$		EtOH (abs)	1.1×10^{10}	[8805]
Anthracene	ET	Cyclohexane	9×10^9	[8120]
149 Indole, 5-methoxy-				
O$_2$		Cyclohexane	2.1×10^{10}	[8805]
150 Indole, 1-methyl-				
O$_2$		Benzene	1.4×10^{10}	[7704]
9,10-Anthraquinone	CT	Benzene	7.8×10^9	[7705]
151 β-Ionone				
Di-tert-butylnitroxide		MCH	4.2×10^9	[8418]
O$_2$		Toluene	5.1×10^9	[8515]
β-Carotene	ET	Toluene	3.1×10^9	[8515]

Table 6c Triplet-State Quenching of Organic Molecules

No. Quencher	Mechanism	Solvent	K_q (L mol^{-1} s^{-1})	Ref.
152 Isobenzofuran, 1,3-diphenyl-				
Azulene	ET	Benzene	6.5×10^7	[8125]
153 (*all-E*)-Lycopene				
O$_2$		Hexane	5.4×10^9	[7304]
154 Mesoporphyrin, dimethyl ester				
O$_2$		Benzene	1.4×10^9	[8015]
155 Methylene Blue cation				
O$_2$		MeCN	1.7×10^9	[8420]
Fe(acac)$_3$	ET	Benzene/EtOH (9:1)	5.4×10^8	[7611]
Benzenediazonium cation	OT	MeCN	6.7×10^6	[8615]
Allylthiourea	RT	MeCN	8.3×10^6	[8615]
DABCO	RT	MeCN	1.9×10^8	[8615]
Triethylamine	RT	MeOH	4.7×10^7	[7619]
Triphenylamine	RT	MeCN	3.0×10^9	[7720]
156 Methylene Blue cation, conjugate monoacid				
EDTA	RT	H$_2$O (pH 4.5)	5×10^8	[7405]
Fe^{II}(CN)$_6^{4-}$	RT	H$_2$O (pH 4.4)	1.4×10^{10}	[8008]
Ferrocene	RT	EtOH/H2O (19:1)	4.9×10^9	[8008]
Methylene Blue cation	RT	H$_2$O	1.1×10^8	[8109]
Thionine cation	RT	MeCN/H$_2$O (1:1)	9.0×10^7	[8219]
157 2-Naphthaldehyde				
MV^{2+}	OT	MeCN/H$_2$O (9:1)	3.8×10^9	[8108]
Triethylamine	RT	Benzene	1.3×10^7	[7812]
158 Naphthalene				
O$_2$		Benzene	2.1×10^9	[9401]
O$_2$		Cyclohexane	1.6×10^9	[0201]
O$_2$		MeCN	2.6×10^9	[0201]
O$_2$		Toluene	2.33×10^9	[0301]
TEMPO		MeCN	9.6×10^8	[8014]

Table 6c Triplet-State Quenching of Organic Molecules

No. Quencher	Mechanism	Solvent	K_q (L mol^{-1} s^{-1})	Ref.
158 Naphthalene (continued)				
Anthracene	ET	EtOH	1.3×10^{10}	[7406]
Cr(hfac)$_3$	ET	Benzene	8.2×10^9	[8315]
Ferrocene	ET	EtOH	7.0×10^9	[7408]
(*all-E*)-Retinal	ET	Benzene	5.5×10^9	[8116]
MV^{2+}	OT	MeCN/H$_2$O (9:1)	8.5×10^9	[8310]
Diethyl sulfide	RT	MeCN	1.9×10^6	[8410]
159 Naphthalene, 2-acetyl-				
Di-*tert*-butylnitroxide		Benzene	7.6×10^8	[8409]
O$_2$		Benzene	1.7×10^9	[8409]
O$_2$		H$_2$O	2.5×10^9	[0002]
Azulene	ET	Benzene	9.7×10^9	[8409]
Azulene	ET	MeCN	1.4×10^{10}	[8123]
Cu(acac)$_2$	ET	Benzene	1.9×10^9	[6602]
Ferrocene	ET	Benzene	7.6×10^9	[8409]
β-Ionone	ET	Benzene	2.2×10^9	[8409]
Quadricyclane	ET	Benzene	1.0×10^6	[7815]
MV^{2+}	OT	MeCN/H$_2$O (9:1)	5.4×10^9	[8108]
Phenol	RT	H$_2$O (pH 8)	3.3×10^7	[0002]
Phenoxide ion	RT	MeCN/H$_2$O (1:1)	2.6×10^9	[8110]
Triethylamine	RT	Cyclohexane	1.2×10^5	[7812]
L-Tyrosine	RT	H$_2$O (pH 7)	3.7×10^7	[0002]
160 Naphthalene, 1-bromo-				
O$_2$		MeCN	1.8×10^9	[0201]
Biacetyl	ET	Benzene	3.4×10^9	[6402]
161 Naphthalene, 2-bromo-				
O$_2$		MeCN	1.7×10^9	[0201]
162 Naphthalene, 1-chloro-				
O$_2$		MeCN	1.9×10^9	[0201]
Biacetyl	ET	Cyclohexane	4.7×10^9	[7314]

Table 6c Triplet-State Quenching of Organic Molecules

No. Quencher	Mechanism	Solvent	K_q (L mol^{-1} s^{-1})	Ref.
163 Naphthalene, 1-cyano-				
O$_2$		MeCN	1.4×10^9	[0201]
164 Naphthalene, 1,4-dicyano-				
O$_2$		MeCN	2.1×10^9	[8414]
2,5-Dimethyl-2,4-hexadiene	ET	MeCN	1.7×10^9	[8414]
165 Naphthalene, 2,6-dimethoxy-				
O$_2$		MeCN	9.6×10^9	[0201]
166 Naphthalene, 2,7-dimethoxy-				
O$_2$		MeCN	6.6×10^9	[0201]
167 Naphthalene, 2,3-dimethyl-				
O$_2$		MeCN	3.9×10^9	[0201]
168 Naphthalene, 2,6-dimethyl-				
O$_2$		MeCN	4.1×10^9	[0201]
169 Naphthalene, 1,4-dinitro-				
O$_2$		EtOH	2.0×10^9	[8117]
O$_2$		Hexane	1.7×10^9	[8117]
Tetracene	ET	Benzene	1.0×10^{10}	[7618]
170 Naphthalene, 1,8-dinitro-				
O$_2$		EtOH	2.1×10^9	[7708]
Tetracene	ET	Benzene	6.7×10^9	[7708]
Triethylamine	RT	MeCN	4.6×10^9	[7708]
171 Naphthalene, 1-fluoro-				
O$_2$		MeCN	2.2×10^9	[0201]
172 Naphthalene, 2-hydroxy-				
O$_2$		Cyclohexane	2.6×10^9	[7310]
Pyridine	OT	Cyclohexane	1.5×10^9	[7310]
173 Naphthalene, 1-methoxy-				
O$_2$		MeCN	7.2×10^9	[0201]
174 Naphthalene, 2-methoxy-				
O$_2$		MeCN	5.3×10^9	[0201]

Table 6c Triplet-State Quenching of Organic Molecules

No.	Quencher	Mechanism	Solvent	K_q (L mol^{-1} s^{-1})	Ref.
175	**Naphthalene, 1-methyl-**				
	O_2		MeCN	3.2×10^9	[0201]
	β-Ionone	ET	Toluene	2.7×10^9	[8515]
	MV^{2+}	OT	MeCN/H$_2$O (9:1)	7.8×10^9	[8310]
176	**Naphthalene, 2-methyl-**				
	O_2		MeCN	3.1×10^9	[0201]
177	**Naphthalene, 1-[2-(1-naphthyl)ethenyl]-, (*E*)-**				
	O_2		Benzene	4.4×10^9	[8422]
	Azulene	ET	Benzene	2.9×10^9	[8422]
178	**Naphthalene, 1-[2-(2-naphthyl)ethenyl]-, (*E*)-**				
	O_2		Benzene	3.8×10^9	[8422]
	Azulene	ET	Benzene	3.7×10^9	[8422]
179	**Naphthalene, 2-[2-(2-naphthyl)ethenyl]-, (*E*)-**				
	O_2		Benzene	6.0×10^9	[8422]
	Azulene	ET	Benzene	4.4×10^9	[8422]
180	**Naphthalene, 1-[1-(1-naphthyl)ethenyl]-**				
	O_2		Benzene	3.7×10^9	[8411]
	Ferrocene	ET	Benzene	4.4×10^9	[8411]
181	**Naphthalene, 1-nitro-**				
	O_2		EtOH	3.3×10^9	[8117]
	O_2		Hexane	1.3×10^9	[8117]
	Triphenylethylene	ET	Benzene	1.8×10^9	[8218]
182	**Naphthalene, 2-nitro-**				
	O_2		EtOH	1.6×10^9	[8117]
	O_2		Hexane	1.7×10^9	[8117]
	Azulene	ET	Benzene	7×10^9	[8123]
	Azulene	ET	MeCN	1.4×10^{10}	[8123]
	Tetracene	ET	Benzene	7.4×10^9	[7617]
183	**2-Naphthalenesulfonate ion**				
	Fe(CN)$_6^{4-}$	CT	H$_2$O	1.4×10^7	[7615]

Table 6c Triplet-State Quenching of Organic Molecules

No.	Quencher	Mechanism	Solvent	K_q (L mol^{-1} s^{-1})	Ref.
184	**1,4-Naphthoquinone, 2-methyl-**				
	O_2		H_2O (pH 7.0)	1.2×10^9	[8317]
	Thymine	RT	H_2O (pH 7.0)	2.7×10^9	[8317]
185	**1,3,5,7-Octatetraene, 1,8-diphenyl-**				
	O_2		Benzene	6.0×10^9	[8222]
	O_2		Cyclohexane	4.8×10^9	[8221]
	O_2		MeCN	6.4×10^9	[8222]
186	**Orotate ion**				
	O_2		H_2O (pH 6.3)	3.0×10^9	[7102]
187	**Orotic acid**				
	O_2		H_2O (pH 1.1)	2.2×10^9	[7102]
188	**1,3,4-Oxadiazole, 2,5-diphenyl-**				
	O_2		Benzene	1.6×10^9	[7714]
189	**Oxazole, 2,5-bis(4-biphenylyl)-**				
	O_2		Benzene	1.7×10^9	[7714]
190	**Oxazole, 2,5-diphenyl-**				
	O_2		Benzene	2.5×10^9	[7714]
	Anthracene	ET	Cyclohexane	4.7×10^9	[8016]
191	**Oxazole, 2-(1-naphthyl)-5-phenyl-**				
	O_2		Benzene	2.3×10^9	[7714]
192	**Oxirane, 2,3-di-(2-naphthyl)-, (Z)-**				
	O_2		Benzene	1.5×10^9	[8409]
	Azulene	ET	Benzene	9.7×10^9	[8409]
193	**Pentacene**				
	O_2		Benzene	1.7×10^9	[7307]
	Cr(hfac)$_3$	CT	Benzene	6.0×10^8	[8315]
194	**1,3-Pentadiene, 2-methyl-, (E)-**				
	β-Carotene	ET	Toluene	5.0×10^9	[8417]
	MV^{2+}	OT	MeOH	4×10^8	[8224]

Table 6c Triplet-State Quenching of Organic Molecules

No.	Quencher	Mechanism	Solvent	K_q (L mol^{-1} s^{-1})	Ref.
195	**Pentahelicene**				
	O$_2$		MCH	5.7×10^8	[7908]
196	**2-Pentanone**				
	O$_2$		Hexane	3.4×10^{10}	[8413]
	O$_2$		MeCN	3.2×10^{10}	[8413]
	O$_2$		MeOH	2.9×10^{10}	[8413]
197	**Perylene**				
	TEMPO		MeCN	2.1×10^6	[8014]
	Cr(acac)$_3$	ET	Benzene	4.7×10^7	[8126]
	Ferrocene	ET	EtOH	1.3×10^9	[7408]
	MV^{2+}	OT	MeCN/H$_2$O (9:1)	1.9×10^9	[8310]
198	**Phenalene, 2,3-dihydro-**				
	O$_2$		EPA	2×10^9	[8132]
199	**Phenanthrene**				
	O$_2$		Benzene	2.0×10^9	[7307]
	Azulene	ET	Benzene	6.6×10^9	[8123]
	Azulene	ET	MeCN	1.0×10^{10}	[8123]
	Cu(acac)$_2$	ET	MeOH	2.2×10^9	[8321]
	Cu^{2+}	OT	MeOH/H$_2$O (9:1)	1.0×10^8	[7811]
	tert-Butyl hydroperoxyde	RT	Benzene	2.3×10^7	[8309]
200	**Phenazine**				
	O$_2$		Benzene	2.0×10^9	[9401]
	Ferrocene	ET	EtOH	4.6×10^9	[7408]
201	**Phenol**				
	O$_2$		H$_2$O (pH 7.1)	6.1×10^9	[7511]
202	**Phenol, 4-methyl-**				
	O$_2$		H$_2$O (pH 7.5)	5.3×10^9	[7511]
203	**Phenothiazine**				
	O$_2$		MeOH	2.4×10^{10}	[7515]
	Cu^{2+}	OT	MeOH	6.0×10^9	[7515]

Table 6c Triplet-State Quenching of Organic Molecules

No.	Quencher	Mecha-nism	Solvent	K_q (L mol^{-1} s^{-1})	Ref.
204	**Phenothiazine, 10-methyl-**				
	Cu^{2+}	OT	H$_2$O/EtOH (2:1)	1.0×10^9	[7919]
205	**Phenothiazinium, 3-(dimethylamino)-7-(methylamino)-**				
	EDTA	RT	H$_2$O (pH 8.2)	5.5×10^7	[7405]
206	**Phenothiazinium, 3-(dimethylamino)-, conjugate monoacid**				
	EDTA	RT	H$_2$O (pH 4.5)	9×10^8	[7405]
207	**Phenoxazinium, 3,7-bis(diethylamino)-**				
	O$_2$		EtOH	8.3×10^8	[8223]
	1,3,5,7-Cyclooctatetraene	ET	EtOH	1.7×10^7	[8223]
208	**Phenoxazinium, 3,7-bis(diethylamino)-, conjugate monoacid**				
	O$_2$		EtOH	3.0×10^8	[8223]
	1,3,5,7-Cyclooctatetraene	ET	EtOH	8.7×10^6	[8223]
209	**Phenoxazinium, 3,7-diamino-**				
	Allylthiourea	RT	MeOH	1.8×10^6	[7616]
	EDTA	RT	H$_2$O (pH 8.0)	1.5×10^7	[7712]
210	**Phenylalanine**				
	O$_2$		H$_2$O (pH 7.5)	3.3×10^9	[7512]
211	**Phenylalanine, *N*-acetyl-**				
	O$_2$		H$_2$O (pH 8.1)	3.9×10^9	[7512]
212	**Phthalazine**				
	O$_2$		H$_2$O (pH 7.1)	1.4×10^9	[7514]
213	**Picene**				
	O$_2$		Benzene	1.4×10^9	[7307]
214	**Pivalophenone, 4'-methoxy-**				
	2,5-Dimethyl-2,4-hexadiene	ET	Benzene	5.0×10^9	[8517]
215	**Pivalothiophenone, 4-chloro-**				
	Di-*tert*-butylnitroxide		Benzene	2.4×10^9	[8703]
	Ferrocene	ET	Benzene	6.6×10^9	[8703]
	Triethylamine	RT	Benzene	7.8×10^6	[8703]

Table 6c Triplet-State Quenching of Organic Molecules

No.	Quencher	Mechanism	Solvent	K_q (L mol^{-1} s^{-1})	Ref.
216	**Pivalothiophenone, 4-methoxy-**				
	Di-*tert*-butylnitroxide		Benzene	2.0×10^9	[8703]
	Ferrocene	ET	Benzene	5.5×10^9	[8703]
	Triethylamine	RT	Benzene	4.1×10^6	[8703]
217	**Porphyrin, tetrakis(4-sulfonatophenyl)-**				
	O$_2$		H$_2$O (pH 7)	1.9×10^9	[8121]
	1,4-Benzoquinone	OT	H$_2$O (pH 7)	3.4×10^9	[8121]
	Nitrobenzene	OT	H$_2$O (pH 7)	3.2×10^6	[8121]
218	**Porphyrin, tetraphenyl-**				
	O$_2$		Cyclohexane	2.1×10^9	[0301]
	Fullerene C$_{60}$	ET	Benzene	3.5×10^7	[9101]
	Fullerene C$_{70}$	ET	Benzene	4.4×10^7	[9102]
219	**Porphyrin-2,18-dipropanoic acid, 7,12-diethenyl-3,8,13,17-tetramethyl-, dimethyl ester**				
	O$_2$		Benzene	2.7×10^9	[7706]
	β-Carotene	ET	Benzene	1.8×10^9	[7706]
	Tetracene	ET	Benzene	5×10^8	[7706]
220	**Porphyrin-2,18-dipropanoic acid, 3,7,12,17-tetramethyl-, dimethyl ester**				
	O$_2$		Benzene	2.3×10^9	[8015]
221	**Propiophenone**				
	Di-*tert*-butylnitroxide		Benzene	2.3×10^9	[7911]
	Cu(acac)$_2$	CT	MeOH	7.6×10^9	[8321]
	1-Methylnaphthalene	CT	Benzene	1.4×10^{10}	[8416]
	MV^{2+}	OT	MeCN/H$_2$O (9:1)	5.5×10^9	[8416]
222	**Propiophenone, 2,3-epoxy-4'-methoxy-3-phenyl-**				
	O$_2$		Benzene	3.9×10^9	[8514]
	Azulene	ET	Benzene	8.3×10^9	[8514]
	(*E*)-Stilbene	ET	Benzene	6.0×10^9	[8514]
	4-Methoxyphenol	RT	Benzene	4.9×10^9	[8514]

Table 6c Triplet-State Quenching of Organic Molecules

No. Quencher	Mecha-nism	Solvent	K_q $(L\,mol^{-1}\,s^{-1})$	Ref.
223 Psoralen				
O_2		H_2O (pH 8)	3.3×10^9	[8312]
(*all-E*)-Retinol	ET	EtOH	1.1×10^9	[7917]
224 Psoralen, 3-carbethoxy-				
O_2		H_2O	3.3×10^9	[8318]
(*all-E*)-Retinol	ET	EtOH	9.6×10^8	[8217]
Tryptophan	RT	H_2O	3.6×10^9	[8217]
225 Psoralen, 3-carbethoxy-4′,5′,-dihydro-				
O_2		H_2O	3×10^9	[8217]
Tryptophan	RT	H_2O	2.6×10^9	[8217]
226 Psoralen, 8-methoxy-				
O_2		H_2O	3.3×10^9	[7907]
Tryptophan		MeOH	3.5×10^8	[7912]
(*all-E*)-Retinol	ET	EtOH	5.3×10^9	[7917]
227 Psoralen, 4,5′,8-trimethyl-				
Tryptophan		MeOH	6.9×10^8	[7912]
228 4-Pteridinone, 2-amino-				
O_2		H_2O (pH 9.2)	1.3×10^9	[8122]
Tryptophan	RT	H_2O (pH 9.2)	4.9×10^9	[8122]
229 Pyranthrene				
O_2		Toluene	4.2×10^9	[8320]
Rubrene	ET	Toluene	2.3×10^9	[8128]
230 Pyrazine				
O_2		H_2O (pH 7.1)	3.2×10^9	[7409]
I^-	RT	H_2O (pH 7.1)	1.0×10^{10}	[7514]
231 Pyrene				
O_2		Benzene	2.2×10^9	[7307]
O_2		Cyclohexane	1.6×10^9	[7004]
TEMPO		MeCN	4.0×10^7	[8014]
1,4-Benzoquinone	CT	Benzene	9.9×10^9	[7906]

Table 6c Triplet-State Quenching of Organic Molecules

No.	Quencher	Mecha-nism	Solvent	K_q (L mol^{-1} s^{-1})	Ref.
231	**Pyrene** (continued)				
	Cu(acac)$_2$	CT	MeOH	1.9×10^9	[8321]
	2,5-Dimethyl-2,4-hexadiene	CT	Benzene	4.0×10^5	[8124]
	Ferrocene	CT	EtOH	6.0×10^9	[7408]
	MV^{2+}	OT	MeCN/H$_2$O (9:1)	7.0×10^9	[8310]
232	**1-Pyrenecarboxaldehyde**				
	O$_2$		Benzene	1.9×10^9	[8430]
	O$_2$		Cyclohexane	1.8×10^9	[8430]
	Azulene	ET	Benzene	8.6×10^9	[8422]
233	**Pyrimidine**				
	O$_2$		H$_2$O (pH 7.0)	4.6×10^9	[7514]
	2-Propanol	RT	H$_2$O (pH 7.0)	8.6×10^7	[7514]
234	**Pyruvic acid**				
	Ethanol	RT	Benzene	2.3×10^6	[7205]
235	**Pyruvic acid, ethyl ester**				
	1-Methylnaphthalene	ET	MeCN	8.0×10^9	[8608]
236	**Quinoxaline**				
	O$_2$		H$_2$O (pH 7.0)	4.6×10^9	[7514]
237	**(*all-E*)-Retinal**				
	O$_2$		Cyclohexane	3.7×10^9	[8430]
	O$_2$		MeOH	4.6×10^9	[8430]
	β-Carotene	ET	Toluene	6.2×10^9	[8418]
	Ferrocene	ET	Toluene	3.6×10^7	[8418]
	Tetracene	ET	Toluene	2.6×10^9	[8418]
238	**(*all-E*)-Retinoic acid**				
	O$_2$		MeOH	1.4×10^9	[8211]
239	**(*all-E*)-Retinol**				
	Di-*tert*-butylnitroxide		MCH	8.5×10^8	[8418]
	O$_2$		Hexane	4.7×10^9	[7304]
	Azulene	ET	Toluene/EtI	1×10^7	[8418]

Table 6c Triplet-State Quenching of Organic Molecules

No.	Quencher	Mechanism	Solvent	K_q (L mol^{-1} s^{-1})	Ref.
239	**(*all-E*)-Retinol** (continued)				
	β-Carotene	ET	Toluene/EtI	8.0×10^9	[8418]
	Ferrocene	ET	Toluene/EtI	6.5×10^7	[8418]
	Tetracene	ET	Toluene/EtI	2.4×10^9	[8418]
240	**Retinyl acetate**				
	O_2		MeOH	1.0×10^9	[8211]
241	**Rhodamine 6G cation**				
	O_2		EtOH	1.6×10^9	[7407]
	1,4-Benzoquinone	OT	H_2O (pH 6)	2.5×10^9	[7809]
	Ascorbic acid	RT	H_2O (pH 6)	8.0×10^8	[7809]
	p-Phenylenediamine	RT	H_2O (pH 6)	1.0×10^9	[7809]
242	**Riboflavin**				
	Aniline	RT	MeOH	4.3×10^9	[0303]
	DABCO	RT	MeOH	1.5×10^9	[0303]
	p-Dimethoxybenzene	RT	MeOH	3.2×10^9	[0303]
	Indole	RT	MeOH	5.1×10^9	[0303]
	TMPD	RT	MeOH	6.9×10^9	[0303]
	Triethylamine	RT	MeOH	3.4×10^8	[0303]
243	**Rubrene**				
	O_2		Toluene	3.4×10^9	[8216]
244	**Stilbene**				
	O_2		Benzene	9.0×10^9	[8123]
	O_2		MeCN	9.5×10^9	[8123]
	O_2		MeOH	8.6×10^9	[8123]
245	**(*E*)-Stilbene**				
	Anthracene	ET	Benzene	2.6×10^9	[7202]
	β-Carotene	ET	Toluene	3.8×10^9	[8417]
246	**(*Z*)-Stilbene**				
	O_2		Benzene	3.8×10^9	[7704]

Table 6c Triplet-State Quenching of Organic Molecules

No.	Quencher	Mechanism	Solvent	K_q (L mol^{-1} s^{-1})	Ref.
247	**(*E*)-Stilbene, 4,4'-dinitro-**				
	Ferrocene	ET	Benzene	9.1×10^9	[7921]
	DABCO	RT	MeCN	3.5×10^9	[8607]
248	**(*E*)-Stilbene, 4-nitro-**				
	Ferrocene	ET	Benzene	4.8×10^9	[7921]
	DABCO	RT	MeCN	3.6×10^9	[8607]
249	**Styrene, β-methyl-, (*E*)-**				
	MV^{2+}	OT	MeOH	6.5×10^8	[8225]
250	**Styrene, α-phenyl-**				
	O$_2$		Benzene	9.0×10^9	[8218]
	O$_2$		MeCN	6.9×10^9	[8218]
251	***p*-Terphenyl**				
	O$_2$		Benzene	1.2×10^9	[7816]
	2-Nitrothiophene	ET	Acetone	1.0×10^9	[8209]
	MV^{2+}	OT	MeCN/H$_2$O (9:1)	9.2×10^9	[8310]
252	**α-Terthienyl**				
	Anthracene	ET	Cyclohexane	1.7×10^9	[8613]
	MV^{2+}	OT	MeOH	6.8×10^9	[8613]
253	**Testosterone**				
	O$_2$		EtOH	2.2×10^9	[8012]
254	**Testosterone acetate**				
	Diphenylamine	RT	MeCN	3.9×10^9	[8804]
255	**Tetracene**				
	Di-*tert*-butylnitroxide		Hexane	3.5×10^7	[7308]
	O$_2$		Benzene	3.1×10^9	[7307]
	Ferrocene	ET	Benzene	3.6×10^7	[7508]
	Rubrene	ET	Toluene	2.0×10^9	[8128]
	MV^{2+}	OT	MeCN/H$_2$O (9:1)	1.5×10^8	[8310]

Table 6c Triplet-State Quenching of Organic Molecules

No. Quencher	Mechanism	Solvent	K_q (L mol^{-1} s^{-1})	Ref.
256 Thiobenzophenone				
2,5-Dimethyl-2,4-hexadiene		Benzene	4.2×10^8	[8407]
O$_2$		Benzene	2.9×10^9	[8407]
Azulene	ET	Benzene	4.2×10^9	[8407]
β-Carotene	ET	Benzene	1.1×10^{10}	[8407]
257 Thiobenzophenone, 4,4′-bis(dimethylamino)-				
2,5-Dimethyl-2,4-hexadiene		Benzene	1.7×10^8	[8407]
Azulene	ET	Benzene	9.3×10^9	[8407]
258 Thiobenzophenone, 4,4′-dimethoxy-				
2,5-Dimethyl-2,4-hexadiene		Benzene	2.7×10^8	[8407]
O$_2$		Benzene	3.7×10^9	[8407]
Azulene	ET	Benzene	4.6×10^9	[8407]
β-Carotene	ET	Benzene	1.0×10^{10}	[8407]
259 Thiocoumarin				
1,4-Cyclohexadiene		Benzene	1.1×10^9	[8610]
O$_2$		Benzene	3.0×10^9	[8610]
N,N-Dimethylaniline	RT	MeCN	1.0×10^9	[8610]
Tributylstannane	RT	Benzene	3.6×10^9	[8610]
260 Thioindigo				
O$_2$		Benzene	3.2×10^9	[7818]
261 Thionine cation				
O$_2$		H$_2$O (pH 8)	4.5×10^8	[7006]
Crystal Violet	ET	MeOH	1.9×10^8	[6904]
Allylthiourea	RT	MeOH	8.0×10^6	[7713]
Azulene	RT	MeOH	2.0×10^9	[7713]
DABCO	RT	MeOH	8.0×10^8	[7713]
Diphenylamine	RT	MeCN	5.4×10^9	[7814]
I$^-$	RT	H$_2$O (pH 7.1)	4.7×10^9	[7410]
L-Phenylalanine	RT	H$_2$O (pH 7.1)	2.2×10^7	[7410]
Triphenylamine	RT	MeOH	6.4×10^9	[7620]

Table 6c Triplet-State Quenching of Organic Molecules

No.	Quencher	Mechanism	Solvent	K_q (L mol^{-1} s^{-1})	Ref.
262	**Thionine cation, conjugate monoacid**				
	O$_2$		H$_2$O (pH 4.62)	2.6×10^8	[7006]
	Allylthiourea	RT	MeOH	7.0×10^8	[7713]
	Azulene	RT	MeOH	4.0×10^9	[7713]
263	**Thiophene, 2-nitro-**				
	O$_2$		Acetone	1.0×10^9	[8209]
	Azulene	ET	MeCN	1.4×10^{10}	[8209]
	CCl$_4$	OT	MeCN	4.4×10^6	[8209]
	I$^-$	RT	H$_2$O (pH 7)	8.6×10^9	[8210]
	1,3,5-Trimethoxybenzene	RT	MeCN	5.1×10^9	[8209]
264	**4-Thiouridine**				
	Cysteine		H$_2$O	4×10^8	[8319]
	I$^-$		H$_2$O	6×10^9	[8319]
	Methionine		H$_2$O	1.5×10^9	[8319]
	Tryptophan		H$_2$O	1.5×10^9	[8319]
265	**Thioxanthen-9-one**				
	Styrene		Benzene	3×10^9	[8115]
	1,1-Diphenylethylene	ET	Benzene	5.5×10^9	[8218]
	I$^-$	ET	MeCN/H$_2$O (3:2)	6.6×10^9	[8516]
	Ethyl 4-(dimethylamino)benzoate	RT	Benzene	6×10^9	[8115]
266	**Thioxanthione**				
	O$_2$		Benzene	2.8×10^9	[8407]
	Azulene	ET	Benzene	5.8×10^9	[8407]
	β-Carotene	ET	Benzene	1.0×10^{10}	[8407]
267	**Thioxanthylium, 3,6-bis(dimethylamino)-**				
	1,4-Benzoquinone	OT	H$_2$O (pH 7.2)	2.8×10^9	[8226]
	Allylthiourea	RT	MeOH	5.4×10^5	[8011]
	N,N-Dimethylaniline	RT	MeCN	8.2×10^9	[8011]
	EDTA	RT	H$_2$O (pH 7.2)	1.4×10^6	[8226]

Table 6c Triplet-State Quenching of Organic Molecules

No.	Quencher	Mechanism	Solvent	K_q (L mol^{-1} s^{-1})	Ref.
268	**Thymidine**				
	(*all-E*)-Retinol	ET	MeCN	6×10^9	[7914]
	Metronidazole	OT	MeCN	7.3×10^9	[8702]
269	**Thymidine 5'-monophosphate**				
	(*all-E*)-Retinol	ET	EtOH	2×10^9	[7914]
270	**Thymine**				
	(*all-E*)-Retinol	ET	MeCN	$\sim 6 \times 10^9$	[7516]
	Metronidazole	OT	MeCN	4.3×10^9	[8702]
	Tetracyanoethylene	OT	MeCN	7.6×10^9	[9501]
271	**1,3,5-Triazine**				
	O_2		MeCN	5.0×10^9	[7510]
	2-Propanol	RT	MeCN	1.4×10^8	[7510]
272	**Triphenylene**				
	Di-*tert*-butylnitroxide		Hexane	2.2×10^9	[7308]
	O_2		Benzene	1.1×10^9	[7307]
	Azulene	ET	Benzene	7×10^9	[8123]
	Cu(acac)$_2$	ET	Benzene	1.3×10^9	[6602]
	1,3-Cyclohexadiene	ET	Benzene	1.3×10^9	[6602]
	Ferrocene	ET	EtOH	6.5×10^9	[7408]
	Naphthalene	ET	Hexane	1.3×10^9	[6102]
	MV^{2+}	OT	MeCN/H$_2$O (9:1)	7.1×10^9	[8310]
273	**Triphenylethylene**				
	O_2		Benzene	4.8×10^9	[8218]
274	**Tryptamine**				
	O_2		H$_2$O (pH 7.5)	5.7×10^9	[7513]
275	**Tryptophan**				
	O_2		H$_2$O (pH 7.5)	5.0×10^9	[7513]
	Anthracene	ET	EtOH	4.0×10^9	[7513]
276	**Tryptophan, *N*-methyl-**				
	O_2		H$_2$O	5×10^9	[8805]

Table 6c Triplet-State Quenching of Organic Molecules

No.	Quencher	Mechanism	Solvent	K_q (L mol^{-1} s^{-1})	Ref.
277	L-Tryptophanamide, *N*-acetyl-				
	O_2		H_2O	5×10^9	[8805]
278	**Tyrosine**				
	O_2		H_2O (pH 7.5)	4.8×10^9	[7511]
	Tryptophan	ET	H_2O (pH 7.3)	6.0×10^9	[7511]
	Cysteine	OT	H_2O (pH 7.0)	5.2×10^8	[7511]
	Hydroxide ion	RT	H_2O (pH 7-10.3)	2.0×10^{10}	[7511]
279	**Uracil**				
	(*all-E*)-Retinol	ET	MeCN	~8×10^9	[7516]
	Fumaronitrile	OT	MeCN	5.3×10^9	[8702]
	Metronidazole	OT	MeCN	1.3×10^{10}	[8702]
	Tetracyanoethylene	OT	MeCN	1.7×10^{10}	[9501]
280	**Uridine**				
	(*all-E*)-Retinol	ET	MeCN	6×10^9	[7914]
281	**Uridine 5'-monophosphate**				
	O_2		H_2O	3×10^9	[7914]
	(*all-E*)-Retinol	ET	EtOH	5×10^9	[7914]
282	**9-Xanthione**				
	O_2		Benzene	2.8×10^9	[8407]
	Azulene	ET	Benzene	8.2×10^9	[8407]
	Ethanol	RT	EtOH	1.1×10^4	[7918]
	Triethylamine	RT	Benzene	2.1×10^8	[8703]
283	**Xanthone**				
	O_2		Benzene	5.6×10^9	[7612]
	Azulene	ET	Benzene	8.5×10^9	[8130]
	Biphenyl	ET	MeOH	8.8×10^9	[8517]
	I$^-$	ET	MeCN/H_2O (3:2)	7.1×10^9	[8516]
	Naphthalene	ET	Benzene	9.5×10^9	[7612]
	(*E*)-Stilbene	ET	Benzene	7×10^9	[8123]
	Toluene	CT	MeCN	5.8×10^6	[0001]

Table 6c Triplet-State Quenching of Organic Molecules

No.	Quencher	Mechanism	Solvent	K_q (L mol^{-1} s^{-1})	Ref.
283	**Xanthone** (continued)				
	p-Xylene	CT	MeCN	5.9×10^7	[0001]
	Cyclohexane	RT	CCl$_4$	7.7×10^6	[8010]
	Indole	RT	Benzene	1.1×10^{10}	[7807]
	2-Propanol	RT	CCl$_4$	1.1×10^8	[8010]
	Tributylstannane	RT	CCl$_4$	1.5×10^9	[8010]

Table 6d Excited-State Quenching of Transition Metal Complexes

No.	Quencher	Mechanism	Solvent	T (°C)	K_q (L mol^{-1} s^{-1})	Ref.
1	**[Al(phthalocyanine)Cl]**					
	Ferrocene	RT	DMSO/H$_2$O (9:1)	~28	6.6×10^8	[8301]
	1,4-Benzoquinone	OT	DMA/H$_2$O (2.3:1)	~15	9.2×10^6	[8301]
	Phenotiazine	RT	DMSO/H$_2$O (4:1)	~28	1×10^6	[8301]
2	**[Cr(bpy)$_3$]$^{3+}$**					
	Ferrocene	RT	MeCN/H$_2$O (2.3:1)	~15	6.2×10^9	[8401]
	I$^-$	RT	H$_2$O, $\mu = 1.0$ (NaCl)	22	1.2×10^9	[7801]
	O$_2$		H$_2$O, $\mu = 1.0$ (NaCl)	22	1.7×10^7	[7801]
	[Ru(bpy)$_3$]$^{2+}$	RT	H$_2$O, $\mu = 0.2$	23	4.0×10^8	[7501]
	Aniline	RT	MeCN, 0.02 mol L^{-1} TEAP	~22	9.9×10^9	[7802]
	Triethylamine	RT	MeCN, 0.02 mol L^{-1} TEAP	~22	7.4×10^8	[7802]
3	**[Cr(phen)$_3$]$^{3+}$**					
	I$^-$	RT	H$_2$O, pH 10.5, 1 mol L^{-1} NaCl	15	2.8×10^9	[8302]
	O$_2$		H$_2$O, 1 mol L^{-1} HCl	25	4.9×10^7	[7803]
	[Ru(bpy)$_3$]$^{2+}$	RT	H$_2$O, 1 mol L^{-1} H$_2$SO$_4$	25	8.3×10^8	[7803]
4	**[Cu(2,9-Ph$_2$phen)$_2$]$^+$**					
	Anthracene-9-carboxylate ion	ET	EtOH	40	1×10^9	[8402]
	MV^{2+}	OT	EtOH	40	3×10^9	[8402]
5	**Eu^{3+}**					
	[Fe(CN)$_6$]$^{4-}$	RT	H$_2$O, 1 mol L^{-1} KCl	20	6.0×10^8	[8601]
	Acridine	ET	Acetone	20	1.2×10^6	[7001]
6	**[Eu(crypt)]$^{3+}$**					
	[Fe(CN)$_6$]$^{4-}$		H$_2$O, 1 mol L^{-1} KCl	20	5.0×10^8	[8601]
7	**[Eu(phen)$_3$]$^{3+}$**					
	Anthracene	ET	Acetone	20	2×10^6	[7001]
8	**[Ir(bpy)$_2$(C^3,N'-Hbpy)$_3$]$^+$**					
	O$_2$	ET	MeOH		3.4×10^8	[7701]
	Biacetyl	ET	H$_2$O, 0.05 mol L^{-1} HClO$_4$	23	2.6×10^8	[7901]

Table 6d Excited-State Quenching of Transition Metal Complexes

No.	Quencher	Mechanism	Solvent	T (°C)	K_q (L mol^{-1} s^{-1})	Ref.
8	**[Ir(bpy)$_2$(C^3,N'-Hbpy)$_3$]$^+$** (continued)					
	MV^{2+}	OT	H$_2$O, pH 1.4	20	1.3×10^6	[8602]
9	**[Ir(phen)$_3$]$^{3+}$**					
	O$_2$	ET	MeOH		2.8×10^8	[7701]
10	**[Os(bpy)$_3$]$^{2+}$**					
	Fe^{3+}	OT	H$_2$O, 5 mol L^{-1} H$_2$SO$_4$		5.5×10^9	[8001]
	O$_2$		H$_2$O	25	5.2×10^9	[7601]
11	**[Os(phen)$_3$]$^{2+}$**					
	Fe^{3+}	OT	H$_2$O, 5 mol L^{-1} H$_2$SO$_4$		4.0×10^9	[8001]
	O$_2$	ET	MeOH		5.7×10^9	[7701]
	MV^{2+}	OT	MeCN, 0.1 mol L^{-1} LiClO$_4$	~23	3.1×10^9	[8501]
	Phenotiazine	RT	MeCN, 0.1 mol L^{-1} LiClO$_4$	~23	8.3×10^6	[8501]
12	**[Os(terpy)$_2$]$^{2+}$**					
	3-Methoxy-phenothiazine	RT	MeCN, 0.1 mol L^{-1} LiClO$_4$	~23	9.5×10^8	[8501]
13	**[Pd(octaethylporphyrin)]**					
	MV^{2+}	OT	Butyronitrile		1.5×10^9	[7602]
14	**[Pd(tetraphenylporphyrin)]**					
	1,4-Benzoquinone	OT	EtOH		2.6×10^9	[8303]
15	**[Pt(8-hydroxyquinolinate)$_2$]**					
	1,4-Benzoquinone	OT	MeCN	25	2.1×10^{10}	[8603]
	MV^{2+}	OT	MeCN 0.01 mol L^{-1} TEAP	25	1.1×10^{10}	[8603]
16	***fac*-[Re(CO)$_3$(phen)Cl]**					
	Aniline	RT	MeCN 0.01 mol L^{-1} TBAP	25	5.8×10^7	[7804]
	MV^{2+}	OT	MeCN 0.01 mol L^{-1} TBAP	25	3.1×10^9	[7804]
17	**[Rh(4,7-Ph$_2$phen)$_3$]$^{3+}$**					
	1,2-Diaminobenzene	RT	MeCN/H$_2$O		8.1×10^9	[8502]

Table 6d Excited-State Quenching of Transition Metal Complexes

No.	Quencher	Mechanism	Solvent	T (°C)	K_q (L mol^{-1} s^{-1})	Ref.
18	**[Rh(phen)$_3$]$^{3+}$**					
	Fe^{2+}	RT	H$_2$O, $\mu = 1$		4.0×10^9	[8403]
	1,4-Dimethoxybenzene	RT	H$_2$O		4.0×10^9	[8502]
19	**[Rh(tetraphenylporphyrin)Cl]**					
	MV^{2+}	OT	EtOH		3.0×10^9	[8604]
20	**[Ru(4,4$'$-Me$_2$bpy)$_3$]$^{2+}$**					
	Fe^{3+}	OT	H$_2$O, 0.5 mol L^{-1} H$_2$SO$_4$	25	2.9×10^9	[7603]
	O$_2$	OT	H$_2$O	25	4.2×10^9	[7603]
	[Rh(bpy)$_3$]$^{2+}$	OT	H$_2$O, 0.5 mol L^{-1} H$_2$SO$_4$	25	1.1×10^9	[8101]
	DQ^{2+}	OT	H$_2$O 0.1 mol L^{-1} KCl	25	2.5×10^9	[8503]
21	**cis-[Ru(bpy)$_2$(CN)$_2$]**					
	O$_2$	ET	H$_2$O	~21	4.5×10^9	[7301]
	MV^{2+}	OT	H$_2$O, $\mu = 0.024$ (NaCl)	~23	5.3×10^9	[7902]
22	**[Ru(bpy)$_2$(4,4$'$(-COOH)$_2$bpy)]$^{2+}$**					
	MV^{2+}	OT	H$_2$O, 0.1 mol L^{-1} HCl, $\mu = 0.137$ (NaCl)	~23	6.1×10^8	[7902]
23	**[Ru(bpy)$_2$(4,4$'$(-COO)$_2$bpy)]**					
	MV^{2+}	OT	H$_2$O, $\mu = 0.029$ (NaCl)	~23	2.5×10^9	[7902]
24	**[Ru(phen)$_3$]$^{2+}$**					
	I$^-$		H$_2$O, $\mu = 0$ (calc'd)	~21	1×10^6	[7604]
	O$_2$	ET	H$_2$O	~21	4.7×10^9	[7301]
	Ascorbate ion	RT	H$_2$O, pH 5, $\mu = 0.7$	25	2.3×10^8	[8201]
	N,N'-Dibenzyl-4,4$'$-bipyridinium	OT	H$_2$O		3.1×10^9	[7805]
	MV^{2+}	OT	MeCN, 0.01 mol L^{-1} TBAP		2.9×10^9	[7602]
	MV^{2+}	OT	H$_2$O		2.4×10^9	[7805]
25	**cis-[Ru(phen)$_2$(CN)$_2$]**					
	O$_2$	ET	H$_2$O	~21	5.5×10^9	[7301]

Table 6d Excited-State Quenching of Transition Metal Complexes

No.	Quencher	Mechanism	Solvent	T (°C)	K_q (L mol^{-1} s^{-1})	Ref.
26	**[Ru(phthalocyanine)(DMF)$_2$]**					
	Fe^{3+}		MeCN/H$_2$O, 0.01 mol L^{-1} TEAP		7.0×10^9	[8304]
	N,N'-Dibenzyl-4,4'-bipyridinium	OT	MeCN, 0.01 mol L^{-1} TEAP		1.7×10^9	[8304]
26	**[Ru(phthalocyanine)(pyridine)$_2$]**					
	Fe^{3+}	OT	MeCN/H$_2$O, 0.01 mol L^{-1} TEAP		7.4×10^9	[8304]
	MV^{2+}	OT	MeCN, 0.01 mol L^{-1} TEAP		2.0×10^9	[8304]
27	**[Ru(tetraphenylporphyrin)(CO)]**					
	MV^{2+}	OT	DMSO	~23	6.6×10^8	[8101]
28	**[Ru(bpy)$_3$]$^{2+}$**					
	[Co(NH$_3$)$_6$]	ET, OT	H$_2$O, pH 4.7, $\mu = 0.01$	25	1.5×10^7	[8504]
	[Co(sep)$_3$]$^+$	OT	H$_2$O, pH 5, $\mu = 0.1$	20	1.5×10^8	[8505]
	[Cr(bpy)$_3$]$^{3+}$	OT	H$_2$O, $\mu = 0.1$	23	3.3×10^9	[7501]
	trans-[Cr(cyclam)(CN)$_2$]$^+$	ET	H$_2$O, 1 mol L^{-1} CF$_3$SO$_3$Na	15	2.4×10^7	[8605]
	Cu^{2+}	OT	H$_2$O, pH 0.6, $\mu = 0.8$	25	6.2×10^7	[8002]
	Eu^{3+}	OT	H$_2$O, pH 0.3, $\mu = 2.8$	25	3.6×10^5	[8002]
	Fe^{3+}	OT	H$_2$O, 0.1 mol L^{-1} HClO$_4$	18	1.5×10^9	[7903]
	Ferrocene	ET	EtOH	25	5.9×10^9	[7502]
	[Fe(CN)$_6$]$^{3-}$	OT	H$_2$O 0.5 mol L^{-1} NaCl	25	6.5×10^9	[7601]
	[Fe(CN)$_6$]$^{3-}$	RT	H$_2$O, $\mu = 0.1$ (NaCl)	25	7.2×10^9	[8103]
	Hg^{2+}	OT	H$_2$O 0.5 mol L^{-1} HClO$_4$		1.5×10^8	[8202]
	I$^-$		H$_2$O, $\mu = 0$ (calc'd)	~21	1×10^6	[7604]
	O$_2$	ET	MeCN, 0.1 mol L^{-1} TBAC		6.8×10^8	[8506]
	O$_2$	OT?	H$_2$O, pH 4.6	25	3.2×10^9	[8104]
	O$_2$	ET	MeOH	~21	1.7×10^9	[7301]
	[Os(bpy)$_3$]$^{2+}$	ET	H$_2$O, $\mu = 0.1$ (NaCl)	25	1.5×10^9	[8003]
	[PtCl$_4$]$^{2-}$	ET	H$_2$O, $\mu = 2$ (HClO$_4$)		1.4×10^9	[8305]

Table 6d Excited-State Quenching of Transition Metal Complexes

No.	Quencher	Mechanism	Solvent	T (°C)	K_q (L mol^{-1} s^{-1})	Ref.
28	**[Ru(bpy)$_3$]$^{2+}$** (continued)					
	[Rh(bpy)$_3$]$^{3+}$	OT	H$_2$O, 0.5 mol L^{-1} H$_2$SO$_4$	25	6.2×10^8	[8101]
	Anthracene	ET	Benzene/EtOH (15:1)	25	2.2×10^9	[7302]
	Anthracene-9-carboxylate ion	ET	H$_2$O, pH 5		5.0×10^9	[8306]
	1,4-Benzoquinone	OT	H$_2$O, pH 6.9, $\mu = 0.04$		3.7×10^9	[8105]
	N,N'-Dibenzyl-4,4'-bipyridinium	OT	H$_2$O, pH 5, $\mu = 0.1$	20	1.4×10^9	[8505]
	DQ^{2+}	OT	H$_2$O, pH 5	22	1.7×10^9	[8203]
	4-Methoxyphenol	RT	EtOH/H$_2$O (3:1), 0.002 mol L^{-1} NaOH	23	5.0×10^9	[8204]
	Nile Blue A	ET	MeOH	22	2.5×10^9	[8307]
	Phenothiazine	RT	MeOH		5.6×10^9	[7702]
	Pirydine		H$_2$O, pH 4-5, $\mu = 0.50$	25	$<1 \times 10^7$	[7703]
	PirydineH$^+$	OT	H$_2$O, pH 5.8, $\mu = 0.1$	25	$<1 \times 10^7$	[8004]
	Rhodamine 101	ET	MeOH	22	8.5×10^8	[8307]
	Tetrathiafulvalene	RT	MeOH		1.1×10^{10}	[8205]
	MV^{2+}	OT	H$_2$O, $\mu = 0.52$ (NaCl)	~23	2.0×10^9	[7902]
29	**Tb^{3+} [^{5}D$_4$]**					
	Acridine	ET	Acetone	20	1.6×10^6	[7001]
	Anthracene	ET	MeCN	20	1.4×10^5	[7503]
	Eosin	ET	H$_2$O	20	2.8×10^8	[7303]
	Fuchsin	ET	H$_2$O	20	8.6×10^7	[7303]
30	**UO$_2$$^{2+}$**					
	Ag$^+$	RT	H$_2$O, $\mu = 0.1$	25	1.6×10^9	[8005]
	Br$^-$	RT	H$_2$O, 1 mol L^{-1} HClO$_4$	25	4.8×10^9	[7605]
	Cl$^-$	RT	H$_2$O, 1 mol L^{-1} HClO$_4$	25	1.7×10^9	[7605]
	I$^-$	RT	H$_2$O, 1 mol L^{-1} HClO$_4$	25	6.9×10^9	[7605]
	Methanol	RT	H$_2$O, 1 mol L^{-1} H$_2$SO$_4$		1.5×10^9	[7606]
31	**[Zn(octaethylporphyrin)]** [singlet]					
	9,10-Anthraquinone		Toluene		2.0×10^{10}	[8404]
	1,4-Benzoquinone	OT	MeCN		1.8×10^{10}	[8404]

Table 6d Excited-State Quenching of Transition Metal Complexes

No.	Quencher	Mechanism	Solvent	T (°C)	K_q (L mol^{-1} s^{-1})	Ref.
32	**[Zn(octaethylporphyrin)]** [triplet]					
	1,4-Benzoquinone	OT	Toluene	24	9.0×10^9	[8206]
33	**[Zn(5-phenyl-10,15,20-tris(4-sulfonatophenyl)porphyrin)]$^{3-}$**					
	MV^{2+}	OT	H$_2$O, pH 7.0		1.5×10^{10}	[8507]
34	**[Zn(phthalocyanine)]**					
	Ferrocenium ion	OT	DMA/H$_2$O (2.3:1), 0.01 mol L^{-1} HClO$_4$, μ=1-1.5		1.0×10^9	[8207]
	MV^{2+}	OT	DMF/H$_2$O (9:1)		5.5×10^7	[8405]
35	**[Zn(tetraphenylporphyrin)]** [singlet]					
	Eu^{3+}	OT, ET?	MeCN		1.6×10^{10}	[7904]
	1,4-Benzoquinone	EX	Toluene		2.2×10^9	[7905]
36	**[Zn(tetraphenylporphyrin)]** [triplet]					
	O$_2$	OT	H$_2$O, pH 9.2	20	1.5×10^9	[8606]
	MV^{2+}	OT	EtOH		7×10^8	[8006]

REFERENCES

[5401] Eigen, M. *Z. Physik. Chem.* **1954,** *1,* 176-200.
[5701] Fujimori, E.; Livingston, R. *Nature (London)* **1957,** *180,* 1036-1038.
[6001] Lindqvist, L. *Ark. Kemi,* **1960,** *16,* 79-138.
[6101] Grossweiner, L. I.; Zwicker, E. F. *J. Chem. Phys.* **1961,** *34,* 1411-1417.
[6102] Porter, G.; Wilkinson, F. *Proc. R. Soc. London, Ser. A* **1961,** *264,* 1-18.
[6201] Ware, W. R. *J. Phys. Chem.* **1962,** *66,* 455-458.
[6401] Kasche, V.; Lindqvist, L. *J. Phys. Chem.* **1964,** *68,* 817-823.
[6402] Sandros. K. *Acta Chem. Scand.* **1964,** *18,* 2355-2374.
[6501] Shakhverdov, P. A.; Terenin, A. N. *Dokl. Phys. Chem.* **1965,** *160,* 163-165.
[6601] Chessin, M.; Livingston, R.; Truscott, T. G. *Trans. Faraday Soc.* **1966,** *62,*1519-1524.
[6602] Fry, A. J.; Liu, R. S. H.; Hammond. G. S. *J. Am. Chem. Soc.* **1966,** *88,* 4781-4782.
[6701] Usui, Y.; Koizumi, M. *Bull. Chem. Soc. Jpn.* **1967,** *40,* 440-446.
[6801] Chrysochoos, J.; Grossweiner, L. I. *Photochem. Photobiol.* **1968,** *8,* 193-208.
[6802] Stevens, B.; Algar, B. E. *J. Phys. Chem.* **1968,** *72,* 2582-2587.
[6803] Knibbe, H.; Rehm, D.; Weller, A. *Ber. Bunsenges. Phys. Chem.* **1968,** *72,* 257-263.
[6901] Clark, W. D. K.; Litt, A. D.; Steel, C. *J. Am. Chem. Soc.* **1969,** *91,* 5413-5415.
[6902] Turro, N. J.; Engel, R. *Mol. Photochem.* **1969,** *1,* 143-146.
[6903] Turro, N. J.; Engel, R. *Mol. Photochem.* **1969,** *1,* 235-238.
[6904] Kramer, H. E. A. *Z. Phys. Chem. (Frank-furt Am Main)* **1969,** *66,* 73-85.
[6905] Chibisov, A. K. *Photochem. Photobiol.* **1969,** *10,* 331-347.
[6906] Turro, N. J.; Engel, R. *J. Am. Chem. Soc.* **1969,** *91,* 7113-7121.
[6907] Sykes, A.; Truscott, T.G. *J. Chem. Soc. D* **1969,** 929-930.
[6908] Caldwell, R. A., *Tetrahedron Lett.* **1969,** 2121-2124.
[7001] Ermolaev, V. L.; Tachin, V. S. *Opt. Spectrosc. (USSR)* **1970,** *29,* 49-52.
[7002] Turro, N. J.; Lee, T. -J., *Mol. Photochem* **1970,** *2,* 185-190.
[7003] Algar, B. E.; Stevens, B. *J. Phys. Chem.* **1970,** *74,* 3029-3034.
[7004] Patterson, L. K.; Porter, G.; Topp, M. R. *Chem. Phys. Lett.* **1970,** *7,* 612-614.
[7005] Rehm, D.; Weller, A. *Isr. J. Chem.* **1970,** *8,* 259-271.
[7006] Fischer, H.; Kramer, H. E. A.; Maute, A. *Z. Phys. Chem. (Frankfurt Am Main)* **1970,** *69,* 113-131.
[7007] Porter, G.; Topp, M. R. *Proc. R. Soc. London, Ser. A* **1970,** *315,* 163-184.
[7008] Wettack, F. S.; Renkes, G. D.; Rockley, M. G.; Turro, N. J.; Dalton, J. C. *J. Am. Chem. Soc.* **1970,** *92,* 1793-1794.
[7009] Birks, J. B.; Leite, M. S. S. C. P. *J. Phys. B* **1970,** *3,* 417-424.
[7010] Kelly, J. M.; Porter, G. *Proc. R. Soc. London, Ser. A* **1970,** *319,* 319-329.
[7101] Kemp, D. R.; Porter, G. *Proc. R. Soc. London, Ser. A* **1971,** *326,* 117-130.
[7102] Herbert, M. A.; Johns, H. E. *Photochem. Photobiol.* **1971,** *14,* 693-704.
[7201] Hulme, B. E.; Land, E. J.; Phillips, G. O. *J. Chem. Soc., Faraday Trans. 1* **1972,** *68,* 2003-2012.
[7202] Dainton, F. S.; Robinson, E. A.; Salmon, G. A. *J. Phys. Chem.* **1972,** *76,* 3897-3904.
[7203] Tetreau, C.; Lavalette, D.; Land, E. J.; Peradejordi, F. *Chem. Phys. Lett.* **1972,** *17,* 245-247.
[7204] Yip, R. W.; Loutfy, R. O.; Chow, Y. L.; Magdzinski, L. K. *Can. J. Chem.* **1972,** *50,* 3426-3431.

[7205] Ayscough, P. B.; Sealy, R. C. *J. Photochem.* **1972,** *1,* 83-85.
[7206] Foerster, E. W.; Grellmann. K. H. *Chem. Phys. Lett.* **1972,** *14,* 536-538.
[7301] Demas, J. N.; Diemente, D.; Harris, E.W. *J. Am. Chem. Soc.* **1973,** *95,* 6864-6865.
[7302] Wrighton, M., Markham, J. *J. Phys. Chem.* **1973,** *77,* 3042-3044.
[7303] Shakhverdov, T. A.; Bodunov, E. N. *Opt. Spectrosc. (USSR)* **1973,** *34,* 646-650.
[7304] Truscott, T. G.; Land, E. J.; Sykes, A. *Photachem. Photobiol.* **1973,** *17,* 43-51.
[7305] Lakowicz, J. R.; Weber, G. *Biochemistry* **1973,** *12,* 4171-4179.
[7306] Watkins, A. R. *J. Phys. Chem.* **1973,** *77,* 1207-1210.
[7307] Gijzeman, O. L. J.; Kaufman, F.; Porter, G. *J. Chem. Soc., Faraday Trans. 2,* **1973,** *69,* 708-720.
[7308] Gijzeman, O. L. J.; Kaufman, F.; Porter, G. *J. Chem. Soc., Faraday Trans. 2,* **1973,** *69,* 727-737.
[7309] Nakashima, N.; Mataga, N.; Yamanaka, C. *Int. J. Chem. Kinet.* **1973,** *5,* 833-839.
[7310] Kikuchi, K.; Watarai, H.; Koizumi, M. *Bull. Chem. Soc. Jpn.* **1973,** *46,* 749-754.
[7311] Lutz, H.; Breheret, E.; Lindqvist, L. *J. Phys. Chem.* **1973,** *77,* 1758-1762.
[7312] Porter, G.; Dogra, S. K.; Loutfy, R. O.; Sugamori, S. E.; Yip, R. W. *J. Chem. Soc., Faraday Trans. 1* **1973,** *69,* 1462-1474.
[7313] Mathis, P.; Kleo, J. *Photochem. Photobiol.* **1973,** *18,* 343-346.
[7314] Sandros, K. *Acta Chem. Scand.* **1973,** *27,* 3021-3032.
[7401] Rosenfeld, T.; Alchalel, A.; Ottolenghi, M. *J. Phys. Chem.* **1974,** *78,* 336-341.
[7402] Giering, L.; Berger, M.; Steel, C. *J. Am. Chem. Soc.* **1974,** *96,* 953-958.
[7403] Watkins, A. R. *J. Phys. Chem.* **1974,** *78,* 1885-1890.
[7404] Watkins, A. R. *J. Phys. Chem.* **1974,** *78,* 2555-2558.
[7405] Bonneau, R.; Fornier de Violet, Ph.; Joussot-Dubien, J. *Photochem. Photobiol.* **1974,** *19,* 129-132.
[7406] Dempster, D. N.; Morrow, T.; Quinn, M. F. *J. Photochem.* **1974,** *2,* 329-341.
[7407] Dempster, D. N.; Morrow, T.; Quinn, M. F. *J. Photochem.* **1974,** *2,* 343-359.
[7408] Kikuchi, M.; Kikuchi, K.; Kokubun, H. *Bull. Chem. Soc. Jpn.* **1974,** *47,* 1331-1333.
[7409] Bent, D. V.; Hayon, E.; Moorthy, P. N. *Chem. Phys. Lett.* **1974,** *27,* 544-547.
[7410] Kraljic, I.; Lindqvist, L. *Photochem. Photobiol.* **1974,** *20,* 351-355.
[7411] Chamey, D. R.; Dalton, J. C.; Hautala, R. R.; Snyder, J. J.; Turro, N. J. *J. Am. Chem. Soc.* **1974,** *96,* 1407-1410.
[7501] Bolletta, F.; Maestri, M.; Moggi, L.; Balzani, V. *J. Chem. Soc. Chem. Commun* **1975,** 901-902.
[7502] Wrighton, M. S.; Psunsap, L.; Morse, D. L. *J. Phys. Chem.* **1975,** *79,* 96-71.
[7503] Ermolaev, V. L.; Tachin, V. S. *Opt. Spectrosc. (USSR)* **1975,** *38,* 656-657.
[7504] Brede, O.; Helmstreit, W.; Mehnert, R. *Z. Phys. Chem. (Leipzig)* **1975,** *256,* 505-512.
[7505] Post, M. F. M.; Langelaar, J.; Van Voorst, J. D. W. *Chem. Phys. Lett.* **1975,** *32,* 59-62.
[7506] Marcondes, M. E. R.; Toscano, V. G.; Weiss, R. G. *J. Am. Chem. Soc.* **1975,** *97,* 4485-4490.
[7507] Wamser, C. C.; Medary, R. T.; Kochevar, I. E.; Turro, N. J.; Chang, P. L. *J. Am. Chem. Soc.* **1975,** *97,* 4864-4869.
[7508] Farmilo, A.; Wilkinson, F. *Chem. Phys. Lett.* **1975,** *34,* 575-580.
[7509] Bortolus, P.; Dellonte, S. *J. Chem. Soc., Faraday Trans. 2* **1975,** *71,* 1338-1342.
[7510] Bent, D. V.; Hayon, E. *Chem. Phys. Lett.* **1975,** *31,* 325-327.
[7511] Bent, D. V.; Hayon, E. *J. Am. Chem. Soc.* **1975,** *97,* 2599-2606.
[7512] Bent, D. V.; Hayon, E. *J. Am. Chem. Soc.* **1975,** *97,* 2606-2612.

[7513] Bent, D. V.; Hayon, E. *J. Am. Chem. Soc.* **1975,** *97,* 2612-2619.
[7514] Bent, D. V.; Hayon, E.; Moorthy, P. N. *J. Am. Chem. Soc.* **1975,** *97,* 5065-5071.
[7515] Alkaitis, S. A.; Beck, G.; Graetzel, M. *J. Am. Chem. Soc.* **1975,** *97,* 5723-5729.
[7516] Salet, C.; Bensasson, R. *Photochem. Photobiol.* **1975,** *22,* 231-235.
[7601] Lin, C. -T.; Sutin, N. *J. Phys. Chem.* **1976,** *80,* 97-105.
[7602] Young, R. C.; Meyer, T. J.; Whitte, D. G. *J. Am. Chem. Soc.* **1976,** *98,* 286-287.
[7603] Lin, C. -T.; Boettcher, W.; Chou, M.; Creutz, C.; Sutin, N. *J. Am. Chem. Soc.* **1976,** *98,* 6536-6544.
[7604] Demas, J. N.; Addington, J. W. *J. Am. Chem. Soc.* **1976,** *98,* 5800-5806.
[7605] Yokoyama, Y.; Shepherd, T. M. *J. Chem. Soc. Faraday Trans. 2* **1976,** *72,* 557-564.
[7606] Sergeeva, G. I.; Chibisov, A. K.; Levshin, L. V.; Karyakin, A. V. *J. Photochem.* **1976,** *5,* 253-264.
[7607] Zador, E.; Warman, J. M.; Hummel, A. *J. Chem. Soc., Faraday Trans. 1* **1976,** *72,* 1368-1376.
[7608] Bensasson, R.; Land, E. J.; Lafferty, J.; Sinclair, R. S.; Truscott, T. G. *Chem. Phys. Lett.* **1976,** *41,* 333-335.
[7609] Land, E. J. *Photochem. Photobiol.* **1976,** *24,* 475-477.
[7610] Wallace, W. L.; Van Duyne, R. P.; Lewis, F. D. *J. Am. Chem. Soc.* **1976,** *98,*5319-5326.
[7611] Wilkinson, F.; Farmilo, A. *J. Chem. Soc., Faraday Trans. 2* **1976,** *72,* 604-618.
[7612] Garner, A.; Wilkinson, F. *J. Chem. Soc., Faraday Trans. 2* **1976,** *72,* 1010-1020.
[7613] Alkaitis, S. A.; Graetzel, M. *J. Am. Chem. Soc.* **1976,** *98,* 3549-3554.
[7614] Bensasson, R.; Salet, C.; Balzani, V. *J. Am. Chem. Soc.* **1976,** *98*,3722-3724.
[7615] Treinin, A.; Hayon, E. *J. Am. Chem. Soc.* **1976,** *98,* 3884-3891.
[7616] Vogelmann, E.; Kramer, H. E. A. *Photochem. Photobiol.* **1976,** *23,* 383-390.
[7617] Capellos, C.; Suryanarayanan, K. *Int. J. Chem. Kinet.* **1976,** *8,* 529-539.
[7618] Capellos, C.; Suryanarayanan, K. *Int. J. Chem. Kinet.* **1976,** *8,* 541-548.
[7619] Kayser, R. H.; Young, R. H. *Photochem. Photobiol.* **1976,** *24,* 395-401.
[7620] Vogelmann, E.; Schreiner, S.; Rauscher, W.; Kramer, H. E. A. *Z. Phys. Chem. (Frankfurt Am Main)* **1976,** *101,* 321-336.
[7701] Demas, J. N.; Harris, E. W.; McBride, R. P. *J. Am. Chem. Soc.* **1977,** *99,* 3457-3451.
[7702] Maestri, M.; Graetzel, M. *Ber. Bunsenges. Phys.. Chem.* **1977,** *81,* 504-507.
[7703] Toma, H. E.; Creutz, C. *Inorg. Chem.* **1977,** *16,* 545-550.
[7704] Garner, A.; Wilkinson, F. *Chem. Phys. Lett.* **1977,** *45,* 432-435.
[7705] Wilkinson, F.; Garner, A. *J. Chem. Soc., Faraday Trans. 2* **1977,** *73,* 222-233.
[7706] Chantrell, S. J.; McAuliffe, C. A.; Munn, R. W.; Pratt, A. C.; Land, E. J. *J. Chem. Soc., Faraday Trans. 1* **1977,** 858-865.
[7707] Archer, M. D.; Ferreira, M. I. C.; Porter, G.; Tredwell, C. J. *Nouv. J. Chim.* **1977,** *1,* 9-12.
[7708] Capellos, C.; Suryanarayanan, K. *Int. J. Chem. Kinet.* **1977,** *9,* 399-407.
[7709] Loutfy, R. O.; Yip, R. W.; Dogra, S. K.,*Tetrahedron Lett.* **1977,** 2843-2846.
[7710] Aloisi, G. G.; Mazzucato, U.; Birks, J. B.; Minuti, L. *J. Am. Chem. Soc.* **1977,** *99,* 6340-6347.
[7711] Bensasson, R.; Dawe, E. A.; Long, D. A.; Land, E. J. *J. Chem. Soc., Faraday Trans. 1* **1977,** *73,* 1319-1325.
[7712] Bonneau, R. *Photochem. Photobiol.* **1977,** *25,* 129-132.
[7713] Steiner, U.; Winter, G.; Kramer, H. E. A. *J. Phys. Chem.* **1977,** *81,* 1104-1110.
[7714] Fouassier, J. -P.; Lougnot, D. -J.; Wieder, F.; Faure, J. *J. Photochem.* **1977,** *7,* 17-28.
[7715] Harriman, A.; Liu, R. S. H. *Photochem. Photobiol.* **1977,** *26,* 29-32.

[7716] Chapple, A. P.; Vikesland, J. P.; Wilkinson, F. *Chem. Phys. Lett.* **1977,** *50,* 81-84.
[7717] Amouyal, E.; Bensasson, R. *J. Chem.Soc., Faraday Trans. 1* **1977,** *73,* 1561-1568.
[7718] Brown, R. G.; Porter, G. *J. Chem.Soc., Faraday Trans. 1* **1977,** *73,* 1569-1573.
[7719] Grodowski, M. S.; Veyret, B.; Weiss, K. *Photochem. Photobiol.* **1977,** *26,* 341-352.
[7720] Kikuchi, K.; Tamura, S. -I.; Iwanaga, C.; Kokubun, H.; Usui, Y. *Z. Phys. Chem. (Frankfurt am Main)* **1977,** *106,* 17-24.
[7801] Maestri, M.; Bolletta, F.; Moggi, L.; Balzani, V.; Henry, M. S.; Hofmann, M. *Z. J. Am. Chem. Soc.* **1978,** *100,* 2694-2701.
[7802] Ballardini, R.; Varani, G.; Indelli, M. T.; Scandola, F.; Balzani, V. *J. Am. Chem. Soc.* **1978,** *100,* 7219-7223.
[7803] Brunschwig, B.; Sutin, N. *J. Am. Chem. Soc.* **1978,** *100,* 7568-7577.
[7804] Luong, J. C.; Nadjo, L.; Wrighton, M. S. *J. Am. Chem. Soc.* **1978,** *100,* 5790-5795.
[7805] Takuma, K.; Shuto, Y.; Matsuo, T. *Chem. Lett.* **1978,** 983-986.
[7806] Levin, G. *J. Phys. Chem.* **1978,** *82,* 1584-1588.
[7807] Wilkinson, F.; Garner, A. *Photochem. Photobiol.* **1978,** *27,* 659-670.
[7808] Berger, M.; McAlpine, E.; Steel, C. *J. Am. Chem. Soc.* **1978,** *100,* 5147-5151.
[7809] Korobov, V. E.; Chibisov, A. K. *J. Photochem.* **1978,** *9,* 411-424.
[7810] Schoof, S.; Guesten, H.; von Sonntag, C. *Ber. Bunsenges. Phys. Chem.* **1978,** *82,* 1068-1073.
[7811] Marshall, E. J.; Pilling, M. J. *J. Chem. Soc., Faraday Trans. 2* **1978,** *74,* 579-590.
[7812] Gorman, A. A.; Parekh, C. T.; Rodgers, M. A. J.; Smith, P. G. *J. Photochem.* **1978,** *9,* 11-17.
[7813] Yamamoto, S. -A.; Kikuchi, K.; Kokubun, H. *Z. Phys. Chem. (Wiesbaden)* **1978,** *109,* 47-58.
[7814] Tamura, S. -I.; Kikuchi, K.; Kokubun, H.; Usui, Y. *Z. Phys. Chem. (Wiesbaden)* **1978,** *111,* 7-18.
[7815] Barwise, A. J. G.; Gorman, A. A.; Leyland, R. L.; Smith, P. G.; Rodgers, M. A. J. *J. Am. Chem. Soc.* **1978,** *100,* 1814-1820.
[7816] Gorman, A. A.; Lovering, G.; Rodgers, M. A. J. *J. Am. Chem. Soc.* **1978,** *100,* 4527-4532.
[7817] Brown, R. G.; Harriman, A.; Harris, L. *J. Chem. Soc., Faraday Trans.2* **1978,** *74,* 1193-1199.
[7818] Grellmann, K. H.; Hentzschel, P. *Chem. Phys. Lett.* **1978,** *53,* 545-551.
[7901] Bergeron, S. F.; Watts, R. J. *J. Am. Chem. Soc.* **1979,** *101,* 3151-3156.
[7902] Gaines, G. L., Jr. *J. Phys. Chem.* **1979,** *83,* 3088-3091.
[7903] Ferreira, M. I. C.; Harriman, A. *J. Chem. Soc. Faraday Trans. 2* **1979,** *75,* 874-879.
[7904] Potter, W.; Levin, G. *Photochem. Photobiol.* **1979,** *30,* 225-231.
[7905] Harriman, A.; Porter, G.; Searle, N. *J. Chem. Soc. Faraday Trans. 2* **1979,** *75,* 1515-1521.
[7906] Wilkinson, F.; Schroeder, J. *J. Chem. Soc., Faraday Trans. 2* **1979,** *75,* 441-450.
[7907] Sloper, R. W.; Truscott, T. G.; Land, E. J. *Photochem. Photobiol.* **1979,** *29,* 1025-1029.
[7908] Grellmann, K. -H.; Hentzschel, P.; Wismontski-Knittel, T.; Fischer, E. *J. Photochem.* **1979,** *11,* 197-213.
[7909] George, M. V.; Kumar, Ch. V.; Scaiano, J. C. *J. Phys. Chem.* **1979,** *83,* 2452-2455.
[7910] Heelis, P. F.; Parsons, B. J.; Phillips, G. O. *Biochim. Biophys. Acta* **1979,** *587,* 455-462.
[7911] Encinas, M. V.; Scaiano, J. C. *J. Photochem.* **1979,** *11,* 241-247.

[7912] Beaumont, P. C.; Parsons, B. J.; Phillips, G. O.; Allen, J. C. *Biochim. Biophys. Acta* **1979,** *562,* 214-221.
[7913] Kuz'min, V. A.; Darmanyan, A. P.; Levin, P. P. *Chem. Phys. Lett.* **1979,** *63,* 509-514.
[7914] Salet, C.; Bensasson, R.; Becker, R. S. *Photochem. Photobiol.* **1979,** *30,* 325-329.
[7915] Delouis, J. F.; Delaire, J. A.; Ivanoff, N. *Chem. Phys. Lett.* **1979,** *61,* 343-346.
[7916] Vogelmann, E.; Rauscher, W.; Kramer. H. E. A. *Photochem. Photobiol.* **1979,** *29,* 771-776.
[7917] Sa E Melo, M. T.; Averbeck, D.; Bensasson, R. V.; Land, E. J.; Salet, C. *Photochem. Photobiol.* **1979,** *30,* 645-651.
[7918] Bruhlmann, U.; Huber, J. R. *J. Photochem.* **1979,** *10,* 205-213.
[7919] Moroi, Y.; Braun, A. M.; Graetzel, M. *J. Am. Chem. Soc.* **1979,** *101,* 567-572.
[7920] Encina, M. V.; Lissi, E. A. *J. Polym. Sci. Polym. Chem. Ed.* **1979,** *17,* 1645-1653.
[7921] Schulte-Frohlinde, D.; Goemer, H. *Pure Appl. Chem.* **1979,** *51,* 279-297.
[8001] Ohsawa, Y.; Saji, T.; Aoyagui, S. *J. Electroanal. Chem. Interfacial Electrochem.* **1980,** *106,* 327-338.
[8002] Baggott, J. E.; Pilling, M. J. *J. Phys. Chem.* **1980,** *84,* 3012-3019.
[8003] Creutz, C.; Chou, M.; Netzel, T. L.; Okumura, M.; Sutin, N. *J. Am. Chem. Soc.* **1980,** *102,* 1309-1319.
[8004] Boettcher, W.; Haim, A. *J. Am. Chem. Soc.* **1980,** *102,* 1564-1569.
[8005] Marcantonatos, M. D.; Deschaux M. *Chem. Phys. Lett.* **1980,** *76,* 359-365.
[8006] Pileni, M. *Chem. Phys. Lett.* **1980,** *75,* 540-544.
[8007] Osif, T. L.; Lichtin, N. N.; Hoffman, M. Z.; Ray, S. *J. Phys. Chem.* **1980,** *84,* 410-414.
[8008] Ohno, T.; Lichtin, N. N. *J. Am. Chem. Soc.* **1980,** *102,* 4636-4643.
[8009] Saltiel, J.; Shannon, P. T.; Zafiriou, O. C.; Uriarte, A. K. *J. Am. Chem. Soc.* **1980,** *102,* 6799-6808.
[8010] Scaiano, J.C. *J. Am. Chem. Soc.* **1980,** *102,* 7747-7753.
[8011] Winter, G.; Steiner, U. *Ber. Bunsenges. Phys. Chem.* **1980,** *84,* 1203-1214.
[8012] Bonneau, R. *J. Am. Chem. Soc.* **1980,** *102,* 3816-3822.
[8013] Scaiano, J. C.; Neta, P. *J. Am. Chem. Soc.* **1980,** *102,* 1608-1611.
[8014] Watkins, A. R. *Chem. Phys. Lett.* **1980,** *70,* 262-265.
[8015] Bonnets, R.; Charalambides, A. A.; Land, E. J.; Sinclair, R. S.; Tait, D.; Truscott, T. G. *J. Chem. Soc., Faraday Trans. 1* **1980,** *76,* 852-859.
[8016] Takahashi, T.; Kikuchi, K.; Kokubun, H. *J. Photochem.* **1980,** *14,* 67-76.
[8017] Wilson, T.; Halpern, A. M. *J. Am. Chem. Soc.* **1980,** *102,* 7279-7283.
[8018] Wagner, P. J.; Lam, H. M. H. *J. Am. Chem. Soc.* **1980** , *102,* 4167-4172.
[8019] Turro, NJ.; Tanimoto, Y. *J. Photochem.* **1980,** *14,* 199-203.
[8101] Chan, S. -F.; Chou, M.; Creutz, C.; Matsubara, T.; Sutin, N. *J. Am. Chem. Soc.* **1981,** *103,* 369-379.
[8102] Rillema, D. P.; Nagle, J. K.; Barringer, L. F., Jr.; Meyer, T. J. *J. Am. Chem. Soc.* **1981,** *103,* 56-62.
[8103] Rybak, W.; Haim, A.; Netzel, T. L.; Sutin, N. *J. Phys. Chem.* **1981,** *85,* 2856-2860.
[8104] Kurimura, Y.; Yokota, H.; Muraki, Y. *Bull. Chem. Soc. Jpn.* **1981,** *54,* 2450-2453.
[8105] Darwent, J. R.; Kalyanasundaram, K. *J. Chem. Soc. Faraday Trans. 2* **1981,** *77,* 373-382.
[8106] Griller, D.; Howard, J. A.; Marriott, P. R.; Scaiano, J. C. *J. Am. Chem. Soc.* **1981,** *103,* 619-623.

[8107] Davidson, R. S.; Bonneau, R.; Fornier de Violet, P.; Joussot-Dubien, J. *Chem. Phys. Lett.* **1981**, *78*, 475-478.

[8108] Das, P. K. *Tetrahedron Lett.* **1981**, *22*, 1307-1310.

[8109] Kamat, P. V.; Lichtin, N. N. *J. Phys. Chem.* **1981**, *85*, 814-818.

[8110] Das, P. K.; Bhattacharyya, S. N. *J. Phys. Chem.* **1981**, *85*, 1391-1395.

[8111] Kemp, T. J.; Martins, L. J. A. *J. Chem. Soc., Faraday Trans. 1* **1981**, *77*, 1425-1435.

[8112] Zacharova, G. V.; Lifanov, Yu. I.; Chibisov, A. K. *High Energy Chem.* **1981**, *15*, 56-60.

[8113] Das, P. K.; Encinas, M. V.; Scaiano, J. C. *J. Am. Chem. Soc.* **1981**, *103*, 4154-4162.

[8114] Encinas, M. V.; Scaiano, J. C. *J. Am. Chem. Soc.* **1981**, *103*, 6393-6397.

[8115] Amirzadeh, G.; Schnabel, W. *Makromol. Chem.* **1981**, *182*, 2821-2835.

[8116] Wilbrandt, R.; Jensen, N. -H. *J. Am. Chem. Soc.* **1981**, *103*, 1036-1041.

[8117] Capellos, C. *J. Photochem.* **1981**, *17*, 213-225.

[8118] Monti, S.; Gardini, E.; Bortolus, P.; Amouyal, E. *Chem. Phys. Lett.* **1981**, *77*, 115-119.

[8119] Traber, R.; Vogelmann, E.; Schreiner, S.; Werner, T.; Kramer, H. E. A. *Photochem. Photobiol.* **1981**, *33*, 41-48.

[8120] Klein, R.; Tatischeff, I.; Bazin, M.; Santus, R. *J. Phys. Chem.* **1981**, *85*, 670-677.

[8121] Nahor, G. S.; Rabani, J.; Grieser, F. *J. Phys. Chem.* **1981**, *85*, 697-702.

[8122] Chahidi, C.; Aubailly, M.; Monzikoff, A.; Bazin, M.; Santus, R. *Photochem. Photobiol.* **1981**, *33*, 641-649.

[8123] Goerner, H.; Schulte-Frohlinde, D. *J. Phys. Chem.* **1981**, *85*, 1835-1841.

[8124] Gorman, A. A.; Gould, I. R.; Hamblett, I. *J. Am. Chem. Soc.* **1981**, *103*, 4553-4558.

[8125] Herkstroeter, W. G.; Merkel, P. B. *J. Photochem.* **1981**, *16*, 331-341.

[8126] Wilkinson, F.; Tsiamis, C. *J. Chem. Soc., Faraday Trans. 2* **1981**, *77*, 1681-1693.

[8127] Kossanyi, J.; Sabbah, S.; Chaquin. P.; Ronfart-Haret, J. C. *Tetrahedron* **1981**, *37*, 3307-3315.

[8128] Damranyan, A. P.; Kuz'min, V. A. *Dokl. Phys. Chem.* **1981**, *260*, 938-941.

[8129] Das, P. K; Bobrowski, K. *J. Chem. Soc., Faraday Trans. 2* **1981**, *77*, 1009-1027.

[8130] Goemer, H.; Schulte-Frohlinde, D. *J. Photochem.* **1981**, *16*, 169-177.

[8131] Stevens, B.; Marsh, K. L.; Barltrop, J. A. *J. Phys. Chem.* **1981**, *85*, 3079-3082.

[8132] Demuth, M.; Amrein, W.; Bender, C. O.; Braslavsky, S. E.; Burger, U.; George. M. V.; Lemmer, D.; Schaffner, K. *Tetrahedron* **1981**, *37*, 3245-3261.

[8201] Krishnan, C. V.; Creutz, C.; Mahajan, D.; Schwarz, H. A.; Sutin, N. *Isr. J. Chem.* **1982**, *22*, 98-105.

[8202] Kalyanasundaram, K.; Neumann-Spallart, M. *Chem. Phys. Lett.* **1982**, *88*, 6-12.

[8203] Launikonis, A.; Ioder, J. W.; Mau, A. W. -H.; Sasse, W. H. F.; Summers, L. A.; Wells, D. *Aust. J. Chem.* **1982**, *35*, 1341-1355.

[8204] Miedlar, K.; Das, P. K. *J. Am. Chem. Soc.* **1982**, *104*, 7462-7469.

[8205] Graetzel, C. K.; Graetzel, M. *J. Phys. Chem.* **1982**, *86*, 2710-2714.

[8206] Feitelson, J.; Mauzerall, D. *J. Phys. Chem.* **1982**, *86*, 1623-1628.

[8207] Ohno, T.; Kato, S.; Lichtin, N. N. *Bull. Chem. Soc. Jpn.* **1982**, *55*, 2753-2759.

[8208] Shizuka, H.; Obuchi, H. *J. Phys. Chem.* **1982**, *86*, 1297-1302.

[8209] Martins, L. J. A.; Kemp, T. J. *J. Chem. Soc., Faraday Trans. 1* **1982**, *78*, 519-531.

[8210] Martins, L. J. A. *J. Chem. Soc., Faraday Trans. 1* **1982**, *78*, 533-543.

[8211] Lo, K. K. N.; Land, E. J.; Truscott, T. G. *Photochem. Photobiol.* **1982**, *36*, 139-145.

[8212] Becker, H. G. O.; Jirkovsky, J.; Fojtik, A.; Kleinschmidt, J. *J. Prakt. Chem.* **1982**, *324*, 505-511.

[8213] Scaiano, J. C.; Lee, C. W. B.; Chow, Y. L.; Buono-Core, G. E. *J. Photochem.* **1982,** *20,* 327-334.

[8214] O'Dowd, R. F.; O'Hare, A.; Cooke, J.; Taaffe, J. K. *J. Phys. E* **1982,** *15,* 736-740.

[8215] Guesten, H.; Heinrich, G. *J. Photochem.* **1982,** *18,* 9-17.

[8216] Darmanyan, A. P. *Chem. Phys. Lett.* **1982,** *86,* 405-410.

[8217] Ronfard-Haret, J. C.; Averbeck, D.; Bensasson, R. V.; Bisagni, E.; Land, E. J. *Photochem. Photobiol.* **1982,** *35,* 479-489.

[8218] Goemer, H. *J. Phys. Chem.* **1982,** *86,* 2028-2035.

[8219] Kamat, P. V.; Lichtin, N. N. *J. Photochem.* **1982,** *18,* 197-209.

[8220] Smith, G. J. *J. Chem. Soc., Faraday Trans. 2* **1982,** *78,* 769-773.

[8221] Chattopadhyay, S. K.; Das, P. K.; Hug, G. L. *J. Am. Chem. Soc.* **1982,** *104,* 4507-4514.

[8222] Goemer, H. *J. Photochem.* **1982,** *19,* 343-356.

[8223] Kamat, P. V.; Lichtin, N. N. *Isr. J. Chem.* **1982,** *22,* 113-116.

[8224] Caldwell, R. A.; Singh, M. *J. Am. Chem. Soc.* **1982,** *104,* 6121-6122.

[8225] Caldwell, R. A.; Cao, C. V. *J. Am. Chem. Soc.* **1982,** *104,* 6174-6180.

[8226] Ortmann, W.; Kassem, A.; Hinzmann, S.; Fanghaenel, E. *J. Prakt Chem.* **1982,** *324,* 1017-1025.

[8227] Maharaj, U.; Winnik, M. A. *Tetrahedron Lett.* **1982,** *23,* 3035-3038.

[8228] Simpson, J. T.; Krantz, A.; Lewis, F. D.; Kokel, B. *J. Am. Chem. Soc.* **1982,** *104,* 7155-7161.

[8301] Ohno, T.; Kato, S.; Yamada, A.; Tanno, T. *J. Phys. Chem.* **1983,** *87,* 775-781.

[8302] Bolletta, F.; Maestri, M.; Moggi, L.; Jamieson, M. A.; Serpone, N.; Henry, M. S.; Hofmann, M. Z. *Inorg. Chem.* **1983,** *22,* 2502-2509.

[8303] Harriman, A.; Porter, G.; Wilowska, A. *J. Chem. Soc. Faraday Trans. 2* **1983,** *79,* 807-816.

[8304] Prasad, D. R.; Ferraudi, G. *Inorg. Chem.* **1983,** *22,* 1672-1674.

[8305] Vinogradov, S. A.; Balashev, K. P.; Shagisultanova, G. A. *Koord. Khim.* **1983,** *9,* 949-954.

[8306] Johansen, O.; Mau, A. W. -F.; Sasse, W. H. F. *Chem. Phys. Lett.* **1983,** *94,* 113-117.

[8307] Mandal, K.; Pearson, T. D. L.; Krug, W. P.; Demas, J. N. *J. Am. Chem. Soc.* **1983,** *105,* 701-707.

[8308] Johansen, O.; Mau, A. W. -H.; Sasse, W. H. F. *Chem. Phys. Lett.* **1983,** *94,* 107-112.

[8309] Stewart, L. C.; Carlsson, D. J.; Wiles, D. M.; Scaiano, J. C. *J. Am. Chem. Soc.* **1983,** *105,* 3605-3609.

[8310] Das, P. K. *J. Chem. Soc., Faraday Trans. 1* **1983,** *79,* 1135-1145.

[8311] Loeff. I.; Treinin, A.; Linschitz, H. *J. Phys. Chem.* **1983,** *87,* 2536-2544.

[8312] Craw, M.; Lambert, C. *Photochem. Photobiol.* **1983,** *38,* 241-243.

[8313] Cogdell, R. J.; Land, E. J.; Truscott, TG. *Photochem. Photobiol.* **1983,** *38,* 723-725.

[8314] Maciejewski, A. *Chem. Phys. Lett.* **1983,** *94,* 344-349.

[8315] Wilkinson, F.; Tsiamis, C. *J. Am. Chem. Soc.* **1983,** *105,* 767-774.

[8316] Chattopadhyay, S. K.; Kumar, Ch. V.; Das, P. K. *Chem. Phys. Lett.* **1983,** *98,* 250-254.

[8317] Fisher, G. J.; Land, E. J. *Photochem. Photobiol.* **1983,** *37,* 27-32.

[8318] Craw, M.; Bensasson, R. V.; Ronfard-Haret, J. C.; Sa E Melo, M. T.; Truscott, T. G. *Photochem. Photobiol.* **1983,** *37,* 611-615.

[8319] Salet, C.; Bensasson, R. V.; Favre, A. *Photochem. Photobiol.* **1983,** *38,* 521-525.

[8320] Darmanyan, A. P. *Chem. Phys. Lett.* **1983,** *96,* 383-389.

[8321] Chow, Y. L.; Buono-Core, G. E.; Marciniak, B.; Beddard, C. *Can. J. Chem.* **1983,** *61,* 801-808.

[8401] Ohno, T.; Kato, S. *Bull. Chem. Soc. Jpn.* **1984,** *57,* 1528-1533.

[8402] Edel, A.; Marnot, P. A.; Sauvage, J. -P. *Nouv. J. Chim.* **1984,** *8,* 495-498.

[8403] Indelli, M. T.; Carioli, A.; Scandola, F. *J. Phys. Chem.* **1984,** *88,* 2685-2686.

[8404] Barboy, N.; Feitelson, J. *J. Phys. Chem.* **1984,** *88,* 1065-1068.

[8405] Ohtani, H.; Kobayashi, T.; Ohno, T.; Kato, S.; Tanno, T.; Yamada, A. *J. Phys. Chem.* **1984,** *88,* 4431-4435.

[8406] Scaiano, J. C.; Stewart, L. C.; Livant, P.; Majors, A. W. *Can. J. Chem.* **1984,** *62,* 1339-1343.

[8407] Kumar, C. V.; Qin, L.; Das, P. K. *J. Chem. Soc. Faraday Trans. 2* **1984,** *80,* 783-793.

[8408] Wilkinson, F.; Farmilo, A. *J. Chem. Soc., Faraday Trans. 2* **1984,** *80,* 1117-1124.

[8409] Das, P.K.; Griffin, G. W. *J. Org. Chem.* **1984,** *49,* 3452-3457.

[8410] Beecroft, R. A.; Davidson, R. S.; Goodwin, D.; Pratt, J. E. *Tetrahedron* **1984,** *40,* 4487-4496.

[8411] Lazare, S.; Bonneau, R.; Lapouyade, R. *J. Phys. Chem.* **1984,** *88,* 18-23.

[8412] Baral-Tosh, S.; Chattopadhyay, S. K.; Das, P. K. *J. Phys. Chem.* **1984,** *88,* 1404-1408.

[8413] Naito, I.; Schnabel, W. *Bull. Chem. Soc. Jpn.* **1984,** *57,* 771-775.

[8414] Das, P. K.; Muller, A. J.; Griffin, G. W. *J. Org. Chem.* **1984,** *49,* 1977-1985.

[8415] Hamanoue, K.; Tai, S.; Hidaka, T.; Nakayama, T.; Kimoto, M.; Teranishi, H. *J. Phys. Chem.* **1984,** *88,* 4380-4384.

[8416] Wismontski-Knittel, T.; Kilp, T. *J. Phys. Chem.* **1984,** *88,* 110-115.

[8417] Kumar, C. V.; Chattopadhyay, S. K.; Das, P. K. *Chem. Phys. Lett.* **1984,** *106,* 431-436.

[8418] Chattopadhyay, S. K.; Kumar, C. V.; Das, P. K. *J. Chem. Soc., Faraday Trans. 1* **1984,** *80,* 1151-1161.

[8419] Lee, W. A.; Graelzel, M.: Kalyanasundaram, K. *Chem. Phys. Lett.* **1984,** *107,* 308-313.

[8420] Grajcar, L.: Ivanoff, N.; Delouis, J. F.; Faure, J. *J. Chim. Phys. Phys.-Chim. Biol.* **1984,** *81,* 33-38.

[8421] Davis, H. F.; Chattopadhyay, S. K.; Das, P. K. *J. Phys. Chem.* **1984,** *88,* 2798-2803.

[8422] Wismontski-Knittel, T.; Das. P. K. *J. Phys. Chem.* **1984,** *88,* 2803-2808.

[8423] Chattopadhyay, S. K.; Kumar, C. V.; Das, P. K. *J. Photochem.* **1984,** *26,* 39-47.

[8424] Eftink, M. R.; Ghiron, C. A. *Biochemistry* **1984,** *23,* 3891-3899.

[8425] Kumar, C. V.; Davis, H. F.; Das, P. K. *Chem. Phys. Lett.* **1984,** *109,* 184-189.

[8426] Darmanyan, A. P. *Chem. Phys. Lett.* **1984,** *110,* 89-94.

[8427] Gorman, A. A.; Hamblett, I.; Harrison, R. J. *J. Am. Chem. Soc.* **1984,** *106,* 6952-6955.

[8428] Yip, R. W.; Sharma, D. K.; Giasson, R.; Gravel, D. *J. Phys. Chem.* **1984,** *88,* 5770-5772.

[8429] Masuhara, H.; Shioyama, H.; Saito, T.; Hamada, K.; Yasoshima, S.; Mataga, N. *J. Phys. Chem.* **1984,** *88,* 5868-5873.

[8430] Chattopadhyay, S. K.; Kumar, C. V.; Das, P. K. *J. Photochem.* **1984,** *24,* 1-9.

[8431] Kikuchi, K.; Yamamoto, S.; Kokubun, H. *J. Photochem.* **1984,** *24,* 271-283.

[8432] Scaiano, J. C.; Lissi, E. A.; Stewart, L. C. *J. Am. Chem. Soc.* **1984,** *106,* 1539-1542.

[8433] Das, P. K.; Muller, A. J.; Griffin, G. W.; Gould, I. R.; Tung, C. -H.; Turro, N. J. *Photochem. Photobiol.* **1984,** *39,* 281-285.

[8434] Orr, U.; Traber, R.; Hemmerich, P.; Kramer, H. E. A. *Photochem. Photobiol.* **1984,** *40,* 309-318.

[8435] Sundararajan, K.; Ramakrishnan, V.; Kuriacose, J. C. *Indian J. Chem., Sect. B* **1984,** *23B,* 1068-1070.

[8501] Kober, E. M.; Marshall, J. L.; Dreassick, W. J.; Sullivan, B. P.; Caspar, J. V.; Meyer, T. J. *Inorg. Chem.* **1985,** *24,* 2755-2763.

[8502] Ohno, T. *Coord. Chem. Rev.* **1985,** *64,* 311-320.

[8503] Krishnan, C. V.; Brunschwig, B. S.; Creutz, C.; Sutin, N. *J. Am. Chem. Soc.* **1985,** *107,* 2005-2015.

[8504] Zahir, K.; Boettcher, W.; Haim, A. *Inorg. Chem.* **1985,** *24,* 1966-1968.

[8505] Creaser, I. I.; Gahan, L. R.; Geue, R. J.; Launikonis, A.; Lay, P. A.; Lydon, J. D.; McCarthy, M. G.; Mau, A. W. -H; Sargeson, A. M.; Sasse, W. H. F. *Inorg. Chem.* **1985,** *24,* 2671-2680.

[8506] Ollino, M.; Cherry, W. R. *Inorg. Chem.* **1985,** *24,* 1417-1418.

[8507] Okura, I.; Kaji, N.; Aono, S.; Kita, T.; Yamada, A. *Inorg. Chem.* **1985,** *24,* 451-453.

[8508] Mau, A. W. -H.; Johansen, O.; Sasse, W. H. F. *Photochem. Photobiol.* **1985,** *41,* 503-509.

[8509] Yoshimura, A.; Kato, S. *Bull. Chem. Soc. Jpn.* **1985,** *58,* 1556-1559.

[8510] Chattopadhyay, S. K.; Kumar, C. V.; Das, P. K. *J. Photochem.* **1985,** *30,* 81-91.

[8511] Petrushenko, K. B.; Vokin, A. I.; Turchaninov, V. K.; Gorshkov, A. G.; Frolov, Yu. L. *Bull. Acad. Sci. USSR, Div. Chem. Sci.* **1985,** *34,* 242-246.

[8512] Lougnot, D. J.; Jacques, P.; Fouassier, J. P.; Casal, H. L.; Nguyen, K. -T; Scaiano, J. C. *Can. J. Chem.* **1985,** *63,* 3001-3006.

[8513] Heelis, P. F.; De la Rosa, M. A.; Phillips, G. O. *Photobiochem. Photobiophys.* **1985,** *9,* 57-63.

[8514] Kumar, C. V.; Ramaiah, D.; Das, P. K.; George, M. V. *J. Org. Chem.* **1985,** *50,* 2818-2825.

[8515] Chattopadhyay, S. K.; Kumar, C. V.; Das, P. K. *Photochem. Photobiol.* **1985,** *42,* 17-24.

[8516] Abdullah, K. A.; Kemp, T. J. *J. Chem. Soc. Perkin Trans 2* **1985,** 1279-1283.

[8517] Scaiano, J. C.; Leigh, W. J.; Meador, M. A.; Wagner, P. J. *J. Am. Chem. Soc.* **1985,** *107,* 5806-5807.

[8518] Rohatgi-Mukherjee, K. K.; Bhattacharyya, K.; Das, P. K. *J. Chem. Soc.. Faraday Trans. 2* **1985,** *81,* 1331-1344.

[8601] Sabbatini, N.; Perathoner, S.; Dellonte, S.; Lattanzi, G.; Balzani, V. *J. Less Common Met.* **1986,** *126,* 329-334.

[8602] Slama-Schwok, A.; Rabani, J. *J. Phys. Chem.* **1986,** *90,* 1176-1179.

[8603] Ballardini, R.; Varani, G.; Indelli, M. T.; Scandola, F. *Inorg. Chem.* **1986,** *25,* 3858-3865.

[8604] Hoshino, M.; Seki, H.; Yasufuku, K.; Shizuka, H. *J. Phys. Chem.* **1986,** *90,* 5149-5153.

[8605] Tamilarasan, R.; Endicott, J.F. *J. Phys. Chem.* **1986,** *90,* 1027-1033.

[8606] Bazin, M.; Santus, R *Photochem. Photobiol.* **1986,** *43,* 235-242.

[8607] Goerner, H.; Schulte-Frohlinde, D. *Chem. Phys. Lett.* **1986,** *124,* 321-325.

[8608] Scaiano, J. C.; Encinas, M. V.; Lissi, E. A.; Zanocco, A.; Das, P. K. *J. Photochem.* **1986,** *33,* 229-236.

[8609] Hoshi, M.; Kikuchi, K.; Kokubun, H.; Yamamoto, S. -A. *J. Photochem.* **1986,** *34,* 63-71.

[8610] Bhattacharyya, K.; Das, P. K.; Ramamurthy, V.; Rao, V. P. *J. Chem. Soc., Faraday Trans. 2* **1986**, *82*, 135-147.

[8611] Bhattacharyya, K.; Das, P. K. *J. Phys. Chem.* **1986**, *90*, 3987-3993.

[8612] Herkstroeter, W. G.; Farid, S. *J. Photochem.* **1986**, *35*, 71-85.

[8613] Evans, C.; Weir, D.; Scaiano, J. C.; Mac Eachern, A.; Arnason, J. T.; Morand, P.; Hollebone, B.; Leitch, L. C.; Philogene, B. J. R. *Photochem. Photobiol.* **1986**, *44*, 441-451.

[8614] Wagner, P. J.; Truman, R. J.; Puchalski, A. E.; Wake, R. *J. Am. Chem. Soc.* **1986**, *108*, 7727-7738.

[8615] Becker, H. G. O.; Schuetz, R.; Tillack, B.; Rehak, V. *J. Prakt. Chem.* **1986**, *328*, 661-672.

[8616] Riddick, J. A.; Bunger, W. B.; Sakano, T. K. *Organic Solvents. Physical Properties and Methods of Purification, 4th Ed.*; John Wiley and Sons: New York (USA), 1986, volume II, 1325p.

[8701] Gersdorf, J.; Mattay, J.; Goerner, H. *J. Am. Chem. Soc.* **1987**, *109*, 1203-1209.

[8702] Kemp, T. J.; Parker, A. W.; Wardman, P. *J. Chem. Soc. Perkin Trans. 2* **1987**, 397-403.

[8703] Bhattacharyya, K.; Ramamurthy, V.; Das, P. K. *J. Phys. Chem.* **1987**, *91*, 5626-5631.

[8704] Usui, Y.; Misawa, H.; Sakuragi, H.; Tokumaru, K. *Bull. Chem. Soc. Jpn.* **1987**, *60*, 1573-1578.

[8705] Arai, T.; Oguchi, T.; Wakabayashi, T.; Tsuchiya, M.; Nishimura, Y.; Oishi, S.; Sakuragi, H.; Tokumaru, K. *Bull. Chem. Soc. Jpn.* **1987**, *60*, 2937-2943.

[8801] Yoshimura, A.; Ohno, T. *Photochem. Photobiol.* **1988**, *48*, 561-565.

[8802] Saini, R. D.; Dhanya, S.; Bhattacharyya, P. K. *J. Photochem. Photobiol.* **1988**, *43*, 91-103.

[8803] Kuz'min, V. A.; Levin, P. P. *Bull. Acad. Sci. USSR, Div. Chem. Sci.* **1988**, *37*,429-432.

[8804] Weir, D.; Scaiano, J.C.; Schuster, D. I. *Can. J. Chem.* **1988**, *66*, 2595-2600.

[8805] Ghiron, C. A.; Bazin, M.; Santus, R. *Biochim. Biophys. Acta* **1988**, *957*, 207-216.

[8806] Saltiel, J.; Atwater, B. W. *Adv. Photochem.* **1988**, *14*, 1-90.

[8901] Hoffman, M. Z.; Bolletta, F.; Moggi, L.; Hug, G. L. *J. Phys. Chem. Ref. Data* **1989**, *18*, 219-543.

[9101] Arbogast, J. W.; Darmanyan, A. P.; Foote, C. S.; Rubin, Y.; Diederich, F. N.; Alvarez, M. M.; Anz, S. J.; Whetten, R. L. *J. Phys. Chem.* **1991**, *95*, 11-12.

[9102] Arbogast, J. W.; Foote, C. S. *J. Am. Chem. Soc.* **1991**, *113*, 8886-8889.

[9201] Samanta, A.; Kamat, P. V. *Chem. Phys. Lett.* **1992**, *199*, 635-639.

[9301] McLean, A. J.; Rodgers, M. A. J. **1993**, *115*, 4786-4792.

[9401] Wilkinson, F.; McGarvey, D. J.; Olea, A. F. *J. Phys. Chem.* **1994**, *98*, 3762-3769.

[9501] Zhang, X. Y.; Jin, S.; Ming, Y. F.; Fan, M. G.; Lian, Z. R.; Yao, S. D.; Lin, N. Y. *J. Photochem. Photobiol. A: Chem.* **1995**, *85*, 85-88.

[9601] Parola, J. A.; Pina, F.; Ferreira, E.; Maestri, M.; Balzani, V. *J. Am. Chem. Soc.* **1996**, *118*, 11610-11616.

[9701] Sasaki, Y.; Fujitsuka, M.; Watanabe, A.; Ito, O. *J. Chem. Soc., Faraday Trans.* **1997**, *93*, 4275-4279.

[9702] Fujitsuka, M.; Watanabe, A.; Ito, O.; Yamamoto, K.; Funasaka, H. *J. Phys. Chem. A*, **1997**, *101*, 4840-4844.

[9703] Heisler, L.; Hossenlopp, J. M.; Niu, Y.; Steinmetz, M. G.; Zhang, Y. *J. Photochem. Photobiol. A: Chem.* **1997**, *107*, 125-136.

[9704] Hariharan, C.; Vijaysree, V.; Mishra, A. K. *J. Lumin.* **1997**, *75*, 205-211.

[9801] Grosso, V. N.; Chesta, C. A.; Previtali, C. M. *J. Photochem. Photobiol. A: Chem.* **1998**, *118*, 157-163.

[9802] Thomas, K. G.; Biju, V.; George, M. V.; Guldi, D. M.; Kamat, P. V. *J. Phys. Chem. A* **1998**, *102*, 5341-5348.

[9803] Novaira, A. I.; Borsarelli, C. D.; Cosa, J. J.; Previtali, C. M. *J. Photochem. Photobiol. A: Chem.* **1998**, *115*, 43-47.

[9804] Okada, K.; Yamaji, M.; Shizuka, H. *J. Chem. Soc., Faraday Trans.* **1998**, *94*, 861-866.

[9901] Sasaki, Y.; Konishi, T.; Yamazaki, M.; Fujitsuka, M.; Ito, O. *Phys. Chem. Chem. Phys.* **1999**, *1*, 4555-4559.

[9902] Wilkinson, F.; Abdel-Shafi, A. A. *J. Phys. Chem. A* **1999**, *103*, 5425-5435.

[0001] Coenjarts, C.; Scaiano, J. C. *J. Am. Chem. Soc.* **2000**, *122*, 3635-3641.

[0002] Canonica, S.; Hellrung, B.; Wirz, J. *J. Phys. Chem. A* **2000**, *104*, 1226-1232.

[0101] Biju, V.; Barazzouk, S.; Thomas, K. G.; George, M. V.; Kamat, P. V. *Langmuir* **2001**, *17*, 2930-2936.

[0201] Abdel-Shafi, A. A.; Wilkinson, F. *Phys. Chem. Chem. Phys.* **2002**, *4*, 248-254.

[0202] Encinas, M. V.; Bertolotti, S. G.; Previtali, C. M. *Helv. Chim. Acta* **2002**, *85*, 1427-1437.

[0301] Gerhardt, S. A.; Lewis, J. W.; Kliger, D. S.; Zhang, J. Z.; Simonis, U. *J. Phys. Chem. A* **2003**, *107*, 2763-2767.

[0302] Rivarola, C. R.; Chesta, C. A.; Previtali, C. M. *Photochem. Photobiol. Sci.* **2003**, *2*, 893-897.

[0303] Porcal, G.; Bertolotti, S. G.; Previtali, C. M.; Encinas, M. V. *Phys. Chem. Chem. Phys.* **2003**, *5*, 4123-4248.

[0307] Lide, D. R. (ed.) *CRC Handbook of Chemistry and Physics, 84th Ed.*; CRC Press: Boca Raton, FL (USA), 2003.

[0401] Lorente, C.; Capparelli, A. L.; Thomas, A. H.; Braun, A. M.; Oliveros, E. *Photochem. Photobiol. Sci.* **2004**, *3*, 167-173.

[0402] Chatterjee, S.; Basu, S.; Ghosh, N.; Chakrabarty, M. *Chem. Phys. Lett.* **2004**, *388*, 79-83.

7

Ionization Energies, Electron Affinities, and Reduction Potentials

The tables in this section consist of critical compilations of parameters related to the study of exciplexes and electron-transfer processes. Section 7a contains tables of data derived from gas-phase measurements. The data in this section are ionization energies and electron affinities. Section 7b contains the electrochemistry data on organic solutions. The data in this section consist mainly of oxidation and reduction halfwave potentials.

7a IONIZATION ENERGIES AND ELECTRON AFFINITIES

In the theory of charge-transfer complexes [5001, 0106], the energy of the charge-transfer state is computed by the following set of processes involving the electron donor, D, and the electron acceptor, A:

$$D \rightarrow D^+ + e^- \qquad (7\text{-}I)$$

$$A + e^- \rightarrow A^- \qquad (7\text{-}II)$$

$$D^+ + A^- \rightarrow [D^+ \cdots A^-] \qquad (7\text{-}III)$$

The minimum energy required to remove an electron to infinity in process (7-I) is the ionization energy (formerly called ionization potential), E_i. The adiabatic ionization energy refers to the formation of the molecular ion in its ground vibrational state, and the vertical ionization energy applies to the transition to the molecular ion without change in geometry. The energy released in the formation of the singly charged negative ion, starting with A and e^- at infinity (process 7-II), is the electron affinity, E_{ea}, of A [9704]. The energy associated to process (7-III) is the Coulomb potential energy of the ion-pair (or complex) at the distance of closest approach, $R_{D^+A^-}$. The total energy associated to the formation of the complex starting with D and A is

$$E_{D^+A^-} = E_i - E_{ea} - \frac{e^2}{4\pi \, \varepsilon_0 \varepsilon_r \, R_{D^+A^-}} \tag{7-1}$$

where e is the elementary charge, 1.602×10^{-19} C, ε_r is the relative medium static permittivity (formerly called dielectric constant), and ε_0 is the vacuum permittivity, 8.854×10^{-12} F m^{-1}.

This could represent a first approximation

$$E_{[DA]^*} \approx E_{D^+A^-} \tag{7-2}$$

to the energy of an exciplex (an excited DA complex with no stable ground state) or excited DA complex with a higher local excitation energy.

The ionization energies of some common electron donors are collected in Table 7a-1. Electron affinities of electron acceptors are emphasized, along with their ionization energies, in Table 7a-4. The two intermediate tables contain data on unsaturated organic compounds. Many of the aromatics in Table 7a-3 can act as either donors or acceptors, depending on the nature of the excited state-quencher pair.

Table 7a-1 Ionization Energies of Selected Electron Donors

No.	Compound	Adiabatic (eV)	Vertical (eV)
1	Acridine	–	7.88[a]
2	Ammonia	10.16[b]	10.85[b]
3	*tert*-Amylamine	8.5[b]	9.20[b]
4	Aniline	7.72[c]	8.05[c]
5	Aniline, *N,N*-diethyl-	6.95[d]	7.2[e]
6	Aniline, *N,N*-dimethyl-	7.12[c]	7.37[c]
7	Aniline, *N,N*-diphenyl-	6.80[c]	7.00[c]
8	Aniline, *N*-methyl-	7.33[c]	–
9	Aniline, *N*-phenyl-	7.16[c]	7.44[c]
10	Benzene, 1,2-dimethoxy-	–	7.96[f]
11	Benzene, 1,3-dimethoxy-	–	8.16[f]
12	Benzene, 1,4-dimethoxy-	–	7.8[f]
13	Benzene, methoxy	8.21[c]	8.42[c]
14	Benzene, 1,2,3-trimethoxy-	–	8.3[f]
15	Benzene, 1,2,4-trimethoxy-	–	7.36[f]
16	Benzene, 1,3,5-trimethoxy-	–	8.11[f]
17	Benzidine	6.9[g]	–
18	Benzidine, *N,N,N',N'*-tetramethyl-	6.40[c]	–
19	Benzylamine	8.64[c]	–
20	Butylamine	8.71[b]	9.40[b]
21	*sec*-Butylamine	8.7[b]	9.30[b]
22	*tert*-Butylamine	8.6[b]	9.25[b]
23	Carbazole	7.57	–
24	Cobaltocene	–	5.55[h]
25	Cyclohexylamine	8.6[b]	9.16[b]
26	1,4-Diazabyciclo[2.2.2]octane (DABCO)	7.2[i]	7.52[i]
27	Diethylamine	7.86[b]	8.63[b]
28	Diisopropylamine	7.60[b]	8.40[b]
29	Diisopropylamine, *N*-ethyl-	7.2[j]	7.7[j]
30	Diethylmethylamine	7.5[b]	8.22[b]
31	Diphenylamine	7.16[k]	–
32	Dipropylamine	7.77[b]	8.55[b]
33	Dimethylamine	8.23[b]	8.93[b]

Table 7a-1 Ionization Energies of Selected Electron Donors

No.	Compound	Adiabatic (eV)	Vertical (eV)
34	Ethanamine, N-ethyl-	8.01[b]	8.63[b]
35	Ethylamine	8.86[b]	9.47[b]
36	Ferrocene	–	6.86[h]
37	Isobutylamine	8.7[b]	9.30[b]
38	Isopropylamine	8.64[b]	9.33[b]
39	Methylamine	8.97[b]	9.66[b]
40	Naphthalene, 1-methoxy-	7.70[c]	7.72[c]
41	Naphthalene, 2-methoxy-	7.4[c]	7.87[c]
42	Neopentylamine	8.5[b]	9.25[b]
43	Phenazine	–	8.33[a]
44	Phenoxazine	7.23[l]	–
45	Phenothiazine	7.31[l]	–
45	Phenylalanine	8.5[m]	8.9[m]
46	Phenylalanine, N,N-dimethyl-	7.7[m]	8.2[m]
47	Phenylalanine, N-methyl-	8.5[m]	8.7[m]
48	p-Phenylenediamine, N,N,N',N'-tetramethyl-	6.20[c]	6.75[c]
49	Phthalocyanine, free base	–	6.41[n]
50	Phthalocyanine, Zn(II)	–	6.37[n]
51	Piperidine	8.20[o]	8.66[c]
52	Piperidine, N-methyl-	7.8[p]	8.30[b]
53	Porphyrin, free base	–	6.9[n]
54	Porphyrin, octaethyl-, free base	–	6.39[n]
55	Porphyrin, octaethyl-, Zn(II)	–	6.29[n]
56	Porphyrin, tetraphenyl-, free base	–	6.39[n]
57	Porphyrin, tetraphenyl-, Zn(II)	–	6.42[n]
58	Propylamine	8.55[b]	9.41[b]
59	Pyrrolidine	8.0[o]	8.75[b]
60	Pyrrolidine, N-methyl-	–	8.41[b]
61	Quinuclidine	7.51[i]	8.06[i]
62	Ruthenocene	–	7.45[h]
63	Tetraselenafulvalene	7.14[q]	7.21[q]
64	Tetrathiafulvalene	6.70[r]	6.92[q]
65	Tetrathiafulvalene, bis(ethylenedithio)-	6.7[r]	–

Table 7a-1 Ionization Energies of Selected Electron Donors

No.	Compound	Adiabatic (eV)	Vertical (eV)
66	Toluene, *p*-methoxy-	–	8.09[f]
67	*o*-Toluidine	7.44[d]	7.83[s]
68	Tributylamine	7.4[b]	7.90[b]
69	Triethylamine	7.50[b]	8.08[b]
70	Triisopropylamine	6.95[j]	7.2[j]
71	Trimethylamine	7.82[b]	8.53[b]

[a] From ref. [9702]; [b] From ref. [7601]; [c] From refs. [8201], [8802], [9004]; [d] From ref. [9901]; [e] From ref. [7301]; [f] From ref. [9402]; [g] From ref. [6901]; [h] From ref. [0105]; [i] From ref. [8101]; [j] From ref. [9701]; [k] From ref. [0307]; [l] From ref. [7502]; [m] From ref. [9401]; [n] From ref. [9903]; [o] From ref. [9603]; [p] From ref. [9201]; [q] From ref. [7501]; [r] From ref. [9001]; [s] From ref. [7401].

Table 7a-2 Ionization Energies[a] and Electron Affinities[b] of Alkenes, Dienes, and Alkynes

No.	Compound	Adiabatic E_i (eV)	Vertical E_i (eV)	Electron affinity (eV)
1	Acetylene	11.40	11.43	–1.8
2	Allene	9.65[c]	–	–
3	Bicyclo[2.1.0]pentane	8.7	–	–
4	Bicyclo[2.1.0]pent-2-ene	8.0	8.6	–
5	Butadiene	9.07	9.03	–0.65
6	1,3-Butadiene, 2,3-dimethyl-	8.71	8.72	–
7	(E)-2-Butene	9.10	9.11	–
8	(Z)-2-Butene	9.11	9.11	–
9	1-Butyne	10.18	–	–
10	2-Butyne	9.56	9.79	–
11	1,3-Cyclohexadiene	8.25	8.25	–0.73
12	1,4-Cyclohexadiene	8.82	8.82	–1.75
13	Cyclohexene	8.95	9.12	–2.70
14	1,3-Cyclooctadiene	8.4	–	–
15	Cyclopentadiene	8.56	–	–1.05
16	Cyclopentene	9.01	9.01	–
17	Cyclopropane	9.86	–	–
18	Ethylene	10.51	10.50	–1.55
19	(E,E)-2,4-Hexadiene	8.18	–	–
20	(E,Z)-2,4-Hexadiene	8.24	–	–
21	(Z,Z)-2,4-Hexadiene	8.3	–	–
22	Isobutene	9.23[c]	–	–
23	Isoprene	8.84	8.87	–
24	Norbornadiene	8.35	8.70	–
25	Norbornylene	8.31	8.95	–1.70
26	(E)-Piperylene	8.59	–	–
27	(Z)-Piperylene	8.63	8.60	–
28	Propylene	9.73	9.91	–2.0[d]
29	Propyne	10.36	10.36	–

[a] From refs. [8201], [8802], [9004], unless otherwise noted; [b] From ref. [8402], unless otherwise noted; [c] From ref. [9703]; [d] From ref. [8801].

Table 7a-3 Ionization Energies[a] and Electron Affinities[b] of Aromatics

No.	Compound	Adiabatic E_i (eV)	Vertical E_i (eV)	Electron affinity (eV)
1	Acenaphthene	7.7	7.76	–
2	Acenaphthylene	8.22	–	+0.45
3	Acetylene, phenyl-	8.81	8.82	–
4	Acridine	–	7.88[c]	–
5	Adenine	8.26[d]	8.44[d]	–
6	Anthracene	7.45	7.41	+0.69[e]
7	Anthracene, 9-bromo-	7.47[e]	–	–
8	Anthracene, 9,10-dimethyl-	7.11[e]	–	–
9	Anthracene, 9-methyl-	7.24	7.24	–
10	Anthracene, 9-phenyl-	–	7.25	–
11	(*E*)-Azobenzene	–	8.6[f]	–
12	Azulene	7.41	7.44	+0.78[e]
13	Benz[*a*]anthracene	7.43	7.41	+0.63
14	Benzene	9.24	9.24	–0.74[e]
15	Benzene, hexamethyl-	7.85[g]	–	–
16	Benzene, isopropyl-	8.73	8.75	–1.08
17	Benzene, pentamethyl-	7.92[g]	–	–
18	Benzene, 1,2,4,5-tetramethyl-	8.04	8.05	+0.07
19	Benzene, 1,2,3-trimethyl-	8.42[g]	–	–
20	Benzene, 1,2,4-trimethyl-	8.27[g]	–	–
21	Benzene, 1,3,5-trimethyl-	8.41	8.45	–1.03
22	Benzo[*b*]chrysene	7.14	7.20	+0.89[e]
23	Benzo[*a*]coronene	7.1	7.08	–
24	Benzo[*c*]cinnoline	–	8.3[c]	–
25	Benzo[*rst*]pentaphene	7.0	–	–
26	Benzo[*ghi*]perylene	7.15	7.15	–
27	Benzo[*c*]phenanthrene	7.60	7.60	+0.58[e]
28	Benzo[*a*]pyrene	7.12	7.41	+0.78[e]
29	Benzo[*e*]pyrene	7.41	–	+0.55[e]
30	Benzo[*b*]triphenylene	7.39	7.39	–
31	Biphenyl	7.95	8.34	+0.12[e]
32	Biphenylene	7.56	7.60	–
33	Carbazole	7.57	–	–

Table 7a-3 Ionization Energies[a] and Electron Affinities[b] of Aromatics

No.	Compound	Adiabatic E_i (eV)	Vertical E_i (eV)	Electron affinity (eV)
34	Chrysene	7.59	7.59	+0.42[e]
35	Coronene	7.29	7.29	+0.74[e]
36	*m*-Cresol	–	8.39[h]	–
37	*o*-Cresol	–	8.30[h]	–
38	*p*-Cresol	–	8.22[h]	–
39	Cytosine	8.68[d]	8.94[d]	–
40	Dibenz[*a,c*]anthracene	7.39[e]	–	+0.69[e]
41	Dibenz[*a,h*]anthracene	7.38	7.38	+0.70[e]
42	Dibenz[*a,j*]anthracene	7.40	7.40	+0.67[e]
43	Fluoranthene	7.95	–	+0.80[e]
44	Fluorene	7.89	7.93	+0.22
45	Furan	8.88	–	–1.76
46	Guanine	7.77[d]	8.24[d]	–
47	Indan	8.3	8.45	–
48	Indene	8.14	8.15	+0.16
49	Indole	–	8.20[i]	–
50	Isoquinoline	8.53	8.54	–0.42
51	Isoxazole	–	10.15[i]	–
52	Naphthacene	–	7.01[j]	–
53	Naphthalene	8.14	8.15	+0.14[e]
54	Naphthalene, 1,4-dimethyl	7.78[e]	–	–
55	Naphthalene, 1-methyl-	7.85	8.01	+0.13
56	Naphthalene, 2-methyl-	7.8	8.01	+0.16
57	Oligo(*p*-phenylenevinylene)-1[k]	5.76[p]	–	< +2.92[p]
58	Oligo(*p*-phenylenevinylene)-2[l]	5.55[p]	–	< +2.95[p]
59	Oligo(*p*-phenylenevinylene)-3[m]	5.51[p]	–	< +3.07[p]
60	Oligo(*p*-phenylenevinylene)-4[n]	5.42[p]	–	< +3.03[p]
61	Oligo(*p*-phenylenevinylene)-5[o]	5.36[p]	–	< +3.10[p]
62	Ovalene	6.71	6.71	–
63	Oxazole	–	9.83[i]	–
64	Pentacene	6.61	6.61	+1.37[e]
65	Pentaphene	7.27	7.27	+0.75[e]
66	Perylene	6.9	6.97[c]	+0.98[e]

Table 7a-3 Ionization Energies[a] and Electron Affinities[b] of Aromatics

No.	Compound	Adiabatic E_i (eV)	Vertical E_i (eV)	Electron affinity (eV)
67	Phenanthrene	7.86	7.86[c]	+0.31
68	1,10-Phenanthroline	–	8.35[c]	–
69	4,7-Phenanthroline	–	8.51[c]	–
70	Phenazine	–	8.33[c]	–
71	Phenol	–	8.47[h]	–1.0[q]
72	Poly(*p*-phenylenevinylene)	5.11[p]	–	–
73	Pyrazine	9.29	9.63	–0.80
74	Pyrene	7.41	7.41	+0.62[e]
75	Pyridazine	8.64	9.31	–0.49
76	Pyridine	9.25	–	–0.59
77	Pyrimidine	9.23	9.73	–0.33
78	Pyrrole	–	8.90[i]	–
79	Quinoline	8.62	8.62	–0.60
80	(*E*)-Stilbene	7.70	7.90	+0.38
81	Styrene	8.43	8.50	+0.12[e]
82	*m*-Terphenyl	8.01	–	–
83	*o*-Terphenyl	8.00	–	–
84	*p*-Terphenyl	7.78	–	+0.27[e]
85	Tetrabenz[*a,c,h,j*]anthracene	7.43	7.43	–
86	Tetracene	6.97	6.97	+1.09[e]
87	Thiophene	–	8.91[i]	–
88	Thymine	8.87[d]	9.14[d]	–
89	Toluene	8.82	8.85	–0.4
90	Toluene, *p*-chloro-	–	8.73[h]	–
91	Toluene, *p*-methoxy-	–	8.09[h]	–
92	Triphenylene	7.84	7.88	+0.29[e]
93	*m*-Xylene	8.56	8.55	–1.06
94	*o*-Xylene	8.56	8.57	–1.12
95	*p*-Xylene	8.44	8.43	–1.07

[a] From refs. [8201], [8802], [9004], unless otherwise noted; [b] From ref. [8402], unless otherwise noted; [c] From ref. [9702]; [d] From ref. [0104]; [e] From ref. [9902]; [f] From ref. [0301]; [g] From ref. [0307]; [h] From ref. [9402]; [i] From ref. [0103]; [j] From ref. [9601]; [k] 1,4-Bis(3,5-di-*t*-butylstyryl)benzene; [l] 1,4-Bis(3,5-di-*t*-butylstyryl)stilbene; [m] 1,4-Bis[4-(3,5-di-*t*-butylstyryl)styryl]benzene; [n] 1,4-Bis[4-(3,5-di-*t*-butylstyryl)-styryl]stilbene; [o] 1,4-Bis[4-[4-(3,5-di-*t*-butylstyryl)styryl]styryl]benzene; [p] From ref. [9501]; [q] From ref. [8801].

Table 7a-4 Ionization Energies[a] and Electron Affinities[b] of Selected Electron Acceptors

No.	Compound	Adiabatic E_i (eV)	Adiabatic E_i (eV)	Electron affinity (eV)
1	Anthracene-9-carboxaldehyde	7.69	7.67	+1.02
2	9,10-Anthraquinone	9.25	–	+1.55
3	Benzene, 1-cyano-4-nitro-	10.2	–	+1.82
4	Benzene, 1,2-dicyano-	9.90	10.27	+0.95
5	Benzene, 1-methoxy-4-nitro-	8.8	9.08	+0.81
6	Benzene, nitro-	9.86	9.88	+2.1
7	Benzene, 1,2,4,5-tetracyano-	–	–	+1.6
8	Benzonitrile	9.62	9.71	+0.25
9	Benzophenone	9.05	–	+0.63
10	1,4-Benzoquinone	10.0	9.99	+1.83
11	1,4-Benzoquinone, tetrachloro-	9.74	–	+1.37
12	Biacetyl	9.24	9.55	+0.75
13	Ethylene, 1,2-dicyano-, (*E*)-	11.16	–	+0.96
14	Ethylene, 1,1-dicyano-	–	11.38	+1.54
15	Ethylene, tetracyano-	11.77	11.79	+2.9
16	9-Fluorenone	8.36	–	+1.19
17	Fullerene C_{60}	7.57[c]	–	+2.65[d]
18	Fullerene C_{70}	7.36[c]	–	+2.73[d]
19	Maleic anhydride	10.8	11.45	+1.41
20	1-Naphthaldehyde	8.3	–	+0.68
21	Naphthalene, 2-acetyl	–	8.23	+0.6
22	1,4-Naphthoquinone	9.56	–	+1.71
23	4-Nitrotoluene	9.4	9.54	+0.89
24	9,10-Phenanthrenequinone	8.64	–	+1.83
25	Tetracyano-1,4-benzoquinone	–	–	+1.8
26	7,7,8,8-Tetracyanoquinodimethane	–	–	+2.84

[a] From refs. [8201], [8802], [9004], unless otherwise noted; [b] From refs. [8402], [8501], unless otherwise noted; [c] From ref. [0001]; [d] From refs. [9101], [9602].

7b REDUCTION POTENTIALS

The standard free energy, ΔG^0, associated to the formation of a pair of separated ions from a neutral donor and a neutral acceptor (processes 7-I and 7-II) is

$$\Delta G^0 = N_A \, e \, [E^0(D^+/D) - E^0(A/A^-)] \qquad (7\text{-}3)$$

with $e = 1.602 \times 10^{-19}$ C the elementary charge, $N_A = 6.023 \times 10^{23}$ mol^{-1} the Avogadro constant, $E^0(D^+/D)$ the standard electrode potential of the donor cation resulting from the oxidation process, and $E^0(A/A^-)$ the standard electrode potential of the acceptor (relative to the same reference electrode).[a]

The tables report the halfwave potential values for the half-reactions of interest written as reductions [$E_{1/2}(D^+/D)$ and $E_{1/2}(A/A^-)$], that are approximately equal to the standard potentials, E^0. If the potential values are in V, the following equation gives ΔG^0 in kcal mol^{-1}:

$$\Delta G^0 \approx 23.05 \, [E_{1/2}(D^+/D) - E_{1/2}(A/A^-)] \qquad (7\text{-}4)$$

In the theory of electron-transfer quenching of excited states [0106], the free energies for the processes,

$$D^* + A \;\rightarrow\; [D^+ \cdots A^-] \qquad (7\text{-}IV)$$

$$D + A^* \;\rightarrow\; [D^+ \cdots A^-] \qquad (7\text{-}V)$$

are central for the quenching of excited states of donors and acceptors, respectively. The standard free energy change in either of these electron-transfer quenching processes has been related to the standard potentials of D and A in a theory due to Rehm and Weller [7003]. Making use of the approximation in Eq. 7-4, their equation can be written as

$$\Delta G^0{}_{et} \approx N_A \left\{ e[E_{1/2}(D^+/D) - E_{1/2}(A/A^-)] - \frac{e^2}{4\pi \, \varepsilon_0 \varepsilon_r \, R_{D^+A^-}} \right\} - \Delta E_{0-0} \qquad (7\text{-}5)$$

where ΔE_{0-0} is the excited state energy of the electron donor (process 7-IV) or acceptor (process 7-V), ε_r is the relative medium static permittivity (formerly called dielectric constant), and ε_0 is the vacuum permittivity, 8.854×10^{-12} F m^{-1}.

The electron-transfer rate constant, k_{et}, can be written classically as an activated rate constant [8404, 0106],

$$k_{et} = k_0 \exp\left(\frac{-\Delta G^{\ddagger}}{RT} \right) \qquad (7\text{-}6)$$

where k_0 is the reciprocal of the dielectric relaxation time. Rehm and Weller [7003] proposed an empirical relation,

$$\Delta G^{\ddagger} = \left\{ \left[\frac{\Delta G^0_{\text{et}}}{2} \right]^2 + [\Delta G^{\ddagger}(0)]^2 \right\}^{1/2} + \frac{\Delta G^0_{\text{et}}}{2} \qquad (7\text{-}7)$$

for the free energy of activation of the process of electron-transfer quenching. In Eq. 7-7, $\Delta G^{\ddagger}(0)$ is the free energy of activation when the standard free energy change ΔG^0_{et} for the overall quenching process is zero. The value of ΔG^0_{et} is given by Eq. 7-5. Equations such as these (along with various modifications) have proved fruitful in elucidating electron-transfer processes in excited states.

Four points can be noted with regard to the selection and use of the tables:

1. In the tables we have reported exclusively halfwave potentials.
2. When dealing with a large group of compounds (e.g., amines or transition metal complexes), we selected those that are frequently encountered in photoinduced electron-transfer studies; within a certain series (e.g., the fullerenes), we included only the most representative species.
3. We have uniformly tried to list $E_{1/2}$ with respect to standard calomel electrodes (SCE). There are large differences between SCE and the various Ag electrodes that are popular to use in organic solvents. Only in the case of the nitriles (Table 7b-7) did we quote $E_{1/2}$ vs. Ag electrodes. Unfortunately these $E_{1/2}$ values are not compatible for use in Eq. 7-5 when the other $E_{1/2}$ values has been measured relative to SCE. These values vs. Ag electrodes can still be used to make qualitative correlations between the various nitriles as quenchers.
4. We tried consistently to choose $E_{1/2}(X/X^-)$ values in N,N-dimethylformamide (DMF) solutions and $E_{1/2}(X^+/X)$ values in MeCN. These choices were made since they gave the largest selection of $E_{1/2}$ data in non aqueous solvents. Given the uncertainties in the measurements, mixing DMF and MeCN results are not nearly as serious as mixing $E_{1/2}$ from different reference electrodes. For example, in MeCN, $E_{1/2}$ values measured relative to Ag electrodes can vary anywhere from ±0.3 V compared to SCE [7001].
5. It is not clear whether all the data on nitriles refer to reversible processes.

---

[a] Traditionally this equation has been written as:

$$\Delta G^0 = N_A \, e \, (E^0_{\text{ox}} - E^0_{\text{red}})$$

where E^0_{ox} and E^0_{red} are the electrode potential values at which the oxidation and reduction processes, respectively, occur. Moreover, in some older literature, the standard electromotive forces of oxidation and reduction are called, respectively, "oxidation potential" and "reduction potential". These terms are intrinsically confusing and should be avoided altogether, because they conflate the chemical concept of reaction with the physical concept of electrical potential [0501].

Table 7b-1 Halfwave Reduction Potentials of Amines

No.	Compound	$E_{\frac{1}{2}}(D^+/D)$ (V vs. SCE) MeCN	Ref.
1	Aniline	+0.98	[7503]
2	Aniline, 2,4-dimethoxy-*N,N*-dimethyl-	+0.27	[6403]
3	Aniline, 3,4-dimethoxy-*N,N*-dimethyl-	+0.20	[6403]
4	Aniline, 3,5-dimethoxy-*N,N*-dimethyl-	+0.50	[6403]
5	Aniline, *N,N*-diethyl-	+0.76	[8102]
6	Aniline, *N,N*-dimethyl-	+0.81	[8102]
7	Aniline, *N,N*-diphenyl-	+0.92	[7503]
8	Aniline, 2-methoxy-*N,N*-dimethyl-	+0.48	[6403]
9	Aniline, 3-methoxy-*N,N*-dimethyl-	+0.49	[6403]
10	Aniline, 4-methoxy-*N,N*-dimethyl-	+0.33	[6403]
11	Aniline, 4-methyl-	+0.78	[7503]
12	Aniline, 4-methyl-*N,N*-dimethyl-	+0.72	[9605]
13	Aniline, *N*-phenyl-	+0.86	[7503]
14	Anthracene, 2-amino-	+0.44	[7503]
15	Anthracene, 9-amino-	+0.15	[7503]
16	1-Azabicyclo[2.2.2]octane (Quinuclidine)	+0.82	[8401]
17	Benzidine	+0.55	[9502]
18	Benzidine, *N,N,N',N'*-tetramethyl-	+0.43	[7503]
19	Cyclohexylamine	+1.78	[9605]
20	1,4-Diazabicyclo[2.2.2]octane (DABCO)	+0.56	[0002]
21	Dibenzylamine	+1.38	[7801]
22	Dibutylamine	+1.17	[7801]
23	Dicyclohexylamine	+1.12	[7801]
24	Diethylamine	+1.30	[7801]
25	Di-*i*-propylethylamine	+0.90	[7001]
26	Diphenylamine	+0.94	[9605]
27	Dipropylamine	+1.22	[7801]
28	Indole	+1.21	[0303]
29	Indole, 1,2-dimethyl-	+1.09	[0303]
30	Indole, 1-methyl-	+1.20	[0303]
31	Indole, 2-methyl-	+1.07	[0303]
32	*m*-Phenylenediamine, *N,N,N',N'*-tetramethyl-	+0.32	[6403]

Table 7b-1 Halfwave Reduction Potentials of Amines

No.	Compound	$E_{1/2}(D^+/D)$ (V vs. SCE) MeCN	Ref.
33	*o*-Phenylenediamine	+0.45	[0003]
34	*o*-Phenylenediamine, *N,N,N',N'*-tetramethyl-	+0.28	[6403]
35	*p*-Phenylenediamine	+0.28	[9502]
36	*p*-Phenylenediamine, *N,N,N',N'*-tetramethyl-	+0.13	[9605]
37	Piperidine	+1.12	[9603]
38	Pyrene, 1,6-bis(dimethylamino)-	+0.49	[6701]
39	Pyrrolidine	+0.85	[9603]
40	Tributylamine	+0.92	[7801]
41	Triethanolamine	+0.90	[0303]
42	Triethylamine	+0.96	[7801]
43	Trimethylamine	+1.12	[7001]
44	Triphenylamine	+0.86	[7801]

Table 7b-2 Halfwave Reduction Potentials of Selected Electron Donors

No.	Compound	$E_{1/2}(D^+/D)$ (V vs. SCE) MeCN	Ref.
1	Adenine	+1.70	[9604]
2	Anthracene, 9,10-bis(2,6-dimethoxyphenyl)-	+1.18	[6701]
3	Anthracene, 9,10-bis(methylthio)-	+1.11	[6701]
4	Anthracene, 9,10-dimethoxy-	+0.98	[7001]
5	Anthracene, 9,10-diphenoxy-	+1.20	[6701]
6	Anthracene, 9-methoxy-	+1.05	[7001]
7	Benzene, 1,2-dimethoxy-	+1.45	[6402]
8	Benzene, 1,3-dimethoxy-	+1.50	[9605]
9	Benzene, 1,4-dimethoxy-	+1.34	[6402]
10	Benzene, hexamethoxy-	+1.24	[6402]
11	Benzene, methoxy-	+1.76	[6402]
12	Benzene, pentamethoxy-	+1.07	[6402]
13	Benzene, 1,2,3,4-tetramethoxy-	+1.25	[6402]
14	Benzene, 1,2,3,5-tetramethoxy-	+1.09	[6402]
15	Benzene, 1,2,4,5-tetramethoxy-	+0.81	[6402]
16	Benzene, 1,2,3-trimethoxy-	+1.42	[6402]
17	Benzene, 1,2,4-trimethoxy-	+1.12	[6402]
18	Biphenyl, 2,2'-bis(methylthio)-	+1.39	[6701]
19	Biphenyl, 3,3'-bis(methylthio)-	+1.475	[6701]
20	Biphenyl, 4,4'-bis(methylthio)-	+1.255	[6701]
21	Biphenyl, 2,2'-dimethoxy-	+1.51	[6701]
22	Biphenyl, 3,3'-dimethoxy-	+1.60	[6701]
23	Biphenyl, 4,4'-dimethoxy-	+1.30	[6701]
24	Biphenyl, 4-methoxy-	+1.53	[6701]
25	Carbazole	+1.16	[7001]
26	Carbazole, *N*-ethyl-	+1.14	[0101]
27	Carbazole, *N*-methyl-	+1.10	[7001]
28	Corrole, octaethyl-	+0.38[a]	[0006]
29	Cytosine	+1.88	[9604]
30	Guanine	+1.23	[9604]
31	Phenoselenazine, *N*-methyl-	+0.77	[0401]
32	Phenothiazine	+0.59	[7201]

Table 7b-2 Halfwave Reduction Potentials of Selected Electron Donors

No.	Compound	$E_{1/2}(D^+/D)$ (V vs. SCE) MeCN	Ref.
33	Phenothiazine, N-methyl-	+0.72	[0401]
34	Phenoxazine	+0.59	[7201]
35	Phenoxazine, N-methyl-	+0.63	[0401]
36	Porphyrin, octaethyl-[b]	+0.83	[0005]
37	Porphyrin, tetrabenzo-[c]	+0.55	[0005]
38	Porphyrin, tetramesityl-[b]	+0.91	[0005]
39	Porphyrin, tetraphenyl-[b]	+1.00	[0005]
40	Pyrene, 1,6-bis(methylthio)-	+0.96	[6701]
41	Pyrene, 1,6-dimethoxy-	+0.82	[6701]
42	Tetraselenafulvalene	+0.48	[7501]
43	Tetrathiafulvalene	+0.32	[9904]
44	Tetrathiafulvalene, bis(ethylenedithio)-	+0.49	[9102]
45	Tetrathiafulvalene, bis(pyrrolo)-	+0.38	[9905]
46	Thiophene	+2.15	[7001]
47	Thymine	+1.85	[9604]

[a] In CH_2Cl_2, vs. Ag/AgCl; [b] In CH_2Cl_2; [c] In dimethylsulfoxide.

Table 7b-3 Halfwave Reduction Potentials of Aromatic Hydrocarbons

No.	Compound	$E_{1/2}(X^+/X)$ (V vs. SCE) MeCN	Ref.	$E_{1/2}(X/X^-)$ (V vs. SCE) DMF	Ref.
1	Acenaphthene	+1.21	[6301]	−2.67	[7001]
2	Acetylene, diphenyl-	−		−2.11	[7701]
3	Anthracene	+1.09	[6301]	−1.95	[7701]
4	Anthracene, 9,10-bis(phenylethynyl)-	+1.165	[6701]	−1.29	[6701]
5	Anthracene, 9,10-dimethyl-	+0.95	[6401]	−	
6	Anthracene, 9,10-diphenyl-	+1.22	[7701]	−1.94	[7701]
7	Anthracene, 9-methyl-	+0.96	[6301]	−1.97[a]	[6201]
8	Anthracene, 9-phenyl-	−		−1.86	[7001]
9	Azulene	+0.71	[6301]	−1.65[a]	[6201]
10	Benz[*a*]antracene	+1.18	[6301]	−	
11	Benz[a]anthracene, 7,12-dimethyl-	+0.96	[9906]	−	
12	Benz[*a*]antracene, 10-methyl-	+1.14	[6301]	−2.03	[7001]
13	Benz[*a*]antracene, 11-methyl-	+1.14	[6301]	−2.00	[7001]
14	Benz[*a*]antracene, 12-methyl-	+1.07	[6301]	−1.98	[7001]
15	Benz[*a*]antracene, 1-methyl-	+1.14	[6301]	−2.03	[7001]
16	Benz[*a*]antracene, 2-methyl-	+1.14	[6301]	−2.00	[7001]
17	Benz[*a*]antracene, 3-methyl-	+1.14	[6301]	−2.00	[7001]
18	Benz[*a*]antracene, 5-methyl-	+1.15	[6301]	−2.02	[7001]
19	Benz[*a*]antracene, 6-methyl-	+1.15	[6301]	−2.01	[7001]
20	Benz[*a*]antracene, 7-methyl-	+1.08	[6301]	−1.99	[7001]
21	Benz[*a*]antracene, 8-methyl-	+1.13	[6301]	−2.02	[7001]
22	Benz[*a*]anthracene, 9-methyl-	+1.15	[6301]	-2.02	[7001]
23	Benzene	+2.30	[7701]	−	
24	Benzene, hexamethyl-	+1.46	[6401]	−	
25	Benzene, 1,3,5-trimethyl-	+1.85	[6401]	−	
26	Benzo[*ghi*]perylene	+1.01	[9906]	−	
27	Benzo[*c*]phenanthrene	−		−2.20	[7001]
28	Benzo[*c*]phenanthrene, 1-methyl-	−		−2.17	[7001]
29	Benzo[*c*]phenanthrene, 3-methyl-	−		−2.22	[7001]
30	Benzo[*c*]phenanthrene, 4-methyl-	−		−2.20	[7001]
31	Benzo[*c*]phenanthrene, 5-methyl-	−		−2.22	[7001]
32	Benzo[*c*]phenanthrene, 6-methyl-	−		−2.20	[7001]
33	Benzo[*a*]pyrene	+0.94	[6301]	−2.10	[7001]
34	Benzo[*e*]pyrene	+1.27	[6301]	−2.13[a]	[6201]

Table 7b-3 Halfwave Reduction Potentials of Aromatic Hydrocarbons

No.	Compound	$E_{1/2}(X^+/X)$ (V vs. SCE) MeCN	Ref.	$E_{1/2}(X/X^-)$ (V vs. SCE) DMF	Ref.
35	Benzo[*b*]triphenylene	+1.25	[6301]	−2.08[a]	[6201]
36	Biphenyl	−		−2.55	[7701]
37	Biphenylene	−		−2.28[a]	[6201]
38	1,3-Butadiene	+2.33	[6401]	−	
39	1,3-Butadiene, 2,3-dimethyl-	+2.13	[6401]	−	
40	1,3-Butadiene, 2-methyl-	+2.14	[6401]	−	
41	Cholanthene, 3-methyl-	+0.87	[9906]	−	
42	Chrysene	+1.35	[6301]	−2.25	[7001]
43	Coronene	+1.23	[6301]	−2.07	[7001]
44	Dibenz[*a,c*]anthracene	+1.25	[9906]	−	
45	Dibenz[*a,h*]anthracene	+1.19	[6301]	−2.10[a]	[6201]
46	Dibenz[*a,j*]anthracene	+1.26	[9906]	−	
47	Ethene, 1,1-diphenyl-	−		−2.30	[7001]
48	Ethene, 1,2-(*E*)-diphenyl-	−		−2.21	[7001]
49	Ethene, 1,2-(*Z*)-diphenyl-	−		−2.07	[7001]
50	Fluoranthene	+1.45	[6301]	−1.74	[7001]
51	Fullerene C_{60}	+1.26[b]	[9301]	−0.35[c]	[0304]
52	Fullerene C_{70}	+1.20[b]	[9301]	−0.39[c]	[9103]
53	Indan	+1.59	[9906]	−	
54	Naphthacene	+0.77	[9906]	−1.58	[9906]
55	Naphthalene	+1.54	[6301]	−2.49	[7701]
56	Naphtho[1,2,3,4-*def*]chrysene	+1.01	[6301]	−1.91[a]	[6201]
57	Perylene	+0.85	[6301]	−1.67	[7001]
58	Phenanthrene	+1.50	[6301]	−2.44	[7001]
59	Pyrene	+1.16	[6301]	−2.09	[7001]
60	Rubrene	−		−1.41	[7001]
61	(*E*)-Stilbene	+1.43	[7701]	−2.08	[7701]
62	(*Z*)-Stilbene	−		−2.07	[7701]
63	Styrene	−		−2.60	[7701]
64	Tetracene	+0.77	[6301]	−1.58	[7001]
65	Toluene	+2.28	[6401]	−	
66	Triphenylene	+1.55	[6301]	−2.46[a]	[6201]

[a] Values measured relative to Hg-pool reference electrode and adjusted to SCE by subtracting 0.55 V [6201]; [b] In 1,1,2,2-tetrachloroethane, vs. ferrocenium/ferrocene couple; [c] In tetrahydrofuran.

Table 7b-4 Halfwave Reduction Potentials and Lowest Singlet Excited State Energies of Naphthalenes

No.	Compound	$E_{\frac{1}{2}}(X^+/X)$ (V vs. SCE) MeCN[a]	$E_{\frac{1}{2}}(X/X^-)$ (V vs. SCE) 75%DO[b]	$E_{\frac{1}{2}}(X/X^-)$ (V vs. SCE) DMF[a]	E_{S_1}[c] (eV)
1	Naphthalene	+1.70[d]	−2.437	−	3.99
2	Naphthalene, 1-allyl-	−	−2.447	−	−
3	Naphthalene, 2-allyl-	−	−2.444	−	−
4	Naphthalene, 1-amino-	+0.54[e]	−	−	−
5	Naphthalene, 2-amino-	+0.64[e]	−	−	−
6	Naphthalene, 1,5-bis(dimethylamino)-	+0.585	−	−2.64	−
7	Naphthalene, 2,6-bis(dimethylamino)-	+0.26	−	−2.71	−
8	Naphthalene, 2,7-bis(dimethylamino)-	+0.57	−	−2.77	−
9	Naphthalene, 1,4-bis(methylthio)-	+1.07	−	−2.10	−
10	Naphthalene, 1,5-bis(methylthio)-	+1.265	−	−2.15	−
11	Naphthalene, 1,8-bis(methylthio)-	+1.09	−	−2.22	−
12	Naphthalene, 2,3-bis(methylthio)-	+1.355	−	−2.21	−
13	Naphthalene, 2,6-bis(methylthio)-	+1.10	−	−2.24	−
14	Naphthalene, 2,7-bis(methylthio)-	+1.33	−	−2.25	−
15	Naphthalene, 1-*t*-butyl-	−	−2.493	−	−
16	Naphthalene, 1-cyano-		See table 7b-7		
17	Naphthalene, 1,4-dicyano-		See table 7b-7		
18	Naphthalene, 1,5-disulphonate-	−	−	−2.15[f]	−
19	Naphthalene, 1,3-dimethoxy-	+1.265	−	−2.61	−
20	Naphthalene, 1,4-dimethoxy-	+1.10	−	−2.69	−
21	Naphthalene, 1,5-dimethoxy-	+1.28	−	−2.76	−
22	Naphthalene, 1,6-dimethoxy-	+1.28	−	−2.68	−
23	Naphthalene, 1,7-dimethoxy-	+1.28	−	−2.67	−
24	Naphthalene, 1,8-dimethoxy-	+1.17	−	−2.72	−
25	Naphthalene, 2,3-dimethoxy-	+1.39	−	−2.73	−
26	Naphthalene, 2,6-dimethoxy-	+1.33	−	−2.60	−
27	Naphthalene, 2,7-dimethoxy-	+1.47	−	−2.68	−
28	Naphthalene, 1,2-dimethyl-	−	−2.479	−	3.84
29	Naphthalene, 1,3-dimethyl-	−	−2.483	−	3.85
30	Naphthalene, 1,4-dimethyl-	−	−2.471	−	3.85
31	Naphthalene, 1,5-dimethyl-	−	−2.475	−	3.86

Table 7b-4 Halfwave Reduction Potentials and Lowest Singlet Excited State Energies of Naphthalenes

No.	Compound	$E_{1/2}(X^+/X)$ (V vs. SCE) MeCN[a]	$E_{1/2}(X/X^-)$ (V vs. SCE) 75%DO[b]	$E_{1/2}(X/X^-)$ (V vs. SCE) DMF[a]	E_{S_1}[c] (eV)
32	Naphthalene, 1,6-dimethyl-	–	–2.476	–	3.85
33	Naphthalene, 1,7-dimethyl-	–	–2.469	–	3.85
34	Naphthalene, 1,8-dimethyl-	–	–2.521	–	3.85
35	Naphthalene, 2,3-dimethyl-	+1.38[g]	–2.501	–	3.87
36	Naphthalene, 2,6-dimethyl-	+1.36[g]	–2.476	–	3.83
37	Naphthalene, 2,7-dimethyl-	–	–2.485	–	3.86
38	Naphthalene, 1-dimethylamino-	+0.75	–	–2.58	–
39	Naphthalene, 2-dimethylamino-	+0.67	–	–2.63	–
40	Naphthalene, 2-hydroxy-	–	–	–2.52[g,h]	–
41	Naphthalene, 1-methoxy-	+1.38	–	–2.65	–
42	Naphthalene, 2-methoxy-	+1.52	–	–2.60	–
43	Naphthalene, 1-methyl-	+1.43[g]	–2.458	–	3.91
44	Naphthalene, 2-methyl-	+1.45[g]	–2.460	–	3.87
45	Naphthalene, 1-methylthio-	+1.32	–	–2.25	–
46	Naphthalene, 2-methylthio-	+1.365	–	–2.28	–
47	Naphthalene, 1-nitro-	–	–	–0.97[g]	–
48	Naphthalene, 2-nitro-	–	–	–0.98[g]	–
49	Naphthalene, 1,4,5,8-tetramethoxy-	+0.70	–	–2.69	–
50	Naphthalene, 1,3,7-trimethyl-	–	–2.496	–	–
51	Naphthalene, 1,4,5-trimethyl-	–	–2.529	–	–
52	Naphthalene, 1,6,7-trimethyl-	–	–2.515	–	–
53	Naphthalene, 2,3,6-trimethyl-	–	–2.523	–	–
54	Naphthalene, 1-vinyl-	–	–2.09	–	–
55	Naphthalene, 2-vinyl-	–	–2.15	–	3.66

[a] Supporting electrolyte: 0.1 mol L^{-1} tetrabutylammonium perchlorate; from ref. [6701], unless otherwise noted; [b] In 75% dioxane-water; supporting electrolyte: 0.1 mol L^{-1} tetrabutylammonium iodide; from refs. [6001], [6302]; [c] From refs. [6404], [7002], [7101]; [d] From ref. [7701]; [e] From ref. [7801]; [f] From ref. [7001]; [g] From ref. [7701]; [h] Value measured relative to Hg-pool reference electrode and adjusted to SCE by subtracting 0.55 V [6201].

Table 7b-5 Halfwave Reduction Potentials of Azaaromatics Compounds

No.	Compound	$E_{1/2}(A/A^-)$ (V vs. SCE) MeCN	Ref.
1	Acridine	−1.62	[7503]
2	Benzo[c]cinnoline	−1.55	[7503]
3	Benzo[f]quinoline	−2.14	[7503]
4	Benzo[h]quinoline	−2.21	[7503]
5	Benzo[f]quinoxaline	−1.74	[7503]
6	2,2′-Bipyridine	−2.18	[8901]
7	2,2′-Bipyridine, 4,4′-dimethyl-	−2.24	[8901]
8	2,2′-Bipyridinium, N,N′-dimethyl-	−0.73	[7901]
9	3,3′-Bipyridinium, N,N′-dimethyl-	−0.84	[7901]
10	4,4′-Bipyridinium, N,N′-bis(n-butyl)-	−0.44	[9002]
11	4,4′-Bipyridinium, N,N′-bis(cyanomethyl)-	−0.18	[9002]
12	4,4′-Bipyridinium, N,N′-bis(n-octyl)-	−0.44	[9002]
13	4,4′-Bipyridinium, N,N′-dibenzyl-	−0.35	[9503]
14	4,4′-Bipyridinium, N,N′-dimethyl-	−0.45	[9002]
15	4,4′-Bipyridinium, N,N′-diphenyl-	−0.24	[9002]
16	2,2′-Bipyridimine	−1.80	[8901]
17	2,3-Bis(2-pyridyl)pyrazine	−1.93[a]	[0201]
18	2,5-Bis(2-pyridyl)pyrazine	−1.53[a]	[0201]
19	Cinnoline	−1.69	[7503]
20	Cyclobis(paraquat-p-phenylene)	−0.29	[9503]
21	2,7-Diazapyrenium, N,N′-diphenyl-	−0.41	[0102]
22	2,7-Diazapyrenium, N,N′-dimethyl-	−0.46	[0102]
23	Isoquinoline	−2.22	[7503]
24	Phenanthridine	−2.12	[7503]
25	1,10-Phenanthroline	−2.04	[8901]
26	1,7-Phenanthroline	−2.09	[7503]
27	4,7-Phenanthroline	−2.04	[7503]
28	Phenazine	−1.23	[7503]
29	Phthalazine	−1.98	[7503]
30	Pyrazine	−2.08	[7503]
31	Pyridazine	−2.12	[7503]
32	Pyridine	−2.62	[7503]

Table 7b-5 Halfwave Reduction Potentials of Azaaromatics Compounds

No.	Compound	$E_{1/2}(A/A^-)$ (V vs. SCE) MeCN	Ref.
33	Pyrimidine	−2.08	[7503]
34	Quinazoline	−1.80	[7503]
35	Quinoline	−2.11	[7503]
36	Quinoxaline	−1.70	[7503]
37	2,2′:6′,2″-Terpyridine	−1.99	[0302]

[a] In DMF solution at −54 °C.

Table 7b-6 Halfwave Reduction Potentials of Carbonyl Compounds

No.	Compound	$E_{1/2}(A/A^-)$ (V vs. SCE)	Solvent	$E_{1/2}(A/A^-)$ (V vs. SCE) EtOH/H$_2$O 1:1[a]
1	Acetophenone	-2.14^b	MeCN	-1.66
2	Acetophenone, 4'-chloro-	-2.10^b	MeCN	–
3	Acetophenone, 4'-cyano-	-1.58^b	MeCN	–
4	Acetophenone, 4'-methoxy-	-2.23^b	MeCN	–
5	Acetophenone, 4'-methyl-	-2.19^b	MeCN	–
6	9,10-Anthraquinone	-0.86^c	DMF	–
7	Anthrone	-1.51^c	DMF	–
8	Benzaldehyde	-1.93^c	DMF	–
9	Benzil	$-1.16^{d,e}$	MeCN	-0.71
10	Benzoic acid	-2.24^c	DMF	–
11	Benzoic acid, methyl ester	-2.32^c	DMF	–
12	Benzophenone	-1.83^b	MeCN	-1.55
13	Benzophenone, 4-chloro-	-1.75^b	MeCN	–
14	Benzophenone, 4-cyano-	-1.42^b	MeCN	–
15	Benzophenone, 4,4'-dimethoxy-	-2.02^b	MeCN	–
16	Benzophenone, 4,4'-dimethyl-	-1.90^b	MeCN	–
17	1,4-Benzoquinone	-0.45^c	DMF	–
18	1,4-Benzoquinone, tetrachloro-	$+0.02^f$	MeCN	–
19	1,4-Benzoquinone, tetrafluoro-	-0.04^f	MeCN	–
20	1,4-Benzoquinone, tetramethyl-	-0.69^f	MeCN	–
21	1,4-Benzoquinone, trichloro-	-0.08^f	MeCN	–
22	Biacetyl	$-1.32^{d,e}$	MeCN	-1.03
23	2-Cyclopentenone	-2.16^c	DMF	–
24	9-Fluorenone	-1.29^c	DMF	-1.21
25	Maleic anhydride	-0.85^c	DMF	–
26	2-Naphthaldehyde	–		-1.34
27	Naphthalene, 2-acetyl-	–		-1.72
28	1,4-Naphthoquinone	-0.63^c	DMF	–
29	9,10-Phenanthrenequinone	$-0.67^{c,g}$	DMF	–
30	Phthalic anhydride	-1.27^c	DMF	–
31	Xanthone	$-1.77^{c,g}$	DMF	–

[a] From ref. [7802]; [b] From ref. [8601]; [c] From ref. [7701]; [d] From ref. [8701]; [e] Value measured relative to a Ag/Ag$^+$ reference electrode and adjusted to SCE [7001]; [f] From ref. [7001]; [g] Values measured relative to Hg-pool reference electrode and adjusted to SCE by subtracting 0.55 V [6201].

Table 7b-7 Halfwave Reduction Potentials of Nitriles

No.	Compound	$E_{1/2}(A/A^-)$ (V vs. Ag electrode) DMF[a]	$E_{1/2}(A/A^-)$ (V vs. SCE) MeCN
1	Anthracene, 9-cyano-	–	-1.58^b
2	Anthracene, 9,10-dicyano-	–	-0.98^b
3	Benzene, 1-cyano-3,5-dinitro-	–0.96	–
4	Benzene, 1-cyano-4-nitro-	–1.25	–
5	Benzene, 1,2-dicyano-	–2.12	–
6	Benzene, 1,3-dicyano-	–2.17	–
7	Benzene, 1,4-dicyano-	–1.97	–
8	Benzene, 1,2,4,5-tetracyano-	–1.02	-0.71^a
9	Benzonitrile	–2.74	–
10	Benzonitrile, 4-chloro-	–2.4	–
11	Benzonitrile, 4-methoxy-	–3	–
12	Benzonitrile, 4-methyl-	–2.75	–
13	Benzoylacetonitrile	–2	–
14	4-Cyanobenzoic acid	–1.91	–
15	Ethene, tetracyano-	–	$+0.24^a$
16	Naphthalene, 1-cyano-	–	$-2.33^{c,d}$
17	Naphthalene, 1,4-dicyano-	–	$-1.67^{c,d}$
18	Pyridine, 4-cyano-	–2.03	–
19	7,7,8,8-Tetracyanoquinodimethane	–0.2	$+0.19^a$

[a] From ref. [7001]; [b] From ref. [7804]; [c] From ref. [7602]; [d] Values measured relative to a Ag/Ag+ reference electrode, not adjusted to SCE.

Table 7b-8 Halfwave Reduction Potentials of Halogenated Benzenes[a]

No.	Substituent	Iodobenzenes	Bromobenzenes	Chlorobenzenes
1	H	−1.21	−1.81	−2.13
2	m-Br	−0.96	−1.45	−
3	p-Br	−1.08	−1.54	−
4	m-CF$_3$	−1.00	−1.52	−
5	p-CF$_3$	−1.01	−1.53	−
6	p-CHO	−0.96	−	−1.20
7	m-C$_6$H$_5$	−1.15	−1.58	−
8	p-C$_6$H$_5$	−1.16	−1.56	−
9	m-CN	−	−1.29	−
10	p-CN	−	−1.26	−1.36
11	p-COC$_6$H$_5$	−0.96	−1.06	−
12	m-COMe	−	−1.19	−
13	p-COMe	−1.04	−1.15	−
14	m-Cl	−0.98	−1.53	−
15	p-Cl	−1.06	−1.61	−1.85
16	m-I	−0.92	−	−
17	p-I	−1.01	−	−
18	m-Me	−1.22	−1.85	−2.16
19	p-Me	−1.23	−1.84	−2.16
20	p-NHCOMe	−	−1.88	−2.09
21	p-NH$_2$	−	−1.96	−
22	m-NMe$_2$	−	−	−2.23
23	p-NMe$_2$	−1.35	−1.97	−
24	m-NMe$_3^+$	−	−	−1.48
25	p-NMe$_3^+$	−0.91	−1.34	−
26	p-OC$_6$H$_5$	−	−1.73	−
27	p-OEt	−	−1.82	−
28	m-OMe	−1.19	−1.76	−
29	p-OMe	−1.25	−1.84	−2.15

[a] From ref. [6802]; DMF solutions, values relative to a Ag/AgBr electrode; Supporting electrolyte: 0.02 mol L^{-1} tetraethylammonium bromide.

Table 7b-9 Halfwave Reduction Potentials of Selected Electron Acceptors

No.	Compound	$E_{1/2}(A/A^-)$ (V vs. SCE)	Solvent	Ref.
1	(E)-Azobenzene	−1.36	DMF	[6801]
2	(E)-4,4′-Azobiphenyl	−1.22	DMF	[6801]
3	(E)-1,1′-Azonaphthalene	−1.13	DMF	[6801]
4	(E)-2,2′-Azonaphthalene	−1.22	DMF	[6801]
5	(E)-4,4′-Azopyridine	−0.80	DMF	[6801]
6	Benzaldehyde, 4-nitro-	−0.86	MeCN	[6101]
7	Benzene, 1-chloro-4-nitro-	−1.06	MeCN	[6101]
8	Benzene, 1-fluoro-4-nitro-	−1.13	MeCN	[6101]
9	Benzene, 1,4-dinitro-	−0.69	MeCN	[6101]
10	Benzene, 1-methyl-2-nitro-	−1.28	DMF	[7701]
11	Benzene, 1-methyl-4-nitro	−1.14	DMF	[7701]
12	Benzene, nitro	−1.08	DMF	[7701]
13	Benzene, 1,1′-sulfonylbis-	−2.05	DMF	[7701]
14	BDPY[a]	−1.15[b]	CH_2Cl_2	[0306]
15	1,3-Butadiene, 1,4-diphenyl-	−1.98	Dioxane 75%	[7503]
16	β-Carotene	−1.63	THF	[7803]
17	Cyclopentadiene	−2.91	DMF	[7701]
18	Diazene, diphenyl-	−1.36	DMF	[7701]
19	Eosin Y dianion[c]	−1.38	MeCN	[0402]
20	Fulleropyrrolidine, N-TEG-[d]	−0.44	THF	[0304]
21	Fulleropyrrolidinium, N-methyl-N-TEG-[d]	−0.29[e]	THF	[0304]
22	1,3,5-Hexatriene, 1,6-diphenyl-	−1.76	Dioxane 96%	[7503]
23	Isoprene	−2.70	DMF	[7701]
24	Mellitic triimide, N,N′,N″-tris-(n-butyl)]-[f]	−0.52	CH_2Cl_2	[0305]
25	Methanofullerene, bis(trimethylsilylethynyl)-	−0.52	CH_2Cl_2	[9802]
26	Methyl 4-nitrobenzoate	−0.95	MeCN	[6101]
27	1,4,5,8-Napthalene diimide, N,N′-bis(n-pentyl)-	−0.70	CH_2Cl_2	[0004]
28	1,8-Naphthalimide	−1.30	MeCN	[7001]
29	1,3,5,7-Octatetraene, 1,8-diphenyl-	−1.62	Dioxane 96%	[7503]
30	Oligo(p-phenylenevinylene)-1[g]	−2.29[b]	THF	[9104]
31	Oligo(p-phenylenevinylene)-2[h]	−2.17[b]	THF	[9104]
32	Oligo(p-phenylenevinylene)-3[i]	−2.06[b]	THF	[9104]
33	Oligo(p-phenylenevinylene)-4[j]	−2.01[b]	THF	[9104]

Table 7b-9 Halfwave Reduction Potentials of Selected Electron Acceptors

No.	Compound	$E_{1/2}(A/A^-)$ (V vs. SCE)	Solvent	Ref.
34	Phthalimide	−1.47	DMF	[7001]
35	Phthalocyanine	−0.7	DMF	[9302]
36	Phthalocyanine, tetrabutyl-	−0.7	DMF	[9302]
37	Phthalocyanine, tetraneopenthoxy-	−0.9	DMF	[9302]
38	Phthalocyanine, tetrasulfonato-	−0.53	DMF	[9302]
39	Piperylene	−2.76	DMF	[7701]
40	Porphyrin, octaethyl-	−1.46	CH_2Cl_2	[0005]
41	Porphyrin, tetrabenzo-	−1.13	DMSO	[0005]
42	Porphyrin, tetramesityl-	−1.41	CH_2Cl_2	[0005]
43	Porphyrin, tetraphenyl-	−1.23	CH_2Cl_2	[0005]
44	Pyridine-N-oxide	−2.30	DMF	[7701]
45	Pyromellitic diimide	−0.82	DMF	[7001]
46	Pyromellitic diimide, N,N'-bis(n-pentyl)-	−0.95	CH_2Cl_2	[0004]
47	Retinal	−1.42	THF	[7803]
48	Thiofluorenone	−1.07	MeCN	[8502]
49	Thioxanthione	−1.05	MeCN	[8502]
50	Triphenylphosphine[k]	−1.53	DMF	[7001]

[a] 4,4-Difluoro-1,3,5,7-tetramethyl-4-bora-3a,4a-diaza-s-indacene; [b] Value reported vs. ferrocenium/ferrocene couple and adjusted to SCE (see Table 7b-10 [9203]); [c] Tetrabutylammonium salt; [d] TEG: triethyleneglycol monomethylether; [e] At −60 °C; [f] Value reported vs. decamethylferrocenium/decamethylferrocene couple and adjusted to SCE (see Table 7b-10 [9907]); [g] 1,4-Bis(3,5-di-t-butylstyryl)benzene; [h] 1,4-Bis(3,5-di-t-butylstyryl)stilbene; [i] 1,4-Bis[4-(3,5-di-t-butylstyryl)styryl]-benzene; [j] 1,4-Bis[4-(3,5-di-t-butylstyryl)styryl]stilbene; [k] Value measured relative to Hg-pool reference electrode and adjusted to SCE by subtracting 0.55 V [6201].

Table 7b-10 Halfwave Reduction Potentials of Sandwich Metal Compounds

No.	Compound	$E_{1/2}(X^+/X)$ (V vs. SCE)	Solvent	Ref.
1	Bis(benzene)chromium	−0.83	DMF	[7001]
2	Chromocene	−1.04	MeCN	[0202]
3	Cobaltocene	−0.91	DMF	[0202]
4	Cobaltocene, decamethyl-	−1.43	DMF	[0202]
5	Ferrocene	+0.41[a]	MeCN	[9203]
		+0.51[a]	CH_2Cl_2	[9203]
		+0.51[a]	DMF	[9203]
		+0.57[a]	THF	[9203]
6	Ferrocene, acetamido-	+0.223	MeCN	[7001]
7	Ferrocene, acetyl-	+0.58	MeCN	[7001]
8	Ferrocene, benzyl-	+0.314	MeCN	[7001]
9	Ferrocene, carboxylic acid	+0.550	MeCN	[7001]
10	Ferrocene, decamethyl-	−0.095[a,b]	MeCN	[9907]
		−0.022[a,b]	CH_2Cl_2	[9907]
11	Ferrocene, 1,1′-diacetyl-	+0.796	MeCN	[7001]
12	Ferrocene, 1,1′-dibenzyl-	+0.296	MeCN	[7001]
13	Ferrocene, 1,1′-dimethyl-	+0.241	MeCN	[7001]
14	Ferrocene, 1,1′-diphenyl-	+0.370	MeCN	[7001]
15	Ferrocene, methyl-	+0.281	MeCN	[7001]
16	Ferrocene, phenyl-	+0.281	MeCN	[7001]
17	Ferrocene, vinyl-	+0.325	MeCN	[7001]
18	Osmocene	+0.633	MeCN	[7001]
19	Osmocene, decamethyl-	+0.46	CH_2Cl_2	[0202]
20	Ruthenocene	+0.693	MeCN	[7001]
21	Ruthenocene, decamethyl-	+0.55	CH_2Cl_2	[0202]

[a] Supporting electrolyte: 0.1 mol L^{-1} tetrabutylammonium perchlorate; [b] Value reported vs. ferrocenium/ferrocene couple and adjusted to SCE [9203].

Table 7b-11 **Halfwave Reduction Potentials of Transition Metal Complexes**

No.	Compound[a]	$E_{1/2}(X^+/X)$ (V vs. SCE)	Solvent	$E_{1/2}(X/X^-)$ (V vs. SCE)	Solvent	Ref.
1	$[Au^{III}(TPP)]^+$	+1.78	CH_2Cl_2	−0.62	CH_2Cl_2	[0005]
2	$[Co^{II}(bpy)_3]^{2+}$	+0.22	MeCN	−		[0203]
3	$Co^{III}(OEC)$	+0.11	CH_2Cl_2	−0.30	CH_2Cl_2	[0006]
4	$[Co^{II}(phen)_2]^{2+}$	+0.30	MeCN	−		[0203]
5	$Co^{II}(Pc)$	+0.86	MeCN	−		[9302]
6	$[Co^{II}(TPP)$	+0.81	CH_2Cl_2	−0.95	CH_2Cl_2	[0005]
7	$Co^{II}(TPPc)$	+0.835	MeCN	−		[9302]
8	$[Cr^{III}(bpy)_3]^{3+}$	−		−0.25	MeCN	[7603]
9	$[Cr^{III}(phen)_3]^{3+}$	−		−0.28	MeCN	[7603]
10	$[Cr^{III}(terpy)_2]^{3+}$	−		−0.15	MeCN	[7603]
11	$[Cu^{II}(bpy)_3]^{2+}$	−		+0.03	MeCN	[9804]
12	$[Cu^{I}(bpy)_2]^+$	+0.18	MeCN	−0.38		[9005]
13	$[Cu^{II}(cyclam)]^{2+}$	−		−0.90	MeCN	[0403]
14	$[Cu^{I}(DPphen)_2]^+$	+0.70	MeCN	−1.66	MeCN	[9908]
15	$[Cu^{I}(phen)_2]^+$	+0.04	MeCN	−0.50	MeCN	[9005]
16	$[Cu^{II}(phen)_3]^{2+}$	−		+0.04	MeCN	[9804]
17	$Cu^{II}(TPP)$	+1.00	CH_2Cl_2	−1.20	DMF	[0005]
18	$[Fe^{II}(bpy)_3]^{2+}$	+1.01[b]	DMF	−1.32[b]	DMF	[9205]
19	$[Fe^{II}(phen)_3]^{2+}$	+1.00[b]	DMF	−1.36[b]	DMF	[9205]
20	$Fe^{III}(OEC)(py)$	+0.21	CH_2Cl_2	−1.04	CH_2Cl_2	[0006]
21	$Fe^{II}(TPP)$	-0.31	DCE^c	−1.01	DCE^c	[0005]
22	$[Fe^{II}(terpy)_2]^{2+}$	+1.18[b]	MeCN	−1.24[b]	MeCN	[9207]
23	$[Ir^{III}(bpy)_3]^{3+}$	−		−0.83	MeCN	[7805]
24	$[Ir^{III}(bpy)(ppy)]^{2+}$	+1.24	MeCN	−1.40	MeCN	[0404]
25	$Ir^{III}(ppy)_3$	+0.63[b]	DMF	−2.33[b]	DMF	[9204]
26	$[Ir^{III}(terpy)_2]^{3+}$	−		−0.76	DMF	[9909]
27	$Mg^{II}(TPP)$	+0.60	CH_2Cl_2	−1.35	DMF	[0005]
28	$Mn^{II}(TPP)$	+1.13	CH_2Cl_2	−1.31	DMF	[0005]
29	$[Ni^{II}(bpy)_3]^{2+}$	+1.70	MeCN	−1.24	MeCN	[8403]
30	$[Ni(cyclam)]^{2+}$	+0.54	MeCN	−		[0403]
31	$Ni^{II}(TPP)$	+1.05	CH_2Cl_2	−1.28	CH_2Cl_2	[0005]
32	$[Os^{II}(bpy)_3]^{2+}$	+0.81	DMF	−1.29	DMF	[8902]
33	$[Os^{II}(phen)_3]^{2+}$	+0.82	DMF	−1.21	DMF	[8902]

Table 7b-11 Halfwave Reduction Potentials of Transition Metal Complexes

No.	Compound[a]	$E_{1/2}(X^+/X)$ (V vs. SCE)	Solvent	$E_{1/2}(X/X^-)$ (V vs. SCE)	Solvent	Ref.
34	$[Os^{II}(terpy)_2]^{2+}$	+0.97	MeCN	−1.23	MeCN	[9403]
35	$Pd^{II}(ppy)_2$	+1.38[d]		−2.04	DMF	[9204]
36	$Pd^{II}(TPP)$	+1.20	CH_2Cl_2	−1.24	CH_2Cl_2	[0005]
37	$[Pt^{II}(bpy)_2]^{2+}$	−		−0.73[b]	DMF	[9805]
38	$Pt^{II}(bpy)Cl_2$	−		−1.12[b]	DMF	[9605]
39	$[Pt^{II}(bpy)(py)_2]^{2+}$	−		−0.87[b]	DMF	[9206]
40	$Pt^{II}(ppy)_2$	+0.26[d]	MeCN	−2.16	MeCN	[9204]
41	$Re^I(bpy)(CO)_3Cl$	+1.36	MeCN	−1.32	MeCN	[9803]
42	$Re^I(terpy)(CO)_2Cl$	+1.19	MeCN	−1.40	DMF	[8804]
43	$[Rh^{III}(phen)_3]^{3+}$	−		−0.76	MeCN	[7504]
44	$[Rh^{III}(phen)(ppy)]^{2+}$	−		−1.67	DMF	[9204]
45	$[Ru^{II}(bpy)_3]^{2+}$	+1.29	MeCN	−1.33	MeCN	[8803]
46	$[Ru^{II}(bpy)(ppy)]^+$	+0.51[e]	MeCN	−1.56[e]	MeCN	[9204]
47	$[Ru^{II}(DMbpy)_3]^{2+}$	+1.10	MeCN	−1.45	MeCN	[8803]
48	$[Ru^{II}(phen)_3]^{2+}$	+1.27	MeCN	−1.35	MeCN	[8803]
49	$Ru^{II}(TPP)CO$	+0.85	CH_2Cl_2	−1.35	DMF	[0005]
50	$[Ru^{II}(terpy)_2]^{2+}$	+1.30	MeCN	−1.24	MeCN	[9403]
51	$Zn^{II}(OEP)$	+0.63	CH_2Cl_2	−1.61	CH_2Cl_2	[0005]
52	$Zn^{II}(Pc)$	+0.695	DMF	−0.85	DMF	[9302]
53	$Zn^{II}(TPP)$	+0.78	CH_2Cl_2	−1.32	CH_2Cl_2	[0005]
54	$Zn^{II}(TPPc)$	+0.755	DMF	−		[9302]

[a] bpy: 2,2′-bipyridine; cyclam: 1,4,8,11-tetraazacyclotetradecane; DMbpy: 4,4′-dimethyl-2,2′-bipyridine; DPphen: 2,9-diphenyl-1,10-phenanthroline; OEC: octaethylcorrole; OEP: octaethylporphyrin; Pc: phthalocyanine; phen: 1,10-phenanthroline; ppy: 2-phenylpyridine; TPP: tetraphenylporphyrin; TPPc: tetraphenylphthalocyanine; terpy: 2,2′:6′,2″-terpyridine; [b] Value reported vs. ferrocenium/ferrocene couple and adjusted to SCE (see Table 7b-10 [9203]); [c] 1,2-Dichloroethane; [d] Not fully reversible process; [e] Value measured relative to a Ag/Ag+ reference electrode and adjusted to SCE [7001].

REFERENCES

[5001] Mulliken, R. S. *J. Am. Chem. Soc.* **1950**, *74*, 811-825.

[6001] Klemm, L. H.; Lind, C. D.; Spence, J. T. *J. Org. Chem.* **1960**, *25*, 611-616.

[6101] Maki, A. H.; Geske, D. H. *J. Am. Chem. Soc.* **1961**, *83*, 1852-1860.

[6201] Streitwieser, A., Jr.; Schwager, I. *J. Phys. Chem.* **1962**, *66*, 2316-2320.

[6301] Pysh, E. S.; Yang, N. C. *J. Am. Chem. Soc.* **1963**, *85*, 2124-2130.

[6302] Klemm, L. H.; Kohlik, A. J. *J. Org. Chem.* **1963**, *28*, 2044-2049.

[6401] Neikam, W. C.; Desmond, M. M. *J. Am. Chem. Soc.* **1964**, *86*, 4811-4819.

[6402] Zweig, A.; Hodgson, W.G.; Jura, W. H. *J. Am. Chem. Soc.* **1964**, *86*, 4124-4129.

[6403] Zweig, A.; Lancaster, J. E.; Neglia, M. T.; Jura, W. H. *J. Chem. Phys.* **1964**, *86*, 4130-4136.

[6404] Clar, E. *Polycyclic Hydrocarbons*; Academic Press: New York (USA), 1964, volume 1.

[6701] Zweig, A.; Maurer, A. H.; Roberts, B. G. *J. Org. Chem.* **1967**, *32*, 1322-1329.

[6801] Sadler, J. L.; Bard, A. J. *J. Am. Chem. Soc.* **1968**, *90*, 1979-1989.

[6802] Sease, J. W.; Burton, F. G.; Nickol, S. L. *J. Am. Chem. Soc.* **1968**, *90*, 2595-2598.

[6901] Nakajima, A.; Akamatu, H. *Bull. Chem. Soc. Jpn.* **1969**, *42*, 3030-3032.

[7001] Mann, C. K.; Barnes, K. K. *Electrochemical Reactions in Nonaqueous Systems*; Marcel Dekker: New York (USA), 1970.

[7002] Birks, J. B. *Photophysics of Aromatic Molecules*; Wiley-Interscience: New York (USA), 1970.

[7003] Rehm, D.; Weller, A. *Isr. J. Chem.* **1970**, *8*, 259-271.

[7101] Perkampus, H. -H.; Sandeman, I.; Timmons, C. J. *DMS UV Atlas of Organic Compounds*; Verlag Chemie: Weinheim (Germany); Butterworths: London (UK), 1966-1971, volumes I-V.

[7201] Kowert, B. A.; Marcoux, L.; Bard, A. J. *J. Am. Chem. Soc.* **1970**, *94*, 5538-5550.

[7301] Maier, J. P.; Turner, D. W. *J. Chem. Soc. Faraday Trans. 2* **1973**, *69*, 521-524.

[7401] Kobayashi, T.; Nagakura, S. *Bull. Chem. Soc. Jpn.* **1974**, *47*, 2565-2567.

[7501] Engler, E. M.; Kaufman, F. B.; Green, D. C.; Klots, C. E.; Compton, R. N. *J. Am. Chem. Soc.* **1975**, *97*, 2921-2922.

[7502] Haink, H. J.; Huber, J. R. *Chem. Ber.* **1975**, *108*, 1118-1124.

[7503] Siegerman, H. *Technique of Electroorganic Synthesis*; Wiley: New York (USA), 1975, volume 5, part II, pp. 667-1056.

[7504] Kew, G.; Hanck, K. W.; DeArmond, K. *J. Phys. Chem.* **1975**, *79*, 1828-1835.

[7601] Aue, D. H.; Webb, H. M.; Bowers, M. T. *J. Am. Chem. Soc.* **1976**, *98*, 311-317.

[7602] Arnold, D. R.; Maroulis, A. J. *J. Am. Chem. Soc.* **1976**, *98*, 5931-5937.

[7603] Hughes, M. C.; Macero, D. J. *Inorg. Chem.* **1976**, *15*, 2040-2044.

[7701] Meites, L.; Zuman, P. *CRC Handbook Series in Organic Electrochemistry*; CRC Press: Boca Raton, FL (USA), 1977-1982, volumes I-V.

[7801] Ballardini, R.; Varani, G.; Indelli, M. T.; Scandola, F.; Balzani, V. *J. Am. Chem. Soc.* **1978**, *100*, 7219-7223.

[7802] Barwise, A. J. G.; Gorman, A. A.; Leyland, R. L.; Smith, P.G.; Rodgers, M. A. J. *J. Am. Chem. Soc.* **1978**, *100*, 1814-1820.

[7803] Park, S. -M. *J. Electrochem. Soc.* **1978**, *125*, 216-222.

[7804] Eriksen, J.; Foote, C. S. *J. Phys. Chem.* **1978**, *82*, 2659-2662.

[7805] Kahl, J. L.; Hanck, K. W.; DeArmond, K. *J. Phys. Chem.* **1978**, *82*, 540-545.

[7901] Bock, C. R.; Connor, J. A.; Gutierrez, A. R.; Meyer, T. J.; Whitten, D. G.; Sullivan, B. P.; Nagle, J. K. *J. Am. Chem. Soc.* **1979**, *101*, 4815-4824.

[8101] Alder, R. W.; Arrowsmith, R. J.; Casson, A.; Sessions, R. B.; Heilbronner, E.; Kovac, B.; Huber, H.; Taagepera, M. *J. Am. Chem. Soc.* **1981**, *103*, 6137-6142.

[8102] Nocera, D. G.; Gray, H. B. *J. Am. Chem. Soc.* **1981**, *103*, 7349-7350.

[8201] Levin, R. D.; Lias, S. G. *National Standards Reference Data Series*; National Bureau of Standards: Washington, DC (USA), 1982, volume 71.

[8401] Hub, W.; Schneider, S.; Doerr, F.; Oxman, J. D.; Lewis, F. D. *J. Am. Chem. Soc.* **1984**, *106*, 701-708.

[8402] Christodoulides, A. A.; McCorkle, D. L.; Christoporou, L. G. in *Electron-Molecule Interactions and Their Applications*; Christophorou, L. G. (ed.); Academic Press: New York (USA), 1984, volume 2, pp. 423-641.

[8403] Henne, B. J.; Bartak, D. E. *Inorg. Chem.* **1984**, *23*, 369-373.

[8404] Newton, M. D.; Sutin, N. *Ann. Rev. Phys. Chem.* **1984**, *35*, 437-480.

[8501] Grimsrud, E. P.; Caldwell, G.; Chowdhury, S.; Kebarle, P. *J. Am. Chem. Soc.* **1985**, *107*, 4627-4634.

[8502] Ramamurthy, V. *Org. Photochem.* **1985**, *7*, 231-338.

[8601] Wagner, P. J.; Truman, R. J.; Puchalski, A. E.; Wake, R. *J. Am. Chem. Soc.* **1986**, *108*, 7727-7738.

[8701] Gersdorf, J.; Mattay, J.; Goerner, H. *J. Am. Chem. Soc.* **1987**, *109*, 1203-1209.

[8801] Pearson, R. G. *Inorg. Chem.* **1988**, *27*, 734-740.

[8802] Lias, S. G.; Bartmess, J. E.; Liebman, J. F.; Holmes, J. L.; Levin, R. D.; Mallard, W.G. *J. Phys. Chem. Ref. Data* **1988**, *17*, 861.

[8803] Juris, A.; Balzani, V.; Barigelletti, F.; Campagna, S.; Belser, P.; von Zelewsky, A. *Coord. Chem. Rev.* **1988**, *84*, 85-277.

[8804] Juris, A.; Campagna, S.; Bidd, I.; Lehn, J. -M.; Ziessel, R. *Inorg. Chem.* **1988**, *27*, 4007-4011.

[8901] Kawanishi, Y.; Kitamura, N.; Tazuke, S. *Inorg. Chem.* **1989**, *28*, 2968-2975.

[8902] Kalyanasundaram, K.; Nazeeruddin, M. K. *Chem. Phys. Lett.* **1989**, *158*, 45-50.

[9001] Lichtenberger, D. L.; Johnston, R. L.; Hinkelmann, K.; Suzuki, T.; Wudl, F. *J. Am. Chem. Soc.* **1990**, *112*, 3302-3312.

[9002] Nolan, J. E.; Plambeck, J. A. *J. Electroanal. Chem.* **1990**, *294*, 1-20.

[9003] Bashkin, J. K.; Kinlen, P. J. *Inorg. Chem.* **1990**, *29*, 4507-4509.

[9004] Lias, S. G.; Bartmess, J. E.; Liebman, J. F.; Holmes, J. L.; Levin, R. D.; Mallard, W.G. *NIST Positive Ion Energetics Database*; National Institute of Standards and Technology: Gaithesbug, MD (USA), 1990, version 1.1.

[9005] Federlin, P.; Kern, J. -M.; Rastegar, A.; Dietrich-Buchecker, C.; Marnot, P. A.; Sauvage, J. -P. *New J. Chem.* **1990**, *14*, 9-12.

[9101] Wang, L. -S.; Conceicao, J.; Jin, C.; Smalley, R. E. *Chem. Phys. Lett.* **1991**, *182*, 5-11.

[9102] Hansen, T. K.; Lakshmikantham, M. V.; Cava, M. P.; Metzger, R. M.; Becher, J. *J. Org. Chem.* **1991**, *56*, 2720-2722.

[9103] Allemand, P. M.; Koch, A.; Wudl, F.; Rubin, Y.; Diederich, F.; Alvarez, M. M.; Anz, S. J.; Whetten, R. L. *J. Am. Chem. Soc.* **1991**, *113*, 1050-1051.

[9104] Schenk, R.; Gregorius, H.; Meerholz, K.; Heinze, J.; Müllen, K. *J. Am. Chem. Soc.* **1991**, *113*, 2634-2647.

[9201] Bortolus, P.; Camaioni, N.; Flamigni, L.; Minto, F.; Monti, S.; Faucitano, A. *Photochem. Photobiol.* **1992**, *68*, 239.

[9202] Brüggermann, K.; Kochi, J. K. *J. Org. Chem.* **1992**, *57*, 2956-2960.

[9203] Dubois, D.; Moninot, G.; Kutner, W.; Jones, M. T.; Kadish, K. M. *J. Phys. Chem.* **1992**, *96*, 7137-7145.

[9204] Maestri, M.; Balzani, V.; Deuschel-Cornioley, C.; von Zelewsky, A. *Adv. Photochem.* **1992**, *17*, 1-68.

[9205] Braterman, P. S.; Song, J. -I.; Peacock, R. D. *Inorg. Chem.* **1992**, *31*, 555-559.

[9206] Braterman, P. S.; Song, J. -I.; Wimmer, F.M.; Wimmer, S.; Kaim, W.; Klein, A.; Peacock, R. D. *Inorg. Chem.* **1992**, *31*, 5084-5088.

[9207] Constable, E. C.; Cargill Thompson, A. M. W. *J. Chem. Soc. Dalton Trans.* **1992**, 2947-2950.

[9301] Xie, Q.; Arias, F.; Echegoyen, L. *J. Am. Chem. Soc.* **1993**, *115*, 9818-9819.

[9302] Lever, A. B. P. *Inorg. Chim. Acta* **1993**, *203*, 171-174.

[9401] Campbell, S.; Marzluff, E. M.; Rodgers, M. T.; Beauchamp, J. L.; Rempe, M. E.; Schwinck, K. F.; Lichtenberger, D. L. *J. Am. Chem. Soc.* **1994**, *116*, 5257-5264.

[9402] Yagci, Y.; Schnabel, W.; Wilpert, A.; Bendig, J. *J. Chem. Soc. Faraday Trans.* **1994**, *90*, 287-291.

[9403] Sauvage, J. -P.; Collin, J. -P.; Chambron, J. -C.; Guillerez, S.; Coudret, C.; Balzani, V.; Barigelletti, F.; De Cola, L.; Flamigni, L. *Chem. Rev.* **1994**, *94*, 993-1019.

[9501] Schmidt, A.; Anderson, M. L.; Dunphy, D.; Wehrmeister, T.; Müllen, K.; Armstrong, N. R. *Adv. Mater.* **1995**, *7*, 722-726.

[9502] Cordova, E.; Bissell, R. A.; Kaifer, A. E. *J. Org. Chem.* **1995**, *60*, 1033-1038.

[9503] Ashton, P. R.; Ballardini, R.; Balzani, V.; Credi, A.; Gandolfi, M. T.; Menzer, S.; Pérez–García, L.; Prodi, L.; Stoddart, J. F.; Venturi, M.; White, A. J. P.; Williams, D. J. *J. Am. Chem. Soc.* **1995**, *117*, 11171-11197.

[9601] Zakrzewski, V. G.; Dolgounitcheva, O.; Ortiz, J.V. *J. Chem. Phys.* **1996**, *105*, 8748-8753.

[9602] Boltalina, O. V.; Dashkova, E. V.; Sidorov, L. N. *Chem. Phys. Lett.* **1996**, *256*, 253-260.

[9603] Jonsson, M.; Wayner, D. D. M.; Lusztyk, J. *J. Phys. Chem.* **1996**, *100*, 17539-17543.

[9604] Seidel, C. A. M.; Schulz, A.; Sauer, M. H. M. *J. Phys. Chem.* **1996**, *100*, 5541-5553.

[9605] Yang, L.; Wimmer, F. M.; Wimmer, S.; Zhao, J.; Braterman, P. S. *J. Organomet. Chem.* **1996**, *525*, 1-8.

[9605] Jacques, P.; Burget, D.; Allonas, X. *New J. Chem.* **1996**, *20*, 933-937.

[9701] von Raumer, M.; Suppan, P.; Haselbach, E. *Helv. Chim. Acta* **1997**, *80*, 719-724.

[9702] Dolgounitcheva, O.; Zakrzewski, V. G.; Ortiz, J. V. *J. Phys. Chem. A* **1997**, *101*, 8554-8564.

[9703] De Boelpaep, I.; Vetters, B.; Peeters, J. *J. Phys. Chem. A* **1997**, *101*, 787-793.

[9704] McNaught, A. D.; Wilkinson, A. (eds.) *Compendium of Chemical Terminology - The Gold Book, 2nd Ed.*; Blackwell Science: Oxford (UK), 1997.

[9801] Martín, E.; Weigand, R. *Chem. Phys. Lett.* **1998**, *288*, 52-58.

[9802] Armaroli, N.; Diederich, F.; Dietrich-Buchecker, C. O.; Flamigni, L.; Marconi, G.; Nierengarten, J. -F.; Sauvage, J. -P. *Chem. Eur. J.* **1998**, *4*, 406-416.

[9803] Ashton, P. R.; Balzani, V.; Credi, A.; Kocian, O.; Pasini, D.; Prodi, L.; Spencer, N.; Stoddart, J. F.; Tolley, M. S.; Venturi, M.; White, A. J. P.; Williams, D. J. *Chem. Eur. J.* **1998**, *4*, 590-607.

[9804] Majumdar, P.; Ghosh, A. K.; Falvello, L. R.; Peng, S. -M.; Goswami, S. *Inorg. Chem.* **1998**, *37*, 1651-1654.

[9805] Brown, A. R.; Guo, Z; Mosselmans, F. W. J.; Parsons, S.; Schroeder, M.; Yellowlees, L. J. *J. Am. Chem. Soc.* **1998**, *120*, 8805-8811.

[9901] Ichida, M.; Sodha, T.; Nakamura, A. *Chem. Phys. Lett.* **1999**, *310*, 373-378.

[9902] Chen, E. S.; Chen, E. C. M.; Sane, N.; Talley, L.; Kozanecki, N.; Shulze, S. *J. Chem. Phys.* **1999**, *110*, 9319-9329.

[9903] Piet, D. P.; Danovich, D.; Zuilhof, H.; Sudhoelter, E. J. R. *J. Chem. Soc. Perkin Trans. 2* **1999**, 1653-1661.

[9904] Ashton, P. R.; Balzani, V.; Becher, J.; Credi, A.; Fyfe, M. C. T.; Mattersteig, G.; Menzer, S.; Nielsen, M. B.; Raymo, F. M.; Stoddart, J. F.; Venturi, M.; Williams, D. J. *J. Am. Chem. Soc.* **1999**, *121*, 3951-3957.

[9905] Jeppesen, J. O.; Takimiya, K.; Jensen, F.; Becher, J. *Org. Lett.* **1999**, *1*, 1291-1294.

[9906] Dabestani, R.; Ivanov, I. N. *Photochem. Photobiol.* **1999**, *70*, 10-34.

[9907] Noviandri, I.; Brown, K. N.; Fleming, D. S.; Gulyas, P. T.; Lay, P. A.; Masters, A. F.; Phillips, L. *J. Phys. Chem. B* **1999**, *103*, 6713-6722.

[9908] Miller, M. T.; Gantzel, P. K.; Karpishin, T. B. *Inorg. Chem.* **1999**, *38*, 3414-3422.

[9909] Collin, J. -P.; Dixon, I. M.; Sauvage, J. -P.; Williams, J. A. G.; Barigelletti, F.; Flamigni, L. *J. Am. Chem. Soc.* **1999**, *121*, 5009-5016.

[0001] Boltalina, O. V.; Ioffe, I. N.; Sidorov, L. N.; Seifert, G.; Vietze, K. *J. Am. Chem. Soc.* **2000**, *122*, 9745-9749.

[0002] Inada, T. N.; Kikuchi, K.; Takahashi, Y.; Ikeda, H.; Miyashi, T. *J. Photochem. Photobiol. A: Chem.* **2000**, *137*, 93-97.

[0003] Nad, S.; Pal, H. *J. Phys. Chem. A* **2000**, *104*, 673-680.

[0004] Hamilton, D. G.; Montalti, M.; Prodi, L.; Fontani, M.; Zanello, P.; Sanders, J. K. M. *Chem. Eur. J.* **2000**, *6*, 608-617.

[0005] Kadish, K. M.; Royal, G.; van Caemelbecke, E.; Gueletti, L. in *The Porphyrin Handbook*; K. M. Kadish, K. M. Smith, R. Guilard (eds.); Academic Press: San Diego, CA (USA), 2000, volume 9, pp. 1-219.

[0006] Erben, C.; Will, S.; Kadish, K. M. in *The Porphyrin Handbook*; K. M. Kadish, K. M. Smith, R. Guilard (eds.); Academic Press: San Diego, CA (USA), 2000, volume 2, pp. 233-300.

[0101] Onodera, H.; Araki, Y.; Fujitsuka, M.; Onodera, S.; Ito, O.; Bai, F.; Zheng, M.; Yang, J. -L. *J. Phys. Chem. A* **2001**, *105*, 7341-7349.

[0102] Balzani, V.; Credi, A.; Langford, S. J.; Prodi, A.; Stoddart, J. F.; Venturi, M. *Supram. Chem.* **2001**, *13*, 303-311.

[0103] Albini, A.; Fagnoni, M. in *Electron Transfer in Chemistry*; Balzani, V. (ed.); Wiley-VCH: Weinheim (Germany), 2001, volume 2, pp. 338-378.

[0104] Lewis, F. D. in *Electron Transfer in Chemistry*; Balzani, V. (ed.); Wiley-VCH: Weinheim (Germany), 2001, volume 3, pp. 105-175.

[0105] Hubig, S. M.; Kochi, J. K. in *Electron Transfer in Chemistry*; Balzani, V. (ed.); Wiley-VCH: Weinheim (Germany), 2001, volume 2, pp. 618-676.

[0106] V. Balzani (ed.) *Electron Transfer in Chemistry*; Wiley-VCH: Weinheim (Germany), 2001, volume 1.

[0201] Marcaccio, M.; Paolucci, F.; Paradisi, C.; Carano, M.; Roffia, S.; Fontanesi, C.; Yellowlees, L. J.; Serroni, S.; Campagna, S.; Balzani, V. *J. Electroanal. Chem.* **2002**, *532*, 99-112.

[0202] Baik, M. -H.; Friesner, R. A. *J. Phys. Chem. A* **2002**, *106*, 7407-7412.

[0203] Sapp, S. A.; Elliott, C. M.; Contado, C.; Caramori, S.; Bignozzi, C. A. *J. Am. Chem. Soc.* **2002**, *124*, 11215-11222.

[0301] Schultz, T.; Quenneville, J.; Levine, B.; Toniolo, A.; Martinez, T. J.; Lochbrunner, S.; Schmitt, M.; Shaffer, J. P.; Zgierski, M. Z.; Stolow, A. *J. Am. Chem. Soc.* **2003**, *125*, 8098-8099.

[0302] Yoshikawa, N.; Matsumura-Inoue, T. *Anal. Sci.* **2003**, *19*, 761-765.

[0303] Porcal, G.; Bertolotti, S. G.; Previtali, C. M.; Encinas, M. V. *Phys. Chem. Chem. Phys.* **2003**, *5*, 4123-4128.

[0304] Carano, M.; Da Ros, T.; Fanti, M.; Kordatos, K.; Marcaccio, M.; Paolucci, F.; Prato, M.; Roffia, S.; Zerbetto, F. *J. Am. Chem. Soc.* **2003**, *125*, 7139-7144.

[0305] Carroll, J. B.; Gray, M.; McMenimen, K. A.; Hamilton, D. G.; Rotello, V. M. *Org. Lett.* **2003**, *5*, 3177-3180.

[0306] Wan, C. -W.; Burghart, A.; Chen, J.; Bergstroem, F.; Johansson, L. B. -A.; Wolford, M. F.; Kim, T. G.; Topp, M. R.; Hochstrasser, R. M.; Burgess, K. *Chem. Eur. J.* **2003**, *9*, 4430-4441.

[0307] Lide, D. R. (ed.) *CRC Handbook of Chemistry and Physics, 84th Ed.*; CRC Press: Boca Raton, FL (USA), 2003.

[0401] Elliott, M. (University of Colorado, Fort Collins, USA), private communication.

[0402] Marchioni, F. (Università di Bologna, Italy), private communication.

[0403] Dong, Y.; Lindoy, L. F.; Turner, P.; Wei, G. *Dalton Trans.* **2004**, 1264-1270.

[0404] Lo, K. K. -W.; Chan, J. S.-W.; Lui, L. -H.; Chung, C. -K. *Organometallics* **2004**, *23*, 3108-3116.

[0501] Braslawsky, S. E.; Houk, K. N.; Verhoeven, J. W. *Glossary of Terms Used in Photochemistry, 3rd Ed.*; IUPAC Commission, *Pure Appl. Chem.*, in press.

8

Bond Dissociation Energies

Bond dissociation energies D_e are often useful for the estimation of the enthalpy of a reaction. Three tables are included here that cover many of the bonds commonly encountered in photochemical reactions. All of the energies in these tables are in units of kJ mol^{-1}.

Table 8a Bond Dissociation Energies (kJ mol^{-1}) of Single Bonds[a]

	H—	CH$_3$—	C$_2$H$_5$—	iso-Pr—	tert-Bu—	C$_6$H$_5$—	C$_6$H$_5$CH$_2$—	CH$_3$CO—	CF$_3$—
H—	436	439	423	409	404	473	376	374	450
CH$_3$—	439	377	371	367	366	434	332	354	423
C$_2$H$_5$—	423	371	343	335	322	410	294	349	–
iso-Pr—	409	367	335	324	305	404	282	339	–
tert-Bu—	404	366	322	305	282	389	282	332	–
C$_6$H$_5$—	473	434	410	404	389	494	328	415	–
C$_6$H$_5$CH$_2$—	376	332	294	282	268	328	–	297	–
CH$_3$CO—	374	354	349	339	332	415	297	–	–
CF$_3$—	450	423	–	–	–	–	–	–	413
CH$_2$=CHCH$_2$—	362	–	–	–	–	–	–	–	–
cyclo-C$_3$H$_5$—	445	–	–	–	–	–	–	–	–
cyclo-C$_4$H$_7$—	404	–	–	–	–	–	–	–	–
cyclo-C$_5$H$_9$—	404	–	–	–	–	–	–	–	–
cyclo-C$_6$H$_{11}$—	400	–	–	–	–	–	–	–	–
CH$_3$COCH$_2$—	411	–	–	–	–	–	–	–	–
CH$_3$OCH$_2$—	402	–	–	–	–	–	–	–	–
HOCH$_2$—	402	–	–	–	–	–	–	–	–

Table 8a Bond Dissociation Energies (kJ mol^{-1}) of Single Bonds[a]

	H—	CH₃—	C₂H₅—	iso-Pr—	tert-Bu—	C₆H₅—	C₆H₅CH₂—	CH₃CO—	CF₃—
N≡C—	528	510[b]	—	—	—	—	—	—	—
(CH₃)₂C(OH)—	381	—	—	—	—	—	—	—	—
N≡CCH₂—	393	336	322	—	—	—	275	—	—
Cl₃C—	393	362	—	—	—	—	—	—	335
CH₂=CH—	465	406	393	385	372	473	—	—	—
CH≡C—	556	—	—	—	—	—	—	—	—
NH₂CH₂—	390	—	332	—	—	—	285	—	—
Br—	366	293	295	298	296	346	256	292	294
Cl—	432	350	354	353	355	407	310	354	362
F—	570	472	463	460	—	533	—	512	547
I—	298	239	236	234	231	280	215	223	227
H₂N—	453	358	357	359	362	439	298[b]	417	—
ON—	195	167	176	153	165	213	—	—	179
HO—	497	385	393	396	402	472	346	459	—
CH₃O—	436	348	355	356	353	423	—	421	—
C₂H₅O—	438	339	339	343	339	423	—	—	—
tert-BuO—	440	—	—	—	—	—	—	—	—

Table 8a Bond Dissociation Energies (kJ mol^{-1}) of Single Bonds[a]

	H—	CH$_3$—	C$_2$H$_5$—	iso-Pr—	tert-Bu—	C$_6$H$_5$—	C$_6$H$_5$CH$_2$—	CH$_3$CO—	CF$_3$—
C$_6$H$_5$O—	362	238	264	–	–	368	–	–	–
CH$_3$COO—	443	301	301	–	–	389	–	–	–
HOO—	369	–	–	–	–	–	–	–	–

[a] Values selected from ref. [7601, 0301], and ref. [0301], p. 9-67.

Table 8b Bond Dissociation Energies (kJ mol^{-1}) of Small Molecules[a]

	H	F	Cl	Br	I	OH
H	436	570	432	366	298	497
F	570	159	256	280	≤271.5	–
Cl	432	256	243	218	211	251
Br	366	280	218	193	179	234
I	298	≤271.5	211	179	151	234
OH	497	-	251	234	234	213

[a] Values selected from ref. [0301].

Table 8c Bond Dissociation Energies of Peroxides and Multiple Bonds[a]

Type of Bond	D_e/kJ mol^{-1}
N≡N	945
O=O	498
CO	1077
NO	631
C≡C	962
C=C	720
O=C=O	532
HO—OH	213
C_2H_5O—OC_2H_5	159
$(CH_3)_2CHO$—$OCH(CH_3)_2$	158
$(CH_3)_3CO$—$OC(CH_3)_3$	159
$CH_3C(=O)O$—$O(O=)CCH_3$	127
CH_3CH_2O—OH	184
$(CH_3)_3CO$—OH	151

[a] Values selected from ref. [7601, 0301].

REFERENCES

[7601] Benson, S.W. *Thermochemical Kinetics. Methods for the Estimation of Thermo-chemical Data and Rate Parameters*; Wiley and Sons: New York, 1976, 336p.

[0301] Lide, D.R. *Handbook of Physics and Chemistry. Eighty Fourth Edition*; CRC Press: Boca Raton, (FL), 2003-2004.

9

Solvent Properties

This section contains five tables useful for the choice of an appropriate solvent and for an understanding of photochemical and spectroscopic behavior in that solvent.

The first table includes molecular mass, boiling and melting points, density (d), refractive index (n_D), viscosity (η), dielectric constant (ε), and Kosower's Z and Dimroth's E solvent parameters. The second table contains data regarding end absorption of many solvents. Since many studies are done in the presence of oxygen, the concentration of O_2 at 0.2122 bars and 1.013 bars are included to facilitate calculation of quenching constants. In the same table are reported the values of the lifetime of the $^1\Delta_g$ excited state of molecular oxygen in the various solvents. The fourth table contains a list of transparent organic glasses, along with some data on these glasses at low temperature. In the final table are gathered the donor numbers of a selection of solvents.

For a critical compilation of scales of solvent parameters, see also [9901].

Table 9a Physical Properties[a] of Solvents

No.	Solvent	Mol. wt. g mol^{-1}	bp °C	mp °C	d[b] g mL^{-1}	n_D[c]	η[d] 10^{-3} Pas	ε[e]	Kosower Z[f]	Dimroth E[g]
1	Acetic acid	60.05	117.9	16.7	1.0492	1.3719	1.13[h]	6.15	79.2	51.1
2	Acetic acid, ethyl ester	88.11	77.11	−83.55	0.9006	1.37239	0.4508	6.053	64.0	38.1
3	Acetone	58.08	56.2	−95	0.7899	1.35868	0.303[h]	20.7[h]	65.7	42.2
4	Acetonitrile	41.05	81.8	−44	0.7857	1.34411	0.345[h]	35.94[h]	71.3	46.0
5	Benzene	78.12	80	6	0.8765	1.50112	0.649	2.284	54	34.5
6	Benzene, bromo-	157.02	155.91	−30.82	1.4959	1.55680	1.196[i]	5.40[h]	–	37.5
7	Benzene, chloro-	112.56	131.69	−45.58	1.1063	1.5248	0.799	5.621[h]	58.0	37.5
8	Benzene, 1,2-dimethyl-	106.17	144	−25	0.8802	1.50545	0.809	2.568	–	–
9	Benzene, 1,3,dimethyl-	106.17	139	−48	0.8642	1.49722	0.617	2.3742	–	–
10	Benzene, 1,4-dimethyl-	106.17	138	13	0.8611	1.49582	0.644	2.2699	–	–
11	Benzene, isopropyl-	120.19	152	−96	0.8618	1.49145	0.791	2.383	–	–
12	Benzene, methoxy-	108.14	153.60	−37.5	0.9940	1.51700	1.32	4.33[h]	–	37.2
13	Benzene, 1,3,5-trimethyl-	120.19	165	−45	0.8652	1.49937	1.154	2.279	–	–
14	Benzonitrile	103.12	191.1	−12.7	1.0093[i]	1.52823	1.24[h]	25.20[h]	65.0	42.0

Table 9a Physical Properties[a] of Solvents

No.	Solvent	Mol. wt. g mol⁻¹	bp °C	mp °C	d^b g ml⁻¹	n_D^c	η^d 10^{-3}Pa s	ε^e	Kosower Z^f	Dimroth E^g
15	1-Butanol	74.12	118	−90	0.8098	1.3993	2.948	17.51[h]	77.7	50.2
16	2-Butanol	74.12	99.5	−114.7	0.8063	1.3971	3.632	16.56[h]	75.4	–
17	2-Butanone	72.11	79.6	−87	0.8054	1.3788	0.399	18.51	64.0	41.3
18	Carbon disulfide	76.14	46	−112	1.2632	1.62746	0.363	2.64	–	32.6
19	Carbon tetrachloride	153.82	77	−23	1.5940	1.4601	0.969	2.238	–	32.5
20	Chloroform	119.38	61	−64	1.4832	1.4459	0.58	4.806	63.2	39.1
21	Cyclohexane	84.16	81	7	0.7785	1.42623	0.975	2.023	60.1	31.2
22	Cyclohexane, methyl-	98.19	101	−127	0.7694	1.42312	0.734	2.020	–	–
23	Cyclohexanol	100.16	161	25	0.9624	1.4641	68.0	15.0[h]	75.0	–
24	Cyclopentane	70.13	49.3	−93.9	0.7457	1.40645	0.439	1.969	–	–
25	Decahydronaphthalene	138.25	191.7	−124	0.8865	1.4758	2.415[h]	2.1542[h]	–	–
26	Decane	142.29	174.15	−29.64	0.7300	1.41189	0.9284	1.991	–	–
27	Dichloromethane	84.93	39.6	−95	1.3266	1.42416	0.449[i]	8.93[h]	64.2	41.1
28	Diethyl ether	74.12	34	−116	0.7138	1.35243	0.242	4.335	–	34.6
29	1,2-Dimethoxyethane	90.12	85	−58	0.8691	1.37963	0.455[h]	7.20[h]	61.2	38.2

Table 9a Physical Properties[a] of Solvents

No.	Solvent	Mol. wt. g mol⁻¹	bp °C	mp °C	d[b] g mL⁻¹	$n_D{}^c$	η^d 10⁻³Pas	ε^e	Kosower Z^f	Dimroth E^g
30	*N,N*-Dimethylformamide	73.09	153	−60	0.944	1.43047	0.924	36.71[h]	68.5	43.8
31	Dimethylsulfoxide	78.14	189	19	1.1014	1.4793	1.991[h]	46.45[h]	70.2	45.0
32	1,4-Dioxane	88.11	101	12	1.0337	1.42241	1.439[i]	2.209[h]	64.6	36.0
33	Dodecane	170.34	216.32	−9.58	0.7487	1.42167	1.508	2.015	–	–
34	Ethanamine, *N*-ethyl-	73.14	55	−50	0.7056	1.3845	0.306	3.894	–	–
35	Ethane, 1,2-dichloro	98.96	84	−36	1.2351	1.4448	0.800[k]	10.37[h]	63.4	41.9
36	Ethane, 1,1,2,2-tetrachloro-1,2-difluoro-	203.83	93	27	1.6447[h]	1.41297[h]	1.21[h]	2.52[h]	–	–
37	Ethane, 1,1,2-trichloro-1,2,2-trifluoro-	187.38	48	−35	1.5635[h]	1.35572[h]	0.711	2.41[h]	–	–
38	Ethanol	46.07	78	−114	0.7893	1.36143	1.200	24.55[h]	79.6	51.9
39	Ethylene glycol	62.07	197	−13	1.1088	1.4318	19.9	37.7[h]	85.1	56.6
40	Formamide	45.04	210.5	2.55	1.1334	1.44754	3.764	111.0	83.3	56.6
41	Glycerol	92.09	290	18	1.2613	1.4746	1412	42.5[h]	82.7	–
42	Heptane	100.20	98	−91	0.6837	1.38764	0.4181	1.9246	–	–
43	Hexamethylphosphoramide	179.20	233	7.20	1.027	1.4588	3.47	29.3	62.8	40.9

Table 9a Physical Properties[a] of Solvents

No.	Solvent	Mol. wt. g mol^{-1}	bp °C	mp °C	d[b] g ml^{-1}	n_D[c]	η[d] 10^{-3}Pas	ε[e]	Kosower Z[f]	Dimroth E[g]
44	Hexane	86.18	69	-95	0.6548[h]	1.37486	0.3126	1.8863	-	30.9
45	Methanol	32.04	65	-98	0.7914	1.32840	0.5929	32.66[h]	83.6	55.5
46	2-Methylbutane	72.15	27.88	-159.9	0.6193	1.35373	0.225	1.843	-	-
47	2-Methyl-2-butanol	88.15	102.0	-8.8	0.8096	1.4050	3.548[h]	5.78[h]	70.7	-
48	3-Methyl-1-butanol	88.15	130.5	-117.2	0.8104	1.4072	3.738[h]	15.19[h]	77.6	47.0
49	3-Methylpentane	86.18	62.28	glass	0.6643	1.37652	0.307[h]	1.895	-	-
50	2-Methyl-1-propanol	74.12	107.89	-108	0.8016	1.39591	3.333[h]	17.93[h]	77.7	-
51	2-Methyl-2-propanol	74.12	82.35	25.62	0.7887	1.3878	5.942	12.47[h]	71.3	43.9[i]
52	Naphthalene, decahydro-, (E)-	138.25	187.3	-30.3	0.8699	1.46949	2.128	2.172	-	-
53	Naphthalene, decahydro-, (Z)-	138.25	195.8	-42.9	0.8965	1.48098	3.381	2.197	-	-
54	Octane	114.23	126	-57	0.6986[h]	1.39743	0.5466	1.948	60.1	-
55	Pentane	72.15	36	-130	0.6262	1.35748	0.235	1.844	-	-
56	1-Pentanol	88.15	138	-79	0.8144	1.4100	3.5128[h]	13.9[h]	-	-
57	2-Pentanol	88.15	119.0	glass	0.8094	1.4064	5.307[i]	13.71[h]	-	-
58	1-Propanol	60.10	97	-126	0.8035	1.38556	1.9430[h]	20.45[h]	78.3	50.7

Table 9a Physical Properties[a] of Solvents

No.	Solvent	Mol. wt. g mol⁻¹	bp °C	mp °C	d[b] g mL⁻¹	n_D[c]	η[d] 10⁻³Pas	ε[e]	Kosower Z[f]	Dimroth E[g]
59	2-Propanol	60.10	82	−90	0.7855	1.3772	2.0436[h]	19.92[h]	76.3	48.6
60	Pyridine	79.10	115	−42	0.9819	1.51016	0.952	12.92[h]	64.0	40.2
61	Tetrahydrofuran	72.11	65	−108	0.8892	1.40716	0.575	7.58[h]	58.8	37.4
62	Tetramethylene sulfone	120.17	287.3	28.45	1.2604[j]	1.4833[j]	10.286[j]	43.26[j]	77.5	44.0
63	Toluene	92.14	111	−95	0.8669	1.49693	0.5859	2.379[h]	–	33.9
64	Trifluoroacetic acid	114.02	71.78	−15.216	1.4890	1.2850	0.926	8.55	–	–
65	2,2,2-Trifluoroethanol	100.04	74.05	−43.5	1.3826[l]	–	1.995	26.67[l]	–	–
66	2,2,4-Trimethylpentane	114.23	99.238	−107.388	0.69193	1.39145	0.504	1.940	–	–
67	Water	18.015	100.0	0.0	0.9982	1.332988	1.0019	80.16	94.6	63.1

[a]Values for molecular weight, boiling point, melting point, density, refractive index, viscosity and dielectric constant are at 20 °C unless otherwise noted, and have been extracted from ref. [8601, 0301]; Values for Kosower's Z and Dimroth's E solvent parameters are at 25 °C and have been taken from ref. [6801] and [7101], respectively; For a discussion of Z, E, and other solvent parameters see ref. [7901, 8101]; [b]Density at 20 °C unless otherwise noted; [c]Refractive index at 20 °C at the average sodium D line, unless otherwise noted; [d]Viscosity at 20 °C unless otherwise noted; [e]Dielectric constant at 20 °C, unless otherwise noted. Limiting values at low frequencies; [f]Kosower's solvent parameter derived from the wavelength of the charge-transfer band in the visible spectrum of 1-ethyl-4-methoxycarbonylpyridinium iodide ($Z = 2.859 \times 10^4/\lambda$ where λ is the position of the absorption maximum in nanometers); [g]Dimroth's solvent parameter derived from the visible spectrum of two pyridinium betaines; [h]Measured at 25 °C; [i]Measured at 15 °C; [j]Measured at 30 °C; [k]Measured at 19 °C; [l]From ref. [8501].

Table 9b Ultraviolet Transmission[a] of Solvents

No.	Solvent	$\%T$[b] 254 nm	$\%T$ 313 nm	$\%T$ 366 nm	$T = 10\%$[c] at λ(nm) =
1	Acetic acid, ethyl ester	<10	99	100	255
2	Acetone	0	0	100	329
3	Acetonitrile	98	100	100	190
4	Benzene	0	94	100	280
5	Benzonitrile	0	85	100	299
6	Carbon disulfide	0	0	0	380
7	Carbon tetrachloride	0	100	100	265
8	Chloroform	80	100	100	245
9	Cyclohexane	100	100	100	205
10	Cyclohexane, methyl-	100	100	100	207
11	Dichloromethane	98	100	100	232
12	Diethyl ether	84	100	100	215
13	*N,N*-Dimethylformamide	0	93	100	270
14	Dimethyl sulfoxide	0	96	100	262
15	1,4-Dioxane	64	100	100	215
16	Ethane, 1,2-dichloro	97	98	100	226
17	Ethanol	98	100	100	205
18	Heptane	100	100	100	197
19	Hexane	100	100	100	195
20	Methanol	100	100	100	205
21	2-Methylbutane	100	100	100	192
22	2-Propanol	98	100	100	205
23	Pyridine	0	88	100	305
24	Tetrahydrofuran	57	99	100	233
25	Toluene	0	90	100	285
26	2,2,4-Trimethylpentane	100	100	100	197

[a] Values taken from ref. [7102], numbers M1-M19 and 97, 98; [b] Transmission through 1 cm of neat solvent at indicated wavelength; [c] Wavelength (in nm) at which the percent transmission has dropped to 10% or where the absorption is equal to 1 (for 1 cm pathlength of neat solvent).

Table 9c O_2 Concentration in Solvents

No	Solvent	Mol.mass g/mol	T °C	d g/mL[a]	[O2]/mmol L⁻¹ (1.013 bar O2)[b]	[O2]/mmol L⁻¹ (0.213 bar O2)[c]	τ (¹Δg) O2 µs
1	Acetic acid	60.052	20	1.0495	8.11	1.7	
2	Acetic acid, ethyl ester	88.106	20	0.9006	8.89	1.9	
3	Acetone	58.08	20	0.7900	11.4	2.4	
			25	0.7894	11.3	2.4	45[d]
4	Acetonitrile	41.052	24	0.7765[e]	9.1[f]	1.9[f]	61[g]
5	Aniline	93.128	25	1.0175	2.47	0.52	
6	Aniline, N,N-dimethyl-	121.182	25	0.9523	5.64	1.2	
7	Aniline, N-methyl-	107.155	25	0.9822	2.67	0.56	
8	Benzene	78.113	20	0.8790	9.02	1.9	
			25	0.8736	9.06	1.9	31[g]
9	Benzene, bromo-	157.01	25	1.4882	7.09	1.5	
10	Benzene, chloro-	112.559	25	1.1009	8.78	1.8	

Table 9c O₂ Concentration in Solvents

No	Solvent	Mol.mass g/mol	T °C	d g/mL[a]	$[O_2]$/mmol L^{-1} (1.013 bar O_2)[b]	$[O_2]$/mmol L^{-1} (0.213 bar O_2)[c]	$\tau(^1\Delta_g)\ O_2$ μs
11	Benzene, 1,2-dimethyl-	106.167	25	0.8759	9.22	1.9	
12	Benzene, 1,3-dimethyl-	106.167	25	0.8601	9.69	2.0	
13	Benzene, 1,4-dimethyl-	106.167	25	0.8566	10.0	2.1	
14	Benzene, ethyl-	106.167	25	0.8625	9.91	2.1	
15	Benzene, fluoro-	96.104	25	1.022[h]	16.0	3.4	
16	Benzene, hexafluoro-	186.056	20 25	1.6187 1.6073	21.1 20.8	4.4 4.4	
17	Benzene, isopropyl-	120.194	25	0.8574	9.90	2.1	
18	Benzene, nitro-	123.111	25	1.1983	4.82	1.0	
19	1-Butanol	74.122	20 25	0.8096 0.8057	8.77 8.65	1.8 1.8	
20	2-Butanone	72.107	25	0.7997	11.2	2.4	
21	Butylamine	73.138	20	0.7392	11.3	2.4	

544

Table 9c O$_2$ Concentration in Solvents

No	Solvent	Mol.mass g/mol	T °C	d g/mL[a]	[O$_2$]/mmol L^{-1} (1.013 bar O$_2$)[b]	[O$_2$]/mmol L^{-1} (0.213 bar O$_2$)[c]	τ ($^1\Delta_g$) O$_2$ µs
22	Carbon disulfide	76.131	25	1.2555	7.24	1.5	1500[d]
23	Carbon tetrachloride	153.823	20	1.5940	12.4	2.6	900[d]
			25	1.5844	12.4	2.6	
24	Chloroform	119.378	20	1.4891	11.6	2.4	244[d]
25	Cyclohexane	84.161	20	0.7786	11.5	2.4	20[d]
			25	0.7739	11.5	2.4	
26	Cyclohexane, methyl-	98.188	20	0.7694	12.3	2.6	
			25	0.7651	12.5	2.6	
27	Cyclohexanol	100.16	26	0.9684[e]	8.27	1.7	
28	Cyclohexanone	98.144	20	0.9452	6.14	1.3	
			25	0.943	6.11	1.3	
29	Decane	142.284	25	0.7263	11.2	2.3	
30	Dichloromethane	84.933	20	1.3256	10.7	2.2	
31	Diethyl ether	74.122	20	0.7136	14.7	3.1	

Table 9c O₂ Concentration in Solvents

No	Solvent	Mol.mass g/mol	T °C	d g/mL[a]	$[O_2]$/mmol L^{-1} (1.013 bar O_2)[b]	$[O_2]$/mmol L^{-1} (0.213 bar O_2)[c]	$\tau(^1\Delta_g)\,O_2$ μs
32	Dimethoxymethane	76.095	20	0.860	13.5	2.8	
33	Dimethyl sulfoxide	78.129	25	1.0954	2.20	0.46	
34	1,4-Dioxane	88.106	20	1.0336	6.10	1.3	
			25	1.0280	6.28	1.3	
35	Dodecane	170.337	25	0.7452	8.14	1.7	
36	Ethane, 1,2-dichloro	98.96	20	1.2521	7.40	1.6	
37	Ethanol	46.069	20	0.7892	10.1	2.1	
			25	0.7849	9.92	2.1	
38	Ethanol (95%)	-	25	0.8074	7.84[i]	1.64[i]	
39	Ethene, tetrachloro-	165.834	20	1.6228	8.27	1.7	
40	Ethylene glycol	62.068	20	1.1135	0.58	0.12	
41	Glycerol	92.094	20	1.2613	0.3	0.07	
42	Heptane	100.203	20	0.6837	13.4	2.8	
			25	0.6795	13.2	2.8	

545

Table 9c O$_2$ Concentration in Solvents

No	Solvent	Mol.mass g/mol	T °C	d g/mL^a	[O$_2$]/mmol L^{-1} (1.013 bar O$_2$)^b	[O$_2$]/mmol L^{-1} (0.213 bar O$_2$)^c	τ ($^1\Delta_g$) O$_2$ μs
43	Hexane	86.177	20	0.6593	15.0	3.1	
			25	0.6548	14.7	3.1	
44	2-Hexanone	100.16	25	0.8067	9.42	2.0	
45	Methanol	32.042	20	0.7910	10.3	2.2	
			25	0.7864	10.2	2.1	
46	Methyl acetate	74.079	20	0.9342	11.1	2.3	
			25	0.9279	11.2	2.3	
47	2-Methyl-1-propanol	74.122	20	0.8016	9.21	1.9	
			25	0.7978	9.04	1.9	
48	Nitromethane	61.04	25	1.1313	8.60	1.8	
49	Nonane	128.257	25	0.7137	11.2	2.3	
50	Octane	114.23	20	0.7027	13.3	2.8	
			25	0.6986	13.3	2.8	
51	1-Octanol	130.23	20	0.8250	7.22	1.5	
			25	0.8216	7.13	1.5	
52	Pentane	72.15	25	0.6214	17.7	3.7	

Table 9c O₂ Concentration in Solvents

No	Solvent	Mol.mass g/mol	T °C	d g/mL[a]	[O₂]/mmol L⁻¹ (1.013 bar O₂)[b]	[O₂]/mmol L⁻¹ (0.213 bar O₂)[c]	τ (¹Δg) O₂ μs
53	1-Pentanol	88.149	20	0.8145	8.44	1.8	
54	2-Pentanone	86.133	25	0.8015	10.4	2.2	
55	Perfluorodecalin	462.08	25	1.946[j]	16.1	3.4	
56	Perfluoroheptane	388.04	25	1.7333[k]	24.8	5.2	
57	Perfluoro(methylcyclohexane)	350.05	25	1.788[j]	23.3	4.9	
58	Perfluorononane	488.07	25	1.84[j]	20.2	4.2	
59	Perfluorooctane	438.06	25	1.738[j]	21.2	4.4	
60	Piperidine	85.149	20	0.8606	7.52	1.6	
			25	0.8566	7.39	1.5	
61	1-Propanol	60.096	25	0.7996	6.69	1.4	
62	2-Propanol	60.096	20	0.7855	10.3	2.2	22.1[g]
			25	0.7813	10.2	2.1	
63	Propylamine	59.111	20	0.7173	10.1	2.1	

547

Table 9c O$_2$ Concentration in Solvents

No	Solvent	Mol.mass g/mol	T °C	d g/mL[a]	[O$_2$]/mmol L^{-1} (1.013 bar O$_2$)[b]	[O$_2$]/mmol L^{-1} (0.213 bar O$_2$)[c]	τ($^1\Delta_g$) O$_2$ μs
64	Pyridine	79.101	20	0.9832	5.66	1.2	
			25	0.9782	5.66	1.2	
65	Pyrrolidine	71.122	20	0.8586	7.29	1.5	
			25	0.8538	7.27	1.5	
66	Tetrahydrofuran	72.107	20	0.8892	9.90	2.1	
			25	0.8842[l]	10.0	2.1	
67	Toluene	92.14	20	0.8668	8.63	1.8	25[d]
			25	0.8622	9.88	2.1	
68	2,2,4-Trimethylpentane	114.23	25	0.6878	15.5	3.3	
69	Water	18.0152	20	0.9982	1.39	0.29	
			25	0.9970	1.27	0.27	

[a] Values selected from ref. [8601]; [b] Values often taken from evaluated data of ref. [8102]. These values and others chosen (from individual measurements compiled in ref. [8102] were almost always from experiments done at 1.013 bar partial pressure of O$_2$; [c] Values for 0.213 bar partial pressure of O$_2$ calculated from b, assuming Henry's Law holds; [d] From ref. [8301]; [e] At 25°C; [f] From ref. [7103]; [g] From ref. [8901]; [h] At 20°C; [i] From ref. [4601]; [j] From ref. [7701]; [k] From ref. [4701]; [l] From ref. [8502].

Table 9d Low-Temperature Organic Glasses for Spectroscopy[a]

Glass[a]	Ratio	Ref.	% crack[b]	V_{77}/V_{293}[c]	η/η_{3MP}[d]
Hydrocarbons					
Pentane (tech)	–	[6301]	0	–	–
Petroleum ether (30-60)	–	[6301]	10	–	–
2-Methylpentane (2MP)	–	[6501]	–	–	11
		[6802]			
3-Methylpentane (3MP)	–	[6201]	–	–	1
		[6401]			
		[6802]			
3-Ethylpentane	–	[6803]	–	–	1.1×10^6
2,3-Dimethylpentane	–	[6803]	–	–	1.6×10^4
3-Methylhexane	–	[6803]	–	–	1.5×10^6
4-Methylpentane	–	[6803]	–	–	9.1×10^{12}
3-Methyloctane	–	[6803]	–	–	5.5×10^{18}
Ethylcyclohexane	–	[6803]	–	–	2.0×10^{10}
Methylcyclohexane (MeCH)	–	[5301]	–	–	$\sim 4 \times 10^4$
		[6202]			
		[6701]			
		[6802]			
Pentane / Heptane	1:1	[6301]	0	–	–
2MP / MeCH	1:1	[7201]	–	–	0.90
3MP / isopentane (IP)	1:0	[5301]	–	0.784	1
		[6202]			
		[6401]			
		[6802]			
3MP / IP	9:1	[6401]	–	–	0.25
3MP / IP	8.1:1	[6802]	–	–	0.13
3MP / IP	4:1	[6401]	–	–	6.6×10^{-2}
		[6802]			

Table 9d Low-Temperature Organic Glasses for Spectroscopy[a]

Glass[a]	Ratio	Ref.	% crack[b]	V_{77}/V_{293} [c]	η/η_{3MP} [d]
3MP / IP	7:3	[6401]	–	–	1.5×10^{-2}
3MP / IP	2:1	[6802]	–	–	6.0×10^{-3}
3MP / IP	3:2	[5301]	–	0.771	5.4×10^{-3}
		[6202]			
		[6401]			
3MP / IP	1:1	[6401]	–	–	6.3×10^{-4}
3MP / IP	2:3	[6401]	–	–	1.9×10^{-4}
3MP / IP	3:7	[6401]	–	–	5.5×10^{-5}
3MP / IP	1:3	[6802]	–	–	1.6×10^{-5}
3MP / IP	1:4	[6401]	–	–	1.9×10^{-5}
3MP / IP	1:6	[5301]	–	0.760	–
		[6202]			
3MP / IP	1:9	[6401]	–	–	3.8×10^{-6}
3MP / IP	1:32	[6401]	–	–	9.5×10^{-7}
MeCH / Pentane	4:1 - 3:2	[6301]	0	–	–
MeCH / IP	4:1	[5301]	–	0.810	–
		[6202]			
MeCH / IP	1:3	[5301]	–	0.773	1.1×10^{-2}
		[6202]			
		[6501]			
MeCH / IP	1:4	[5301]	–	0.769	–
		[6202]			
MeCH / IP	1:5	[5301]	–	0.767	–
		[6202]			
MeCH / Methylcyclopentane	1:1	[6501]	–	–	2.8
		[7201]			
MeCH / Methylcyclopentane	3:2	[7201]	–	–	5.4

Table 9d Low-Temperature Organic Glasses for Spectroscopy[a]

Glass[a]	Ratio	Ref.	% crack[b]	V_{77}/V_{293}[c]	η/η_{3MP}[d]
Ethers					
Diethyl ether (Et$_2$O)	–	[6301]	10	–	–
2-Methyltetrahydrofuran	–	[6201]	–	–	6.9×10^5
		[6501]			
		[6803]			
Et$_2$O / IP	1:1, 2:1	[6203]	–	–	–
		[6201]			
Et$_2$O / MeCH	2:3	[6701]	–	–	–
Propyl ether / Pentane	2:1	[6301]	10	–	–
Alcohols					
Ethanol (EtOH)	–	[6301]	10	–	–
Glycerol	–	[7201]	–	–	1.2×10^{45}
1-Propanol	–	[6301]	20	–	1.6×10^7
		[7201]			
Propylene glycol	–	[7201]	–	–	4.8×10^{31}
EtOH / Methanol (MeOH)	5:1 - 9:1	[6301]	15	–	–
EtOH / MeOH	4:1	[5301]	–	0.802	–
		[6202]			
EtOH / MeOH	1:1	[6501]	–	–	1.7×10^{11}
		[7201]			
2-Propanol / IP	3:7	[6201]	–	–	–
1-Butanol / IP	3:7	[6201]	–	–	–
EtOH / Et$_2$O	1:1	[6203]	–	–	–
1-Propanol / 2-Propanol	1:1	[7201]	–	–	1.2×10^9
1-Propanol / 2-Propanol	3:2	[7201]	–	–	1.1×10^8
1-Propanol / Et$_2$O	2:5	[6203]	–	–	–
2-Propanol / Et$_2$O	1:3	[6201]	–	–	–
1-Butanol / Et$_2$O	2:5	[6203]	–	–	–

Table 9d Low-Temperature Organic Glasses for Spectroscopy[a]

Glass[a]	Ratio	Ref.	% crack[b]	V_{77}/V_{293}[c]	η/η_{3MP}[d]
EtOH / MeOH / Et$_2$O	8:2:1	[6301]	10	–	–
EPA (EtOH / IP / Et$_2$O)	2:5:5	[5301]	0	0.778	5.5×10^{-2}
		[6202]			
		[6301]			
2-Propanol / IP / Et$_2$O	2:5:5	[6203]	–	–	–
Miscellaneous					
Perfluorodimethylcyclohexane	–	[6803]	–	–	1.1×10^{35}
CH$_2$Cl$_2$ / MeOH	1:1	[0501]			
Et$_2$O / EtOH / Toluene	2:1:1	[6201]	–	–	
Ethyl iodide / Et$_2$O / IP	1:2:1	[6201]	–	–	–
Ethyl iodide / EtOH / MeOH	1:16:4	[6201]	–	–	–
cis-2-Pentene / Et$_2$O	1:2	[6203]	–	–	–
Cyclohexene / IP	2:1	[6701]	–	–	–
Cyclohexene / Et$_2$O	3:2	[6701]	–	–	–
Piperylene / Et$_2$O / MeCH	1:2:2	[6701]	–	–	–
EtOH / NH$_3$ (28% aq.)	<20:1	[6301]	30	–	–
Et$_2$O / EtOH / NH$_3$ (28% aq.)	10:9:1	[6301]	10	–	–
Triethylamine / Et$_2$O / Pentane	2:5:5	[6301]	0	–	–
Triethylamine / Et$_2$O / IP	2:5:5	[6201]	–	–	–
EtOH / conc. HCl	19:1	[6301]	30	–	–

[a] Organic solvent mixtures that form clear glasses at 77 K with a low cracking frequency. For other glasses see listed references; [b] From ref. [6301], frequency of cracking in percent; [c] From ref. [5301, 6202], volume at 77 K relative to volume at 293 K; [d] Viscosities, η, relative to 3MP at 77 K. (1) [6501], η's relative to a value of 11 for 2MP. Value of 11 was adopted from ref. [6802] to put the η's on a common scale. Values reported have been extrapolated on the graph in ref. [6501] and are subject to considerable error. For an absolute scale, a value for 1:3 MeCH / IP of 1.26×10^7 Pa·s at 77 K can be extracted from the data. (2) [6802], using an absolute value of 2.2×10^{11} Pa·s for 3MP at 77 K. The other values for this reference have been extrapolated to 77 K. (3) [6401], an absolute value of 9.4×10^{11} Pa·s is reported for MP at 77 K.

9e DONOR NUMBER

Of the solvents, aromatic and olefinic hydrocarbons are π-donors (π-EPD); alcohols, ethers, amines, carboxamides, nitriles, ketones, sulfoxides and N- and P-oxides are n-donors (n-EPD); and haloalkanes are σ-donors (σ-EPD). Boron and antimony trihalides are acceptor solvents (v-EPA), as are halogens and mixed halogens (σ-EPA), and liquid sulfur dioxide (π-EPA). In principle, all solvents are amphoteric in this respect, *i.e.* they may act as donor (nucleophile) and acceptor (electrophile) simultaneously. For example, water can act as a donor (by means of the oxygen atom) as well as an acceptor (by forming hydrogen bonds) [9001].

n-Donor solvents are particularly important for the solvation of cations. Examples are hexamethylphosphoric triamide, pyridine, dimethyl sulfoxide, N,N-dimethylformamide, acetone, methanol, and water. Their specific EPD-properties make them excellent cation solvators, and they are, therefore, good solvents for salts. They are also known as *coordinating solvents* [6501]. The majority of inorganic reactions are carried out in coordinating solvents.

An empirical semiquantitative measure of the nucleophilic properties of EPD-solvents is provided by the so-called *donor number DN* (or donicity) of Gutmann [7501, 7601, 7801, 7902]. This donor number has been defined as the negative ΔH-values for 1:1 adduct formation between antimony pentachloride and electron pair donor solvents (D) in dilute solution in the non-coordinating solvent 1,2-dichloroethane, according to Eq. 9-1 [9001].

$$D: \quad + \quad SbCl_5 \quad \xrightarrow{\quad RT, \; ClCH_2CH_2Cl \quad} \quad \overset{\oplus}{D}\text{---}\overset{\ominus}{SbCl_5} \tag{9-1}$$

$$\text{Solvent Donor Number } DN = -\Delta H_{D-SbCl_5} / (kcal \cdot mol^{-1})$$

The linear relationship between $-\Delta H_{D-SbCl_5}$ and the logarithm of the corresponding equilibrium constant (Log K_{D-SbCl_5}) shows that the entropy contributions are equal for all the studied acceptor/donor solvent reactions. Therefore, one is justified in considering the donor numbers as semiquantitative expression for the degree of coordination interaction between EPD solvents and antimony pentachloride. Antimony pentachloride [9001] is regarded as an acceptor on the borderline between hard and soft Lewis acids.

A list of alphabetically ordered organic solvents, selected from ref [8401] is given in table 9e. The higher the donor number, the stronger the interaction between solvent and acceptor.

Unfortunately, donor numbers have been defined in the non-SI unit *kcal mol⁻¹*. Marcus has presented a scale of dimensionless, normalized donor numbers DN^N, which are defined according to $DN^N = DN/(38.8 \; kcal \; mol^{-1})$[8401]. The non-donor solvent 1,2-dichloroethane ($DN = DN^N = 0.0$) and the strong donor solvent hexamethylphosphoric triamide (HMPT: $DN = 38.8$ *kcal mol^{-1}*; $DN^N = 1.0$) are used to fix the scale. It should be noted however that a much higher DN value of 50.3 *kcal mol^{-1}* for HMPT has been measured by Bollinger et al [8302].

Table 9e Solvent Donor Numbers[a]

No.	Solvent	DN^b
1	Acetic Acid	20
2	Acetic Anhydride	10.5
3	Acetil Chloride	0.7
4	Acetone	17.0
5	Acetonitrile	14.1
6	Acetophenone	15
7	Acrylonitrile	11
8	Ammonia	(59) (42)
9	Aniline	35
10	Anisole	9
11	Benzaldehyde	16
12	Benzene	0.1 (3)
13	Benzonitrile	11.9
14	Benzophenone	18
15	Benzyl Alcohol	23
16	Benzyl cyanide	15.1
17	Bromobenzene	3
18	Butyronitrile	16.6
19	1-Butanol	29
20	2-Butanone	17.4
21	Carbon Disulfide	2
22	Carbon Tetrachloride	0
23	Cyclohexane	0
24	Chlorobenzene	(3.3)
25	Chloroform	4
26	Cumene	6
27	Cyclohexanone	18
28	Cyclopentanone	18
29	Dichloroethane	(0)
30	Diacetyl	11
31	Diethylamine	50

Table 9e Solvent Donor Numbers[a]

No.	Solvent	DN^b
32	Diethylether	19.2
33	N,N-Dimethylformamide	26.6
34	Dimethylsulfoxide	29.8
35	1,4-Dioxane	14.8
36	Ethanol	32
37	Ethyl acetate	17.1
38	Ethyl benzoate	15
39	Ethylamine	40
40	Ethylbenzene	6
41	Ethylene Glycol	20
42	Ethylenediamine	(55)
43	Fluorbenzene	3
44	Formamide	36
45	Formic Acid	19
46	Furan	6
47	Glycerol	19
48	Hexamethylphosphor Triamide	38.8
49	n-Hexane	0
50	Hydrazine	(44)
51	Iodobenzene	4
52	Mesitylene	10
53	Methanol	(19)(30)
54	Methyl acetate	16.3
55	Methylamine	39
56	Methylene Chloride	1
57	N-Methylformamide	49
58	2-Methyltetrahydrofuran	18
59	Nitrobenzene	4.4 (11)
60	Nitromethane	2.7 (6)
61	1-Pentanol	25
62	3-Pentanone	15

Table 9e Solvent Donor Numbers[a]

No.	Solvent	DN^b
63	Phenol	11
64	Piperidine	40
65	Piridine-*N*-oxide	32
66	2-Propanol	36
67	1-Propylamine	38
68	Propylene Carbonate	15.1
69	Propionitrile	16.1
70	Pyridine	33.1
71	Quinoline	32
72	Styrene	5
73	Sulfuryl Chloride	0.1
74	Tetrahydrofuran	20.0
75	Thionyl Chloride	0.4
76	Toluene	0.1 (4)
77	Triethyl Phosphate	26
78	Triethylamine	(61)
79	Tri-*i*-butylamine	50
80	Trimethyl Phosphate	23.0
81	Trimethylamine	64
82	Water	(18.0)
83	Xylene	5

[a] For the definition of donor number, see Eq. 9-1. For conversion into SI units: 1 kcal mol^{-1} = 4.184 kJ mol^{-1}; [b] The values in parenthesis are considered less reliable by Marcus [8401].

REFERENCES

[4601] Kretschmer, C. B.; Nowakowska, J.; Wiebe, R. *Ind. Eng. Chem.* **1946**, *38*, 506-509.

[4701] Fowler, R. D.; Hamilton, J. M., Jr.; Kasper, J. S.; Weber, C. E.; Burford, W. B., III; Anderson, H. C. *Ind. Eng. Chem., Ind. Ed.* **1947**, *39*, 375-378.

[5301] Passerini, R.; Ross, I. G. *J. Sci. Instrum.* **1953**, *30*, 274-276.

[6201] Smith, F. J.; Smith, J. K.; McGlynn, S. P. *Rev. Sci. Instrum.* **1962**, *33*, 1367-1371.

[6202] Rosengren, K. J. *Acta Chem. Scand.* **1962**, *16*, 1421-1425.

[6203] Scott, D. R.; Allison, J. B. *J. Phys. Chem.* **1962**, *66*, 561-562.

[6301] Winefordner, J. D.; St. John, P. A. *Anal. Chem.* **1963**, *35*, 2211-2212.

[6401] Lombardi, J. R.; Raymonda, J. W.; Albrecht, A. C. *J. Chem. Phys.* **1964**, *40*, 1148-1156.

[6501] Greenspan, H.; Fischer, E. *J. Phys. Chem.* **1965**, *69*, 2466-2469.

[6502] Drago, R. S.; Purcell, K. F.; in *Non Aqueous Solvent Systems*; Waddington, T. C. (ed.); Academic Press: London, UK, 1965, 211-251.

[6701] Murov, S. L. *Ph.D., Thesis*; Univ. Chicago: Chicago (IL), 1967, 227-227.

[6801] Kosower, E. M. *An Introduction to Physical Organic Chemistry*; Wiley and Sons: New York, 1968, 503p.

[6802] Ling, A. C.; Willard, J. E. *J. Phys. Chem.* **1968**, *72*, 1918-1923.

[6803] Ling, A. C.; Willard, J. E. *J. Phys. Chem.* **1968**, *72*, 3349-3351.

[7101] Fowler, F. W.; Katritzky, A. R.; Rutherford, R. J. D. *J. Chem. Soc. B* **1971**, 460-469.

[7102] Perkampus, H. -H.; Sandeman, I.; Timmons, C. J. *DMS UV Atlas of Organic Compounds*; Verlag Chemie, Weinheim: Butterworths, London (England), 1966-1971, Vol. I-V.

[7103] Clark, W. D. K.; Steel, C. *J. Am. Chem. Soc.* **1971**, *93*, 6347-6355.

[7201] Hutzler, J. S.; Colton, R. J.; Ling, A. C. *J. Chem. Eng. Data* **1972**, *17*, 324-327.

[7501] Gutmann, V. *Coord. Chem. Rev.* **1975**, *15*, 207-237

[7601] Gutmann, V. *Coord. Chem. Rev.* **1976**, *18*, 225-255.

[7701] Wesseler, E. P.; Iltis, R.; Clark, L. C. Jr. *J. Fluorine Chem.* **1977**, *9*, 137-146.

[7801] Gutmann, V. *The Donor-Acceptor Approach to Molecular Interactions*; Plenum: New York, 1978, 279p.

[7901] Griffiths, T. R.; Pugh, D. C. *Coord. Chem. Rev.* **1979**, *29*, 129-211.

[7902] Gutmann, V. *Pure Appl. Chem.* **1979**, *51*, 2197-2210.

[8101] Kamlet, M. J.; Abboud, J. L. M.; Taft, R. W. in *Progress in Physical Organic Chemistry*; Taft, R. W. (Ed.); Wiley and Sons: New York, 1981, vol. 13, 485-630.

[8102] Battino, R. *Solubility Data Series: Volume 7, Oxygen and Ozone*; Pergamon: Oxford (England), 1981, 519p.

[8301] Acs, A.; Schmidt, R.; Brauer, H.-D. *Photochem. Photobiol.* **1983**, *38*, 527-531.

[8302] Bollinger, J.-C.; Yvernault, G.; Yvernault, T. *Thermochim. Acta* **1983**, *60*, 137-147.

[8401] Marcus, Y. *J. Sol. Chem.* **1984**, *13*, 599-624.

[8501] Marcus, Y. *Ion Solvation*; Wiley and Sons: New York, 1985.

[8601] Riddick, J. A.; Bunger, W. B.; Sakano, T. K. *Organic Solvents. Physical Properties and Methods of Purification. Fourth Edition*; Wiley and Sons: New York, 1986, Vol. II, 1325p.

[8901] Clennan, E. L.; Noe, L. J.; Wen, T.; Szneler, E. *J. Org. Chem.* **1989**, *54*, 3581-3584.

[9001] Reichardt, C. *Solvents and Solvent Effects in Organic Chemistry*; VCH: Weinheim (D), 1990, p. 534.

[9901] Abboud, J.-L. M.; Notario, R. *Pure Appl. Chem.* **1999**, *71*, 645-718.

[0301] Lide, D.R. *Handbook of Physics and Chemistry. Eighty Fourth Edition*; CRC Press: Boca Raton, (FL), 2003-2004.

[0501] Credi, A. unpublished results.

10

Luminescence Spectroscopy Measurements

10a CORRECTION OF LUMINESCENCE INTENSITY MEASUREMENTS IN SOLUTION

With a spectrofluorimeter, the types of measurements that can be performed are mainly: i) *emission spectra*, where the emission intensity as a function of the emission wavelength is observed; ii) *excitation spectra*, where the emission intensity as a function of the excitation wavelength is observed; iii) qualitative and quantitative measurements; iv) luminescence *quantum yields* determinations.

Unfortunately, luminescence measurements are often not as easy as they could seem at a first glance [7101, 8101, 8201, 8701, 8702, 9801, 9901]. In fact, while the electric signal produced by a spectrophotometer represents a physical quantity (the absorbance) that can be expressed in an absolute scale, the electric signal produced by a spectrofluorimeter is related to the total luminescence intensity (i.e., to the number of emitted photons) through a number of instrumental factors (intensity of the exciting source, instrument optics, signal amplification) and to the solution characteristics. The observed intensity signal can be related to sample concentration only if corrected in order to take into account these instrumental and solution factors. It should be stressed that luminescence intensity measurements carried out with standard spectrofluorimeters are never absolute, and the intensity values must be expressed in a relative scale even after corrections.

10a-1 Corrected Excitation Spectra

Corrected excitation spectra are often used in analytical chemistry to reveal a luminescent compound which can be present, also only as a trace, in a complex matrix. Reliable excitation spectra can be obtained only if the absorbance of the sample is <0.1 (vide infra), over all the examined wavelength region, because only in these conditions the emission intensity value is directly proportional to the absorbance value at each different excitation wavelength; in this case, the excitation spectrum reproduces the absorption spectrum of the emitting species. This is true if the spectrofluorimeter used is corrected for the spectral intensity of the excitation lamp, which varies with the excitation wavelength. Most of the modern instruments are equipped in this sense.

Alternatively, the most common method to obtain the spectral irradiance of the excitation system (exciting lamp plus monochromator) as a function of the excitation wavelength is the use of a *quantum counter*, at a concentration suitable to absorb all the incident photons. For the wavelength region 220-600 nm, the recommended standard is Rhodamine 101 [8001]. This compound is suitable for both 90° and front surface emission observations [8101, 9901]. Rhodamine 101 is preferred [8001, 8901] with respect to Rhodamine B [7501, 9901], because the second one is sensitive to temperature fluctuations, and to light below 340 nm (if not of high grade purity). Quinine sulfate is a good quantum counter from 220 to 340 nm [8301, 8901]. A correct excitation spectrum, using a quantum counter in a front face geometry, may be obtained by scanning the excitation monochromator and measuring the ratio of the fluorescence intensity from the sample to that from the quantum counter at a fixed emission wavelength. A list of quantum counters is reported in ref [9901], p. 50.

10a-2 Corrected Emission Spectra

In recording the emission spectra one must consider that the sensitivity of the revealing system (emission monochromator plus photomultiplier) depends on the emission wavelength, so corrections are necessary to compare spectra performed with different instruments. These corrections can be performed in two ways:

a) use of standard compounds with known corrected emission spectra: in this simple and reliable method the emission correction factors are obtained by comparing the observed emission spectrum of a standard fluorophore with its corrected emission spectrum tabulated in the literature. For emission spectra of standards covering the range 300-800 nm see ref. [9901] (Chapter 2, Appendix 1, and refs. therein).

b) use of a standard lamp: commercially available tungsten standard lamps of known color temperature can be used as follows: the wavelength integrated intensity of this lamp is measured using the detection system of the spectrofluorimeter. At each wavelength, the sensitivity of the detection system can be calculated by the ratio:

$$S(\lambda) = I(\lambda)_{exp}/I(\lambda)_{std}$$

where $I(\lambda)_{exp}$ is the emission intensity measured and $I(\lambda)_{std}$ is the known intensity, certified by the selling company. The corrected spectra are obtained by dividing experimental intensity value by these sensitivity factors. This method, which has the advantage of covering an emission range from far UV to near IR, needs a more delicate practical procedure; furthermore, the spectral output of the lamp changes with the age and usage, and recalibration with time is needed.

10a-3 Quantitative Measurements at Fixed Emission Wavelength

This paragraph focuses on quantitative steady-state luminescence intensity determination in solution, at a fixed emission wavelength, using a commercial spectrofluorimeter with right-angle excitation (perpendicular geometry) [9801]. This kind of measurements are particularly important in analyte detection, titrations, quenching and sensitization experiments, photoreaction and photoluminescence quantum yield determinations, and whenever a luminescence signal is used to monitor a chemical process.

As a matter of fact, the observed luminescence intensity has no need of corrections only for very dilute solutions, where absorption effects can be ignored [8701]. Although this condition can sometimes be achieved (an analyte with a high luminescence quantum yield – e.g., anthracene – can be detected at concentrations as low as 10^{-10} M), more concentrated solutions are normally to be examined. Other problems arise if the solution contains other chromophores absorbing at the excitation and/or emission wavelengths.

The determination of the "corrected" emission intensity of a luminophore from the luminescence values measured in an experiment implies the knowledge of (i) *the fraction of incident photons absorbed by the emitting species* and (ii) *the fraction of the actually emitted light that is detected by the instrument*, or, in equivalence, the part of emitted photons lost on the way to the photomultiplier. The corrections introduced below take into account all the factors that affect the measured luminescence intensity, thus answering to the above questions.

A correction factor $\Im$ can be defined such that, multiplied by the luminescence intensity I_ℓ^{obs} observed for the luminophore at a fixed emission wavelength λ_{em} upon excitation at λ_{ex}, gives a value of the corrected luminescence intensity I_ℓ^0 that is independent on the experimental conditions (except for the concentration of the considered emitting species):

$$I_\ell^0(\lambda_{ex},\lambda_{em}) = \Im(\lambda_{ex},\lambda_{em}) \cdot I_\ell^{obs}(\lambda_{ex},\lambda_{em}) \qquad (10\text{-}1)$$

Then, if C_1 and C_2 are the concentrations of a luminescent species in two solutions, whatever are the experimental conditions, and I_1^0 and I_2^0 are the corresponding corrected luminescence intensities, in absence of quenching and sensitization phenomena, one obtains

$$\frac{I_1^0(\lambda_{ex},\lambda_{em})}{I_2^0(\lambda_{ex},\lambda_{em})} = \frac{C_1}{C_2} \qquad (10\text{-}2)$$

This relationship implies that the corrected luminescence intensity is linearly correlated to the luminophore concentration, as needed. In this way, possible quenching or sensitization processes can be quantitatively assessed if C_1 and C_2 are known.

The procedure that should be followed to make corrections on observed luminescence intensities can be divided in two steps: (i) correction for the (non-linear) luminescence intensity response of the instrument versus absorbance and (ii) correction for inner filter effects (coabsorption of the excitation light and reabsorption of the emitted light).

Instrument calibration curve

Even in the simplest case, when only one species is present in solution, the relationship between the observed luminescence intensity and the concentration (i.e., the absorbance) of the solution at the excitation wavelength is not linear. This is due to two reasons, one of physical and one of "geometrical" nature.

The first reason is trivial. The emitted light is proportional to the number of the excited states present in solution, which are proportional to the light intensity absorbed by the target species at the excitation wavelength, $I_a(\lambda_{ex})$. The intensity of light absorbed by a pathlength d of solution containing a luminophore of concentration C_ℓ and molar absorptivity $\varepsilon_\ell(\lambda_{ex})$ is given by the well-known Beer's law:

$$I_a(\lambda_{ex}) = I_i(\lambda_{ex}) - I_t(\lambda_{ex})$$
$$= I_i(\lambda_{ex}) - I_i(\lambda_{ex}) \cdot 10^{-\varepsilon_\ell(\lambda_{ex})C_\ell d}$$
$$= I_i(\lambda_{ex}) \cdot (1 - 10^{-\varepsilon_\ell(\lambda_{ex})C_\ell d}) \tag{10-3}$$

where $I_i(\lambda_{ex})$ and $I_t(\lambda_{ex})$ are the incident and transmitted light intensity, respectively. By definition, the luminescence quantum yield (Φ_ℓ) is defined by the ratio between the emitted ($I_\ell(\lambda_{ex}, \lambda_{em})$) and absorbed ($I_a(\lambda_{ex})$) photons; then, the total luminescence intensity can expressed as

$$I_\ell(\lambda_{ex}, \lambda_{em}) = \Phi_\ell \cdot I_a(\lambda_{ex}) = \Phi_\ell \cdot I_i(\lambda_{ex}) \cdot (1 - 10^{-\varepsilon_\ell(\lambda_{ex})C_\ell d}) \tag{10-4}$$

From Eq. 10-4 it is evident that the relationship between I_ℓ and the luminophore concentration is not linear. The exponential term in Eq. 10-4 can be expanded in series, and orders higher than 1 can be neglected if $\varepsilon_\ell(\lambda_{ex}) \cdot C_\ell \cdot d$ is reasonably small (< 0.1). Under this condition, Eq. 10-4 can be approximated to Eq. 10-5

$$I_\ell(\lambda_{ex}, \lambda_{em}) = \ln(10) \cdot I_i(\lambda_{ex}) \cdot \Phi_\ell \cdot \varepsilon_\ell(\lambda_{ex}) \cdot C_\ell \cdot d \tag{10-5}$$

showing that I_ℓ is linearly related to C_ℓ for low absorbance values at λ_{ex}.

The second factor determining the non-linearity of the I_ℓ vs. A response of the instrument is related to the geometry of the slits/monochromators/detector ensemble. In commercial spectrofluorimeters, the light emitted by the sample is not collected over the whole space surrounding the cuvette, but only through a "window" in front of the cell surface at 90° with respect to the excitation beam. This means that the light detected by the photomultiplier depends on *"where"* the luminescence is originated *"inside"* the cuvette. If the solution is sufficiently dilute, one can assume that the excitation beam is able to cross the entire cuvette before being considerably attenuated, and that the excited states are uniformly produced along the excitation beam path. Under this condition, the luminescence intensity is related to the absorbance by Eq. 10-4, and no geometrical effect is present. When the absorbance at the excitation wavelength is high, the luminophore absorbs and emits in the first layers of the solution. Because the detector monitors the central part of the observed surface it will lose most of the emitted light leading to the paradox that the luminescence intensity decreases with increasing the luminophore concentration for high values of the absorbance at λ_{ex}. It is then evident that for solutions with large absorbance at the excitation wavelength (typically, $A > 0.1$) the observed luminescence intensity is affected by a "geometrical factor" that must be taken into account. Obviously, any species (not only the luminophore) absorbing at the excitation wavelength enhances this effect.

While the first reason of the non-linear I_ℓ vs. A behavior could be accounted for by calculations, the geometrical factor has to be measured since it depends on the instrument characteristics. A simple correction formula has been proposed [9901] with the assumption that the luminescence is originated in the central point of the cuvette. It is recommended, however, to construct a calibration curve for the instrument [8701, 8702] by plotting the luminescence intensity values (at a chosen λ_{em}) obtained for several solutions of one emitting species with different concentration (i.e., with different absorbance at λ_{ex}) versus the corresponding absorbance at the excitation wavelength. Of course, such a calibration graph depends on the width of the window monitored by the detector, i.e. on the emission slit width. A smaller dependence of the calibration curve on the excitation slit can also be observed. Fig. 10a-1 shows the calibration graph for a Perkin-Elmer LS-50 spectrofluorimeter obtained measuring the luminescence intensity ($\lambda_{em} = 450$ nm; bandpass = 10 nm) of quinine sulfate in H_2SO_4 0.5 M [8301] at different concentrations. Notice that the relationship is practically linear until $A \approx 0.1$ (see Eq. 10-5 and above in the text).

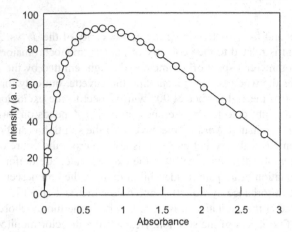

Fig. 10a-1. Calibration graph for a Perkin-Elmer LS-50 spectrofluorimeter.

Generally speaking, if $F_g(A(\lambda_{ex}))$ is the function containing both the mathematical and the geometrical dependence of I_ℓ vs. A, derived from the above described calibration curve, Eq. 10-4 can be rewritten as

$$I_\ell^{obs}(\lambda_{ex}, \lambda_{em}) = \Phi_\ell \cdot I_i(\lambda_{ex}) \cdot F_g(A(\lambda_{ex}))$$
(10-6)

A correction factor $\Im$ that, substituted in Eq. 10-1 gives corrected luminescence intensity values that satisfy Eq. 10-2, can be obtained taking

$$\Im(\lambda_{ex}) = \frac{A(\lambda_{ex})}{F_g(A(\lambda_{ex}))}$$
(10-7)

Then, according to Eqs. 10-1 and 10-6, one obtains

$$I_\ell^0(\lambda_{ex}, \lambda_{em}) = \frac{A(\lambda_{ex})}{F_g(A(\lambda_{ex}))} I_\ell^{obs}(\lambda_{ex}) = \Phi_\ell \cdot I_i(\lambda_{ex}) \cdot A(\lambda_{ex})$$
(10-8)

Eq. 10-8 shows that the corrected luminescence intensity $I_\ell^0(\lambda_{ex}, \lambda_{em})$ is linearly related to the luminophore concentration and satisfies Eq. 10-2.

From Eq. 10-8 it is clear that the total absorbance of the solution at λ_{ex} must be measured for obtaining the corrected intensity. In some analytical application, the absorbance of the solution can be too small to be measured with good confidence. In this case Eq. 10-8 cannot be used but, as discussed above and shown in Fig. 10a-1, the I_ℓ vs. A trend is linear (Eq. 10-5). This means that luminescence intensities observed for solutions whose absorbance at λ_{ex} lies in this region can be directly compared, according to Eq. 10-2, without any correction. This is also the reason why a corrected excitation spectrum (i.e. in which the emission intensity recorded at a fixed λ_{em} is proportional to the absorbance of the species, so that it can be compared with the absorption spectrum) is obtained only for solutions with low absorbance ($A < 0.1$) ([8101], Chapter 7, [8201], Chapter 3).

Inner filter effects

The presence of different substances in the same solution can affect the observed luminescence intensity in different ways (besides quenching and sensitization phenomena), namely by absorbing the incident light or by reabsorbing the emitted light. For the same reasons, correction of the observed luminescence intensity values might be necessary also for a molecule or a supramolecular system [9101] that contains many chromophoric and/or emitting moieties.

Coabsorption of the exciting light: This phenomenon takes place when the incident light is absorbed not only by the luminophore of interest, but also by other chromophores present in solution, acting as filters at the excitation wavelength. If the various species do not interact in the ground state, the absorption spectrum of the mixture is exactly the sum of the spectra of the separated species. At the excitation wavelength λ_{ex} the fraction of light $F_\ell(\lambda_{ex})$ absorbed by the luminophore is given by

$$F_\ell(\lambda_{ex}) = \frac{\varepsilon_\ell(\lambda_{ex}) \cdot C_\ell}{\sum_i A_i(\lambda_{ex})} = \frac{A_\ell(\lambda_{ex})}{A(\lambda_{ex})} \tag{10-9}$$

where $A_\ell(\lambda_{ex})$ is the absorbance of the luminophore and $A(\lambda_{ex})$ is the total absorbance of the solution at the excitation wavelength. The term $F_\ell(\lambda_{ex})$ of Eq. 10-9 should be introduced in Eq. 10-6 when the coabsorption of the exciting light is effective:

$$I_\ell^{obs}(\lambda_{ex}, \lambda_{em}) = \Phi_\ell \cdot I_i(\lambda_{ex}) \cdot F_g\big(A(\lambda_{ex})\big) \cdot \frac{A_\ell(\lambda_{ex})}{A(\lambda_{ex})} \tag{10-10}$$

The correction factor $A(\lambda_{ex})/F_g(A(\lambda_{ex}))$ introduced in Eq. 10-8, however, takes already into account this effect. In fact, by substituting in Eq. 10-8 the above expression of $I_\ell^{obs}(\lambda_{ex}, \lambda_{em})$ (Eq. 10-10), one obtains

$$I_\ell^0(\lambda_{ex}, \lambda_{em}) = \frac{A(\lambda_{ex})}{F_g\big(A(\lambda_{ex})\big)} \Phi_\ell \cdot I_i(\lambda_{ex}) \cdot F_g\big(A(\lambda_{ex})\big) \cdot \frac{A_\ell(\lambda_{ex})}{A(\lambda_{ex})}$$

$$= \Phi_\ell \cdot I_i(\lambda_{ex}) \cdot \varepsilon_\ell(\lambda_{ex}) \cdot C_\ell \tag{10-11}$$

Eq. 10-11 demonstrates that the luminescence intensity corrected with the factor $A(\lambda_{ex})/F_g(A(\lambda_{ex}))$ (Eq. 10-7) is effectively proportional to the concentration of the examined luminophore even in case of coabsorption of exciting light (Eq. 10-8).

It should be noted that the calculations done so far are valid for any emission wavelength, since the "physical" and "geometrical" factors contained in the $F_g(A(\lambda_{ex}))$ function do not alter the spectral characteristics of the luminophore emission detected by the spectrofluorimeter.

Reabsorption of the emitted light: In luminescence measurements it is rather common that the absorbance of the solution in the spectral region of the luminophore emission is not zero. In this case, part of the light emitted by the luminophore is reabsorbed by the luminophore itself and/or by other species. This problem is of great importance in the determination of luminescence quantum yields and of the correct shape of a luminescence band [7101, 8101, 8201, 8701, 8702]. Of course, in our fixed emission wavelength measurements, the best precaution is to read the luminescence intensity at a wavelength where the absorbance of the solution is negligible. This favorable situation, however, is sometimes impossible to achieve, and corrections for the reabsorption of the emitted light must be done. The reabsorption effect depends on the geometry of the cell, that in our case (perpendicular geometry and vertical slits) is that depicted in Fig. 10a-2.

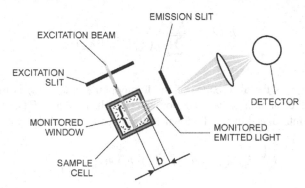

Fig. 10a-2. Schematic representation of the geometrical arrangement of the cell compartment and detector section in a spectrofluorimeter with perpendicular geometry and vertical slits.

The excitation beam, whose thickness will be considered negligible with respect to the width of the cell, incises perpendicularly to the excitation window of the cuvette and passes through the solution keeping parallel to the emission window. It can therefore be considered that the luminophore emission takes place from a narrow strip of solution. The pathlength that the emitted photons have to cross for getting out of the cuvette is not unique and depends on the emission optics (Fig. 10a-2). According to the Beer's law, the fraction of emitted light transmitted through the pathlength b at the wavelength λ_{em} is given by

$$T_\ell(\lambda_{em}) = 10^{-A(\lambda_{em}) \cdot b} \qquad (10\text{-}12)$$

where $A(\lambda_{em})$ is the absorbance of the solution per unit length at the monitored emission wavelength. It is then obtained:

$$I_\ell^{obs}(\lambda_{ex}, \lambda_{em}) = I_\ell^{0'}(\lambda_{ex}, \lambda_{em}) \cdot T_\ell(\lambda_{em}) \qquad (10\text{-}13)$$

$$I_\ell^{0'}(\lambda_{ex}, \lambda_{em}) = \frac{I_\ell^{obs}(\lambda_{ex}, \lambda_{em})}{T_\ell(\lambda_{em})} \qquad (10\text{-}14)$$

where $I_\ell^{0'}$ is the luminescence intensity made free from reabsorption effects (but has to be corrected for the other effects, see above), and $1/T_\ell(\lambda_{em})$ is the correction factor for the reabsorption of the emitted light, or what could be named the emission inner filter (EIF) correction.

As one can see from Eq. 10-12, the b value must be known in order to calculate the EIF correction. This parameter is averaged over all possible pathlengths, and would correspond to the "geometrical" b only if parallel beams were detected (Fig. 10a-2). In most instruments with perpendicular geometry the cell is illuminated centrally; then, in case of 1 cm × 1 cm cuvette, the "geometrical" b is supposed to be 0.5 cm. With this assumption, Lakowicz has proposed a simple correction formula [9901] that, however, gives only a rough estimate of the reabsorption effect. The value of the distance b can be estimated with good confidence by means of a simple experiment. What is needed is a solution containing the luminophore and a non-interacting species absorbing in the spectral region where the luminophore emits. The luminophore is excited, and the luminescence intensity is measured at a wavelength where the second species absorbs; the measurement is repeated for solutions with increasing concentration of absorbing species (increasing absorbance at λ_{em}). The observed intensity values have to be corrected for the instrumental response and for the coabsorption of the exciting light (see above), i.e. for all effects but internal emission filter. According to Eqs. 10-12 and 10-13, from the plot of the logarithm of these partially corrected luminescence values versus the absorbance of the solution at λ_{em} one should obtain a linear correlation, whose slope (with the sign changed) is the actual width of the layer of solution that the emitted light must cross to exit the cuvette. In principle, a dependence

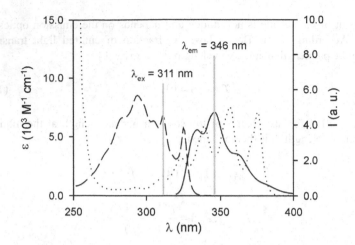

Fig. 10a-3. Absorption spectra of 1,5-dimethoxynaphthalene (dashed line) and anthracene (dotted line), and fluorescence spectrum of 1,5-dimethoxynaphthalene (full line; λ_{ex} = 311 nm) in MeCN solution at room temperature.

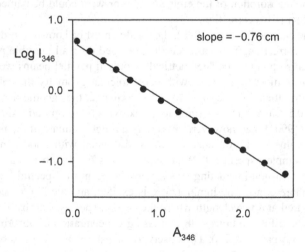

Fig. 10a-4. Plot of the logarithm of the 1,5-dimethoxynaphthalene luminescence intensity at 346 nm, corrected for the instrument response and for the coabsorption of exciting light versus the absorbance of the solution at the same wavelength.

of b on $A(\lambda_{ex})$ can be expected; however, no substantial change in the reabsorption pathlength is normally observed for a wide range of absorbance values. Suitable candidates for such experiment are, for istance, 1,5-dimethoxynaphthalene (2×10^{-4} M) as luminophore and anthracene (from 0 to 10^{-3} M) as an emission inner filter, in MeCN solution at room temperature. Good choices for the excitation and reading wavelength are 311 and 346 nm, respectively (Fig. 10a-3). The fitting of the experimental data is shown in Fig. 10a-4.

Another problem related to the reabsorption of the emitted light is the secondary emission from the examined luminophore. In fact, if there is overlap between the absorption and luminescence spectra of the luminophore, a certain amount of its luminescence is reabsorbed by itself and then reemitted, and so on. The secondary emission is important for concentrated solutions of strongly luminescent molecules with large overlap between absorption and emission spectra. A more complicated situation arises when a chromophore absorbing at λ_{ex} is luminescent, and its emission spectrum overlaps the absorption spectrum of the examined luminophore. Then, some of the light absorbed by the "inner filter" at λ_{ex} will be emitted and re-absorbed by the luminophore (a case of "trivial energy transfer" phenomenon). The final result is that part of the light not directly absorbed by the luminophore actually feeds the luminophore emission. This effect is difficult to take into account because the amount of emitted light reabsorbed by the luminophore is rather complicated to estimate; however, if the luminophore absorbance in the region where the "inner filter" molecule emits is very high, one can consider that all the emitted light is reabsorbed by the luminophore. In this limit case $F_\ell(\lambda_{ex})$ would become

$$F_\ell(\lambda_{ex}) = \frac{A_\ell(\lambda_{ex}) + A_{if}(\lambda_{ex}) \cdot \Phi_{if}}{A(\lambda_{ex})} \qquad (10\text{-}15)$$

where $A_{if}(\lambda_{ex})$ is the absorbance of the emitting species acting as "inner filter" effects at λ_{ex} and Φ_{if} is its luminescence quantum yield.

It should be emphasized, however, that corrections for the emission inner filter effect are always critical and introduce a rather high error on the corrected luminescence values, mainly because of the exponential relationship between $I_\ell^{0'}$ and $A(\lambda_{em})$ (Eqs. 10-12 and 10-14) and of the uncertainty on the b value. For example, if one takes $A(\lambda_{em})$ = 2.0 and $b = 0.5$ cm, according to Eqs. 10-12 and 10-14, the EIF correction $1/T_\ell(\lambda_{em})$ is 10; if $b = 0.52$ cm, the EIF correction becomes 11. This means that a 4% error on the b value causes a 10% error on the corrected luminescence intensity. The reabsorption of the emitted light should always be minimized with a wise choice of the monitored emission wavelength; data corrected using Eqs. 10-12 and 10-14 in condition of strong absorption at the emission wavelength (typically, $A(\lambda_{em}) > 1.5$) should be very carefully examined.

General formula

The general equation that can be employed to get a value of the luminescence intensity $I_\ell^0(\lambda_{em})$ proportional to the luminophore concentration from the observed luminescence intensity $I_\ell^{obs}(\lambda_{ex}, \lambda_{em})$ is:

$$I_\ell^0(\lambda_{em}) = \frac{A(\lambda_{ex})}{F_g(A(\lambda_{ex}))} \cdot \frac{1}{T_\ell(\lambda_{em})} \cdot I_\ell^{obs}(\lambda_{ex}, \lambda_{em}) \qquad (10\text{-}16)$$

In Eq. 10-16, the first term $A(\lambda_{ex})/F_g(A(\lambda_{ex}))$ represents the correction factor for the non-linear instrumental calibration curve described above. It also takes into account the coabsorption of the exciting light by other chromophores, when present (inner filter effect at λ_{ex}). The second term $1/T_\ell(\lambda_{em})$ is the correction factor for the reabsorption of the emitted light at the monitored emission wavelength, if any, by the chromophores present in solution (inner filter effects at λ_{em}; Eq. 10-12).

The propagation of errors on Eq. 10-16 gives the following expression for the relative error on the corrected luminescence intensity:

$$\frac{\delta I_\ell^0}{I_\ell^0} = \sqrt{\left(\frac{\delta A}{A(\lambda_{ex})} + \frac{\delta F_g}{F_g}\right)^2 + \left[\ln(10) \cdot d \cdot \delta A\right]^2 + \left[\ln(10) \cdot A(\lambda_{em}) \cdot \delta d\right]^2 + \left(\frac{\delta I_\ell^{obs}}{I_\ell^{obs}}\right)^2}$$

$$(10\text{-}17)$$

the symbols used have the same meaning as in Eq. 10-16 and preceding ones; for the sake of clarity, the identifications (λ_{ex}) and (λ_{em}) have been omitted where not explicitly required. The error on the function $F_g(A(\lambda_{ex}))$ is related to δA through its derivative with respect to $A(\lambda_{ex})$. It is worth noting that the largest contribution to the error on the corrected luminescence intensity is usually given by that on the emission inner filter correction (third and fourth terms in Eq. 10-17).

10b FLUORESCENCE QUANTUM YIELD STANDARDS

The easiest way to determine the quantum yield of a fluorophore is by comparison with standards of known quantum yield. The most used standards are listed in Table 10b-1; similar compilations are reported in refs. [8801, 8901, 9901]. The emission quantum yields of these reference compounds are mostly independent of the excitation wavelength, so the standards can be used in their full absorption range. The fluorescence spectra of several reference compounds are shown in Figures 10b-1–10b-3 (from ref. [0301]).

Practically, the quantum yield is generally determined by comparison of the wavelength integrated intensity of the emission spectrum of the fluorophore under examination to that of a suitable standard. The emission wavelength range and the shape of the spectra under comparison should match as much as possible; otherwise, when this is not feasible, the comparison can be performed using corrected emission spectra ([9401, 9901] and Section 10a of this handbook). The absorbance values should be kept below 0.1 to avoid inner filter effects, which have to be considered when there is an overlap between the lowest energy absorption band and the emission spectrum. In these conditions, using the same excitation wavelength, the unknown quantum yield is calculated using [7101, 8901, 9901]:

$$\Phi = \Phi_r \frac{I}{I_r} \frac{A_r}{A} \frac{n^2}{n_r^2} \qquad (10\text{-}18)$$

where Φ is the quantum yield, I is the integrated emission intensity, A is the absorbance at the excitation wavelength, and n is the refractive index of the solvent. The subscript r refers to the reference fluorophore of known quantum yield. When dilute solutions cannot be used ($A \geq 0.1$), the previous expression should include the function containing both the mathematical and geometrical dependence of I vs. A (see Section 10a-3):

$$\Phi = \Phi_r \frac{I}{I_r} \frac{F_{gr}}{F_g} \frac{n^2}{n_r^2} \qquad (10\text{-}19)$$

In any case, if the absorbance of the reference and unknown compound are matched at the excitation wavelength, the formula reduces to:

$$\Phi = \Phi_r \frac{I}{I_r} \frac{n^2}{n_r^2} \qquad (10\text{-}20)$$

The ratio of the refractive indices has its origin in consideration of the intensity observed from a point source in a medium of refractive index n_i by a detector in a medium of refractive index n_o. The observed intensity is modified by the ratio $(n_i/n_o)^2$. For more details, see ref. [9901] (page 53, and references therein). Solvents should be of spectral grade and must be checked for spurious emissions. An important precaution must be observed in choosing the excitation wavelength, i.e. spectral absorption ranges with a steep slope must be avoided. This is to prevent large errors deriving from possible different bandwidths in the absorption and emission/excitation systems of the instruments.

Table 10b-1 Recommended Fluorescence Quantum Yield Reference Compounds.

Emission range (nm)	Compound	Solvent	$\Phi^{a,b}$	Ref.
270–320	Benzene	Hexane	0.05	[6801]
280–330	Phenol	H_2O	0.14	[6701, 9901]
		Cyclohexane	0.08	[7102]
300–400	Naphthalene	Cyclohexane	$0.23; 0.036^c$	[7102; 0301]
		EtOH	0.21	[6601, 6801]
310–400	Terphenyl	Cyclohexane	$0.93; 0.82^c$	[7102; 0301]
315–480	2-Aminopyridine	H_2SO_4 0.05 M	0.60^d	[6802]
		H_2SO_4 0.05-0.5 M	0.66^e	[8301]
360–480	Anthracene	EtOH	0.27	[6101, 7102, 6801]
			0.21^f	[0301]
390–500	9,10-Diphenyl anthracene	Cyclohexane	0.90; 0.95	[8301, 8302; 7701]
		Benzene	0.84	[6101]
		EtOH	0.88	[7701]
380–580	Quinine sulfate^{g,h}	H_2SO_4 0.5 M	0.55	[6101, 6801, 8301]
		H_2SO_4 0.05 M	0.53	[7702]
400–600	β-Carboline^i	H_2SO_4 0.5 M	0.60	[8501, 9901]
430–560	Perylene	EtOH	0.92	[7101]
		Benzene	0.89; 0.99	[6101, 6801]
480–650	Fluorescein^{g,j}	NaOH 0.1 M	0.87; 0.92; 0.95	[6801; 8902; 7801]
550–700	Rhodamine 101	EtOH	1.00	[8001]
550–700	Rhodamine 6G	EtOH	0.94^g	[9601]
550–700	$[Ru(bpy)_3]^{2+\,k}$	H_2O	$0.042, 0.028^l$	[8202]
		MeCN	0.06^m	[8204]
580–780	Cresyl violet	MeOH	0.54^l	[7901]
600–850	$[Os(bpy)_3]^{2+\,k}$	MeCN	0.005	[8601]
600–900	$[Os(phen)_3]^{2+\,k}$	MeCN	0.021	[8601]

[a] Room temperature, absolute values or obtained using well established standards; deoxygenated solution, unless otherwise noted; [b] For the errors, see individual references; [c] Not absolute value, measured in aerated solution using deoxygenated naphthalene as a standard; [d] Relative value, using both quinine sulfate and 9,10-diphenylanthracene as standards; [e] Measured with integration sphere method using quinine sulfate as a standard; [f] Aerated solution, strongly oxygen quenched [7102]; for other comments, see ref. [7101]; [g] No quenching by oxygen [7101]; [h] Sensitive to halogen quenching, so that only the highest purity sulphuric acid should be used [8101]; [i] 9H-Pyrido[3,4-b]indole; [j] Degassed by nitrogen bubbling to avoid photodegradation [7101]; [k] Although the emission is a phosphorescence from a ^{3}MLCT state, its relatively short lifetime (see Table 10d-1) makes this compound more similar to fluorescence emitters; for these reasons it is included in this table; [l] Aerated solution; [m] For other polar solvents, see ref. [8204].

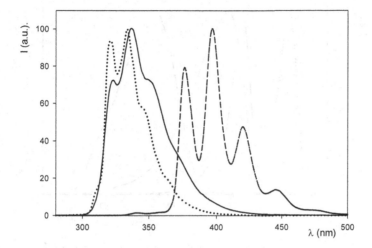

Fig. 10b-1. Uncorrected emission spectra (RT) of reference compounds for the range 300-500 nm: naphthalene in cyclohexane (·········); terphenyl in cyclohexane (————); anthracene in ethanol (– – – –). For the absorption spectra, see Section 3e.

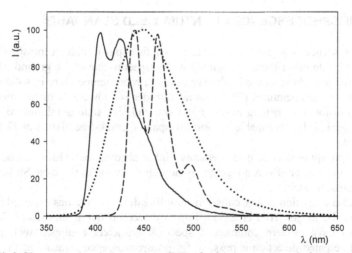

Fig. 10b-2. Uncorrected emission spectra (RT) of reference compounds for the range 350-650 nm: 9,10-diphenylanthracene in cyclohexane (————); quinine sulfate in 0.5 M H_2SO_4 (·········); perylene in cyclohexane (– – – –). For the absorption spectra, see Section 3e.

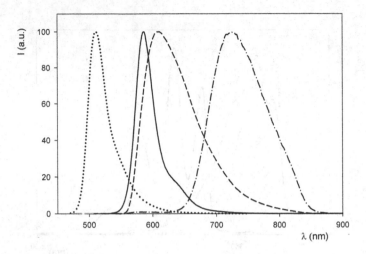

Fig. 10b-3. Uncorrected emission spectra (RT) of reference compounds for the range 450-900 nm: fluorescein in 0.5 M NaOH (·········); rhodamine 101 in ethanol (———); Ru(bpy)$_3$Cl$_2$ in water (— — — —); Os(bpy)$_3$(PF$_6$)$_2$ in acetonitrile (— · — · — ·). For the absorption spectra, see Section 3e.

10c PHOSPHORESCENCE QUANTUM YIELD STANDARDS

Phosphorescence is an emission due to a radiative spin-forbidden transition from an excited state of different multiplicity with respect to that of the ground state. In most cases phosphorescence is observed only at room temperature in solid matrices or at low temperatures (77 K, liquid nitrogen, or lower) in frozen matrices, where collisional quenching of the long-lived emissive state is minimized. For a list of organic solvents capable to form transparent, crack-free glasses at 77 K, see Section 9d.

In phosphorescence measurements higher absorbance values can be used, since the inner filter effects are generally negligible owing to the large Stokes shift of the emission band.

Pulse excitation and/or emission sampling delay techniques are used to discriminate the longer lived emission from any accompanying short-lived fluorescence. In most modern commercial spectrofluorimeters equipped with pulsed lamps time gated detection is possible for phosphorescence measurements. Table 10c-1 list some useful secondary standards for approximate (± 15%) quantum yield determination [8901]. The reference compounds span a reasonable range of wavelengths and can be used to check the operation of spectrometers.

Table 10c-1 Recommended Phosphorescence Quantum Yield Reference Compounds

Emission range (nm)	Compound	Solvent	Φ	Ref.
340–450	Benzene[a]	EPA (77 K)	0.23; 0.18	[5201, 5501; 7201]
400–550	Benzophenone[a]	EPA (77 K)	0.74	[5201, 5501, 7001]
		Freon 113 (298 K)	~0.01	[7802]
400–550	Xanthone[a]	EtOH-ether	0.44	[7001]
460–560	Naphthalene[a]	EPA (77 K)	0.04	[7201]
500–650	Biacetyl[a]	Hexane (298 K)	~0.05	[7802]
1062	Tris(1,1,1,5,5,5-hexafluoroacetylacetonato)neodymium(III)	THF-d_8	0.003	[0001]

[a] These standards are taken from ref. [8901].

10d LUMINESCENCE LIFETIME STANDARDS

For lifetime measurements different techniques can be used. Intensity decay profiles can be observed after excitation by sources such us flash-lamps or pulsed lasers. The use of single-photon counting apparatus with mathematical deconvolution represents the best method for accurate work. With phase-shift fluorimetric technique, the presence of multi-component decay can be revealed, even when observing at only one modulation frequency, by comparing the measured lifetime from the phase-shift to that from the modulation ratio. Homogeneity of the two values is evidence for a single lifetime [8901, 9901]. For phosphorescence lifetime measurements in the range from microseconds (more than ~30 µs) to seconds, modern commercial fluorimeters equipped with pulsed lamps and time-gate detection system can be used. For an updated excellent overview on time- and frequency-domain lifetime measurements, see [9901], chapters 4 and 5. For a review on recommended methods for fluorescence decay analysis, see [9001].

Suitable standards recommended to check apparatus for various decay ranges are given in Table 10d-1. For stilbene-derivatives used as picosecond lifetime standards, see [9102] or [9901], p. 648. Solution standards must be prepared using the highest grade purity compounds and solvents, considering the presence and concentration of any known quenchers (e.g., oxygen). This is particularly important for lifetimes longer than ~100 ns. For lifetime longer than ~1 µs only a few standards are reported in the table, both in fluid solution at room temperature and in glasses at 77 K.

Table 10d-1 Recommended Lifetime Reference Compounds, Ordered According to Increasing Lifetime Value.

Compound	Solvent	λ_{em} (nm)	τ^a	Ref.
Erythrosin	H_2O	580	66 ps[b]	[8203]
Rose bengal	MeOH	580	550 ps[b]	[8203]
Rose bengal	EtOH	580	800 ps[b]	[8203]
2,5-Diphenyloxazole	Cyclohexane	370	1.28 ns[b-d]	[8303]
2,5-Diphenyloxazole	Cyclohexane	440	1.42 ns[c]	[8303]
N,N-dimethyl-1-naphthylamine	CH_2Cl_2	375	2.40 ns[c]	[8303]
Rhodamine B	EtOH	600	2.85 ns[b]	[8303]
Coumarin 450	EtOH	460	4.3 ns[b]	[8203]
3-Methylindole	Cyclohexane	330	4.36 ns[c]	[8303]
Anthracene	EtOH	>400	5.1 ns[b]	[8303]
Anthracene	Cyclohexane	>400	5.23[b,c,e]; 5.52 ns	[8303; 8501]
1,2-Dimethylindole	EtOH	330	5.71 ns[c]	[8303]
1-Methylindole	Cyclohexane	330	6.24 ns[c]	[8303]
9,10-Diphenyl-anthracene	Cyclohexane	>400	7.7[b]; 7.9 ns	[8301; 7701]
3-Methylindole	EtOH	330	8.17 ns[c]	[8303]
9,10-Diphenyl-anthracene	EtOH	f	8.8 ns	[7701]
N-methylcarbazole	EtOH	370	16.0 ns[b]	[7102]
1-Cyanonaphthalene	Hexane	345	18.2 ns[b,c]	[8303]
N-methylcarbazole	Cyclohexane	370	18.3 ns[b]	[7102]
Quinine	0.5 M H_2SO_4	f	20.0 ns[d,g-i]	[6301, 6401, 7102, 8301, 8303, 8501]
β-Carboline[j]	0.5 M H_2SO_4	450	22.0 ns	[8501]
2-Methylnaphthalene	Cyclohexane	320	59 ns[b]	[7102]
Naphthalene	Cyclohexane	320	96 ns[b]	[7102]
Pyrene[k]	Cyclohexane	390	450 ns	[6302, 7002, 7003]
$[Ru(bpy)_3]^{2+}$	H_2O	610	650 ns	[8204]
	MeCN[l]	610	1.10 μs	[8102]
Tris(1,1,1,5,5,5-hexa-fluoroacetylaceto-nato)neodymium(III)	THF-d_8	1062	2.1 μs	[0001]

Table 10d-1 Recommended Lifetime Reference Compounds, Ordered According to Increasing Lifetime Value.

Compound	Solvent	λ_{em} (nm)	τ [a]	Ref.
Biacetyl	Hexane	550	0.5 ms[m]	[7802]
Biacetyl	CH_2Cl_2	518	0.6 ms	[9602]
Benzophenone	EPA, 77 K	450	6 ms[m]	[6702]
Anthracene	EPA, 77 K	690	~40 ms[m]	[7902]
Naphthalene	EPA, 77 K	-[f]	2.35; 2.6 s	[7201; 7902]
Benzene	EPA, 77 K	350	6.3 s[m]	[7201]

[a] Room temperature, deaerated solution, unless otherwise noted; for the exact values of T and the errors, see individual references; [b] Recommended by Eaton [8801]; [c] Recommended by O'Connor and Phillips [8401]; [d] Air equilibrated solution; [e] For air equilibrated solution, $\tau = 4.1$ ns [8303, 8501]; [f] See individual references; [g] For comments on non-exponential decay behavior, see ref. [8501]; [h] Average of the values quoted in the references; [i] For 0.05 M H_2SO_4 see ref. [7401]; [j] 9H-Pyrido[3,4-b]indole; [k] Very dilute solution is recommended ($<10^{-5}$ mol L^{-1}), to avoid excimer formation; [l] For other polar solvents, see ref. [8204]; [m] Recommended by Eaton in ref. [8901].

REFERENCES

[5201] Gilmore, E. H.; Gibson, G. E.; McClure, D. S. *J. Chem. Phys.* **1952**, *20*, 829-836.

[5501] Gilmore, E. H.; Gibson, G. E.; McClure, D. S. *J. Chem. Phys.* **1955**, *23*, 399-399.

[6101] Melhuish, W. H. *J. Phys. Chem.* **1961**, *65*, 229-235.

[6301] Birks, J. B.; Dyson, D. J. *Proc. Roy. Soc.* **1963**, *A275*, 135-148.

[6302] Birks, J. B.; Dyson, D. J.; Munro, I. H. *Proc. Roy. Soc.* **1963**, *A275*, 575-587.

[6401] Ware, W. R.; Baldwin, B. A. *J. Chem. Phys.* **1964**, *40*, 1703-1705.

[6601] Parker, C. A.; Joyce, T. A. *Trans. Faraday Soc.* **1966**, *62*, 2785-2792.

[6701] Chen, R. F. *Anal. Lett.* **1967**, *1*, 35-42.

[6702] Murov, S. L. *Ph. D. Thesis*; Univ. Chicago, Chicago (IL); 1967, p. 227.

[6801] Dawson, W. R.; Windsor, M. W. *J. Phys. Chem.* **1968**, *72*, 3251-3260.

[6802] Rusakowicz, R.; Testa, A. C. *J. Phys. Chem.* **1968**, *72*, 2680-2681.

[7001] Hunter, T. F. *Trans. Faraday Soc.* **1970**, *66*, 300-309.

[7002] Birks, J. B. *Photophysics of Aromatic Molecules*; Wiley: New York (NY), 1970, 704p.

[7003] Patterson, L. K.; Porter, G.; Topp, M. R. *Chem. Phys. Lett.* **1970**, *7*, 612-614.

[7101] Demas, J. N.; Crosby, G. A. *J. Phys. Chem.* **1971**, *75*, 991-1024.

[7102] Berlman, I. B. *Handbook of Fluorescence Spectra of Aromatic Molecules*; Academic Press: New York (NY) 1971, 2nd ed, 415p.

[7201] Li, R.; Lim, E. C. *J. Chem. Phys.* **1972**, *57*, 605-612.

[7401] Chen, R. F. *Anal. Biochem.* **1974**, *57*, 593-604.

[7501] Melhuish, W. H. A*ppl. Optics*. **1975**, *14*, 26-27.

[7701] Mardelli, M.; Olmsted, J. III *J. Photochem.* **1977**,*7*, 277-285.

[7702] Adams, M. J.; Highfield, J. G.; Kirkbright, G. F. *Anal. Chem.* **1977**, *49*, 1850-1852.

[7801] Brannon, J. H.; Magde, D. *J. Phys. Chem.* **1978**, *82*, 705-709.

[7802] Turro, N. J. (ed.) *Modern Molecular Photochemistry*; Benjamin: Menlo Park (CA), 1978, p. 117-118.

[7901] Magde, D.; Brannon, J. H.; Cremers, T. L.; Olmsted, J.III *J. Phys. Chem.* **1979**, *83*, 696-699.

[7902] Heinrich, G.; Güsten, H. *Z. Phys. Chem. (Wiesbaden)* **1979**, *118*, 31-41.

[8001] Karstens, T.; Kobs, K. *J. Phys. Chem.* **1980**, *84*, 1871-1872

[8101] Miller, J. N. (ed.) *Standards in Fluorescence Spectrometry*; Chapman & Hall: London (UK), 1981, 112p.

[8102] Juris, A.; Balzani, V.; Belser, P.; von Zelewsky, A. *Helv. Chim. Acta* **1981**, *64*, 2175-2182.

[8201] Mielenz, K. L. (ed.) *Optical Radiation Measurements*; Academic: New York (NY), 1982, vol. 3, 319p.

[8202] Nakamaru, K. *Bull. Chem. Soc. Japan* **1982**, *55*, 2697-2705.

[8203] Zimmerman, H. E.; Penn, J. H.; Carpenter, C. W. *Proc. Natl. Acad. Sci. USA* **1982**, *79*, 2128-2132.

[8204] Nakamaru, K. *Bull. Chem. Soc. Japan* **1982**, *55*, 1639-1640.

[8301] Meech, S. R.; Phillips, D. *J. Photochem.* **1983**,*23*, 193-217.

[8302] Hamai, S.; Hirayama, F. *J. Phys. Chem.* **1983**, *87*, 83-89.

[8303] Lampert, R. A.; Chewter, L. A.; Phillips, D.; O'Connor, D. V.; Roberts, A. J.; Meech, S. R. *Anal. Chem.* **1983**, *55*, 68-73.

[8401] O'Connor, D. V.; Phillips, D. *Time-correlated Single Photon Counting*; Academic Press: London (UK), 1984, 288p.

[8501] Ghiggino, K. P.; Skilton, P. F.; Thistlethwaite, P. J. *J. Photochem.* **1985**, *31*, 113-121.

[8601] Kober, E. M.; Caspar, J. V.; Lumpkin, R. S.; Meyer, T. J. *J. Phys. Chem.* **1986**, *90*, 3722-3734.

[8701] Rendell, D. *Fluorescence and Phosphorescence*; Wiley: Chichester (UK), 1987, 419p.

[8702] Harris, D. A., Bashford, C. L. (eds.) *Spectrophotometry and Spectrofluorimetry*; IRL: Oxford (UK), 1987, 176p.

[8801] Eaton, D. F. *Pure Appl. Chem.* **1988**, *60*, 1107-1114.

[8901] Scaiano, J. C. (ed.) *Handbook of Organic Photochemistry*; CRC Press: Boca Raton (FL), 1989, volume I.

[8902] Shen, J.; Snook, R. D. *Chem. Phys. Lett.* **1989**, *155*, 583-586.

[9001] Eaton, D. F. *Pure Appl. Chem.* **1990**, *62*, 1631-1648.

[9101] Balzani, V.; Scandola, F. *Supramolecular Photochemistry*; Horwood: Chichester (UK), 1991, 427p.

[9102] Lakowicz, J. R.; Gryczynski, I.; Laczko, G.; Gloyna, D. *J. Fluor.*, **1991**, *1*, 87-93.

[9201] Pant, D.; Tripathi, H. B.; Pant, D. D. *J. Lumin.* **1992**, *51*, 223-230.

[9401] Hofstraat, J. W.; Latuhihin, M. J. *Appl. Spectrosc.* **1994**, *48*, 436-447.

[9601] Fischer, M.; Georges, J. *Chem. Phys. Lett.* **1996**, *260*, 115-118.

[9602] Parola, A. J.; Pina, F.; Ferreira, E.; Maestri, M.; Balzani, V. *J. Am. Chem. Soc.* **1996**, *118*, 11610-11616.

[9801] Credi, A.; Prodi, L. *Spectrochim. Acta A* **1998**, *54*, 159-170.

[9901] Lakowicz, J. R. *Principles of Fluorescence Spectroscopy, 2nd Ed.*; Kluwer: New York (NY), 1999, 698p.

[0001] Iwamuro, M.; Wada, Y; Kitamura, T.; Nakashima, N.; Yaganida, S. *Phys. Chem. Chem. Phys.* **2000**, *2*, 2291-2296.

[0301] Biagini, M. *Diploma Thesis*; Dept. of Chemistry, University of Bologna (Italy); 2003.

11

Light Sources and Filters

11a SPECTRAL DISTRIBUTION OF PHOTOCHEMICAL SOURCES

This section gives a brief survey of the spectral distribution in a variety of photon sources that are useful for photochemical studies; see also ref. [8201] (Chapters 3 and 16) and [8901] (Chapters 5-7).

11a-1 Conventional Light Sources

Tungsten and tungsten-halogen lamps: *tungsten lamps* emit in a continuous manner from the near UV into the IR; in the *tungsten-halogen lamps*, in which the halogen (usually bromine or iodine) adds chemically to the evaporated tungsten at the bulb wall, the gas mixture decomposes with redeposition of the tungsten at the hot filament in a regenerative cycle. The bulb temperature must be high (~250°C) to maintain this cycle; thus, these lamps have small bulbs of quartz or fused silica. Individually calibrated tungsten-halogen lamps are used as standards for calibrating the spectral response of spectrometers.

Arc lamps: another important class of photon sources is that of *arc lamps*. *Deuterium lamps* are UV sources used for spectroscopic purposes, that provide an essentially line-free continuum from 180 to 400 nm. The spectral output of *xenon lamps* consists of a smooth continuum starting from the UV, with a weak superimposition of lines in the visible, with strong lines in the near IR. Most xenon lamps operate at high pressure (~20 bar) with short arc configurations and use a DC power supply for stable operation, and can be easily used in a pulsed way, such as in flash-photolysis spectroscopic studies. The spectral distribution of a xenon flash lamp depends upon the electrical discharge conditions, however in all cases the output in the UV is increased with respect to the output from a steady state lamp. *Xenon-mercury lamps* are short-arc lamps with a pressure of ~ 1 bar xenon, which adds a continuum background to the spectral output, which is mainly that of conventional high-pressure mercury lamps (see below). The qualitative emission spectrum of a Xenon arc is given in Fig. 11a-1. Comparative spectra are given in Fig. 11a-2 for a variety of arc lamps. These plots show the spectral irradiance as a function of the wavelength (energy or power spectra). In order to convert these spectra to photon spectra, the ordinates must be multiplied by factors proportional to the wavelength. According to ref. [8901], page 156, the number of photons emitted by a lamp per second, at a particular wavelength (λ, nm) is related to the optical power as follows:

$$\text{number of photons (einstein s}^{-1}) = \text{power (watts)} \times \lambda \times 8,359 \times 10^{-9}$$

Thus, the reported spectra are greatly exaggerated in the shorter wavelength range if the reader is concerned with number of photons (see, for example, Fig. 11a-1).

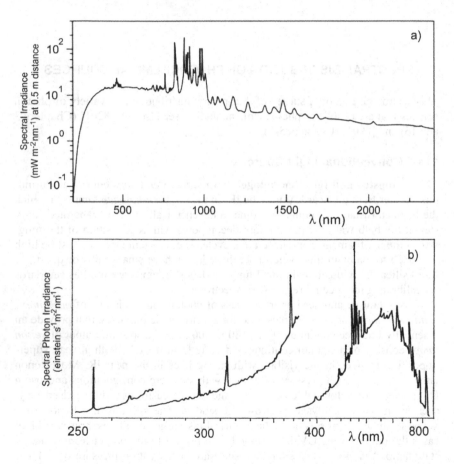

Fig. 11a-1. Typical spectral irradiance of a xenon lamp: a) power spectrum of a 150 W lamp (adapted by permission from LOT-Oriel catalogue); b) photon spectrum of a 350 W lamp (reprinted from ref. [6801], page 165, by permission of Elsevier).

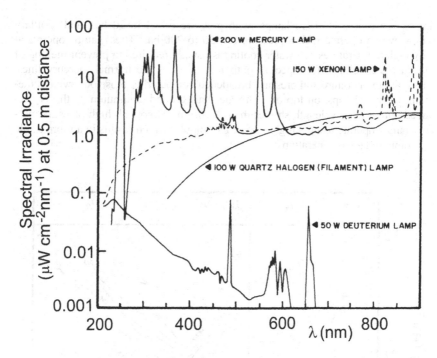

Fig. 11a-2. Spectral irradiance of some arc lamp sources. Reprinted by permission from LOT-Oriel catalogue.

For their relevance in photochemical studies, it is of interest to concentrate some additional attention to the various *mercury lamps* available. *Low-pressure* lamps operate with a vapor pressure below 1 bar, and close to room temperature. Lamps operating at about 10^{-6} bar emit mainly radiation (mercury resonance lines) at 184.9 and 253.7 nm (intensity ratio=1:10), although the spectral output varies with the bulb temperature, mercury pressure, and arc current. The 184.9 nm line is not observed, unless a bulb of suitable UV quartz is used. Almost all low-pressure arcs operate using an AC voltage supply with a low arc current. *Medium-pressure* mercury lamps operate at vapor pressures grater than 1 bar, but are commonly called high-pressure lamps, especially in Europe. The medium pressure lamps considered here operate at a pressure in the range 1-10 bar and are distinguished from high-pressure lamps by the distance between the electrodes. The spectral distribution consists of lines, together with a very weak continuum background. The 253.7 nm line is in most cases absent, owing to the self-absorption by the high concentration of mercury atoms near the bulb walls. Among the commercial lamps differences exist in design and operating conditions, such us temperature and pressure, which lead to small differences in spectral output. Almost all medium-pressure lamps operate using an AC supply. *High-pressure* mercury lamps are of two types, short-arc lamps, with pressure of ~2 to 20 bar, and capillary lamps, which operate with a pressure of ~50 to 200 bar. These lamps operate at very high temperatures and water cooling is usually required to prevent melting of the quartz envelope. The spectral output consists of the normal mercury lines, which are temperature and pressure broadened compared to those observed in medium-pressure lamps, on top of a stronger background continuum, with very reduced output at wavelength shorter than ~280 nm. Almost all high-pressure mercury lamps operate using a high current DC supply, which results in a more stable arc compared to AC operation.

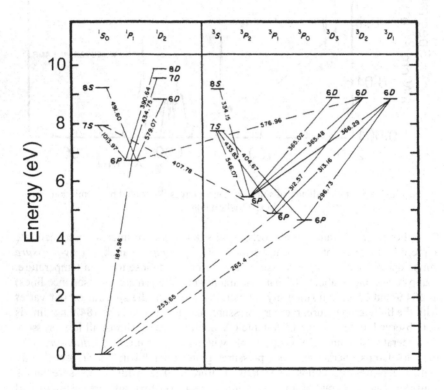

Fig. 11a-3. Grotrian energy-level diagram for mercury. The 184.96 and 253.65 nm lines are resonance lines and their intensities decrease quickly with increasing mercury pressure due to reabsorption. The 265.4 nm line is not observed in absorption due to its highly forbidden character. Adapted from ref. [6601], page 52, by permission of Wiley.

Since mercury has an emission line spectrum, it is often used as a convenient wavelength calibration standard. Fig. 11a-3 gives the energy level diagram for mercury, with the indication of the wavelengths corresponding to the various lines that can be used. Figures 11a-4 and 11a-5 give the emission spectra for low-and medium-pressure mercury lamps, while Table 11a-1 gives the corresponding intensity of the most important lines.

Finally, the solar power spectrum is shown in Fig. 11a-6.

For terms used to express the emission intensities of the various light sources and a glossary of terms used in photochemistry see refs. [0401, 0501].

Table 11a-1 Relative Spectral Energy Distribution of Helios Italquartz Mercury Lamps.

Low-pressure 15 W		Medium-pressure 125 W	
λ (nm)	Spectral irradiance (arbitrary units)	λ (nm)	Spectral irradiance (arbitrary units)
253.7	100	248.2	5.7
296.7	0.74	253.7	9.1
312.9	2.2	265.3	7.4
365.4	3.0	280.4	3.5
404.7	2.2	289.4	2.7
435.8	8.0	296.7	10
546.1	4.8	302.3	16
		312.9	33
		334.2	7.4
		365.4	100
		390.6	1.3
		404.7	24
		407.8	5.4
		435.8	37
		491.6	1.3
		546.1	33
		577.0	20
		579.1	29

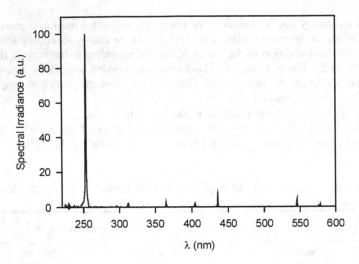

Fig. 11a-4. Emission spectrum of a low-pressure mercury lamp Helios Italquartz 15 W.

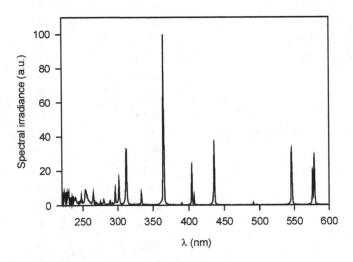

Fig. 11a-5. Emission spectrum of a medium-pressure mercury lamp Helios Italquartz 125W.

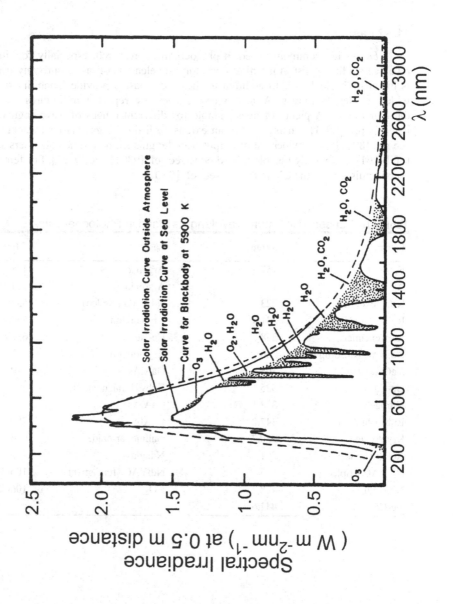

Fig. 11a-6. Solar spectral irradiance. From ref. [6501], reprinted by permission of McGraw-Hill.

11a-2 Lasers

Lasers are commonly used in photochemical research, especially for time resolved studies. A list of the most common wavelengths of lasers currently used is given in Table 11a-2. In addition to these, dye lasers provide tunability over wide wavelength ranges. A few tuning curves are reported in Figures 11a-7 through 11a-10. A plenty of dyes suitable for different ranges of wavelengths is listed in ref. [8901], Chapter 7. For an extensive listing of gas laser wavelengths see ref. [8001]; for a wider list of output wavelengths from commercial lasers see ref. [9701]. For widely tunable laser diodes see ref. [9801] and [9901]. For femtosecond pulses from the UV to the IR, see ref. [9301].

Table 11a-2 Lasers: Common Wavelengths Used in Photochemistry

Laser	λ (nm)	Laser	λ (nm)
F_2	157	Argon ion	488
ArF (excimer)	193	Argon ion	514.5
KrCl (excimer)	223	Nd:YAG (dupled)	532
Ruby (tripled)	231.4	Krypton ion	568.2
KrF (excimer)	248	He-Ne	632.8
Nd:YAG (quadrupled)	266	Krypton ion	647.1
XeCl (excimer)	308	GaAlAs	670
He-Cd	325	Ruby (fundamental)	694.3
Nitrogen	337.1	GaAlAs	750
Ruby (dupled)	347.2	GaAlAs	780
Krypton ion	350.7	Gallium arsenide	904
XeF (excimer)	351	Nd: glass	1060
Nd:YAG (tripled)	355	Nd:YAG (fundamental)	1064
Nitrogen	428	CO_2	10600
He-Cd	441.6		

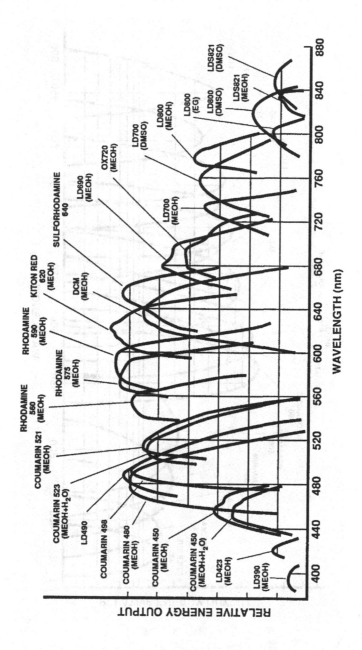

Fig. 11a-7. Tuning curves for flash-lamp pumped dye lasers. Reprinted by permission of Exciton. Dyes names are those of Exciton.

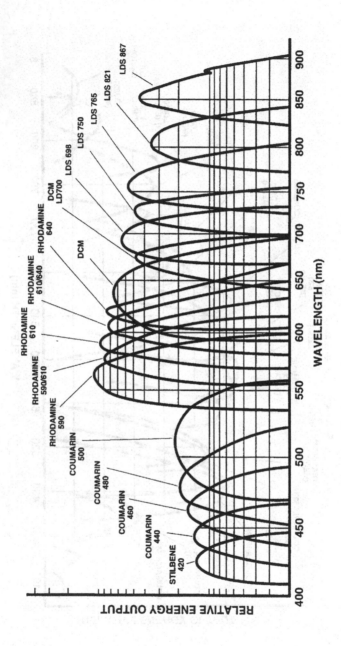

Fig. 11a-8. Tuning curves for Nd:YAG pumped dye lasers. Reprinted by permission of Exciton. Dyes names are those of Exciton.

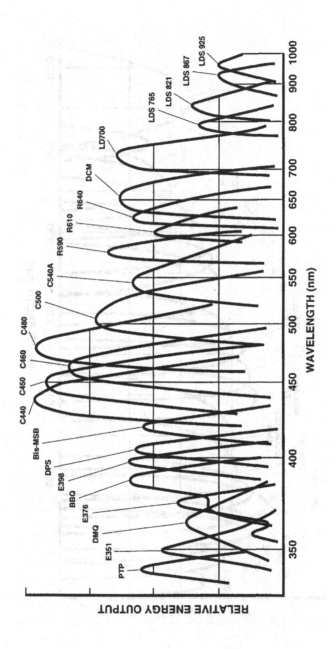

Fig. 11a-9. Tuning curves for dye lasers pumped with XeCl excimer lasers. Reprinted by permission of Exciton. Dyes names are those of Exciton.

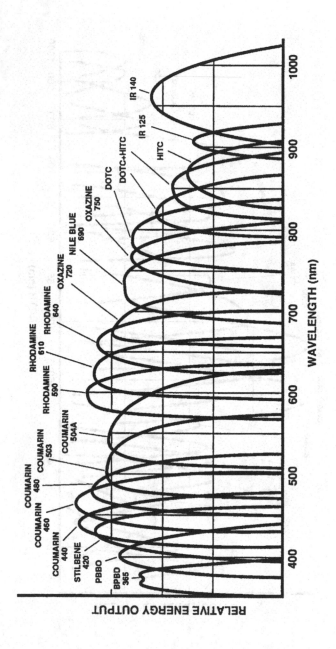

Fig. 11a-10. Tuning curves for nitrogen pumped dye lasers. Reprinted by permission of Exciton. Dyes names are those of Exciton.

11b TRANSMISSION CHARACTERISTICS OF LIGHT FILTERS AND GLASSES

Data and spectra included in this section should help facilitate the selection of appropriate filters and glasses for photophysical and photochemical studies.

Filters are an inexpensive substitute for a monochromator and may be used for excluding a region of wavelengths, *cut-off* filters, or for isolating a more or less wide range of wavelengths, *band-pass* filters and *interference* filters.

A variety of glass filters are commercially available and the reader is referred to the manufacturers' catalogues for details of their transmission curves. Examples of the most important available types of filters are reported in Fig. 11b-1.

Many *solution* filters have been described in the literature [4801, 6401, 6601, 6801, 8101, 8201, 8901]. Their use is convenient in irradiation experiments, if monochromatic light or a band of wavelengths of noticeable intensity are required, particularly when photochemical reactors are used. The transmission curves of some useful liquid cut-off filters are shown in Fig. 11b-2 and their composition is reported in Table 11b-1. Some variations in the precise cut-off wavelength may be achieved by varying the concentration and/or the optical path. The user himself is recommended to measure the transmission of the chosen filter, particularly if very low transmission is required at all the wavelengths shorter than the cut-off point.

A variety of solutions of organic dyes and transition-metal salts are suitable as band-pass filters for wavelengths in the UV and visible region (Fig. 11b-3). To isolate a sufficiently narrow band of wavelengths it is customary to combine a broad band-pass filter, having the required long wavelength cut-off, with a suitable short wavelength cut-off filter. This last combination is particularly convenient for isolating lines of a mercury lamp. An excellent compilation is reported in ref. [6601]. Typical transmittance curves of various types of quartz and glass commercially available are reported in Fig. 11b-4.

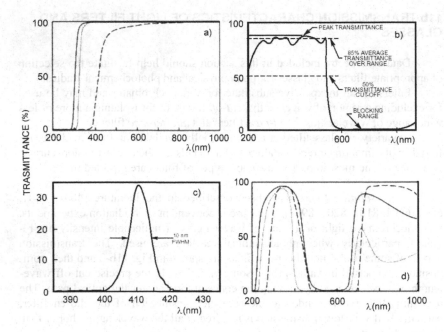

Fig. 11b-1. Typical transmittance curves of the most important types of filters commercially available (adapted by permission from LOT-Oriel catalogue). Long-pass (cut-off) (a); short-pass (cut-off) (b); interference (c); broad-band (d).

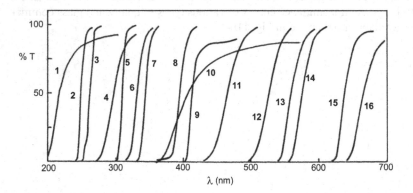

Fig. 11b-2. Transmission curves of short-wavelength cut-off filters in solution. For the composition, see Table 11b-1.

Table 11b-1 Short-Wavelength Cut-Off Filters in Solution

Filter	Compound	Conc.[b,c]	Notes[d]
1	methanol	pure	
2	acetic acid	4 M	
3	KI	0.17% w/v	%T decreases with irradiation
4	$CuSO_4.5H_2O$	1.5% w/v	%T sharply decreases beyond 500 nm; slight %T increase with irradiation
5	Potassium hydrogen phtalate	0.5% w/v	inconsistent, but generally significant decrease of %T with irradiation
6	KNO_3	0.4 M	
7	KNO_3	2 M[e]	
8	$NaNO_2$	1% w/v	
9	$NaNO_2$	75% w/v	slight %T increase with irradiation
10	$Fe_2(SO_4)_3$	3% w/v	
11	K_2CrO_4	0.1% w/v	
12	$K_2Cr_2O_7$	0.5% w/v	
13	$K_2Cr_2O_7$	10% w/v	
14	$Na_2Cr_2O_7.2H_2O$	50% w/v	
15	Rhodamine B	0.2% w/v	
16	Methyl violet	0.02% w/v	also transmits below 460 nm

[a] All filters taken from ref. [6801], except for filters 1 and 10, taken from ref. [6401]; [b] The solvent is in all cases water; [c] 1 cm pathlength, unless otherwise noted; [d] Notes on stability from ref. [6601]; for more details, see this ref.; [e] 2 cm optical path.

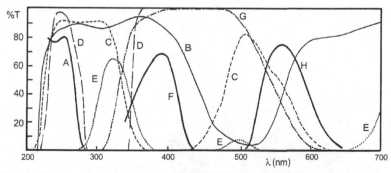

Fig. 11b-3. Transmission curves of band-pass filters in solution. For the composition, see Table 11b-2.

Table 11b-2 Band-Pass Filters in Solution

Filter	Compound	Conc., solvent	d[a] (cm)	Notes[b]
A	Cl_2[c]	1.013 bar	4	also transmits beyond 380 nm
B	$CoSO_4.7H_2O$[d]	7.5% w/v, water	1	do not mix with $NiSO_4$; if pre-irradiated, %T increases to almost stable value
C	$NiSO_4.6H_2O$[d]	50% w/v, water	1	do not mix with $CoSO_4$; if pre-irradiated, %T increases to almost stable value
D	2,7-dimethyl-3,6-diazacyclohepta-2,6-diene perchlorate[e]	0.02% w/v, water,	1	constant transmission with irradiation
E	$KCr(SO_4)_2.12H_2O$[d]	15% w/v, 0.5 M H_2SO_4	1	also has a window of low transmission in the visible
F	I_2[c]	0.75% w/v, in CCl_4,	1	also transmits beyond 650 nm; slight %T increase with irradiation
G	$CuSO_4.5H_2O$[d]	10% w/v, water	5	
H	$CuCl_2.H_2O$[c]	5% w/v, 8 M HCl	1	

[a] Optical path; [b] Notes on stability from ref. [6601], for more details, see this ref.; [c] From ref. [6801]; [d] From ref. [4801]; [e] From ref. [9301].

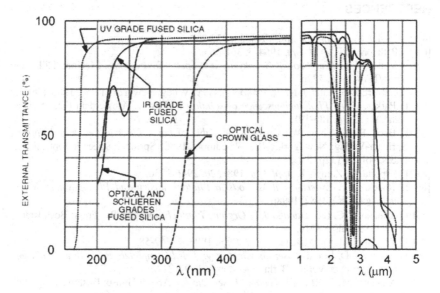

Fig. 11b-4. Typical transmittance curves of various types of quartz and glass commercially available (adapted by permission from LOT-Oriel catalogue). The external transmission includes the surface reflection losses.

REFERENCES

[4801] Kasha, M. *J. Opt. Soc. Am.* **1948**, *38*, 929-935.

[6401] Pellicori, S. F. *Appl. Opt.* **1964**, *3*, 361-366.

[6501] Valley, S. L. *Handbook of Geophysics and Space Enviroments,* McGraw-Hill, New York (NY), 1965, 522p.

[6601] Calvert, J. G.; Pitts, J. N., Jr. *Photochemistry*; Wiley: New York (NY), 1966, 899p.

[6801] Parker, C. A. *Photoluminescence of solutions,* Elsevier: Amsterdam (The Netherlands), 1968, p. 186-191.

[8001] Beck, R.; Englisch, W.; Guers, K. in *Table of Laser Lines in Gases and Vapors*, Springer-Verlag: New York (NY), 3rd edition, 1980 (Springer Series in Optical Sciences, vol. 2), 132p.

[8101] Laporta, P.; Zaraga, F. *Appl. Opt.* **1981**, *20*, 2946-2950.

[8201] Rabek, J. F. *Experimental Methods in Photochemistry and Photophysics*; Wiley: Chichester (UK), 1982, 1098p.

[8901] Scaiano, J. C. (ed) *Handbook of Organic Photochemistry*; CRC Press: Boca Raton, (FL), 1989, vol. 1, 451p.

[9301] Van Driel, H. M.; Mak, G. *Can. J. Phys.* **1993**, *71*, 47-58.

[9701] Andrews, D. L. (ed) *Lasers in Chemistry: Third Completely Revised and Enlarged Edition*; Springer-Verlag: Berlin (Germany), 1997, 212p.

[9801] Anmann, M.-C.; Buus, J. *Tunable Laser Diodes*, Artech House: Boston (Ma), 1998, 289.

[9901] Hou, X.; Zhou, J. X.; Yang, K. X.; Stchur, P.; Michel, R. G. *Adv. Atomic Spectrosc.* **1999**, *5*, 99-143.

[0401] Kuhn, H. J.; Braslavsky, S. E.; Schmidt, R. *Pure Appl. Chem.* **2004**, *76*, 2105-2146.

[0501] Braslawsky, S. E.; Houk, K. N.; Verhoeven, J. W. *Glossary of Terms Used in Photochemistry, 3rd Ed.*; IUPAC Commission, *Pure Appl. Chem.*, in press.

12

Chemical Actinometry

The photochemist is interested in the reactivity of substances under light excitation. The rate of a photochemical reaction can be quantified by the *quantum yield*, also called quantum efficiency, defined as:

$$\Phi = \frac{\text{number of reacted molecules per time unit}}{\text{number of photons absorbed per time unit}}$$

Actinometry allows determination of the photon flux for a system of specific geometry and in a well defined spectral domain; the most favorable case is when the incident light is monochromatic. The term *actinometer* commonly indicates devices used in the UV and visible spectral range. In absolute actinometric measurements a physical device (such as a photomultiplier, a photodiode, a bolometer) converts the energy or the number of the incident photons in a quantifiable electrical signal [9101, 0401]. However, the most commonly utilized method is based on a chemical actinometer, a reference substance undergoing a photochemical reaction whose quantum yield is known, calibrated against a physical device, well studied actinometers or by calorimetric methods.

In principle, any photoactive compound whose quantum yield is known could be used as an actinometer. For a good actinometer, this quantum yield should be, as much as possible, independent of excitation wavelength, temperature, concentration, trace impurities, and oxygen; moreover, the number of reacted molecules should be determined with a convenient and quick analytical method, but none of the numerous actinometers proposed in literature meets all the given criteria. A general report on chemical actinometry was prepared by the Iupac Commission on Photochemistry [0401]. Surveys on the most commonly used actinometers can be found in references [8901, 9101, 9301, 9801].

In the following, we will briefly discuss only the most extensively studied and widely used actinometers, with the aim of covering the entire UV and visible range of wavelengths. A table of the actinometers satisfying at least some of the above mentioned requisites will also be given.

12a FERRIOXALATE ACTINOMETER

This is the most reliable and practical actinometer for UV and visible light up to 500 nm, proposed first by Hatchard and Parker in 1956 [5601]. Under light excitation the potassium ferrioxalate decomposes according to the following equations:

$$Fe(C_2O_4)_3^{3-} \xrightarrow{h\nu} Fe^{2+} + C_2O_4^{\cdot -} + 2C_2O_4^{2-}$$

$$Fe(C_2O_4)_3^{3-} + C_2O_4^{\cdot -} \xrightarrow{\Delta} Fe^{2+} + 2CO_2 + 3C_2O_4^{2-}$$

The quantity of ferrous ions formed during an irradiation period is monitored by conversion to the colored tris-phenanthroline complex ($\varepsilon = 11100$ L mol^{-1} cm^{-1} at $\lambda_{max} = 510$ nm). The original ferric ions are not appreciably complexed by phenanthroline and the complex does not absorb at 510 nm. The moles of ferrous ions formed in the irradiated volume are given by

$$moles\ Fe^{2+} = \frac{V_1 \times V_3 \times \Delta A(510\ nm)}{10^3 \times V_2 \times l \times \varepsilon(510\ nm)}$$

where V_1 is the irradiated volume, V_2 is the aliquot of the irradiated solution taken for the determination of the ferrous ions, V_3 is the final volume after complexation with phenanthroline (all in mL), l is the optical pathlength of the irradiation cell, $\Delta A(510\ nm)$ the optical difference in absorbance between the irradiated solution and that taken in the dark, $\varepsilon(510\ nm)$ is that of the complex Fe(phen)$_3^{2+}$. Thus, the moles of photons absorbed by the irradiated solution per time unit ($Nh\nu/t$) are:

$$Nh\nu/t = \frac{moles\ of\ Fe^{2+}}{\Phi_\lambda \times t \times F}$$

where Φ_λ is the quantum yield of ferrous ion production at the irradiation wavelength, t is the irradiation time, and F is the mean fraction of light absorbed by the ferrioxalate solution (this is important when irradiation is carried out in visible region, where the actinometric solution doesn't absorb all the incident light). The dependence of the quantum yield on the irradiation wavelength has been extensively studied. Table 12a-1 gives most of the available data.

Procedure

A lot of variations of the basic Hatchard and Parker's procedure [5601] have been proposed. The ferrioxalate solution can be prepared just before irradiation mixing ferric sulfate and potassium oxalate [9301], in red light, but is preferable to use solid ferrioxalate [5601, 7701, 9101], whose very easy preparation [5601] must be followed by recrystallization (three times).

Table 12a-1 Quantum Yields of Production of Ferrous Ions from Potassium Ferrioxalate as a Function of the Excitation Wavelength[a]

λ (nm)	$[Fe(C_2O_4)_3^{3-}]$ (mol L^{-1})	$\Phi_{Fe^{2+}}$	Ref.
222	0.006[b]	0.50	[7901]
230	0.006[b]	0.67	[7901]
238-240	0.006[b]	0.68	[7901]
248	0.006[b]	1.35	[7901]
254	0.006	1.25[c]	[5601]
254	0.005	1.25	[5501]
297/302	0.006	1.24[c]	[5601]
313	0.006	1.24	[5601]
334	0.006	1,23[c]	[5601]
358	0.006	1.25	[6601]
363.8[d]	0.006	1.28	[8101]
365.6	0.006	1.21[c]	[5601]
365.6	0.005	1.20	[5501]
365.6	0.006	1.26	[6401], [6601]
392	0.006	1.13	[6601]
405-407	0.006	1.14[c]	[5601]
406.7[d]	0.006	1.19	[8101]
416	0.006	1.12	[6601]
436	0.006	1.11[c]	[5601]
436	0.15	1.01	[5601]
436	0.005	1.07	[5501]
458[d]	0.15	0.85	[8101]
458[d]	0.015	1.10	[8102]
468	0.15	0.93	[5601]
480	0.15	0.94	[5601]
488[d]	0.08	1.08	[8102]
509	0.15	0.86	[5601]
514.5[d]	0.16-0.20	0.90	[8102]
520[d]	0.15	0.65[e]	[8301]
530[d]	0.15	0.53[e]	[8301]
546	0.15	0.15	[5601]
550[d]	0.15	0.15[e]	[8301]

[a] Room temperature, mercury lamp lines, unless otherwise noted; [b] Not stated, assumed from the procedure used, according to ref. [5501]; [c] Recommended value, obtained by different methods; see ref. [5601], page 526; [d] Laser source; [e] Experimental value, but derived from a graph of Φ vs. λ_{ex}.

We recommend the procedure suggested by Fisher [8401], a "micro-version" of the Hatchard and Parker's method: 3 mL of a 0.012 M solution of fer-rioxalate in a spectrophotometric cell is irradiated at any wavelength between 254 and 436 nm, while an identical sample is mantained in the dark. At the end of the irradiation, 0.5 mL of buffered phenanthroline solution is added in the cells and the absorbance at 510 nm measured immediately. Waiting for an hour between irradiation and addition of phenanthroline, or after the addition, does not make any difference. Oxygen does not have to be excluded, because the quantum yield of the ferrioxalate actinometer is independent of the presence of oxygen [5601, 8101, 8901]. Furthermore, the quantum yield of ferrioxalate actinometer does not show a strong temperature dependence, as first pointed out by Hatchard e Parker [5601], and subsequently carefully investigated by Nicodem et al. [7701]. Stirring (with a suitable magnetic stirrer or a flux of nitrogen) is not necessary, but recommended [9301, 0301]. The irradiation time must be short enough in order to avoid more than 10% ferrioxalate decomposition.

This "micro-version" procedure is currently used in our laboratory, but for irradiation at 436 and 464 nm a 0.15 M ferrioxalate solution is used, as suggested in the classical procedure [5601]. The results are in good agreement with those obtained with the more dilute solution (0.012 M) or using the Aberchrome 540 or the Reinekate actinometer [0301]. For these two last actinometers, see below.

Preparation of solutions

Ferrioxalate 0.012 M: 6 g of potassium ferrioxalate in 1 liter of H_2SO_4 0.05 M. Buffered phenanthroline 0.1%: 225 g $CH_3COONa \cdot 3H_2O$, 1 g of phenanthroline in 1 liter of H_2SO_4 0.5 M. Both solutions can be kept for at least a year in dark bottles in a closed cupboard [8401]. When phenanthroline is stored in the light for a long period, its photodecomposition products inhibit color development [7601, 8302].

12b PHOTOCHROMIC ACTINOMETERS

This kind of actinometers are based on a photochromic reaction, photoreversible or thermoreversible; therefore, the actinometric solution can be regenerated by irradiating or heating, with the consequent advantages of ease of handling, accuracy and reusability.

12b-1 Azobenzene

Azobezene undergoes a characteristic *trans-cis* photoisomerization:

Table 12b-1 Quantum Yields for Isomerization of Azobenzene as a Function of the Excitation Wavelength[a]

λ (nm)	[azobenzene][b] (mol L^{-1})	T (°C)	Solvent	$\Phi_{t \to c}$	$\Phi_{c \to t}$	Ref.
254	2.5×10^{-5}	25	Isooctane	0.13	0.44	[5801]
254	6×10^{-4}	RT	Methanol	0.26	0.31	[8501]
280	6×10^{-4}	RT	Methanol	0.12	0.34	[8501]
313	6×10^{-4}	RT	Methanol	0.13	0.30	[8501]
313	$4 \times 10^{-5} - 2 \times 10^{-3}$	RT	Methanol	0.14		[7401]
313	$2 \times 10^{-5} - 5 \times 10^{-4}$	20	Methanol		0.37	[8701]
313	$10^{-5} - 10^{-4}$	RT	Ethanol	0.12	0.31	[8402]
313	$1 \times 10^{-5} - 2 \times 10^{-3}$	15-25	Ethanol	0.22	0.69	[5401]
313	$4 \times 10^{-5} - 2 \times 10^{-3}$	RT	Isopropanol	0.10	0.50	[7401]
313	$2 \times 10^{-5} - 5 \times 10^{-4}$	20	H$_2$O-EtOH		0.40	[8701]
313	$2 \times 10^{-5} - 5 \times 10^{-4}$	20	MeCN		0.35	[8701]
313	$2 \times 10^{-5} - 5 \times 10^{-4}$	20	THF		0.40	[8701]
313	$2.5 \times 10^{-5} / 1 \times 10^{-3}$	25	Isooctane	0.10	0.41/0.42	[5801]
313	$4 \times 10^{-5} - 2 \times 10^{-3}$	RT	Cyclohexane	0.10	0.42	[7401]
313	$2 \times 10^{-5} - 5 \times 10^{-4}$	20	Cyclohexane		0.40	[8701]
313	$2 \times 10^{-5} - 5 \times 10^{-4}$	20	n-hexane		0.44	[8701]
334	6×10^{-4}	RT	Methanol	0.15	0.30	[8501]
365	6×10^{-4}	RT	Methanol	0.15	0.35	[8501]
365	$1 \times 10^{-5} - 2 \times 10^{-3}$	15-25	Ethanol	0.20	(0.6	[5401]
365	$2.5 \times 10^{-5} / 1 \times 10^{-3}$	25	Isooctane	0.12	0.48	[5801]
365	$4 \times 10^{-5} - 2 \times 10^{-3}$	RT	Cyclohexane	0.14		[7401]
405	6×10^{-4}	RT	Methanol	0.20	0.57	[8501]
405	$2.5 \times 10^{-5} / 10^{-3}$	25	Isooctane	0.23/0.21	0.55/0.51	[5801]
436	$4 \times 10^{-5} - 2 \times 10^{-3}$	RT	Methanol	0.28		[7401]
436	6×10^{-4}	RT	Methanol	0.22	0.63	[8501]
436	$1 \times 10^{-5} - 2 \times 10^{-3}$	15-25	Ethanol	0.36	0.45	[5401]
436	$10^{-5} - 10^{-4}$	RT	Ethanol	0.24	0.53	[8402]
436	$4 \times 10^{-5} - 2 \times 10^{-3}$	RT	Isopropanol	0.26	0.42	[7401]
436	$2.5 \times 10^{-5} / 10^{-3}$	25	Isooctane	0.28/0.27	0.55	[5801]
436	$4 \times 10^{-5} - 2 \times 10^{-3}$	RT	Cyclohexane	0.28	0.55	[7401]
436	10^{-3}	25	Benzene	0.26	0.46	[5801]

[a] Room temperature, mercury lamp lines; [b] Total concentration of irradiated azobenzene.

The thermal or photochemical back reaction can be use for regenerating the actinometric solution, making it reusable, but implies that temperature must be kept as low as possible [9801], although between 15° and 25°C the quantum yield values are unaffected within the experimental error [5401].

For the thermal *cis* → *trans* reaction in isooctane a rate constant $k = 1.37 \times 10^{-4}$ min^{-1} was obtained [5801]. Both the forward (*trans* → *cis*) and back (*cis* → *trans*) photoreactions have been extensively studied in a lot of solvents (see Table 12b-1), and are used as actinometric reactions, the first in the 275-340 nm range, the second in the 350-440 nm range and also for 254 nm mercury line. In the UV region conditions of total absorbed light can be used, while in the visible, when *cis* isomer is irradiated, a pre-irradiation of the *trans* isomer is required to produce a photostationary state, but the solution in this case only partly absorbs the incident light, which implies a more complicated procedure [0302]. The kinetic analysis can be done from the spectrophotometric data using two different methods [8701]: i) the tangent at the origin of the isomerization and ii) the method originally proposed by Mauser [7501] for the linearization of the kinetic curves and used by Gauglitz and Hubig [7602, 8103]. A convenient method using concentrated solution of azobenzene in the UV region was also suggested by the same authors [8501]. A practical procedure that can be used for irradiation of azobenzene in the different ranges of wavelengths, together with some suggestions for the kinetic treatment of the absorbance data, are also given in ref. [0401].

12b-2 Fulgide Aberchrome 540

The pale yellow fulgide (*E*)-2-[1-(2,5-dimethyl-3-furyl)-ethylidene]-3-isopropyl-idene succinic anhydride (hereafter indicated A), whose commercial name was Aberchrome 540, is considered a convenient actinometer in the near UV and visible region, because of its reversible photocyclization into the deep red cyclized valence isomer 7,7a-dihydro-2,4,7,7,7a-pentamethylbenzo(b)furan-5,6-dicarboxylic anhydride (C) [8104, 8105]. The quantum yield of this photocyclization (0.2) is independent of the temperature ranging from 10 to 40 °C and it is also claimed to be independent on the "cycling" for photocoloration [8105]; see, however, comments below. Photocoloration at 254 nm is accompanied by marked photodegradation and the use of the actinometer below 300 nm is not recommended [8105].

A(Z)   A(E)   C

This compound is no longer available from Aberchromics Ltd; however, it can be synthesized following the method by Darcy et al. [8104]. Some hundreds of milligrams will last for many years.

According to Heller and Langan, one of the main advantages of this fulgide is the chemical stability and reversibility of its photocyclization reaction [8105]. This actinometer can be used in the range 310-370 nm and 435-545 nm. Nevertheless, the reversible photocyclization competes with the $A(E) \rightleftarrows A(Z)$ photoisomerization reaction [8801], whose quantum yields (toluene, RT) at 365.6 nm are 0.13 ($E \rightarrow Z$) and 0.12 ($Z \rightarrow E$), respectively [9601]. The photostationary state depends on the irradiation wavelength. So, it is suggested not to reuse the irradiated solution when the actinometer is used in the range 310-370 nm [9302]. To overcome this problem one could use the isopropyl derivative, which doesn't give the $Z \rightleftarrows E$ photoisomerization reactions [8801, 9302]. The problem does not concern the use of Aberchrome 540 in the range 435-545 nm, which corresponds only to the bleaching of the cyclic form C.

Procedure for UV region (310-370 nm) [8105, 9101, 0401]

An approximately 5×10^{-3}M solution is prepared by dissolving the fulgide A (25 mg) in dry distilled toluene (20 ml). A known volume of this solution is put in a cell, a magnetic bar is added. Bubbling Ar for about 20 min, or degassing is desired, while not necessary, then the cell is sealed. Read the absorbance (if any) at 494 nm. The stirred solution is then irradiated with monochromatic light for a known period of time, taking care that the magnetic bar does not enter the light beam. After irradiation measure the absorbance at 494 nm. The increase in absorbance at 494 nm enables the photon flux (moles of incident photons in the irradiated volume per time unit) to be calculated using the following equation:

$$ Nh\nu/t = \frac{\Delta A \times V}{\Phi \times \varepsilon \times t} $$

where ΔA is the increase in absorbance value at 494 nm, V is the volume of the irradiated solution (L), Φ is the quantum yield of the photocyclization reaction (0.2 in the range 310-370 nm), ε is the molar absorption coefficient of C (8200 L mol^{-1} cm^{-1} at 494 nm), and t is the irradiation time.

Procedure for the visible region (435-545 nm) [8105, 9101, 9201]

The reverse photoreaction $C \rightarrow A(E)$ can be easily exploited for actinometry in the visible region by measuring the decrease in absorbance at 494 nm of a known volume of the red solution, obtained irradiating for a suitable period of time the fulgide A with UV light (up to an absorbance value of about 2). The quantum yield for the bleaching reaction shows a remarkable linear dependence on the irradiation wavelength, on the temperature, and on the solvent. In toluene, at 21°C, the following expression is valid [9201]:

$$\Phi_\lambda = 0.178 - 2.4 \times 10^{-4} \cdot \lambda$$

where λ is in nm. The numbers of moles of incident photons can be calculated using an equation analogous to that used for the UV region:

$$Nh\nu/t = \frac{\Delta A \times V}{\Phi_\lambda \times \varepsilon \times t \times F}$$

where ΔA is the decrease in absorbance at 494 nm, the other symbols have the above given meaning, and F is the mean fraction of light absorbed at the irradiation wavelength.

Comments on the use of the Aberchrome 540

Many criticisms have been reported in the literature about the use of this fulgide as actinometer since it was proposed, concerning the repeated use in the UV region, maily due to the $A(E) \rightleftarrows A(Z)$ competing photorections, which could cause underestimation of the number of incident photons [8801, 9302]. *Practically, it is suggested not to reuse in the UV region the irradiated solutions of Aberchrome 540*, since the photostationary state and, as a consequence, the measured light intensity are affected by the number of cycles [9302]. This last observation was also confirmed by experiments performed in our laboratory. All these problems could be overcome using the isopropyl derivative, which does not lead to the $A(E) \rightleftarrows A(Z)$ photoisomerizations, because of the steric hindrance of the bulky groups [8801, 9001].

Taking into consideration one of the most recent criticism [9601], the conclusion of the authors is that the photoisomerization reactions $A(E) \rightleftarrows A(Z)$ do not appreciably influence the quantum yield of the photocyclization, as the high concentration of Aberchrome 540 used for this purpose leads only to a low conversion and negligible concentration of the Z isomer. On the other hand, they measured the molar absorption coefficient of the cyclized isomer C obtained by photocoloration of A, after careful chromatographic purification, and found $\varepsilon = 8840$ L mol^{-1} cm^{-1} at 494 nm, which is smaller than the value previously reported by Heller and Langan [8105]. They also determine for the photocoloration reaction a smaller quantum yield value, 0.18 instead of 0.20 [8105]. In the calculations for obtaining the light intensity using the equation reported above, this means that the product $\Phi \times \varepsilon$ changes from 1640 to 1590 (in toluene); the consequent underestimation amounts only to 3%.

In the visible, irradiating the solution obtained by photocoloration, the cyclized compound will be reconverted to the open isomer A(E). As both the A(E) and A(Z) isomers do not absorb visible light, the ring opening photoreaction occur without any side reaction and can be repeatedly used for the determination of the photons of monochromatic light in the range 435-545 nm [0301], as proposed by Heller et al. [8105, 9201].

12c REINECKE'S SALT ACTINOMETER

This actinometer is based on the photosubstitution reaction undergone by Reinecke's salt, $[Cr(NH_3)_2(SCN)_4]^-$, in aqueous solution. Irradiation with UV and visible light causes the substitution of a SCN^- ligand by a water molecule [6601]:

$$[Cr(NH_3)_2(SCN)_4]^- + H_2O \xrightarrow{h\nu} [Cr(NH_3)_2(SCN)_3(H_2O)] + SCN^-$$

The number of photons are determined from the SCN^- released. These ions are complexed by addition of ferric nitrate and the absorbance of the resulting blood-red complex is measured at $\lambda_{max} = 450$ nm ($\varepsilon = 4300$ L mol^{-1} cm^{-1}):

$$Fe^{3+} + SCN^- \rightarrow Fe(SCN)^{2+}$$

The quantum yield of the photoaquation reaction shows only a little dependence on the irradiation wavelength in the range 316–750 nm (Table 12c-1). The Reinecke's salt may be used with ease out to 600 nm, and with more difficulty out to 735 nm. Since the quantum yield depends on the pH [6601] it is necessary before irradiation to verify whether the pH of the aqueous solution of Reinecke's salt is between 5.3 and 5.5. To take into account the analogous thermal aquation reaction [5802], the actinometer needs a correction. Wegner and Adamson [6601] recommend that all the spectrophotometric measurements of the SCN^- concentration should be performed relative to a sample kept in the dark.

Materials and procedure

Reinecke's salt is commercially available as ammonium salt. This can easily converted in the potassium salt [6601]. Alternatively, it can be prepared starting from chromium(III) sulfate according to the method suggested by Szychlinski et al. [8902]. It may be stored indefinitely if kept away from light.

The recommended actinometric procedure is the following [6601]: fresh aqueous solutions in distilled water are prepared for each run and, if more than 0.01 M, filtered. Read the absorbance at the chosen wavelength, before and after irradiation, to account for possible incomplete absorption of the incident light. An aliquot of the same solution is kept in the dark at the same temperature of the irradiated one (this second is better if stirred during illumination), and the temperature should be as low as possible to minimize the thermal reaction contribution. At the end of the irradiation period, aliquots are taken from both the irradiated and dark solutions and analyzed for degree of photoaquation. The extent of either thermal or photochemical aquation of $[Cr(NH_3)_2(SCN)_4]^-$ can be determined by analysis of the thiocyanate ion. An aliquot of the solution to be analyzed is accurately diluted (at least 4:1) into the reagent consisting of 0.1 M Fe(NO$_3$)$_3$ in 0.5 M HClO$_4$. The resulting iron(III) thiocyanate complex has an absorption maximum at 450 nm, with a molar absorption coefficient of 4300 L mol^{-1} cm^{-1}. There is no noticeable effect of pH on quantum yields at 23 °C [6601], so long the solution are acid (up to pH = 5.3-5.5, natural value of pH on dissolving the Reinekate's salt in water).

The almost constancy of the quantum yields of photoaquation allows the use of Reineckate's salt as actinometer also for polychromatic light, but in this case the concentration of the solution must be chosen so that all the incident light of the examined spectral domain is absorbed. This actinometer can be used for poly-chromatic light in the range 390-600 nm, but beyond 600 nm the molar absorption coefficients are too small to reach total absorption. For the experimental procedure and calculations in the case of polychromatic light, see ref. [9101].

Table 12c-1 Quantum Yield Values of SCN⁻ Production by Photolysis of Aqueous Reinecke's Salt (23 °C, pH=5.3)[a]

λ (nm)	ε (L mol^{-1}cm^{-1})	[Reinecke's salt] (mol L^{-1} × 10^3)	Φ_{SCN^-} [b]
316	11000	1.1	0.29[c]
350	>100	3.1	0.39
366			0.32[d]
392[e]	93.5	5.0	0.32
415	67.5	8.0	0.31[c]
452[f]	31.2	10.0	0.31
504	97.5	5.0	0.30
520[e]	106.5	4.0	0.29
545	90.5	5.5	0.28
585	43.8	10	0.27
600	29.0	25	0.28
676	0.75	45	0.27
713	0.35	46	0.28
735	0.27	45	0.30
750	0.15	48	0.27

[a] Data from ref. [6601], unless otherwise noted; [b] Mean values from at least four different runs, except for 316 nm (2 runs) and 366 nm; the original values were reported with three significant figures; [b] Value confirmed by Szychlinski et al. [8902]; [c] From ref. [8403]; [d] λ_{max}; [e] λ_{min}.

12d URANYL OXALATE ACTINOMETER

This was the standard actinometer in solution before the introduction of potassium ferrioxalate [3001]. The photochemical reaction exploited is:

$$H_2C_2O_4 \xrightarrow{\quad hv,\ UO_2^{2+} \quad} CO_2 + CO + H_2O$$

where UO_2^{2+} acts as photosensitizer. The spectral range covered is 208-426 nm, with a quantum yield almost constant ($\Phi_{CO} = 0.5$-0.6) [3001, 6301, 6402, 9301]. The influence of the pH on the photolysis of uranyl oxalate was also carefully examined [7001]. The quantum yield for the consumption of oxalate was found to be independent of the pH between 1 and 5, but decreases outside this range; Φ_{CO} strongly depends on pH. The number of oxalate ions trasformed in the photochemical process is classically determined by titrating the actinometric solution, before and after irradiation, with potassium permanganate. The uranyl oxalate actinometer shows lack of sensitivity for two reasons: the long path lengths needed for complete light absorption in the visible and near UV, and the differential titrimetry method to determine the oxalate consumption, which implies that a significant loss of reactant is needed [5601, 9101, 9301]. Other analytical methods have been attempted, as gas cromatography for CO production [6402], or back titration after addition of an known excess of Ce(IV) [7001], but the GC method should be not useful, considering the above mentioned influence of the pH on the CO production, and the second does not overcome the complexity of the titration method [8901, 9101].

12e OTHER ACTINOMETERS

Other selected actinometers in liquid phase, in order to cover the widest wavelength range, are collected in Table 12e-1. The choice criteria were the reliability of the procedure used for irradiation and the simplicity of the analytical methodology.

Polychromatic light: methodologies for actinometric measurements of polychromatic light sources are reported in ref. [8202] and [9101] (pages 88-93 and 101-104).

Solid and gas phase: for a list of actinometers (with appropriate references) in solid and gas phase see ref. [0401].

Heterogeneous phase: a standardization protocol for relative photonic efficiencies in heterogeneous photocatalysis (solid/liquid or solid/gas systems) is extensively treated by Serpone et al. in refs. [9602, 9702, 9903, 9904].

Drug photostability: a general discussion on actinometers particularly suitable for studies of photostability of drugs are reported in ref [9801], pages 295-304.

Laser actinometry: various actinometers have been proposed for laser sources, e.g., the actinometric compounds labeled *g* in Table 12e-1. For the use of potassium ferrioxalate with lasers see refs. [8002, 8101, 9103]. Azobenzene also is applicable for lasers [8103]. $Ru(bpy)_3^{2+}/$ $Fe(ClO_4)_3$ was used for laser intensity at 353.3 nm [8203]. Demas et. al proposed a photooxigenation actinometer based on the system $Ru(bpy)_3^{2+}/TME$, which has a quantum flat response over the 280-560 nm region [7604]. Of the same author a review on the measurements of laser intensities by chemical actinometry has been published [7605]. For laser pulse actinometry via standard transients see compilation in ref. [0401]. $Ru(bpy)_3(PF_6)_2$ and $Os(bpy)_3(PF_6)_2$ immobilized in PMMA thin films have been proposed as actinometers for transient absorption spectroscopy in the 300-550 and 300-700 nm ranges, respectively [0304].

Table 12e-1 Selected Liquid Phase Actinometers

Compound	λ (nm)	Solvent	Analytical method	Ref.
cis-Cyclooctene[a]	172, 185	n-Pentane	GC[b]	[8106], [8404], [9802]
Ethanol	185	Water	H_2 production, GC[b]	[7603], [7702]
Iodide/iodate	214-330	Water	Abs. 352 nm	[9701], [9901], [0303]
Malachite green leucocyanide[c]	225-289	EtOH	Abs. 620 nm	[5201], [6701]
Heterocoerdianthrone endoperoxide[d,e,f]	248-334	CH_2Cl_2	Abs. 572 nm	[8303], [8405], [9101]
1,2,3,4-Tetraphenyl-cyclobutane[g]	250-270	Methylcyclo-hexane	Abs. 295 nm	[7902], [9102]
2,4-Dimethoxy-6-phenoxy-s-triazine[g]	250-270	Methylcyclo-hexane, EtOH	Abs. 331 nm	[9102]
Azoxybenzene[h]	250-350	EtOH	Abs. 458 nm	[7002], [8406]
1,2-Dimethoxy-4-nitrobenzene[g]	254-366	Water	Abs. 450 nm[i]	[8601]
Stilbenes	254-366	Various	Abs.[i],GC	[7703], [7903], [8407], [8802]
o-Nitrobenzal-dehyde[d,h,j,k]	300-410	CH_2Cl_2, $H_2O/EtOH$	GC, pH-metry	[3401], [9902], [0001], [0002]
Hexan-2-one	313	Various	GC	[6602], [6801], [9301]

Table 12e-1 Selected Liquid Phase Actinometers

Compound	λ (nm)	Solvent	Analytical method	Ref.
trans-2-Nitrocinna-maldehyde[d,k]	313, 366	MeOH	Abs. 440, HPLC	[9803]
Penta-1,3-diene / benzophenone	313, 366	Benzene	GC	[7101], [9301]
Cyclohexa-1,3-diene/benzophenone	313, 366	benzene	GC	[7301]
9,10-Dimethyl-anthracene[d]	334-395	Freon 113	Abs. 324 nm[b]	[8803]
2,2',4,4'-Tetraiso-propylazobenzene	350-390	*n*-heptane	Abs. 365 nm[i]	[7704], [9101]
Ru(bpy)$_3^{2+}$/ Co(NH$_3$)$_5$Cl^{2+}	360-540	water	Photocurrent	[8201]
Ru(bpy)$_3^{2+}$/ peroxydi-sulphate	366, 405, 436	water	Abs. 450	[8001], [8702]
Hematophorphy-rin/TAN[il]	366-546	water	EPR	[7904]
5,12-Diphenyl-tetracene[d]	405-500	Freon 113	Abs. 383[b]	[8803]
DCM styrene dye[g]	410-540	CHCl$_3$, MeOH	HPLC	[9303]
meso-Diphenylhelian-threne[d,f]	475-610	toluene	Abs. 429 nm	[8304], [8305], [8405], [9101]
meso-Diphenylhelian-threne[d,f] / methylene blue	610-670	CHCl$_3$	Abs. 405	[8903]
meso-Diphenylhelian-threne[d,f] / HITC	670-795	CHCl$_3$	Abs. 405	[9002]

[a] Xe-excimer source; [b] For standard procedure, see also ref. [0401]; [c] Especially useful for low fluences; [d] Suitable as polychromatic quantum counter; [e] Reusable; [f] No longer commercially available; [g] Also proposed for lasers; [h] Also proposed in solid phase; [i] Kinetic analysis; [j] For the determination of solar UV radiation and penetration in waters [9304]; [k] Proposed also for drugs photostability testing; [l] Actinometry in an EPR cavity.

REFERENCES

[3001] Leighton, W. G.; Forbes, G. S. *J. Am. Chem. Soc.* **1930**, *52*, 3139-3152.

[3401] Leighton, P. A.; Lucy, F. A. *J. Phys. Chem.* **1934**, *2*, 756-759.

[5201] Calvert, J. G.; Rechen, H. J. L. *J. Am. Chem. Soc.* **1952**, *74*, 2101-2103.

[5401] Birnbaum, P. P.; Style, D. W. G. *Trans. Faraday. Soc.* **1954**,*50*, 1192-1196.

[5501] Baxendale, J. H.; Bridge, N. K. *J. Phys. Chem.* **1955**, *59*, 783-788.

[5601] Hatchard, C. G.; Parker, C. A. *Proc. Roy. Soc. (London)* **1956**, *A235*, 518-536.

[5801] Zimmerman, G.; Chow, L.-Y.; Paik, U.-J. *J. Am. Chem. Soc.* **1958**, *80*, 3528-3531.

[5802] Adamson, A. W. *J. Am. Chem. Soc.* **1958**, *80*, 3183-3189.

[6301] Discher, C. A.; Smith, P. F.; Lippman, I.; Turse, R. *J. Phys. Chem.* **1963**, *67*, 2501-2503.

[6401] Lee, J.; Seliger, H. H. *J. Chem. Phys.* **1964**, *40*, 519-523.

[6402] Volman, D. H.; Seed, J. R. *J. Am. Chem. Soc.* **1964**, *86*, 5095-5098.

[6601] Wegner, E. E.; Adamson, A. W. *J. Am. Chem. Soc.* **1966**, *88*, 394-404.

[6602] Coulson, D. R.; Yang, N. C. *J. Am. Chem. Soc.* **1966**, *88*, 4511-4513.

[6701] Fisher, G. J. ; LeBlanc, J. C.; Johns, H. E. *Photochem. Photobiol.* **1967**, *6*, 757-767.

[6801] Wagner, P. J. *Tetrahedron Lett.* **1968**, *52*, 5385-5388.

[7001] Heidt, L. J.; Tregay, G. W.; Middleton, F. A. Jr. *J. Phys. Chem.* **1970**, *74*, 1876-1882, and refs. therein.

[7002] Mauser, H.; Gauglitz, G.; Stier, F. *Liebigs Ann. Chem.* **1970**, *739*, 84-94.

[7101] Vesley, G. F. *Mol. Photochem.* **1971**, *3*, 193-200.

[7301] Vesley, G. F.; Hammond, G. S. *Mol. Photochem.* **1973**, *5*, 367-369.

[7401] Ronayette, J.; Arnaud, R.; Lebourgeois, P.; Lemaire, J. *Can. J. Chem.* **1974**, *52*, 1848-1857.

[7501] Mauser, H. *Z. Naturforsch., Teil C* **1975**, *30*, 157-160.

[7601] Bowman, W. D.; Demas, N. J. *J. Phys. Chem.* **1976**, *80*, 2434-2435.

[7602] Gauglitz, G. *J. Photochem* **1976**, *5*, 41-47.

[7603] Davies, A. K.; Khan, K. A.; McKellar, J. F.; Phillips, G. O. *Mol. Photochem.* **1976**, *7*, 389-398.

[7604] Demas, J. N.; McBride, R. P.; Harris, E. W. *J. Phys. Chem.* **1976**, *80*, 2248-2253.

[7605] Demas, J. N. in *Creation and Detection of the Excited States*; Ware, W. R., ed.; Marcel Dekker: New York (N.Y.), 1976, vol. 4, 320 p, Chapter 1.

[7701] Nicodem, D. E.; Cabral, M. L. P. F.; Ferreira, J. C. N. *Mol. Photochem.* **1977**, *8*, 213-238.

[7702] Von Sonntag, C.; Schuchmann, H.-P. in *Adv. Photochem.*, Pitts, J. N., Jr.; Hammond, G. S.; Gollnick, K. eds.; Vol. 10, p. 59-145, 80.

[7703] Lewis, F. D.; Johnson, D. E. *J. Photochem.* **1977**, *7*, 421-423.

[7704] Gauglitz, F. *J. Photochem.* **1977**, *7*, 355-357.

[7901] Fernández, E.; Figuera, J. M.; Tobar, A. *J. Photochem.* **1979**, *11*, 69-71.

[7902] Takamuku, S.; Beck, G.; Schnabel, W. *J. Photochem.* **1979**, *11*, 49-52.

[7903] Saltiel, J.; Marinari, A.; Chang, D. W.-L., Mitchner, J. C.; Megarity, E. D. *J. Am. Chem. Soc.* **1979**, *101*, 2982-2996.

[7904] Moan, J.; Høvik, B.; Wold, E. *Photochem. Photobiol.* **1979**, *30*, 623-624.; erratum: *ibidem*, *30*, 625.

[8001] Bolletta, F.; Juris, A.; Maestri, M.; Sandrini, D. *Inorg. Chim. Acta* **1980**, *44*, L175-L176.

[8002] Grueter, H. *J. Appl. Phys.* **1980**, *51*, 5204-5206.

[8101] Demas, J. N.; Bowman, W. D.; Zalewski, E. F.; Velapoldi, R. A. *J. Phys. Chem.* **1981**, *85*, 2766-2771.

[8102] Langford, C. H.; Holubov, C. A. *Inorg. Chim. Acta* **1981**, *53*, L59-L60.

[8103] Gauglitz, G.; Hubig, S. *J. Photochem.* **1981**, *15*, 255-257.

[8104] Darcy, P. J.; Heller, H. G.; Strydom, P. J.; Wittall, J. *J. Chem. Soc. Perkin I*, **1981**, 202-205.

[8105] Heller, H. G.; Langan, J. R. *J. Chem. Soc. Perkin II*, **1981**, 341-343.

[8106] Schuchmann, H.-P.; Von Sonntag, C.; Srinivasan, R. *J. Photochem.* **1981**, *15*, 159-162.

[8201] Dressick, W. J.; Meyer, T. J.; Durham, B. *Isr. J. Chem.* **1982**, *22*, 153-157.

[8202] Gandin, E.; Lion, Y. *J. Photochem.* **1982**, *20*, 77-81.

[8203] Rosenfeld-Grunwald, T.; Brandels, M.; Rabani, J. *J. Phys. Chem.* **1982**, *86*, 4745-4750.

[8301] Hamai, S.; Hirayama, F. *J. Phys. Chem.* **1983**, *87*, 83-89.

[8302] Kirk, A. D.; Namasivayam, C. *Anal. Chem.* **1983**, 55, 2428-2429.

[8303] Brauer, H.-D.; Schmidt, R. *Photochem. Photobiol.* **1983**, *37*, 587-591.

[8304] Brauer, H.-D.; Schmidt, R.; Gauglitz, G.; Hubig, S. *Photochem. Photobiol.* **1983**, *37*, 595-598.

[8305] Acs, A.; Schmidt, R.; Brauer, H.-D. *Photochem. Photobiol.* **1983**, *38*, 527-531.

[8401] Fisher, E. *EPA Newsletters*, **1984**, *21*, 33-34.

[8402] Rau, H. *J. Photochem.* **1984**,*26*, 221-225.

[8403] Szychlinski, J.; Martuszewski, K.; Blazejowski, J.; Bilski, P. *Stud. Mater. Oceanol. (Pol. Akad. Nauk, Kom. Badan Morza)* **1984**, *45*, 217-234.

[8404] Waldemar, A.; Oppenländer, T. *Photochem. Photobiol.* **1984**, *39*, 719-723.

[8405] Schmidt, R.; Brauer, H.-D. *J. Photochem.* **1984**, *25*, 489-499.

[8406] Bunce, N. J.; LaMarre, J.; Vaish, S. P. *Photochem. Photobiol.* **1984**, *39*, 531-533.

[8407] Görner, H. *Ber. Bunsenges. Phys. Chem.* **1984**, *88*, 1199-1208.

[8501] Gauglitz, G.; Hubig, S. *J. Photochem.* **1985**, *30*, 121-125.

[8601] Pavlíckova, L.; Kuzmic, P.; Soucek, M. *Collect. Czech. Chem. Commun.* **1986**, *51*, 368-374.

[8701] Siampiringue, N.; Guyot, G.; Monti, S.; Bortolus, P. *J. Photochem* **1987**, *37*, 185-188.

[8702] Görner, H.; Kuhn, H. J.; Shulte-Frohlinde, D. *EPA Newsletters* **1987**, *31*, 13-19.

[8801] Yokoyama, Y.; Goto, T.; Inoue, T.; Yokoyama, M.; Kurita, Y. *Chem. Lett.* **1988**, *2*, 1049-1052.

[8802] Ho, T.-I.; Su, T.-M.; Hwang, T.-C. *J. Photochem. Photobiol., A: Chem.* **1988**, *41*, 293-298.

[8803] Adick, H.-J.; Schmidt, R.; Brauer, H.-D. *J. Photochem. Photobiol., A: Chem.* **1988**, *45*, 89-96.

[8901] Bunce, N. J. in *CRC Handbook of Organic Photochemistry*; Scaiano, J. C., ed.; CRC Press: Boca Raton, Florida (USA), 1989, vol. 1, 451 p, Chapter 9.

[8902] Szychlinski, J.; Bilski, P.; Martuszewski, K.; Blazejowski, J. *Analyst* **1989**, *114*, 739-741.

[8903] Adick, H.-J.; Schmidt, R.; Brauer, H.-D. *J. Photochem. Photobiol., A: Chem.* **1989**, *49*, 311-316.

[9001] Yokoyama, Y.; Iwai, T.; Kera, N.; Hitomi, I.; Kurita, Y. *Chem. Lett.* **1990**, 263-264.

[9002] Adick, H.-J.; Schmidt, R.; Brauer, H.-D. *J. Photochem. Photobiol., A: Chem.* **1990**, *54*, 27-30.

[9101] Braun, A. M.; Maurette, M.-T.; Oliveros, E. *Photochemical Technology*; Wiley: Chichester (U. K.), 1991, 559 p, Chapter 2.

[9102] Murata, K.; Yamaguchi, Y.; Shizuka, H.; Takamuku, S. *J. Photochem. Photobiol., A: Chem.* **1991**, *60*, 207-214.

[9103] Yoshinobu, I.; Kimiko, E.; Tadashi, Y.; Shunichi, K.; Yuichi, S.; Shunichi, S.; Nobutake, S. *Reza Kenkyu* **1991**, *19*, 247-253; CAN **115**: 193378.

[9201] Glaze, A. P.; Heller, H. G.; Whittall, J. *J. Chem. Soc. Perkin II*, **1992**, 591-594.

[9301] Murov, S. L., Carmichael, I., Hug, G. L. *Handbook of Photochemistry*; Marcel Dekker: New York (NY), 1993, 420 p, Chapter 13.

[9302] Boule, P.; Pilichowski, J. F. *J. Photochem. Photobiol., A: Chem.* **1993**, *71*, 51-53.

[9303] Mialocq, J. C.; Armand, X.; Marguet, S. *J. Photochem. Photobiol., A: Chem.* **1993**, *69*, 351-356.

[9304] Morales, R. G. E.; Jara, G. P.; Cabrera, S. *Limnol. Oceanogr.* **1993**, *38*, 703-705.

[9601] Uhlmann, E.; Gauglitz, G. *J. Photochem. Photobiol. A: Chem.* **1996**, *98*, 45-49.

[9602] Serpone, N.; Sauvé, G.; Koch, R.; Tahiri, H.; Pichat, P.; Piccinini, P.; Pelizzetti, E.; Hidaka, H. *J. Photochem. Photobiol. A: Chem.* **1996**, *94*, 191-203.

[9701] Rahn, R. O. *Photochem. Photobiol.* **1997**, *66*, 450-455.

[9702] Serpone, N. *J. Photochem. Photobiol. A: Chem.* **1997**, *104*, 1-12.

[9801] Favaro, G. in *Drugs: Photochemistry and Photostability*, Albini, A. and Fasani, E., eds.; Spec. Publ. Royal Soc. Chem: Cambridge (UK), **1998**, *225*, 295-304.

[9802] Heit, G.; Neuner, A.; Saugy, P.-Y.; Braun, A. M. *J. Phys. Chem, A* **1998**, *102*, 5551-5561.

[9803] Bovina, E.; De Filippis, P.; Cavrini, V.; Ballardini, R. in *Drugs: Photochemistry and Photostability*, Albini, A. and Fasani, E. eds., Spec. Publ. Royal Soc. Chem.: Cambridge (UK), **1998**, *225*, 305-316.

[9901] Rahn, R. O.; Xu, P.; Miller, S. L. *Photochem. Photobiol.* **1999**, *70*, 314-318.

[9902] Allen, J. M.; Allen, S. A.; Dreiman, J.; Baertschi, S. W. *Photochem. Photobiol.* **1999**, *69*, 17S-18S.

[9903] Serpone, N.; Salinaro, A. *Pure Appl. Chem.* **1999**, *71*, 303-320.

[9904] Salinaro, A.; Emeline, A. V.; Zhao, J.; Hidaka, H.; Ryabchuk, V. K.; Serpone, N. *Pure Appl. Chem.* **1999**, *71*, 321-335.

[0001] Willett, K. L.; Hites, R. A. *J. Chem. Educ.* **2000**, *77*, 900-902.

[0002] Allen, J. M.; Allen, S. K.; Baertschi, S. W. *J. Pharm. Biom. Anal.* **2000**, *24*, 167-178.

[0301] Valitutti, G., Diploma Thesis, July 2003, Lab. of Photochemistry, University of Bologna (Italy).

[0302] Gauglitz, G. in *Photochromism: Molecules and Systems: Revised Edition*, Dürr, H. and Bouas-Laurent, H. eds., Elsevier: Amsterdam (NL), 2003, 1044 p, Chapter 25 and refs. therein.

[0303] Rahn, R. O.; Stephan, M. I.; Bolton, J. R.; Goren, E.; Shaw, P.-S.; Lykke, K. R. *Photochem. Photobiol.* **2003**, *78*, 146-152.

[0304] Bergeron, B. V.; Kelly, C. A.; Meyer, G. J. *Langmuir*, **2003**, *19*, 8389-8394.

[0401] Kuhn, H. J.; Braslavsky, S. E.; Schmidt, R. *Pure Appl. Chem.* **2004**, *76*, 2105-2146.

13

Miscellanea

13a SPIN-ORBIT COUPLING AND ATOMIC MASSES

Heavy-atom effects on spectroscopic and photochemical properties have traditionally been attributed to the spin-orbit interaction [6901]. Some of the most widely studied areas are internal [4901] and external heavy-atom effects [5201] in photophysics. See [8101] and [8102] for an alternative framework.

For one-electron systems, in a central field, the relativistic effects can be studied using the spin-orbit interaction given by [6201]

$$H_{so} = \tfrac{1}{2}\alpha^2 < \frac{1}{r}\frac{\partial V}{\partial r} > \boldsymbol{l} \cdot \boldsymbol{s} \tag{13-1}$$

In Eq. 13-1, α is the fine structure constant, $V(r)$ is the potential energy operator of the central field, $\boldsymbol{l} \cdot \boldsymbol{s}$ is the vector dot product of the electron's orbital angular momentum and spin operators, and the expectation value $<O>$ is over radial wavefunctions. In many-electron systems, problems arise concerning the nature of $V(r)$ and the extent of spin/other-orbit interactions. The interaction of an outer-electron (orbital angular momentum l) with the core is given in terms of a coupling parameter ζ_l which gives a measure of the strength of the spin-orbit interaction. For a one-electron, central-field system, ζ_l is just $\alpha^2 <O>$ in Eq. 13-1.

Experimentally the usual way of finding spin-orbit parameters is from the multiplet splittings in atomic spectra. The calculation of one-electron ζ_l's for various l-electrons has been reviewed in the classic treatise on atomic spectra by Condon and Shortley [3501]. Examples of these calculations have appeared in the photophysics literature [4901, 6101].

In Table 13a, values of ζ_l for a large number of atoms have been listed. The theoretical estimates [7601] were obtained from single-configuration, spin-restricted, numerical Hartree-Fock wavefunctions. These results are supplemented with values computed from experimental spectra in cases where the ground electronic configurations contain (1) only totally-filled shells or (2) only a half-filled s-shell. In case 1, the ζ_l's were estimated by finding the ζ_p for an excited sp^n configuration. Either a p-electron (e.g. rare gases) or an s-electron (e.g. mercury) can be promoted to form such an excited configuration. Case 2 (e.g. alkali metals) was treated by promoting the outer-shell s-electron to a p-shell. Experimental values are also given for the halogens since they are frequently used as heavy-atom quenchers. In the halogens, ζ_p's were calculated from the doublet splitting of the lowest configuration, ns^2np^5.

In considering heavy-atom effects, it seems best to correlate the interaction of the considered system with the *core* of the heavy atom. The parameter ζ_l is a measure of just such an interaction; even though it is actually related to the spin-orbit interaction of an electron in a *specific open-shell orbit* of the heavy atom with its core which is in *some specific electronic configuration*.

Table 13a Atomic Masses and Spin-Orbit Coupling Constants[a,b]

Element	Symbol	Atomic Number	Atomic Mass/u[c]	ζ_l/cm^{-1}
Actinium	Ac	89	227*	1290
Aluminium (Aluminum)	Al	13	26.981538	62
Americium	Am	95	243*	3148
Antimony (Stibium)	Sb	51	121.760	2593
Argon	Ar	18	39.948	*940*
Arsenic	As	33	74.92160	1202
Astatine	At	85	210*	10608
Barium	Ba	56	137.327	*830*
Berkelium	Bk	97	247*	3852
Beryllium	Be	4	9.012182	*2.0*
Bismuth	Bi	83	208.98038	6831
Bohrium	Bh	107	264*	−
Boron	B	5	10.811	10
Bromine	Br	35	79.904	*2460*
Cadmium	Cd	48	112.411	*1140*
Calcium	Ca	20	40.078	*105*
Californium	Cf	98	251*	4238
Carbon	C	6	12.0107	32
Cerium	Ce	58	140.116	687
Cesium (Caesium)	Cs	55	132.90545	*370*
Chlorine	Cl	17	35.453	*587*
Chromium	Cr	24	51.9961	248
Cobalt	Co	27	58.933200	550
Copper	Cu	29	63.546	857
Curium	Cm	96	247*	3488

Table 13a Atomic Masses and Spin-Orbit Coupling Constants[a,b]

Element	Symbol	Atomic Number	Atomic Mass/u[c]	ζ_i/cm^{-1}
Darmstadtium[d]	Ds	110	271*	–
Dubnium	Db	105	262*	–
Dysprosium	Dy	66	162.500	2074
Einsteinium	Es	99	252*	4645
Erbium	Er	68	167.259	2564
Europium	Eu	63	151.964	1469
Fermium	Fm	100	257*	5076
Fluorine	F	9	18.9984032	*269*
Francium	Fr	87	223*	–
Gadolinium	Gd	64	157.25	1651
Gallium	Ga	31	69.723	464
Germanium	Ge	32	72.64	800
Gold	Au	79	196.96655	5104
Hafnium	Hf	72	178.49	1578
Hassium	Hs	108	265*	–
Helium	He	2	4.002602	*0.7*
Holmium	Ho	67	164.93032	2310
Hydrogen	H	1	1.00794	*0.24*
Indium	In	49	114.818	1183
Iodine	I	53	126.90447	*5069*
Iridium	Ir	77	192.217	3909
Iron	Fe	26	55.845	431
Krypton	Kr	36	83.798	*3480*
Lanthanum	La	57	138.9055	556
Lawrencium	Lr	103	262*	–

Table 13a Atomic Masses and Spin-Orbit Coupling Constants[a,b]

Element	Symbol	Atomic Number	Atomic Mass/u[c]	ζ_l/cm^{-1}
Lead	Pb	82	207.2	5089
Lithium	Li	3	6.941	*0.23*
Lutetium	Lu	71	174.967	1153
Magnesium	Mg	12	24.3050	*40.5*
Manganese	Mn	25	54.938049	334
Meitnerium	Mt	109	268*	–
Mendelevium	Md	101	258*	5533
Mercury	Hg	80	200.59	*4270*
Molybdenum	Mo	42	95.94	678
Neodymium	Nd	60	144.24	967
Neon	Ne	10	20.1797	*520*
Neptunium	Np	93	237*	2488
Nickel	Ni	28	58.6934	691
Niobium	Nb	41	92.90638	524
Nitrogen	N	7	14.0067	78
Nobelium	No	102	259*	–
Osmium	Os	76	190.23	3381
Oxygen	O	8	15.9994	154
Palladium	Pd	46	106.42	1504
Phosphorus	P	15	30.973761	230
Platinum	Pt	78	195.078	4481
Plutonium	Pu	94	244*	2810
Polonium	Po	84	209*	8509
Potassium (Kalium)	K	19	39.0983	*38*
Praseodymium	Pr	59	140.90765	824

Table 13a Atomic Masses and Spin-Orbit Coupling Constants[a,b]

Element	Symbol	Atomic Number	Atomic Mass/u[c]	ζ_l/cm^{-1}
Promethium	Pm	61	145*	1119
Protactinium	Pa	91	231.03588	1888
Radium	Ra	88	226*	–
Radon	Rn	86	222*	–
Rhenium	Re	75	186.207	2903
Rhodium	Rh	45	102.90550	1259
Rubidium	Rb	37	85.4678	*160*
Ruthenium	Ru	44	101.07	1042
Rutherfordium	Rf	104	261*	
Samarium	Sm	62	150.36	1286
Scandium	Sc	21	44.955910	77
Seaborgium	Sg	106	263*	
Selenium	Se	34	78.96	1659
Silicon	Si	14	28.0855	130
Silver	Ag	47	107.8682	1779
Sodium (Natrium)	Na	11	22.989770	*11.5*
Strontium	Sr	38	87.62	*390*
Sulfur	S	16	32.065	365
Tantalum	Ta	73	180.9479	1970
Technetium	Tc	43	98*	853
Tellurium	Te	52	127.60	3384
Terbium	Tb	65	158.92534	1853
Thallium	Tl	81	204.3833	3410
Thorium	Th	90	232.0381	1591
Thulium	Tm	69	168.93421	2838

Table 13a Atomic Masses and Spin-Orbit Coupling Constants[a,b]

Element	Symbol	Atomic Number	Atomic Mass/u[c]	ζ_l/cm^{-1}
Tin	Sn	50	118.710	1855
Titanium	Ti	22	47.867	123
Tungsten (Wolfram)	W	74	183.84	2433
Uranium	U	92	238.02891	2184
Vanadium	V	23	50.9415	179
Xenon	Xe	54	131.293	*6080*
Ytterbium	Yb	70	173.04	–
Yttrium	Y	39	88.90585	260
Zinc	Zn	30	65.409	*390*
Zirconium	Zr	40	91.224	387

[a] The ζ_l values in Roman type are calculated spin-orbit coupling constants for an l-electron ($l = p, d, f, ...$) in an open shell of a neutral atom in the ground electronic configuration. Values were the results of single-configuration, numerical calculations using spin-restricted Hartree-Fock wavefunctions [7601]. All ζ_l values are for an open-shell l-electron (other than an s-electron) interacting with the core.

[b] The ζ_l values in bold italics were calculated from atomic spectra [7101] using the theoretical models in Chapters VII (LS-coupling), XI (intermediate coupling, especially the Eqs. on p. 271), and XIII (intermediate coupling in the rare gases) of Condon and Shortley [3501]. These experimental values are for excited configurations, except in the case of the halogens. All ζ_l values are for an open-shell p-electron interacting with the core, which itself may or may not be excited. All experimental values are ζ_p's. The values for light atoms are rough, since the usual approximation for the relativistic effects breaks down; see discussion of helium in reference [3501], pp. 210-212.

[c] Values taken from the Table of Standard Atomic Weights 1997 [9901] and updated in [0101, 0301].

[d] [0302].

* No stable isotope exists. The mass number is give for the isotope of longest half-life [0401].

13b HAMMETT σ CONSTANTS

The Hammett equation can be expressed by

$$\rho\,\sigma = \log k_X - \log k_H$$

where k_X is a reaction parameter (either a rate constant or an equilibrium constant) for a process with the reactant, having a substituent X, and k_H is the reaction parameter for the same process involving the corresponding unsubstituted reactant. ρ depends on the type of reaction. It was originally taken to have a value of 1 for the ionizations of substituted benzoic acids. For many types of reactions, $\rho \sim 1$. The σ-values are used to make correlations between the reaction parameters, k_X and the electronic nature of the substituents, X.

For fixed ρ it can be seen from the defining equation above that electron-releasing substituents will have negative σ's and that electron-withdrawing substituents will have positive σ's. The goal of this analysis is for typical reactions to give values for the σ's such that trends can be predicted for other types of reactions depending on the nature of the substituent.

This has not always worked well, so a variety of σ's have been defined and used with varying success. σ's have been found to be very position dependent; so *meta*- and *para*-values are compiled separately as σ_m, and σ_p, respectively. Electronic effects of *ortho*-substituents are often swamped by steric hindrance. The values of σ_m, and σ_p in Table 13b are the preferred values chosen by the compilers of reference [7901] using pK_a values of substituted benzoic acids.

When the given substituent and the reaction center are conjugated, the normal σ-scales do not yield good linear correlations. For substituents that can involve direct conjugation (through resonance) with a positively charged reaction center, σ^+ has been defined. It has been based on the solvolysis of the cumyl chlorides, taking $\rho = 4.54$ from the hydrolysis of the *meta*-cumyl chlorides. For substituents that can participate in direct conjugation with a negatively charged reaction center, σ^- has been defined. Usually reactions of substituted phenols or anilines are used for defining this scale, e.g. the basicity of phenoxide ions.

The effect of polar substituents is also of interest. One σ-scale used to correlate these effects is that of σ_I. The values reported in Table 13b are based on the ionization of substituted acetic acids. Attempts have been made to separate these inductive (polar) effects, correlated by σ_I, with resonance effects correlated by various σ_R's, which are defined as

$$\sigma_R = \sigma_p - \sigma_I$$

The σ_R's reported in Table 13b are from the σ_R^0 scale which is based on NMR and IR spectra.

Table 13b Hammett σ Constants

Substituent	σ_m	σ_p	σ_m^+	σ_p^+	σ_m^-	σ_p^-	σ_I	σ_R^0
H	0.00	0.00	0.00	0.00	0.00	0.00	0.00	0.00
CH_3	-0.07	-0.17	-0.07	-0.31	-0.03	-0.15	-0.04	-0.11
CH_2CH_3	-0.07	-0.15	-0.06	-0.30	–	–	-0.05	-0.14
$CH(CH_3)_2$	-0.07	-0.15	–	-0.28	–	-0.09	-0.03	-0.12
$CH_2CH(CH_3)_2$	–	-0.12	–	–	–	–	-0.03	–
$C(CH_3)_3$	-0.10	-0.20	-0.06	-0.26	–	-0.13	-0.07	-0.13
$CH=CH_2$	0.05	-0.02	–	0.10	–	–	0.09	-0.03
C_6H_5	0.06	-0.01	0.11	-0.18	–	0.10	0.10	-0.11
CH_2CN	0.16	0.01	–	0.16	–	–	0.18	-0.08
$CH_2C_6H_5$	-0.08	-0.09	-0.07	-0.20	–	-0.12	-0.08	-0.12
CH_2Cl	0.11	0.12	0.14	-0.01	–	–	0.15	-0.03
CH_2OH	0.00	0.00	–	-0.04	0.08	0.08	0.05	0.00
CH_2NH_2	–	–	–	–	–	–	0.00	-0.15
$CH_2NH_3^+$	–	–	–	–	–	–	0.36	0.00
$CH_2Si(CH_3)_3$	-0.16	-0.21	–	-0.22	–	-0.22	-0.07	-0.20
$CH=CHC_6H_5$	0.03	-0.07	–	-1.00	–	0.62	0.02	–
$CH=CHNO_2$	0.32	0.26	–	–	0.37	0.88	0.24	0.13

Table 13b Hammett σ Constants

Substituent	σ_m	σ_p	σ_m^+	σ_p^+	σ_m^-	σ_p^-	σ_I	σ_R^0
COCH$_3$	0.38	0.50	–	–	0.32	0.87	0.29	0.20
CHO	0.35	0.42	–	0.73	0.48	1.13	0.25	0.24
COCF$_3$	0.63	0.80	–	0.85	–	–	0.45	0.33
COOH	0.37	0.45	0.32	0.42	0.56	0.77	0.39	0.29
COOCH$_3$	0.37	0.45	0.37	0.49	–	0.64	0.34	0.16
COOC$_2$H$_5$	0.37	0.45	0.37	0.48	–	0.64	0.21	0.18
COO$^-$	–0.10	0.00	–0.03	–0.02	–	0.24	–0.17	–
CONH$_2$	0.28	0.36	–	–	–	0.63	0.27	0.01
CN	0.56	0.66	0.56	0.66	0.68	1.00	0.56	0.13
CF$_3$	0.43	0.54	0.52	0.61	0.41	0.65	0.42	0.10
C$_6$F$_5$	0.34	0.41	–	–	–	–	0.25	0.02
N$_2^+$	1.76	1.91	–	1.88	–	3.04	1.34	0.64
NH$_2$	–0.16	–0.66	–0.16	–1.30	–0.02	–0.15	0.12	–0.48
NHCH$_3$	–0.30	–0.84	–	–	–	–	–	–0.52
N(CH$_3$)$_2$	–0.15	–0.83	–	–1.70	0.04	–0.12	0.06	–0.52
NH$_3^+$	0.86	0.60	–	–	–	–	0.61	–0.04
N(CH$_3$)$_3^+$	0.88	0.82	0.36	0.41	0.85	0.70	0.93	–0.15

Table 13b Hammett σ Constants

Substituent	σ_m	σ_p	σ_m^+	σ_p^+	σ_m^-	σ_p^-	σ_I	σ_R^0
NHNH$_2$	-0.02	-0.55	—	—	—	—	0.14	-0.43
NHOH	-0.04	-0.34	—	—	—	—	0.12	-0.22
NHCOCH$_3$	0.21	0.00	—	-0.60	—	—	0.24	-0.23
N=N-C$_6$H$_5$	0.32	0.39	—	-0.19	—	0.69	0.25	0.06
NO	0.62	0.91	—	—	—	1.60	0.34	0.32
NO$_2$	0.71	0.78	0.67	0.79	—	1.24	0.76	0.15
O$^-$	-0.47	-0.81	—	-2.30	—	-0.81	-0.16	-0.60
OH	0.12	-0.37	-0.04	-0.92	—	—	0.29	-0.43
OCH$_3$	0.12	-0.27	0.05	-0.78	0.13	-0.14	0.27	-0.43
OCH$_2$CH$_3$	0.10	-0.24	—	—	—	—	0.27	-0.44
OC$_6$H$_5$	0.25	-0.03	0.10	-0.50	—	-0.10	0.39	-0.34
OCF$_3$	0.38	0.35	—	—	0.47	0.27	0.39	-0.04
OCOCH$_3$	0.39	0.31	0.27	-0.06	—	—	0.33	-0.21
SH	0.25	0.15	—	—	—	—	0.26	-0.15
SCH$_3$	0.15	0.00	0.16	-0.60	0.19	0.17	0.23	-0.17
S=O(CH$_3$)	0.52	0.49	—	—	0.45	0.62	0.49	0.00
SO$_2$CH$_3$	0.60	0.72	—	—	0.52	1.05	0.59	0.12

Table 13b Hammett σ Constants

Substituent	σ_m	σ_p	σ_m^+	σ_p^+	σ_m^-	σ_p^-	σ_I	σ_R^0
SO_2NH_2	0.46	0.57	–	–	0.65	0.94	0.46	0.05
SCF_3	0.40	0.50	–	–	0.46	0.57	0.42	0.06
$S(CH_3)_2^+$	1.00	0.90	–	–	1.00	1.16	0.89	0.17
$Si(CH_3)_3$	–0.04	–0.07	0.01	–0.03	0.00	0.17	–0.13	0.06
F	0.34	0.06	0.35	–0.07	–	0.05	0.52	–0.34
Cl	0.37	0.23	0.40	0.11	–	0.27	0.47	–0.23
Br	0.39	0.23	0.41	0.15	–	0.28	0.44	–0.19
I	0.35	0.18	0.36	0.14	–	–	0.39	–0.16

13c FUNDAMENTAL CONSTANTS AND CONVERSION FACTORS

Included in this section are the fundamental constants and conversion factors commonly needed by photochemists and spectroscopists. The International System of Units (SI, Système International d'Unités) is used as much as possible throughout this book. Conversions are given in this section to more traditional units, especially energy units.

Table 13c-1 Fundamental Physical Constants[a]

Quantity	Symbol	Value
Atomic mass constant	$m_u = \dfrac{m(^{12}C)}{12}$	$1.6605387 \times 10^{-27}$ kg
Avogadro constant	N_A	6.0221420×10^{23} mol^{-1}
Bohr magneton	$\mu_B = \dfrac{eh}{4\pi m_e}$	$9.2740090 \times 10^{-24}$ J T^{-1}
Bohr radius	$a_0 = \dfrac{\alpha}{4\pi R_\infty}$	$0.529177208 \times 10^{-10}$ m
Boltzmann constant	$k = \dfrac{R}{N_A}$	1.380650×10^{-23} J K^{-1}
Electric constant (permittivity of vacuum)	$\varepsilon_0 = \dfrac{1}{\mu_0 c^2}$	$8.854187817 \times 10^{-12}$ F m^{-1}
Electron g-factor	g_e	-2.0023193043737
Electron rest mass	m_e	$9.10938188 \times 10^{-31}$ kg
Elementary charge	e	$1.60217646 \times 10^{-19}$ C
Faraday constant	F	9.6485342×10^4 C mol^{-1}
Fine structure constant	$\alpha = \dfrac{e^2}{2\varepsilon_0 hc}$	7.2973525×10^{-3}
Magnetic constant (permeability of vacuum)	$\mu_0 = 4\pi \times 10^{-7}$	$1.2566370614 \times 10^{-6}$ N A^{-2}
Molar gas constant	R	8.31447 J mol^{-1} K^{-1}
Neutron rest mass	m_n	$1.6749272 \times 10^{-27}$ kg
Nuclear magneton	$\mu_N = \dfrac{eh}{4\pi m_p}$	$5.0507832 \times 10^{-27}$ J T^{-1}
Planck constant	h	$6.6260688 \times 10^{-34}$ J s
Proton rest mass	m_p	$1.6726216 \times 10^{-27}$ kg
Rydberg constant	$R_\infty = \dfrac{m_e c \alpha^2}{2h}$	1.0973731568×10^7 m^{-1}
Speed of light in vacuum	c	2.99792458×10^8 m s^{-1}
Stefan-Boltzmann constant	$\sigma = \dfrac{2\pi^5 k^4}{15 h^3 c^2}$	5.670400×10^{-8} W m^{-2} K^{-4}

[a] From reference [0303].
A = ampere, C = coulomb, F = farad, J = joule, K = kelvin, N = newton, T = tesla, W = watt

Table 13c-2 Energy Conversion Factors[a,b]

	J	cm^{-1}	Hz	eV	kJ mol^{-1} [c]	kcal mol^{-1} [c]
1 J	1	5.0342×10^{22}	1.5092×10^{33}	6.2414×10^{18}	6.0223×10^{20}	1.4393×10^{20}
1 cm^{-1}	1.9864×10^{-23}	1	2.9979×10^{10}	1.2398×10^{-4}	1.1963×10^{-2}	2.8592×10^{-3}
1 Hz	6.6261×10^{-34}	3.3356×10^{-11}	1	4.1357×10^{-15}	3.9903×10^{-13}	9.5371×10^{-14}
1 eV	1.6022×10^{-19}	8065.5	2.4180×10^{14}	1	96.488	23.061
1 kJ mol^{-1} [c]	1.6605×10^{-21}	83.593	2.5061×10^{12}	1.0364×10^{-2}	1	2.3901×10^{-1}
1 kcal mol^{-1} [c]	6.9477×10^{-21}	349.75	1.0485×10^{13}	4.3363×10^{-2}	4.1840	1

[a] Adapted from references [8601, 0303] with factors rounded to five significant figures. [b] To find the wavelenght, λ, of a photon having an energy in the units shown in the table, it is easiest to convert to either wavenumber ($\tilde{\nu}$ in cm^{-1}) or frequency (ν in Hz) and then use

$$\lambda = \frac{1}{\tilde{\nu}} = \frac{c}{\nu} = \frac{hc}{E}$$

[c] Unit of energy per amount of substance (mol).

Other Important Conversion Factors

Other energy units often useful are

$$1 \text{ L atm} = 1.013 \times 10^2 \text{ J}$$
$$1 \text{ cal} = 4.1840 \text{ J}$$

In these non standard units, the gas constant ($R = 8.31447 \text{ J mol}^{-1} \text{ K}^{-1}$) is

$$R = 0.0820575 \text{ L atm mol}^{-1} \text{ K}^{-1} = 1.9872 \text{ cal mol}^{-1} \text{ K}^{-1}$$

$$RT = 2.4790 \text{ kJ mol}^{-1} \text{ K}^{-1} \text{ at } 298.15 \text{ K } (25°\text{C})$$

$$kT = 4.1164 \times 10^{-21} \text{ J} = 2.569 \times 10^{-2} \text{ eV at } 298.15 \text{ K } (25°\text{C})$$

The SI unit of pressure is pascal ($1 \text{ Pa} = 1 \text{ N m}^{-2}$). A till used traditional unit of pressure is

$$\text{standard atmosphere} = 101.325 \text{ kPa.}$$

REFERENCES

[3501] Condon, E. U.; Shortley, G. H. *The Theory of Atomic Spectra*; Cambridge U. Pr.: Cambridge (U.K.), 1935, 441p.

[4901] McClure, D. S. *J. Chem. Phys.* **1949**, *17*, 905-913.

[5201] Kasha, M. *J. Chem. Phys.* **1952**, *20*, 71-74.

[6101] Robinson, G. W. *J. Mol. Spectrosc.* **1961**, *6*, 58-83.

[6201] Blume, M.; Watson, R. E. *Proc. Phys. Soc., London* **1962**, *A270*, 127-143.

[6901] McGlynn, S. P.; Azumi, T.; Kinoshita, M. *Molecular Spectroscopy of the Triplet State*; Prentice Hall: Englewood Cliffs (NJ), 1969, 434p.

[7101] Moore, C. E. *Atomic Energy Levels as Derived from the Analyses of Optical Spectra*; Natl. Bur. Stand.: Washington DC, 1971, Vol. I-III.

[7601] Fraga, S.; Saxena, K. M. S.; Karwowski, J. *Handbook of Atomic Data. Physical Sciences Data*; Elsevier: Amsterdam (The Netherlands), 1976, Vol. 5, 551p.

[7901] Hansch, C.; Leo, A. *Substituent Constants for Correlation Analysis in Chemistry and Biology*; Wiley: New York, 1979, 339p.

[8101] Strek, W.; Wierzchaczewski, M. *Chem. Phys.* **1981**, *58*, 185-193.

[8102] Strek, W.; Wierzchaczewski, M. *Acta Phys. Pol. A* **1981**, *A60*, 857-865.

[8601] Cohen, E. R.; Taylor, B. N. *CODATA Bull.* **1986**, 1-36.

[9901] Vocke, R. D., Jr. *Pure Appl. Chem.* **1999**, *71*, 1593-1607.

[0101] Coplen, T. B. *Pure Appl. Chem.* **2001**, *73*, 667–683.

[0301] Loss, R. D. *Pure Appl. Chem.* **2003**, *75*, 1107-1122.

[0302] Carish, J.; Rosenblatt, G. M. *Pure Appl. Chem.* **2003**, *75*, 1613-1615.

[0303] Lide, D. R. (ed.) *CRC Handbook of Chemistry and Physics, 84th Ed.*; CRC Press: Boca Raton, FL (USA), 2003.

[0401] Lide, D. R. (ed.) *CRC Handbook of Chemistry and Physics, 85th Ed.*; CRC Press: Boca Raton, FL (USA), 2004-2005.

Subject index

j

k

l

n

o

p

Z

Printed in the United States
by Baker & Taylor Publisher Services